c1773

DENNIS S. HILL

5/3/1974

Dr. Dennis S. Hill
'Haydn House'
20 Saxby Avenue
Skegness
Lincs.

SERIES

ENTOMOLOGICA

EDITOR

E. SCHIMITSCHEK, GÖTTINGEN

VOLUME 9

DR. W. JUNK B.V. — THE HAGUE — 1973

Text set in 9/11 pt. Monotype Times New Roman,
printed by letterpress, and bound in Great Britain at
The Pitman Press, Bath

AGROMYZIDAE (DIPTERA) OF ECONOMIC IMPORTANCE

by

KENNETH A. SPENCER

DR. W. JUNK B.V. — THE HAGUE — 1973

CONTENTS

FOREWORD

The original stimulus which started KENNETH SPENCER on a study of the Agromyzid flies was an invitation, which he accepted, to translate from the German the monograph on Leaf Miners by Professor E. M. HERING. From this developed nearly 20 years of collaboration until Professor HERING's death in 1967.

Dr. SPENCER has himself described over 600 new species in the family, many of which he collected and reared from known host plants during his extensive travels to all the five main continents. Largely as a result of his work, the number of species known in Britain has increased from 90 in 1945 to 313 today. He is thus uniquely qualified to write this book about the hundred and fifty or so species which are regularly associated with cultivated plants.

Much of the taxonomic detail provided here will be of value primarily to specialists; but with the help of a microscope and the botanical host list (Chapter 2) and the numerous illustrations (mostly prepared by ANN SPENCER) those in agricultural institutes and elsewhere should now be able to identify the majority of species found attacking crops in any part of the world.

Chapter 10 will be of interest to horticulturists because it discusses the main species of Agromyzidae which damage ornamentals in nurseries and private gardens. Those who wish to know what measures can be used to reduce damage to crops will find a useful summary in Chapter 12. The problem of devising an effective insecticide treatment is complicated because these Agromyzids are preadapted to survive modern conditions by having larvae and often pupae which are protected within the living plant. Many instances are known where leaf mining insects have increased rather than decreased following the use of modern insecticides. Most Agromyzids are heavily attacked by parasites, but if the parasites have to search over leaf surfaces which are contaminated with insecticides, it is the beneficial parasite which is killed and the pest which survives. Examples of this effect are cited from several parts of the U.S.A. Perhaps in England the recent rise in importance of *Phytobia cambii* which tunnels in the timber of the cricket bat willow, and of *Phytobia carbonaria* in apple trees in Denmark, is to be explained in this way. A better understanding of their ecology may eventually lead to more successful methods for control.

In previous centuries, but nowadays to a diminishing extent, most of the useful specialized works on insects were written by expert amateur scientists. KENNETH SPENCER began his studies of these flies as a relaxation from his business. As his reputation grew he was invited to undertake a series of special studies for different governments, and he must be one of the very few whose first degree was in an Arts Faculty to be awarded the degree of Doctor of Science by London University. This review of all known Agromyzidae attacking cultivated plants now fills a further important gap in the literature.

G. C. VARLEY

PREFACE

A number of major Agromyzid pests, such as *Ophiomyia phaseoli* (Tryon), *Melanagromyza obtusa* (Malloch), *Liriomyza huidobrensis* (Blanchard) (=*langei* Frick) and *Phytomyza horticola* Goureau are well known both to plant growers and to specialists who frequently receive them for identification. Many other serious pests, further species which occasionally cause significant damage, and an even larger number which feed on cultivated crops with the potential at any time to occur in pest proportions are less well known. Information on the 156 pest species considered here is inaccessibly scattered in hundreds of journals in many languages, is frequently incomplete and in many cases based on inaccurate identification.

In this book I have collated this scattered information and provided essential references, and have included a synoptic re-description of all species. Illustrations are given of male genitalia, early stages and leaf-mines. In addition, 19 new species are described and 21 new synonymies are established. Parasites are recorded where known but the gaps in our knowledge of this important aspect of pest ecology have now become very apparent. No attempt has been made to revise the nomenclature of the parasites and these are referred to as recorded in the literature. In the final chapter, the few beneficial Agromyzidae which have been used or considered for biological control of weeds are briefly discussed.

The polyphagy and oligophagy of a number of important species has made it impractical to arrange the treatment of all species under a botanical sequence of host plants, as this would have led to unnecessary repetition and cross-referencing. Instead, where possible, certain natural groups of host plants have been included in separate chapters, such as stem-borers on leguminous crops (Chapter 3), leaf-miners on cereals (Chapter 9) and species attacking trees (Chapter 10). Inevitably this approach has necessitated a somewhat arbitrary grouping of unrelated plants attacked by stem-borers (Chapter 5) and by leaf-miners (Chapter 6). However, the general index and the pest and host lists given in Chapter 2 will permit ready reference both to host plants and to pest species.

Identification of Agromyzidae is frequently difficult without also having information on early stages and in particular the host plant. In many cases caught specimens can be satisfactorily determined only after examination of male genitalia. However, when species are found infesting cultivated plants, knowledge of the host plant and the way in which the larva feeds will invariably indicate the species concerned. In doubtful cases a positive identification will always be possible from the male genitalia.

I have not attempted to include references to all instances where minor damage has been caused to cultivated crops but essential information is provided on all known species of economic importance. Damage to local crops by Agromyzidae will not yet in all cases have come to the notice of entomologists and new species certainly await discovery, particularly in Africa, Asia and South America.

Valuable collections, including types, have been made available for study by the following individuals and Institutions:

Argentina: DR. L. DE SANTIS, Universidad Nacional de la Plata, DR. N. H. ROSSI DE SIMONS, Inta-Cica, Castelar, DR. P. WYGODZINSKY, American Museum of Natural History, New York.

Austria: DR. A. KALTENBACH, Naturhistorisches Museum, Vienna.

Britain: MR. A. J. BIDDLE, Pea Growing Research Organization Ltd., Peterborough, MR. B. COGAN, British Museum (Natural History), London, DR. R. G. FENNAH, Commonwealth Institute of Entomology, London, DR. N. W. HUSSEY, Miss B. GURNEY and Dr. N. E. SCOPES, Glasshouse Crops Research Institute, Worthing.

France: Dr. L. TSACAS, Musée d'Histoire Naturelle, Paris.

German Democratic Republic: Dr. HABIL. G. MORGE, Custodian of Strobl collection, Admont, Austria, Dr. H. SCHUMANN, Zoologisches Museum, Berlin.

Ghana: Mr. G. SCHEIBELREITER, C.I.B.C., Kumasi.

Hungary: Dr. F. MIHÁLYI, Hungarian Natural History Museum, Budapest.

Iraq: Dr. G. S. AL-SAFI, Division of Entomology, Institute of Agriculture.

Israel: Dr. D. GERLIN, Tel-Aviv University.

Japan: Dr. N. FUKUHARA, National Institute of Agricultural Sciences, Tokyo.

Netherlands: Dr. TH. VAN LEEUWEN, Zoological Museum, Amsterdam.

Pakistan: Dr. M. A. GHANI, C.I.B.C., Rawalpindi.

Peru: Ing. Agr. C. A. KORYTKOWSKI, Lambayeque.

Sweden: Dr. P. I. PERSSON, Naturhistoriska Riksmuseum, Stockholm.

Thailand: Dr. B. MEKSONGSEE, Department of Agriculture, Bangkok.

U.S.A.: Dr. D. E. HARDY, Honolulu, Dr. G. JENSEN, University of Massachusetts, Mr. D. E. MILLER, University of Massachusetts, Dr. E. A. OATMAN, University of Wisconsin, Mr. A. G. WHEELER, Department of Agriculture, Harrisburg, Penn.

Venezuela: Dr. L. CERMELI, Fundacion Shell, Cagua, Dr. J. R. LABRADOR, Universidad del Zulia, Maracaibo.

In addition the following persons have provided material assistance in the form of reprints or information during correspondence or personal discussions: Dr. A. AL-AZAWI, Unesco, Philippines, Dr. L. A. ANDRES, Albany, Calif., Dr. A. R. BRIMBLECOMBE, Brisbane, Australia, Dr. W. E. VAN DEN BRUEL, Gembloux, Belgium, Dr. R. H. FOOTE, Beltsville, Md, U.S.A., Dr. K. E. FRICK, Southern Weed Science Lab., Mississippi, Prof. S. M. HAMMAD, Tripoli, Libyan Arab Republic, Dr. C. S. KOEHLER, Berkeley, Calif., U.S.A., Prof. W. H. LANGE, Davis, Calif., U.S.A., Mr. C. PARKER and Mr. L. KASASIAN, Weed Research Organization, Yarnton, England, Dr. L. E. VAN'T SANT, Wageningen, Netherlands, Dr. F. F. SMITH, Beltsville, Md., U.S.A., Mr. G. STEYSKAL, Washington, D.C., Dr. L. SÜSS, Milan University, Prof. F. VENTURI, Pisa, Dr. D. O. WOLFENBARGER, Homestead Sub-Tropical Experiment Station, Florida.

I would like to thank all the above for their co-operation, which has appreciably assisted the preparation of this book, and also the many further correspondents who are too numerous to mention individually.

During my visit to East Africa in October, 1971, I received hospitality and assistance with collecting from Mr. J. B. GILLETT, Nairobi, and Dr. D. J. GREATHEAD, Kawanda, Uganda. Furthermore, Mr. GILLETT and the staff of the East Africa Herbarium kindly identified a number of plants collected in Kenya and Uganda. Additional assistance with transport was provided by Dr. B. N. MAJISU, Director, E.A.F.F.R.O., Mr. OLOO, National Agricultural Laboratories, Nairobi, Mr. C. MANN, Kapsabet, Kenya, and Mr. T. P. HEATON, Nairobi. This generous help was greatly appreciated.

I would particularly like to thank the Librarians of the Commonwealth Institute of Entomology, the Royal Entomological Society and the Entomology Library of the British Museum (Natural History) for their patience and efficiency in locating the numerous and sometimes obscure papers which have been consulted.

The great majority of the drawings have been done by my wife, ANN SPENCER, to whom special thanks are due for the meticulous care with which these have been prepared.

Finally, I would like to thank the Directors of Shell International Petroleum Company Limited, and the Overseas Development Administration of the Foreign and Commonwealth Office, London, for grants which have provided substantial support during the year I have spent preparing this book.

I. INTRODUCTORY

Early pest species—natural control by parasites—insecticides—insecticidal resistance—types of damage—distribution—systematics—key to genera

The Agromyzidae are known primarily as a family of leaf-miners but in fact only 75% of the known 1800 species feed in this way. Individual species may attack any part of a plant, from the root and the stem to the seed-heads or pods of herbaceous plants, and twigs or stems of trees. Injurious species are found among all these groups with differing larval feeding habits.

Leaf-mines were first recorded in the literature towards the end of the seventeenth century when J.C. BECKMANN (1681) discussed and illustrated the strange forms which had appeared in great numbers the previous year on cherry trees in the Frankfurt/Oder area in Germany. The popular conception at this time was that these mines represented little serpents which had descended from the skies or emanated from the foul air of the local swamps. However, BECKMANN was able to show that the mines were caused by insects and illustrated the lepidopterous larvae reponsible. Fifty years later RÉAUMUR (1737) for the first time discussed and illustrated Agromyzid leaf-miners on *Sonchus oleraceus* L., *Trifolium*, *Ranunculus* and *Lonicera*. These species were not given names but are readily identifiable from their host-plants.

Many species were described in the nineteenth century but only a few isolated cases of damage by Agromyzidae are recorded during this period. The earliest accounts of damage to cultivated plants are given by CURTIS (1844, 1845, 1846) referring to *Viola*, *Phlox*, *Cineraria* and *Ilex aquifolium* L. (holly) in England (cf. Chapter 11). RATZEBURG (1868) reported larval feeding in the trunks of *Acer platanoides* L. in the St. Petersburg area of Russia (cf. Chapter 10). The first record of damage to crops was given by RONDANI (1875), when describing *Liriomyza cicerina* attacking *Cicer arietinum* L. in Italy. Damage to wheat and barley in the Moscow area was described by LINDEMANN (1880) and the species was correctly identified as *Cerodontha* (*Poemyza*) *lateralis* (Macquart). In 1895 TRYON first reported damage to beans by *Ophiomyia phaseoli* in the Brisbane area, Australia, and three years later similar damage was reported in New South Wales (ANON., a, 1898). Damage to asparagus by *Ophiomyia simplex* (Loew) was first described by SAJÓ (1896) in Hungary, the species being misidentified as *Agromyza maura* Meigen.

In the United States BURGESS (1880) described *Liriomyza trifolii* mining *Trifolium repens* L. and in 1884 RILEY described *Liriomyza brassicae* mining cabbage leaves. RILEY, showing some foresight, wrote: 'Although this insect seems rare at present, it may at any time become sufficiently numerous to prove injurious, and it is well to study it in advance.' From 1886 onwards glasshouse-grown chrysanthemums were severely damaged in the New York area by a species which was described as *Phytomyza chrysanthemi* Kowarz (*in* LINTNER, 1891) but this has recently been confirmed as an importation of the European species, *P. syngenesiae* Hardy (Chapter 11, p. 322).

After World War I, when the Agromyzidae were becoming better known and with a more scientific study of agricultural pests, damage by leaf-miners and stem-feeders was reported with increasing frequency.

The degree of damage caused to plants by Agromyzidae depends on the way the larvae feed, the part of the plant attacked and the stage of its growth when the infestation occurs, and above all on the size of the population. All species feeding regularly on crops have the potential to become pests. Some regularly do so, while many others in most seasons remain just below the threshold of economic damage. If a species consistently occurs in large numbers and larval feeding takes place in the seedling stage of the host, such a species will inevitably be a serious pest. Two such species are *Ophiomyia phaseoli* occurring throughout the Old World tropics and *O. spencerella* (Greathead) in East Africa, both attacking beans.

Under normal conditions most species are reasonably well controlled by their own parasite complex. They may be present on crops but not in sufficient numbers to be of any economic significance. Periodically this balance is disrupted, either from natural causes such as loss of the essential synchronism in the emergence times of the parasite and its host or reduction in the parasite population due to disease or unfavourable climatic conditions, and build-up in the Agromyzid population will then follow, as happened with *Ophiomyia pinguis* (Fall.) attacking chicory in Belgium in 1934/5 and with *Agromyza megalopsis* Hering attacking barley in Germany in 1965. Such outbreaks are normally of short duration, lasting at most for two seasons, before the Agromyzids are again reduced to insignificant levels by the re-establishment of normal parasite populations.

A more serious situation has now arisen in a number of areas where long-term use of modern insecticides appears to have critically disrupted natural parasite control of Agromyzid species in several different genera. In the United States and Venezuela *Liriomyza huidobrensis* (Blanchard) and *L. sativae* Blanchard in most seasons now cause severe damage to a wide range of crops (cf. Chapter 8). The earliest known case of the actual increase of a leaf-mining population following the use of DDT and BHC occurred with the tomato leaf-miner, *Liriomyza bryoniae* (Kaltenbach), in glasshouses in the Lea Valley in Hertfordshire, England, in 1948. It was then proved experimentally that Braconid parasites (*Opius* sp. and *Pachysema* sp.) were inherently more susceptible to DDT than the Agromyzid, and the use of these insecticides was thus proving actively counter-productive (SPEYER and PARR, 1948). *Liriomyza sativae* Blanchard became a serious pest in Florida on tomatoes at about the same time and WOLFENBARGER (1948) wrote: 'Coincidentally or otherwise, the leaf-miner was not a serious pest before DDT was used so extensively for control of other insects.'

In 1950 in the Salt River Valley of Arizona, a *Liriomyza* sp., probably *sativae* Blanchard, caused substantial damage to cantaloups and melons. HILLS and TAYLOR (1951) investigated the parasitism of this species and also the effects of DDT. They wrote: 'Insecticide tests on cantaloups in experimental plots in 1950 showed that repeated applications of DDT were ineffective against the leaf-miners but reduced the parasite population, resulting in increased leaf-miner populations.' WENE (1953) investigated what was probably the same species (referred to as *Liriomyza subpusilla* Frost) which was severely damaging peppers in the Lower Rio Grande Valley of Texas. Many experiments were carried out with various insecticides. In his summary WENE commented: 'The experiments conducted during 1951 showed that DDT, TDE and methoxychlor applied as 5 per cent dusts actually increased the number of miner-infested leaves.' WENE (1955) conducted further experiments on this problem (see Chapter 12) and concluded: 'The leaf-miner population was greater on the dieldrin-, endrin-, and lindane-treated plots indicating that these insecticides create an environment favourable for leaf-miner development by reducing the parasite population and not being effective against the leaf-miner.

A large-scale experiment showed that parasites were effective in controlling the leaf-miner.

In California it has been suggested that the unusually heavy leaf-miner infestations on alfalfa in February 1968 at Holtville, Imperial County, might be correlated with the previous treatment of the crop with diazinon (JENSEN and KOEHLER, 1970).

In view of the greater toxicity to hymenopterous parasites of many of the chlorinated hydrocarbons and organophosphorus compounds currently recommended for use against leaf-miners, the search for effective selective insecticides clearly demands the highest priority. GETZIN (1960), with this problem specifically in view, tested a number of organophosphorus insecticides with *Liriomyza munda* Frick in the Lower Rio Grande Valley. He was able to confirm that parathion, ethion and diazinon which are all effective against leaf-miners are particularly toxic to one of the main parasites in this area of Texas, *Derostenus variipes* Crawford. The greatest selectivity both for controlling *L. munda* and permitting survival of *D. variipes* was shown by DELNAV and dimethoate. This significant work has received no organized follow-up, and indeed has been ignored or is apparently unknown in many countries.

In Venezuela an appreciable increase in leaf-mining pests coincided with the widespread use of modern insecticides from about 1950 and the correlation has become apparent (L. CERMELI, personal communication). A leaf-miner on alfalfa in New York and Michigan, *Agromyza frontella* Rond. discussed in Chapter 4, is now present in large numbers and is being investigated by local entomologists. It has no doubt been present all the time but has only now come to attention with a significant increase in population, due in all probability to the destruction of its own parasites by insecticides employed against other pests.

In Europe two similar cases have come to notice with the large tree-boring species of the genus *Phytobia*. *P. cambii* feeds in twigs and trunks of willow and in recent years the larval attack has increased to such an extent in *Salix alba* L., which is used for manufacturing cricket bats in England, that few trees are now found free from this pest. In Denmark an apple orchard has become heavily infested with *P. carbonaria* (Zetterstedt). In the case of *P. cambii* its increase is probably due to destruction of its parasite by insecticides used on other crops in the vicinity; the apple orchard in Denmark is regularly sprayed 12 times a year against other pests and the reason for the increase of *P. carbonaria* seems clear.

It has been established that several species have developed some degree of resistance to the insecticides currently in use. Wolfenbarger (1958) reviewed the history of control measures in Florida during the previous 10 years against *Liriomyza sativae* (referred to as *pusilla* Meig.). He concluded that certain of the chlorinated hydrocarbons such as chlordane, Lindane, Toxaphene and aldrin which had been effective in initial experiments became ineffective even after one to three seasons. Of the phosphorus compounds there is some evidence that parathion is not as effective after 10 years as it was initially. In England Gurney and Hussey (1967) were able to confirm field reports of resistance to BHC and also to diazinon in *Phytomyza syngenesiae* and this was discussed further by Hussey (1969).

Although the primary cause of damage by Agromyzidae is feeding by the larvae, the habit of the females to form a series of punctures in the leaves prior to oviposition can result in significant damage. These feeding punctures were first observed by Curtis (1844) in England in *Liriomyza strigata* (described as *Agromyza violae*), when somewhat unusually the females made numerous punctures in the petals of pansies. As many as 100 feeding punctures have been observed in a single leaf of spinach in California (Wilcox and Howland, 1955). Leaves damaged to this extent can be totally destroyed.

The more primitive species of Agromyzidae feed internally in trees, and in roots, stems or seed-heads of herbaceous plants. The tree-feeders belong to the genus *Phytobia*. The mature larvae can be up to 2 cm in length and long channels are eaten out in the cambium of previous year's wood, spoiling it for subsequent industrial use. Although over 50 species have been described throughout the world feeding in this way, the hosts of only eight are known and these are discussed in Chapter 10.

Four species which are primarily root-feeders deserve mention. *Melanagromyza fabae* sp. n., described below from England, hollows out the root of broad beans (Plate 1), and plants can be significantly weakened. *Ophiomyia lappivora* (Koizumi) damages the edible roots of *Arctium lappa* L. in Japan. *Liriomyza braziliensis* (Frost) feeds in potato tubers in Brazil, Venezuela, Ecuador and probably other countries in South America (Fig. 168). Finally, *Napomyza carotae* Spencer damages carrots in Germany, Switzerland and the Netherlands (Fig. 183).

The two serious pests of beans, *Ophiomyia phaseoli* and *O. spencerella*, can feed down into the upper part of the root but the main larval feeding occurs in the stem, particularly near ground level. The main root system is largely destroyed and the plants only survive if adventitious roots can develop fast enough to prevent their total collapse. It is not uncommon to find as many as 20 larvae of these species in seedling bean plants. Less severe damage is caused by other stem-borers which occur in smaller numbers and attack their hosts when more mature. A recently described species, *Napomyza scrophulariae* Spencer, has badly damaged plantations of *Digitalis* in France and in the German Democratic Republic, being grown for medicinal purposes.

As a modification of internal stem-boring a number of species feed in seed-heads or in pods of leguminous plants. One of the most serious of these is *Melanagromyza obtusa* (Malloch) which can heavily infest pods of *Cajanus indicus* Spreng. in India, Ceylon and Malaya. Two related species feeding in the same way, *M chalcosoma* Spencer and *M. vignalis* Spencer, occur in tropical Africa on *Cajanus* and *Vigna*.

Damage by leaf-miners, depending on the severity of the attack, will obviously reduce assimilation by the plant and can lead to desiccation and premature fall of the leaves. If sufficient leaves are attacked in this way, crops can be reduced or seedling plants even totally destroyed. The most serious leaf-mining species are *Liriomyza huidobrensis* and *L. sativae*, both with a wide distribution in South America and extending to California and Florida. Both are highly polyphagous and attack a wide range of crops and can occur in large numbers. In Europe many species have been locally troublesome, such as *Liriomyza cicerina* (Rondani) on chick-peas, *L. bryoniae* (Kaltenbach) on tomatoes and cucurbits, *L. cepae* (Hering) and *L. nietzkei* sp.n. on onions, *Ophiomyia pinguis* (Fallén) and *Napomyza cichorii* Spencer on chicory and *Phytomyza rufipes* Meigen on Cruciferae. *Phytomyza syngenesiae* (Hardy) regularly causes severe damage to chrysanthemums and other cultivated Composites in Europe, and to a lesser extent in Australia, New Zealand and the United States.

Many species in several genera feed on cereals. Damage in most cases is not serious but populations fluctuate substantially and heavy outbreaks have been recorded in species which are generally not injurious, such as *Agromyza megalopsis* Hering in Germany on barley in 1966 and *Cerodontha* (*Poemyza*) *lateralis* (Macq.) in the Ukraine on spring wheat and barley in 1929. Other species cause damage in most years, such as *Pseudonapomyza spicata* (Malloch) on corn (*Zea mays*) in Hawaii.

Attack by Agromyzidae can lead to injurious secondary fungal infection but the correlation is not always certain and this aspect requires further investigation. GEOFFRION and HÉRISSET (1961) suggest that damage to *Digitalis* stems in France facilitates subsequent infestation by the fungi *Verticillium alboatrum* Reinke and Berthe, and *Fusarium* sp. In England it is believed that *Vicia faba* attacked by *Melanagromyza fabae* is more readily infested by *Fusarium solani* and *Fusarium culmorum* (Pea Growing Research Organization, personal communication). A purely physical secondary result of leaf-mining damage is sunburning of the crop due to excessive exposure. This effect has been reported with tomatoes in Venezuela, and with both tomatoes and melons in California (MICHELBACHER, MIDDLEKAUF and GLOVER, 1951; WILCOX and HOWLAND, 1952; MICHELBACHER, BACON and UNDERHILL, 1953).

The parts of the plants damaged by these species can be analysed as follows:

Leaves	107
Stems	24
Pods, seed-heads	8
Trees (trunks, twigs)	6
Trees (galls)	6
Roots	5
	156

The proportion of leaf-miners among these injurious species, 70%, closely follows the proportion of leaf-mining species in the family as a whole, which is approximately 75%.

The systematics of the Agromyzidae have been fully discussed in many books and papers during the past 40 years, of which the following may be particularly mentioned: HENDEL (1931–1936), Europe; FRICK (1952, 1959), North America; SASAKAWA (1961), Japan; NOWAKOWSKI (1962), Europe and (In press), genus Cerodontha; SPENCER (1969a), Canada and (1972b), Britain; GRIFFITHS (1972b), general. Other regional and generic revisions are readily available. It seems

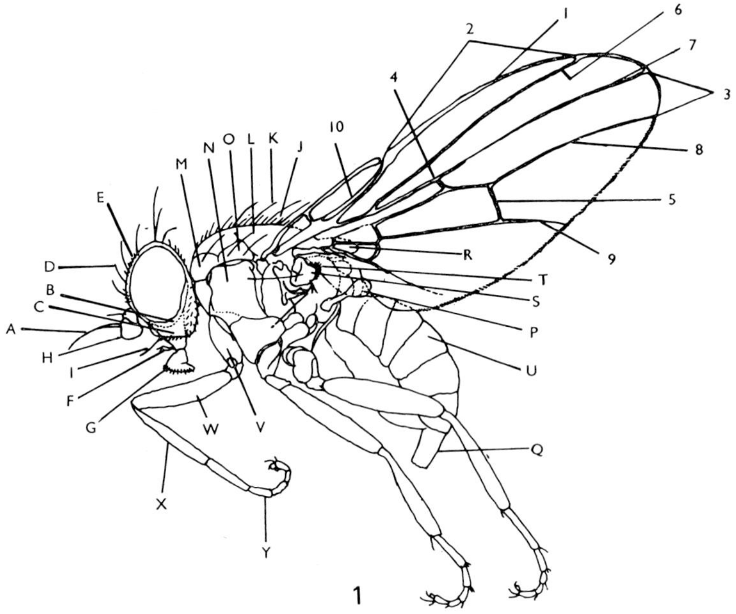

Fig. 1. Side view of typical *Agromyza* sp. (after SASAKAWA): A = arista, B = cheek, C = jowl, D = orbital bristles, E = orbital setulae, F = palp, G = proboscis, H = third antennal segment, I = vibrissa, J = acrostichals, K = dorso-central bristles, L = mesonotum, M = humerus, N = mesopleural area, O = notopleural area, P = haltere, Q = ovipositor sheath, R = scutellum, S = squama, T = squamal fringe, U = tergites, V = coxa, W = femur, X = tibia, Y = tarsi.
1 = costa, 2 = second costal section, 3 = fourth costal section, 4 = first cross-vein, 5 = second cross-vein, 6 = R_1, 7 = R_{4+5}, 8 = M_{1+2}, 9 = M_{3+4}, 10 = sub-costa.

unnecessary here to discuss the systematics and taxonomy of the family in any detail, as identification of species found on crops will in most cases be fairly simple, based on the host association and details of the early stages. Corroboration can be obtained from the more difficult characters of the adults and, if necessary, also from the male genitalia. Nevertheless, it may be helpful to illustrate the main taxonomic characters which experience has shown to be especially useful for the identification of Agromyzids and these are shown in Fig. 1. The genera in the two sub-families Agromyzinae and Phytomyzinae are readily identifiable by the course of the sub-costa, which joins vein R_1 in the Agromyzinae (Fig. 2A), and runs straight to the costa, at least as a fold, in the Phytomyzinae (Fig. 2B).

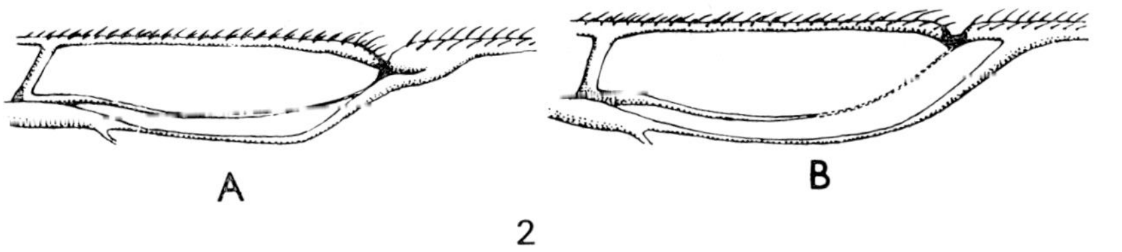

Fig. 2A, B. Sub-costa of A, sub-family Agromyzinae, B, sub-family Phytomyzinae.

No attempt has been made to give detailed descriptions of the complex genitalia structures, as it has seemed of more practical value to provide one or two illustrations of distinctive characters, normally the aedeagus in two views, which generally

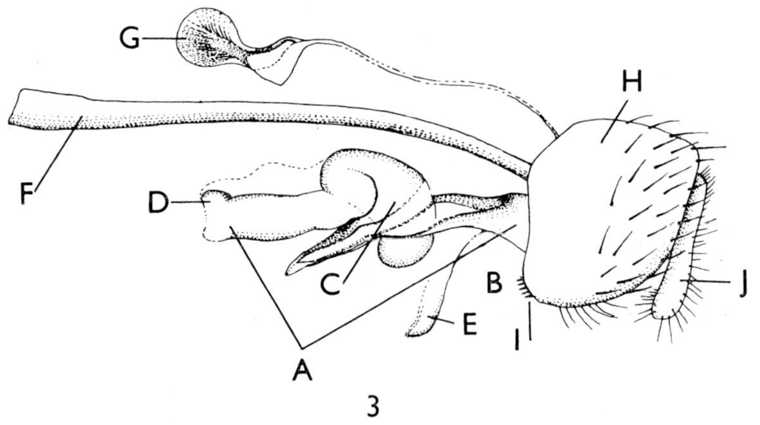

Fig. 3. Male genitalia of typical Agromyzid. A, aedeagus; B, basiphallus; C, mesophallus; D, distiphallus; E. hypophallus; F, aedeagal apodeme; G, ejaculatory apodeme; H, epandrium; I, surstylus; J, cercus.

permits immediate identification. The form of the aedeagus in individual species is remarkably stable, with no significant variation. A simplified illustration of the male genitalia showing the terms used here is given in Fig. 3.

Of 25 genera now known in the family, 16 are recorded in the following chapters damaging crops or ornamentals. The table below shows the significance of these genera in declining order of pest species:

Genus	*Number of species*
Agromyza Fallén	28
Liriomyza Mik	23
Melanagromyza Hendel	20
Phytomyza Fallén	15
Japanagromyza Sasakawa	11
Ophiomyia Braschnikov	10
Tropicomyia gen. nov.	10
Cerodontha Rondani	9
Pseudonapomyza Hendel	6
Hexomyza Enderlein	6
Phytobia Lioy	6
Napomyza Westwood	4
Paraphytomyza Enderlein	4
Amauromyza Hendel	2
Phytoliriomyza Hendel	2
Haplomyza Hendel	1
Ptochomyza Hering	1
	158

A key to those genera which include actual or potential pest species is given below. *Phytoliriomyza* is added, as two undescribed species fiom Bolivia mentioned in Chapter 5 may be referred to this genus (STEYSKAL, personal communication).

Key to genera of pest species

1	Sub-costa developed throughout its length (Fig. 2A), coalescing with R_1 before reaching costa (Sub-family Agromyzinae)	2
—	Sub-costa becoming a fold distally (Fig. 2B) and ending in costa separately and basad of R_1 (Sub-family Phytomyzinae)	7
2	Two pairs of dorso-centrals (dc); if 3 or 4 pairs, halteres black; pre-sutural dc never present	3
—	At least 3 pairs dc, pre-scutellars (prsc) always present; halteres white or yellow; pre-sutural dc frequently present	*Agromyza* Fallén
3	Either prsc or fore-tibial bristle present or halteres yellowish	*Japanagromyza* Sasakawa
—	Halteres black	4
4	Entirely black species; costa ending at vein R_{4+5} or extending to M_{1+2}; mid-tibial bristle frequently absent; many species minute, with wing length less than 2 mm; larval posterior spiracles	

with 3 or more bulbs, never with central horn, frequently on raised stalks; puparium yellow, brown or black | 5

—	Mesonotum or abdomen normally with some metallic coloration, greenish, bluish, or coppery; costa always extending strongly to vein M_{1+2}; mid-tibiae with 1 to 3 strong lateral bristles; frequently large, stout species, wing length 2.5–3.5 mm; larval posterior spiracles with numerous bulbs, from 6 to 20, normally surrounding a strong black horn (rarely atrophied to a mere scar); antennae not separated by raised keel, male never with vibrissal horn	*l*
	Biology: internal feeders in roots, stems or pods	*Meanagromyza* Hendel
5	Antennae normally separated by conspicuous facial keel; male frequently with distinct vibrissal horn (Fig. 53); larval posterior spiracular processes normally with between 6 and 12 bulbs	
	Biology: stem- or leaf-miners	*Ophiomyia* Braschnikov
—	Antennae adjoining; vibrissal horn never present; larval posterior spiracles normally with 3 bulbs on each process	6
6	Minute species, wing length from 1.7–2.3 mm	
	Biology: leaf-miners	*Tropicomyia* gen. nov.
—	Larger species, wing length from 3 mm	
	Biology: gall-causers	*Hexomyza* Enderlein
7	Orbital setulae erect or reclinate, or absent	8
—	Orbital setulae distinctly proclinate	15
8	Wing with costa extending to apex of vein M_{1+2}	9
—	Costa extending only to R_{4+5}	14
9	Scutellum dark, concolorous with mesonotum	11
—	Scutellum yellow; vein M_{1+2} ending nearest wing tip	10
10	Second cross-vein lacking; single ors	*Haplomyza* Hendel
—	Second cross-vein present; 2 ors	*Liriomyza* Mik
11	Halteres with knob white or yellow	12
—	Halteres with knob black or at least partially darkened	*Amauromyza* Hendel
12	Vein R_{4+5} ending nearest wing tip	*Phytobia* Lioy
—	Vein M_{1+2} ending nearest wing tip	13
13	Third antennal segment with a spine, scutellum with only 2 bristles (sub-genus *Cerodontha*); or lunule conspicuously higher than a semicircle, either narrow or triangular (sub-genus *Poemyza*); or lunule conspicuously large and broad, antennal bases widely separated and third	

	antennal segment in male normally greatly enlarged (sub-genus *Dizygomyza*)	*Cerodontha* Rondani
—	Third antennal segment normal, not enlarged in male, frons yellow or brown (*Am. flavifrons* Meigen)	*Amauromyza* (sub-genus *Trilobomyza* Hendel)
14	Second cross-vein either absent, or if present, well beyond first; second costal section at least twice length of fourth; third antennal segment rounded	*Paraphytomyza* Enderlein
—	Second cross-vein basal to first; second costal section conspicuously short, less than $1\frac{1}{2}$ times length of fourth; third antennal segment angulate (Fig. 392)	*Pseudonapomyza* Hendel
15	Costa extending only to vein R_{4+5}	16
—	Costa extending to vein M_{1+2}; orbital setulae conspicuously proclinate (or rarely absent, cf. new species from Bolivia, p. 102)	*Phytoliriomyza* Hendel
16	Second cross-vein present; frons always strongly projecting; second costal section short, always less than twice length of fourth	*Napomyza* Westwood
—	Second cross-vein normally absent; if present, frons not projecting	17
17	Head, wing and aedeagus as in Figs. 238–240	*Ptochomyza* Hering
—	Not so	*Phytomyza* Fallén

II. DISTRIBUTION, EVOLUTION AND HOST SPECIFICITY. PEST AND HOST LISTS

Agromyzidae occur virtually throughout the world from the tropics to the high arctic but the distribution of genera differs significantly and almost certainly reflects their evolutionary development. The main serious pests in the Old World tropics belong to the primitive genera *Melanagromyza* and *Ophiomyia* which evolved in these areas. Leaf-mining genera, such as *Liriomyza* and *Phytomyza* are dominant in more temperate areas and are reduced to insignificant numbers in the tropics. They are of more recent origin and evolved during the main Tertiary explosion of herbaceous plants, which themselves are derived from woody ancestors in the tropics. The tropical leaf-mining genus, *Tropicomyia* gen. nov., is widespread from Africa to the Pacific and northern Australia but the number of species is small. The primitive tree-feeders in the genus *Phytobia* are widely distributed throughout the world. They are particularly well represented in the tropics of the Old World and of Central America but no hosts are known outside Europe and North America. The intriguing question of the probable evolution of the Agromyzidae in close association with that of their host-plants is discussed further below, as this evolutionary background appears to explain their remarkable degree of host specificity.

A distributional analysis of pest species fully reflects the distribution of the family as a whole, as seen in the following table.

Table I. Distribution of pest species by areas

Area	*Number of species*
Europe, Japan, temperate Asia	75
North America	35
Tropical Asia, Pacific	25
Africa	19
South America	17
Australia, New Zealand	14

The majority of these species are limited to a single zoogeographical region but a number are holarctic and are found both in Eurasia and North America (for the most part naturally but in a few cases as a result of recent introduction); a slight overlap occurs between Africa and tropical Asia, between Asia and Australia and between North and South America. There is only a single largely cosmopolitan species, *Liriomyza brassicae* (Riley).

The list of Agromyzid pests below is arranged systematically, as it is felt that this may be helpful for reference purposes. The host-plant, part of the plant attacked and the main areas of distribution are also included.

List of Agromyzid pests

Species	*Host*	*Part of plant affected*	*Distribution*
Sub-family Agromyzinae			
Melanagromyza Hendel			
apii Hg.	*Apium*	stalks	Australia
azawii Sp.	*Vicia faba*, *Sesamum*	stems	Iraq
bonavistae Greathead	*Dolichos*, *Lablab*	pods	East Africa
caerulea (Mall.)	*Ipomoea batatas*	seeds	Florida, Caribbean
candidipennis (Lamb)	*Vigna*	pods	E. & W. Africa
chalcosoma Sp.	*Cajanus*, *Flemingia*, *Vigna*	pods	E. Africa
dolichostigma de Meij.	*Glycine*, *Phaseolus*	stems	Asia
fabae Sp.	*Vicia faba*	roots	England
hibisci Sp.	*Hibiscus*	stems	India
koizumii Kato	*Glycine*	stems	Japan
lini Sp.	*Linum*	stems	Peru
obtusa (Mall.)	*Cajanus*	pods	India, Malaya
phaseolivora Sp.	*Phaseolus*	stems	Ecuador
ricini de Meij.	*Ricinus*	seeds	Indonesia
sojae (Zehnt.)	*Glycine*	stems	Asia, Africa
splendida Frick	*Apium*, *Daucus*; *Lactuca*	stems	Caribbean, Hawaii, California
vignalis Sp.	*Vigna*	pods	E. & W. Africa
sp. (*Brassica*)	*Brassica*	stems	Rhodesia, Singapore
Ophiomyia Braschnikov			
centrosematis de Meij.	*Glycine*	stems	Asia, Africa
cicerivora Sp.	*Cicer*	stems	India
lappivora (Koizumi)	*Arctium lappa*	roots	Japan
orbiculata (Hd.)	*Pisum*	stems	Europe
phaseoli (Tryon)	*Phaseolus*	stems	Australia, Pacific, Asia, Africa
pinguis (Fall.)	*Cichorium*, *Lactuca*	leaves	Europe, Egypt
shibatsuji (Kato)	*Glycine*	roots	Japan
simplex (Lw.)	*Asparagus*	stems	Europe, N. America
spencerella (Greathead)	*Phaseolus*, *Lablab*, *Vigna*	stems	E. Africa
vignivora Sp.	*Vigna*	leaves	Pakistan
Hexomyza Enderlein			
albicula Spencer	*Salix*	twigs (galls)	Canada
cecidogena (Hering)	*Salix*	twigs (galls)	Europe
schineri (Giraud)	*Populus*	twigs (galls)	Europe, N. America
simplicoides (Hd.)	*Salix*	twigs (galls)	Europe
tiliae (Couden)	*Tilia*	twigs (galls)	N. America
websteri (Mall.)	*Wistaria*	twigs (galls)	Japan, N. America

Species	*Host*	*Part of plant affected*	*Distribution*
Tropicomyia Spencer			
capeneri (Hering)	*Vigna*	leaves (epidermal)	Senegal, S. Africa
coffeae (Kon.)	*Coffea*	,,	Java
flacourtiae (Séguy)	Polyphagous	,,	Africa, Madagascar
nigriclava (Bezzi)	Unknown	,,	Rodrigues
pisi (Kleinschmidt)	*Pisum*	,,	Australia
polyphaga (Sp.)	Polyphagous	,,	India
polyphyta (Kleinschmidt)	Polyphagous	,,	Australia
styricicola (Sas.)	Polyphagous (?*Camellia*)	,,	Japan
theae (Cotes)	*Camellia*	,,	Ceylon
vigneae (Séguy)	Leguminosae	,,	Africa, Madagascar
Japanagromyza Sasakawa			
aequalis Sp.	*Vigna*	leaves	Florida
centrosemae (Frost)	*Centrosema*	leaves	Panama
desmodivora Sp.	*Desmodium*	leaves	Florida, Peru
eucalypti Sp.	*Eucalyptus*	leaves	Australia
inaequalis (Mall.)	*Vigna*	leaves	Caribbean, Florida
quercus Sasakawa	*Quercus*	leaves	Japan
salicifolii (Collin)	*Populus*, *Salix*	leaves	Asia (temperate), Mediterranean
tristella (Thomson)	*Glycine*	leaves	Asia
viridula (Coq.)	*Quercus*	leaves	N. America
Japanagromyza sp. (Peru)	*Phaseolus*, *Brassica*	leaves	Peru
Japanagromyza sp. (Puerto Rico)	*Phaseolus*	leaves	Puerto Rico
Agromyza Fallén			
albipennis Mg.	cereals	leaves	Europe, Japan
albitarsis Mg.	*Populus*, *Salix*	leaves	Europe, N. America
alnibetulae Hd.	*Betula*	leaves	Europe
alnivora Sp.	*Alnus*	leaves	Europe
ambigua Fall.	cereals	leaves	Europe, N. America
apfelbecki Strobl	*Cynara*	leaves	Mediterranean
aristata Mall.	*Ulmus*	leaves	N. America
cinerascens Macq.	cereals	leaves	Europe
fragariae Malloch	*Fragaria*	leaves	N. America
frontella (Rond.)	*Medicago sativa*	leaves	Europe
hordei Sp.	*Hordeum*	leaves	India
intermittens Beck.	cereals	leaves	Europe
lathyri Hd.	*Pisum*	leaves	Europe
megalopsis Hg.	*Hordeum*	leaves	Europe
morivora Sas.	*Morus*	leaves	Japan
nana Mg.	*Medicago sativa*	leaves	Europe
nigrella (Rond.)	cereals	leaves	Europe
nigrociliata Hd.	cereals	leaves	Europe
oryzae Munakata	*Oryza*	leaves	Japan
parvicornis Loew	cereals	leaves	N. America
penniseti Sp.	*Pennisetum*	leaves	Africa
pennisetivora Sp.	*Pennisetum*	leaves	India

Species	*Host*	*Part of plant affected*	*Distribution*
populoides Sp.	*Populus*	leaves	N. America
rondensis Str.	cereals	leaves	Europe
spiraeae Kalt.	*Fragaria*	leaves	Europe
trebinjensis Strobl	*Celtis*	leaves	Mediterranean
varifrons Coq.	*Celtis*	leaves	N. America
yanonis (Mats.)	cereals	leaves	Japan
Sub-family Phytomyzinae			
Phytobia Lioy			
betulae (Kangas)	*Betula*	twigs, trunks	N. Europe
betulivora Sp.	*Betula*	twigs, trunks	N. America
cambii (Hendel)	*Salix*	twigs, trunks	N. Europe
carbonaria (Zett.)	*Malus*	twigs, trunks	N. Europe
cerasiferae (Kangas)	*Prunus*	twigs, trunks	Europe
pruni (Gross.)	*Prunus*	twigs, trunks	N. America
Amauromyza Hendel			
flavifrons (Mg.)	*Beta*, *Spinacia*, *Dianthus*	leaves	Europe, N. America
maculosa (Mall.)	*Lactuca*, *Chrysanthemum*	leaves	N. and S. America
Liriomyza Mik			
brassicae (Riley)	Cruciferae, *Pisum*	leaves	Cosmopolitan
braziliensis (Frost)	potato	tubers	S. America
bryoniae (Kalt.)	polyphagous	leaves	Europe
caulophaga (Kleinschmidt)	*Beta*	stalks	Australia
cepae (Hg.)	*Allium*	leaves	Germany
chenopodii (Watt)	*Beta*	leaves	Australia, N. Zealand
chinensis (Kato)	*Allium*	leaves	China, Japan, Malaya
cicerina (Rond.)	*Cicer arietinum*	leaves	Europe
clianthi (Watt)	*Clianthus*, *Sophora*	leaves	New Zealand
congesta (Beck.)	Leguminosae	leaves	Europe, Egypt
crucifericola (Hg.)	crucifers, *Vicia faba*	leaves	Germany
flaveola (Fall.)	cereals	leaves	Europe
huidobrensis (Blanchard)	polyphagous	leaves	N. & S. America
marginalis (Mall.)	cereals	leaves	Florida, S. America
nietzkei Sp.	*Allium*	leaves	Germany
phaseolunata (Frost)	*Phaseolus*	leaves	N. America
pisivora Hg.	*Pisum*	leaves	Europe
quadrata (Mall.)	potato, tomato	leaves	S. America
sativae Blanchard	polyphagous	leaves	N. & S. America, Hawaii
strigata (Mg.)	polyphagous	leaves	Europe
trifoliearum Sp.	*Medicago*, *Trifolium*, *Pisum*	leaves	N. America
trifolii (Burgess)	polyphagous	leaves	N. & S. America
urophorina Mik	*Lilium*	seed-heads	Germany
Haplomyza Hendel			
capsici Sp.	*Capsicum*	leaves	Argentina

Species	Host	Part of plant affected	Distribution
Paraphytomyza Enderlein			
dianthicola (Venturi)	*Dianthus*	leaves	S. Europe
heringi (Hd.)	*Fraxinus*	leaves	Europe
populi (Kalt.)	*Populus*	leaves	Europe
populicola (Walker)	*Populus*	leaves	Europe, Canada
Napomyza Westwood			
carotae Sp.	*Daucus*	roots (stems)	Europe
cichorii Sp.	*Cichorium*	leaves (stems)	Europe
lateralis (Fall.)	*Calendula*	stems, seed-heads	Europe, Canada, Japan
scrophulariae Sp.	*Digitalis*	stems, seed-heads	Europe
Pseudonapomyza Hendel			
asiatica Sp.	*Oryza*	leaves	Asia, Africa
atra (Mg.)	cereals	leaves	Europe
dilatata Sasakawa	cereals	leaves	Fiji, Samoa
spicata (Mall.)	cereals	leaves	Pacific, Formosa, Middle East
spinosa Sp.	*Zea mays*, *Hordeum*	leaves	Pacific, Australia, India, Middle East, Africa
zeae Sp.	*Zea mays*	leaves	Ghana
Phytomyza Fallén			
ferulae Hg.	*Daucus*	leaves	Mediterranean
fuscula Zett.	cereals	leaves	N. Europe
heringiana Hd.	*Malus*	leaves	Europe
horticola Goureau	polyphagous	leaves	Europe, Asia, Africa
ilicicola Loew	*Ilex opaca*	leaves	N. America
ilicis Curtis	*Ilex aquifolium*	leaves	Europe, N. America
nigra Mg.	cereals	leaves	Europe, Japan, N. America
penicilla Hd.	*Cichorium*	leaves, stems	Europe
persicae Frick	*Prunus persica*	leaves	N. America
petoei Hg.	*Mentha*	leaves	Europe
rufipes Mg.	crucifers	leaves, stems	Europe, N. America
syngenesiae (Hardy)	polyphagous	leaves	Europe, N. America, Australia, N. Zealand
vomitoriae Kulp	*Ilex vomitoria*	leaves	S. United States
Ptochomyza Hering			
asparagi Hering	*Asparagus*	leaves	Europe, China
Cerodontha Rondani			
angustipennis Harrison	Gramineae	leaves	New Zealand
australis Mall.	cereals	leaves	Australia, N. Zealand
denticornis (Pz.)	cereals	leaves	Europe
dorsalis (Loew)	cereals	leaves	N. & S. America, Mongolia
milleri Sp.	cereals	leaves	Australia, N. Zealand
incisa (Mg.)	cereals	leaves	Europe, N. America
lateralis (Macq.)	cereals	leaves	Europe
orbitona Sp.	*Oryza sativa*	leaves	W. & S. Africa
oryzivora Sp.	*Oryza sativa*	leaves	India

It will be seen from the pest list above that the great majority of species are restricted in their hosts to plants in a single family or genus. Isolated cases occur where a species feeds normally on one genus or family but also in addition is not infrequently found on one other unrelated genus. Hering (1951) refers to this as disjunctive oligophagy. Two such cases are *Japanagromyza* sp. (Peru) feeding on *Phaseolus* as its main host but also on *Brassica* (p. 83), and *Liriomyza brassicae* (Riley) normally restricted to cruciferous hosts but also feeding on *Pisum* (p. 153). Complete polyphagy is exceptional and occurs in only 13 species in the family – in five *Tropicomyia* species (Chapter 7), in five *Liriomyza* and two *Phytomyza* species (Chapter 8) and also in *Liriomyza crucifericola* Hering (Chapter 6).

Apart from these exceptions the Agromyzidae show a high degree of host specificity. When considering pests, it is important to appreciate the normal and probably obligatory host range of any species which may be encountered. The rigid and remarkable degree of host specificity is almost certainly of ancient origin and can only be fully understood as a reflection of the evolution of the family, in association with the rise and spread of the angiosperms, which provide all hosts apart from an insignificant number of ferns. It is possibly not without interest briefly to consider this evolutionary background.

Many different theories have been held by botanists about the evolution of the angiosperms but today there seems to be general agreement on a number of basic facts which seem very relevant to the evolution of the Agromyzidae. Firstly, it is generally accepted that primitive angiosperms were woody, in fact smallish trees. Herbaceous families are derived from woody ancestors, and there are a number of families including both herbaceous species and also more primitive trees, such as Solanaceae with tree Solanums. It is significant that the earliest angiosperm fossils were almost without exception of trees. It is also now agreed that the monocots are derived from the dicots, but this evolutionary division must have started early in the history of the angiosperms. Although no exact date can be set for the start of angiosperm evolution, there are no pre-Cretaceous angiosperm fossils but many of the early Cretaceous fossils belong to fairly advanced genera and clearly their origin must have been substantially earlier – in Jurassic or even Triassic time. By mid-Cretaceous, angiosperms are replacing gymnosperms as the dominant flora in many deposits. By early Tertiary – some 60 million years ago – practically all plant remains belong to modern genera. By the Miocene many species are indistinguishable from modern ones.

A further important aspect of the evolution of the angiosperms which is very relevant to the evolution of the Agromyzidae is the area where this evolution took place. Obviously the presence of a single primitive plant in a given area is no real indication that that is the area where it originally evolved. But if there are not only numerous primitive species but also numerous primitive genera and also families, this cannot unreasonably be considered as indicating evolution in the area. The area where this aggregation of primitive forms is found is in S.E. Asia – particularly the tropical montane forests of W. China and the mountains of N. Vietnam. Takhtajan (1969) has documented this in considerable detail. Significantly, Africa and South America have by contrast relatively few representatives of primitive angiosperms. An important corollary of this tropical origin is that temperate plants have evolved from tropical ancestors – with the original altitudinal zonation giving rise to a secondary latitudinal one. Thus the early tropical evolution of angiosperms in S.E. Asia was in due course followed by the Tertiary explosion of temperate plants in the northern hemisphere, with modern genera already present by mid-Cretaceous.

One of the factors promoting the success and spread of the angiosperms is the complex relationship which existed with insects, particularly in developing more efficient means of pollination. From early Cretaceous onwards there was obviously active co-evolution of insects and plants – one could not have progressed without the other.

In trying to establish a time scale for the evolution of the Agromyzidae, there is very little positive direct evidence to go on, and it is largely a matter of inference. A number of fossil Agromyzidae have been described, particularly from Baltic amber, but when HENNIG (1965) critically reviewed these a few years ago he discovered that in fact none can be definitely ascribed to this family. However, one of the families considered to be closely related to the Agromyzids, the Odiniidae, are represented by a Baltic amber fossil, and other Acalypterate families such as Milichiids are also positively known from amber, both in Europe and from Mexico. Modern genera of these families were clearly well defined by this time, that is Eocene some 35–40 million years ago, and the families as such must have been established well before this date. Assuming the parallel evolution of the Agromyzidae and the angiosperms, it is not unreasonable to postulate that the family as such existed by late Cretaceous time, and the more primitive of modern genera had almost certainly evolved by the beginning of the Tertiary. The dominance of herbaceous angiosperm pollen, which is notable in fossil deposits from Miocene time onwards, indicates that there was a great evolutionary explosion at this time, and it seems probable that this was when many species in the largely leaf-mining genera evolved in association with their hosts.

The main Agromyzid genera have their own specific feeding pattern, and this shows an evolutionary progression which follows very logically the evolution of the angiosperms. The most primitive Agromyzids are in the genus *Phytobia*, which feed in twigs or trunks of trees, particularly Salicaceae, and *Populus* was identified as a genus already in mid-Cretaceous time. The next step would logically be feeding as internal borers in stems of herbaceous plants which, as indicated earlier, evolved from woody ancestors. The majority of species in the genus *Melanagromyza* feed in this way. The genus *Ophiomyia* is slightly more developed, in which the feeding is not inside the stem, but the larvae form external stem-mines. The final evolutionary step is represented by the most highly specialized feeding of all, leaf-miners.

The evolution of the different orders of angiosperms has been neatly illustrated by CRONQUIST (1968), who has grouped the various orders of dicots into six subclasses. The hosts of the world Agromyzidae show a remarkable correlation with CRONQUIST's system, with relatively few species feeding on the primitive subclasses, progressing to 25% in the Rosidae and no less than 50% in the Asteridae which include Compositae and Umbelliferae, the most highly developed families

Table II.

Subclass	*Orders*	*Orders as hosts of Agromyzidae*	*Percentage of Agromyzid species*
Dicotyledons			
Asteridae	9	9	50%
Rosidae	16	11	25%
Dilleniidae	12	7	2%
Caryophyllidae	11	4	2%
Hamamelidae	9	2	2%
Magnoliidae	6	2	2%

of dicots. Table 2 shows CRONQUIST's sub-classes annotated with the percentage of Agromyzids feeding on each. A similar correlation applies to the monocots.

It seems established that the increasing differentiation of the host-plants has been closely followed by a specialization on them of individual Agromyzid species. The exact mechanics of this evolutionary process is not entirely clear but there seem to be two obvious possibilities. Firstly, as a plant population became isolated, resulting ultimately in speciation, the Agromyzid population present throughout this evolutionary change was itself influenced and modified by the genetic and chemical changes in the plant host, leading to reproductive isolation from its own parent population. Alternatively, a plant population may have speciated in isolation, for example from wind-borne seeds and without its normal Agromyzid feeder. Later, being still closely related to its parent population, the new plant species could be relatively easily colonized by the original Agromyzid species, which would develop a close association with it, leading ultimately to its own speciation. Probably both of these processes have occurred, doubtless with many minor modifications.

The rigid association of Agromyzidae with their hosts which is one of the most striking facts in the ecology of the family has recently been confirmed experimentally. Females of *Phytomyza matricariae* Hendel normally feeding on *Tanacetum* were given a wide range of non-hosts but in no case would they oviposit. It was found, however, that larvae transplanted into leaves of plants not normally used as hosts were in many cases able to complete their feeding and transformation, and even produce adults (SEHGAL, 1971). Agromyzid larvae are unable to move from one plant to another or even externally from one part of a plant to another. There is thus no pressure in the larval stage which might tend to erode established host-specificity. The maintenance of inherited host-specificity therefore devolves exclusively on the females and for this they rely on the two powerful mechanisms of smell and taste. The initial finding of the correct host is a function of scent and smell. Plants contain many attractive substances such as glucosides, oils, alkaloids, saponins and tannins, and the females are able unfailingly to detect the incredibly subtle phytochemical differences even between related hosts. Corroboration that the selected plant is the correct host is then obtained by taste. Agromyzid females have the well documented habit of piercing the leaf or stem tissue with their ovipositors and then imbibing the exuding sap.

This appears to have a threefold function, firstly to confirm that the host is correct, second to ingest proteins specific to the host which are necessary for maturation of the ovaries and without which they are unable to oviposit, and finally actual feeding on the available carbohydrates.

It is hoped that this brief discussion of the evolutionary background of the Agromyzidae and their hosts will provide a fuller understanding of the significance and relative permanence of the host specificity found in the family.

In the host list given below only hosts among cultivated plants are included, although some additional wild hosts have been mentioned in the text, where relevant. The ornamentals discussed in Chapter 11 are arranged botanically, and it has therefore not seemed necessary to include them in the host list, apart from the few which regularly suffer significant economic damage. Plant families and genera are arranged alphabetically rather than systematically to facilitate easy reference.

Host list of Agromyzid pests

Host	*Species*
DICOTOLYDONEAE	
BETULACEAE	
Alnus sp.	*Phytobia* sp.
Alnus glutinosa (L.) Gaert.	*Agromyza alnivora* Sp.
Alnus incana (L.) Moench	*Agromyza alnivora* Sp.
Betula nigra L.	*Phytobia betulivora* Spencer
Betula pubescens Ehrh.	*Phytobia betulae* (Kangas)
Betula spp.	*Agromyza alnibetulae* Hendel
CHENOPODIACEAE	
Beta vulgaris L. var. *cicla* L.	*Amauromyza (Tril.) flavifrons* (Mg.)
	Liriomyza caulophaga (Kleinschmidt)
	Liriomyza chenopodii (Watt)
	Liriomyza huidobrensis (Blanchard)
	Liriomyza strigata (Mg.)
	Liriomyza trifolii (Burgess)
Spinacia oleracea L.	*Amauromyza (Tril.) flavifrons* (Mg.)
	Liriomyza chenopodii (Watt)
	Liriomyza huidobrensis (Blanchard)
	Liriomyza trifolii (Burgess)
COMPOSITAE	
Arctium lappa L.	*Ophiomyia lappivora* (Koizumi)
Calendula officinalis L.	*Napomyza lateralis* (Fall.)
Cichorium endivia L.	*Ophiomyia pinguis* (Fallén)
	Napomyza cichorii Spencer
	Phytomyza penicilla Hd.
Cichorium intybus L.	*Ophiomyia pinguis* (Fall.)
	Napomyza cichorii Spencer
	Phytomyza penicilla Hd.
Cynara cardunculus L.	*Agromyza apfelbecki* Str.
Cynara scolymus L.	*Agromyza apfelbecki* Str.
	Phytomyza syngenesiae (Hardy)

Host list of Agromyzid pests

Host	*Species*
Dimorphotheca sp.	*Napomyza lateralis* (Fall.)
Lactuca sativa L.	*Melanagromyza splendida* Frick
	Ophiomyia pinguis (Fall.)
	Amauromyza maculosa (Mall.)
	Liriomyza huidobrensis (Blanchard)
	Liriomyza trifolii (Burgess)
	Phytomyza horticola Gour.
	Phytomyza syngenesiae (Hardy)
Lactuca sp.	*Liriomyza bryoniae* (Kalt.)
	Liriomyza strigata (Mg.)
CONVOLVULACEAE	
Ipomoea Batatas Lam.	*Melanagromyza caerulea* (Malloch)
CRUCIFERAE	
Brassica campestris L.	*Phytomyza horticola* Gour.
Brassica juncea Coss.	*Phytomyza rufipes* Mg.
Brassica Napus L.	*Phytomyza horticola* Gour.
var. *napobrassica* (L.) Rebh.	*Phytomyza rufipes* Mg.
Brassica oleracea L.	*Phytomyza horticola* Gour.
var. *acephala* DC.	*Phytomyza rufipes* Mg.
var. *botrytis* L.	*Phytomyza rufipes* Mg.
var. *capitata* L.	*Phytomyza rufipes* Mg.
var. *gemmifera* Zeker	*Phytomyza rufipes* Mg.
var. *italica* L.	*Phytomyza rufipes* Mg.
var. *juncea* L.	*Phytomyza rufipes* Mg.
Brassica rapa L.	*Phytomyza horticola* Gour.
	Phytomyza rufipes Mg.
Brassica spp.	*Melanagromyza* sp.
	Japanagromyza sp. (Peru)
	Liriomyza brassicae (Riley)
	Liriomyza sativae Blanchard
Nasturtium sp.	*Liriomyza brassicae* (Riley)
Raphanus sp.	*Liriomyza brassicae* (Riley)
CUCURBITACEAE	
Citrullus vulgaris Schrad.	*Liriomyza bryoniae* (Kalt.)
	Phytomyza horticola Gour.
Cucumis melo L.	*Liriomyza bryoniae* (Kalt.)
	Liriomyza huidobrensis (Blanchard)
	Liriomyza sativae Blanchard
	Liriomyza trifolii (Burgess)
Cucumis melo L. var. *aegyptiacus* Sick.	*Liriomyza bryoniae* (Kalt.)
Cucumis sativus L.	*Liriomyza bryoniae* (Kalt.)
	Liriomyza sativae Blanchard
	Liriomyza trifolii (Burgess)
	Phytomyza horticola Gour.
Cucurbita moschata Duch.	*Liriomyza sativae* Blanchard
Cucurbita pepo L.	*Liriomyza sativae* Blanchard
	Liriomyza trifolii (Burgess)
	Phytomyza horticola Gour.

Host list of Agromyzid pests

Host	*Species*
EUPHORBIACEAE	
Ricinus communis L.	*Melanagromyza ricini* de Meij.
FAGACEAE	
Quercus alba L.	*Japanagromyza viridula* (Coq.)
Quercus borealis Michx. var. *maxima* Sarg.	*Japanagromyza viridula* (Coq.)
Quercus glauca Thunb.	*Japanagromyza quercus* (Sas.)
Quercus rubra L.	*Japanagromyza viridula* (Coq.)
	Phytobia sp.
LABIATAE	
Mentha longifolia (L.) Huds.	*Phytomyza petoei* Her.
Mentha spicata L. emend. Huds.	*Phytomyza petoei* Her.
Mentha spp.	*Phytomyza horticola* Gour.
LEGUMINOSAE	
Cajanus cajan (L.) Millsp.	*Liriomyza sativae* Blanchard
Cajanus indicus Spreng.	*Melanagromyza chalcosoma* Spencer
	Melanagromyza obtusa (Malloch)
	Melanagromyza sojae (Zehnt.)
	Ophiomyia phaseoli (Tryon)
Calopogonium mucunoides Desv.	*Melanagromyza dolichostigma* de Meij.
	Ophiomyia centrosematis (de Meij.)
Canavalia ensiformis DC.	*Ophiomyia phaseoli* (Tryon)
Centrosema pubescens Benth.	*Ophiomyia centrosematis* (de Meij.)
	Japanagromyza centrosemae (Frost)
Cicer arietinum L.	*Liriomyza cicerina* (Rondani)
	Phytomyza horticola Gour.
Cicer sp.	*Ophiomyia cicerivora* Spencer
Crotalaria juncea L.	*Melanagromyza dolichostigma* de Meij.
	Ophiomyia phaseoli (Tryon)
Crotalaria mucronata Desv.	*Ophiomyia centrosematis* (de Meij.)
	Ophiomyia phaseoli (Tryon)
Desmodium tortuosum (Sw.) DC.	*Japanagromyza desmodivora* Spencer
Dolichos sesquepedalis L.	*Ophiomyia phaseoli* (Tryon)
	Tropicomyia vigneae (Ség.)
Dolichos uniflorus Lam.	*Tropicomyia vigneae* (Ség.)
Flemingia rhodocarpa Bax.	*Melanagromyza chalcosoma* Spencer
Flemingia sp.	*Melanagromyza obtusa* (Malloch)
	Melanagromyza sojae (Zehnt.)
Glycine javanica L.	*Tropicomyia vigneae* (Ség.)
Glycine max. (L.) Merrill (=*Soja hispida* Moench)	*Melanagromyza dolichostigma* de Meij.
	Melanagromyza koizumii Kato
	Ophiomyia phaseoli (Tryon)
	Ophiomyia shibatsuji (Kato)
Glycine soja Sieb. and Zucc.	*Melanagromyza dolichostigma* de Meij.
	Melanagromyza koizumii Kato
	Melanagromyza sojae (Zehnt.)

Host list of Agromyzid pests

Host	*Species*
	Japanagromyza tristella (Thomson)
	Ophiomyia centrosematis (de Meij.)
	Ophiomyia shibatsuji (Kato)
Glycine sp.	*Melanagromyza vignalis* Spencer
Indigofera suffruticosa Mill.	*Melanagromyza sojae* (Zehnt.)
Indigofera sumatrana Gaertn. (=*tinctoria* L.)	*Melanagromyza sojae* (Zehntner)
Lablab niger L. (=*Dolichos lablab* L.)	*Melanagromyza bonavistae* Greathead
	Ophiomyia phaseoli (Tryon)
	Ophiomyia spencerella (Greathead)
	Tropicomyia vigneae (Ség.)
Lathyrus latifolius L.	*Agromyza lathyri* Hendel
Lens esculenta Moench	*Phytomyza horticola* Gour.
Lupinus sp.	*Liriomyza sativae* Blanchard
Medicago sativa L.	*Agromyza frontella* Rondani
	Agromyza nana Meigen
	Melanagromyza sojae (Zehnt.)
	Liriomyza congesta (Becker)
	Liriomyza sativae Blanchard
	Liriomyza trifoliearum Spencer
Melilotus sp.	*Melanagromyza sojae* (Zehnt.)
	Agromyza nana Mg.
Phaseolus atropurpureus DC.	*Ophiomyia phaseoli* (Tryon)
Phaseolus calcaratus Roxb.	*Melanagromyza dolichostigma* de Meij.
	Melanagromyza sojae (Zehnt.)
	Ophiomyia phaseoli (Tryon)
	Ophiomyia spencerella (Greathead)
Phaseolus lunatus L.	*Ophiomyia centrosematis* (de Meij.)
	Ophiomyia phaseoli (Tryon)
	Ophiomyia spencerella (Greathead)
	Japanagromyza sp.
	Liriomyza sativae Blanchard
	Liriomyza trifolii (Burgess)
Phaseolus mungo L.	*Ophiomyia phaseoli* (Tryon)
	Ophiomyia spencerella (Greathead)
	Tropicomyia vigneae (Ség.)
Phaseolus panduratus Mart. ex Benth.	*Ophiomyia phaseoli* (Tryon)
Phaseolus radiatus L.	*Melanagromyza dolichostigma* de Meij.
	Melanagromyza obtusa (Malloch)
	Melanagromyza sojae (Zehnt.)
	Ophiomyia phaseoli (Tryon)
	Melanagromyza dolichostigma de Meij.
Phaseolus vulgaris L.	*Melanagromyza phaseolivora* Spencer
	Ophiomyia centrosematis (de Meij.)
	Ophiomyia phaseoli (Tryon)
	Ophiomyia spencerella (Greathead)
	Japanagromyza sp. (Peru)

Host list of Agromyzid pests

Host	*Species*
Phaseolus spp.	*Tropicomyia polyphyta* (Kl.)
Pisum arvense L.	*Phytomyza horticola* Gour.
Pisum sativum L.	*Melanagromyza sojae* (Zehnt.)
	Ophiomyia orbiculata (Hendel)
	Tropicomyia pisi (Kl.)
	Agromyza lathyri Hendel
	Liriomyza brassicae (Riley)
	Liriomyza congesta (Becker)
	Liriomyza huidobrensis (Blanchard)
	Liriomyza pisivora Hering
	Liriomyza strigata (Mg.)
	Liriomyza trifoliearum Spencer
	Liriomyza trifolii (Burgess)
	Phytomyza horticola Gour.
	Phytomyza syngenesiae (Hardy)
Pueraria hirsuta Matsum.	*Japanagromyza tristella* (Thomson)
Pueraria javanica Benth. (=*phaseoloides* Benth.)	*Melanagromyza dolichostigma* de Meij.
	Japanagromyza tristella (Thomson)
Pueraria thunbergiana Benth. (=*P. triloba* Makino)	*Japanagromyza tristella* (Thomson)
Swainsonia galegifolia (And.) R.Br.	*Melanagromyza sojae* (Zehnt.)
Trifolium alexandrinum L.	*Phytomyza horticola* Gour.
Trifolium incarnatum L.	*Liriomyza trifoliearum* Spencer
Trifolium repens L.	*Liriomyza trifoliearum* Spencer
	Liriomyza trifolii (Burgess)
Trifolium spp.	*Agromyza nana* Meigen
	Tropicomyia pisi (Kl.)
Vicia faba L.	*Melanagromyza azawii* Spencer
	Melanagromyza fabae Spencer
	Liriomyza congesta (Beck.)
	Liriomyza crucifericola (Hering)
	Liriomyza huidobrensis (Blanchard)
	Phytomyza horticola Gour.
Vigna angivensis Baker	*Tropicomyia vigneae* (Ség.)
Vigna catjang Walp.	*Melanagromyza vignalis* Spencer
	Ophiomyia phaseoli (Tryon)
Vigna repens Baker (=*luteola* Jacq.)	*Japanagromyza aequalis* Spencer
	Japanagromyza inaequalis (Malloch)
	Liriomyza sativae Blanchard
Vigna unguiculata (L.) Walp. (=*V. sinensis* Endl.)	*Melanagromyza candidipennis* (Lamb)
	Melanagromyza chalcosoma Spencer
	Melanagromyza vignalis Spencer
	Ophiomyia centrosematis (de Meij.)
	Ophiomyia phaseoli (Tryon)
	Ophiomyia spencerella (Greathead)
	Ophiomyia vignivora Spencer
	Tropicomyia capeneri (Hering)
Vigna spp.	*Tropicomyia polyphyta* (Kl.)
Wistaria chinensis DC.	*Hexomyza websteri* (Mall.)

Host list of Agromyzid pests

Host	*Species*
LINACEAE	
Linum usitatissimum L.	*Melanagromyza lini* Spencer
	Liriomyza huidobrensis (Blanchard)
	Phytomyza horticola Gour.
MALVACEAE	
Gossypium sp.	*Tropicomyia flacourtiae* (Ség.)
	Liriomyza sativae Blanchard
Hibiscus esculentus L.	*Melanagromyza hibisci* Spencer
	Liriomyza trifolii (Burgess)
MORACEAE	
Morus sp.	*Agromyza morivora* Sasakawa and Fukuhara
MYRTACEAE	
Eucalyptus camaldulensis Dehnh.	*Japanagromyza eucalypti* Spencer
PEDELIACEAE	
Sesamum indicum L.	*Melanagromyza azawii* Spencer
ROSACEAE	
Crataegus spp.	*Phytobia carbonaria* (Zett.)
Fragaria ananasa Duchesne	*Agromyza spiraeae* Kalt.
Fragaria vesca L.	*Agromyza fragariae* Mall.
	Agromyza spiraeae Kalt.
Fragaria virginiana Duchesne	*Agromyza fragariae* Mall.
Malus spp.	*Phytobia carbonaria* (Zett.)
	Phytomyza heringiana Hd.
Prunus avium L.	*Phytobia pruni* (Grossenbacher)
Prunus cerasifera Ehrh.	*Phytobia cerasiferae* (Kangas)
Prunus domestica L.	*Phytobia cerasiferae* (Kangas)
	Phytobia pruni (Grossenbacher)
Prunus insititia L.	*Phytobia cerasiferae* (Kangas)
Prunus persica L.	*Phytomyza persicae* Frick
Rubus idaeus L.	*Agromyza spiraeae* Kalt.
Rubus occidentalis L.	*Agromyza ?fragariae* Mall.
RUBIACEAE	
Coffea arabica L.	*Tropicomyia coffeae* (Kon.)
	Tropicomyia flacourtiae (Ség.)
	Tropicomyia ?polyphaga (Spencer)
	Tropicomyia polyphyta (Kl.)
RUTACEAE	
Citrus sp.	*Tropicomyia flacourtiae* (Ség.)
	Tropicomyia polyphyta (Kl.)
SALICACEAE	
Populus spp.	*Hexomyza schineri* (Giraud)
	Japanagromyza salicifolii (Collin)
	Agromyza albitarsis Mg.
	Agromyza populoides Spencer

Host list of Agromyzid pests

Host	*Species*
	Phytobia ? cambii (Hendel)
	Paraphytomyza populi (Kalt.)
	Paraphytomyza populicola (Hal.)
Salix alba L.	*Phytobia cambii* (Hendel)
Salix fragilis L.	*Phytobia cambii* (Hendel)
Salix purpurea L.	*Phytobia cambii* (Hendel)
Salix triandra L.	*Phytobia cambii* (Hendel)
Salix viminalis L.	*Phytobia cambii* (Hendel)
Salix spp.	*Hexomyza albicula* Spencer
	Hexomyza cecidogena (Hering)
	Hexomyza simplicoides (Hd.)
	Japanagromyza salicifolii (Collin)
	Agromyza albitarsis Mg.
SCROPHULARIACEAE	
Digitalis lanata Ehrh.	*Napomyza scrophulariae* Sp.
Digitalis purpurea L.	*Napomyza scrophulariae* Sp.
SOLANACEAE	
Capsicum annuum L.	*Liriomyza huidobrensis* (Blanchard)
	Liriomyza sativae Blanchard
	Liriomyza trifolii (Burgess)
	Haplomyza capsici Spencer
Lycopersicum esculentum L.	*Liriomyza bryoniae* (Kalt.)
	Liriomyza huidobrensis (Blanchard)
	Liriomyza quadrata (Mall.)
	Liriomyza sativae Blanchard
	Liriomyza trifolii (Burgess)
	Phytomyza horticola Gour.
Nicotiana tabacum L.	*Liriomyza quadrata* (Mall.)
Solanum andigenum Juz. and Buk.	*Liriomyza braziliensis* (Frost)
Solanum melongena L.	*Liriomyza sativae* Blanchard
	Liriomyza trifolii (Burgess)
	Phytomyza horticola Gour.
Solanum tuberosum L.	*Liriomyza braziliensis* (Frost)
	Liriomyza huidobrensis (Blanchard)
	Liriomyza quadrata (Mall.)
	Liriomyza sativae Blanchard
	Liriomyza trifolii (Burgess)
THEACEAE	
Camellia theae Link	*Tropicomyia atomella* (Mall.)
	Tropicomyia flacourtiae (Ség.)
	Tropicomyia polyphyta (Kl.)
	Tropicomyia styricicola (Sas.)
	Tropicomyia theae (Cotes)
TILIACEAE	
Tilia americana Marsh.	*Hexomyza tiliae* (Couden)

Host list of Agromyzid pests

Host	*Species*
ULMACEAE	
Celtis australis L.	*Agromyza trebinjensis* Strobl
Celtis laevigata Willd.	*Agromyza varifrons* Coq.
Celtis occidentalis L.	*Agromyza varifrons* Coq.
Ulmus americana L.	*Agromyza aristata* Mall.
Ulmus rubra Mühl.	*Agromyza aristata* Mall.
UMBELLIFERAE	
Apium graveolens L.	*Melanagromyza apii* Hering
	Melanagromyza splendida Frick
	Liriomyza huidobrensis (Blanchard)
	Liriomyza trifolii (Burgess)
Daucus carota L.	*Melanagromyza splendida* Frick
	Liriomyza trifolii (Burgess)
	Napomyza carotae Spencer
	Phytomyza ferulae Hering
	Phytomyza syngenesiae (Hardy)
MONOCOTYLEDONEAE	
GRAMINEAE	
Arrhenaterum elatius (L.) J. & C. Presl	*Phytomyza fuscula* Zett.
Avena sativa L.	*Agromyza ambigua* Fall.
	Agromyza nigrella (Rond.)
	Liriomyza flaveola (Fall.)
	Liriomyza marginalis (Mall.)
	Pseudonapomyza atra (Mg.)
	Phytomyza fuscula Zett.
	Phytomyza nigra Mg.
	Cerodontha (*Po.*) *incisa* (Mg.)
	Cerodontha (*Cer.*) *denticornis* (Pz.)
	Cerodontha (*Cer.*) *dorsalis* (Lw.)
Coix Lachryma-Jobi L.	*Pseudonapomyza dilatata* Sas.
	Agromyza albipennis Mg.
Hordeum distichon L.	*Agromyza megalopsis* Her.
Hordeum hexastichon L.	*Agromyza yanonis* (Mats.)
Hordeum vulgare L.	*Agromyza albipennis* Mg.
	Agromyza ambigua Fall.
	Agromyza hordei Spencer
	Agromyza intermittens Beck.
	Agromyza megalopsis Her.
	Agromyza nigrella (Rond.)
	Agromyza rondensis Strobl
	Agromyza yanonis (Mats.)
	Liriomyza flaveola (Fall.)
	Pseudonapomyza atra (Mg.)
	Pseudonapomyza spinosa Spencer
	Phytomyza fuscula Zett.
	Phytomyza nigra Mg.
	Cerodontha (*Po.*) *incisa* (Mg.)
	Cerodontha (*Po.*) *lateralis* (Macq.)
	Cerodontha (*Cer.*) *denticornis* (Pz.)
	Cerodontha (*Cer.*) *dorsalis* (Lw.)

Host list of Agromyzid pests

Host	*Species*
Oryza sativa L.	*Agromyza albipennis* Mg.
	Agromyza oryzae (Mun.)
	Cerodontha (*Po.*) *orbitona* (Spencer)
	Cerodontha (*Po.*) *oryzivora* Sp.
Panicum miliaceum L.	*Agromyza parvicornis* Lw.
	Liriomyza marginalis (Mall.)
	Pseudonapomyza spicata (Mall.)
Panicum sp.	*Cerodontha* (*Cer.*) *dorsalis* (Lw.)
Pennisetum purpureum Schum.	*Agromyza penniseti* Spencer
Pennisetum typhoideum Rich.	*Agromyza pennisetivora* Spencer
Phleum pratense L.	*Phytomyza fuscula* Zett.
	Cerodontha (*Cer.*) *dorsalis* (Lw.)
Saccharum officinarum L.	*Pseudonapomyza dilatata* Sas.
	Pseudonapomyza spicata (Mall.)
Secale cereale L.	*Agromyza albipennis* Mg.
	Agromyza ambigua Fall.
	Agromyza cinerascens Macq.
	Agromyza intermittens Beck.
	Agromyza megalopsis Her.
	Agromyza nigrella (Rond.)
	Agromyza nigrociliata Hd.
	Agromyza rondensis Strobl
	Phytomyza fuscula Zett.
	Phytomyza nigra Mg.
	Cerodontha (*Po.*) *incisa* (Mg.)
	Cerodontha (*Po.*) *lateralis* (Macq.)
	Cerodontha (*Cer.*) *dorsalis* (Lw.)
Sorghum halepense (L.) Pers.	*Pseudonapomyza hispanica* Spencer
Triticum aestivum L.	*Agromyza albipennis* Mg.
	Agromyza ambigua Fall.
	Agromyza ? frontosa Becker
	Agromyza intermittens Becker
	Agromyza ? mobilis Mg.
	Agromyza nigrella (Rond.)
	Agromyza nigrociliata Hd.
	Agromyza rondensis Strobl
	Agromyza yanonis (Mats.)
	Pseudonapomyza atra (Mg.)
	Pseudonapomyza spicata (Mall.)
	Pseudonapomyza spinosa Spencer
	Phytomyza nigra Mg.
	Cerodontha (*Po.*) *incisa* (Mg.)
	Cerodontha (*Po.*) *lateralis* (Macq.)
	Cerodontha (*Cer.*) *australis* (Mall.)
	Cerodontha (*Cer.*) *denticornis* (Pz.)
	Cerodontha (*Cer.*) *dorsalis* (Lw.)
Zea mays L.	*Agromyza oryzae* (Mun.)
	Agromyza parvicornis Lw.
	Pseudonapomyza asiatica Spencer
	Pseudonapomyza dilatata (Sas.)
	Pseudonapomyza spicata (Mall.)
	Pseudonapomyza zeae Spencer
	Cerodontha (*Cer.*) *dorsalis* (Lw.)

Host list of Agromyzid pests

Host	*Species*
LILIACEAE	
Allium cepa L.	*Liriomyza cepae* (Hering)
	Liriomyza chinensis (Kato)
	Liriomyza huidobrensis (Blanchard)
	Liriomyza nietzkei Spencer
	Liriomyza trifolii (Burgess)
	Phytomyza horticola Gour.
Allium porrum L.	*Liriomyza nietzkei* Spencer
Allium schoenoprasum L.	*Liriomyza trifolii* (Burg.)
Asparagus officinalis L.	*Ophiomyia simplex* (Loew)
	Ptochomyza asparagi Hering
Lilium indicum L.	*Liriomyza urophorina* Mik
Lilium martagon L.	*Liriomyza urophorina* Mik

III. LEGUMINOUS CROPS 1: INTERNAL FEEDERS

Beans, pulses and legumes are of extreme importance as a source of food, particularly in the tropics, and substantial damage is caused to a number of crops by several different species considered here. Others occur regularly on cultivated leguminous crops without normally having any significant effect on yields, but even these species deserve attention, as with periodical build-ups in population levels local infestations may from time to time prove to be severe.

Seventeen species are discussed below, of which 12 are centred on the Old World tropics of Africa and Asia. Five occur exclusively in Africa, four are limited to the Oriental Region, while three have a more extensive distribution covering Africa, Asia and also northern Australia. Two are from Europe, two from Japan and only one from South America. Further species certainly await discovery, both in the Old and New World tropics.

Melanagromyza species are mainly internal stem-borers but feeding can also take place in the root, in seed-heads or, in the case of legumes, in pods. One pod-feeder, *M. obtusa* (Malloch) is a serious pest on *Cajanus* in the Oriental Region and has particularly come to notice in India, Java and Malaya. A closely related species, *M. chalcosoma* Spencer, occurs in pods of *Cajanus* and *Vigna* in Africa, where there is also a second species on *Vigna*, *M. vignalis* Spencer. A root-feeder on *Vicia faba*, *M. fabae* sp.n., is now causing concern in England, and a stem-feeder on the same host, *M. azawii* sp.n., has been found damaging broad beans in Iraq. This species is unusual, as in addition to what is probably its normal host, *Vicia faba*, it has also been found in stems of *Sesamum indicum* L.

The three species *M. bonavistae* Greathead in East Africa, *M. dolichostigma* de Meij. in Java and Formosa and *M. koizumii* Kato in Japan represent a closely related complex, as seen from the unusually long larval anterior spiracles and the characteristic form of the posterior spiracles. Apart from differences in genitalia, there are also significant differences in biology – the former feeding in pods, the two latter in stems.

Among *Ophiomyia* species, the most serious pest in the family is *O. phaseoli* (Tryon), which is transferred below to this genus from *Melanagromyza*. With this species, the stems of young plants are frequently so heavily attacked that the plants die. *O. phaseoli* has an extensive distribution throughout Asia and Africa, it has reached northern Australia and Hawaii and is also well established in Egypt. It is not yet known from Central and South America and the utmost care is called for to prevent its introduction to these large areas where it would certainly flourish and cause incalculable losses. A closely related species, *O. spencerella* (Greathead), has recently been discovered in East Africa where it is perhaps an even more serious pest than *O. phaseoli*.

All these species are strictly limited to the Leguminosae for their hosts but within this family many species show a considerable degree of oligophagy, the genera *Dolichos*, *Phaseolus* and *Vigna*, for example, each being attacked by a number of species. It has therefore seemed more practical to arrange the discussion systematically by pests rather than alphabetically by crops.

Melanagromyza azawii sp.n. – Iraq

HEAD. Frons narrow, equal to width of eye, not projecting above eye in profile; 2 ors, 2 ori, the lower incurved; orbital setulae in single row, reclinate; ocellar triangle extended, apex reaching level of upper ori; orbits well differentiated but narrow; jowls deepest in centre below eye, about 1/7 vertical height of eye, cheeks visible as narrow ring (antennae missing in both available specimens).
MESONOTUM. Two strong dc, acrostichals in some 6 rows.
WING. Length in male 2.2–2.4 mm, in female up to 2.7 mm, costa extending strongly to vein M_{1+2}, last section of M_{3+4} slightly over two-thirds penultimate, first cross-vein just beyond midpoint of discal cell.
COLOUR. Frons mat black, ocellar triangle conspicuously, orbits distinctly shining black; mesonotum shining olive-green, abdomen more mat green; squamae and fringe white, margin not differentiated.
MALE GENITALIA. Aedeagus as in Figs. 4 and 5.
HOST-PLANTS. *Vicia faba* L.; also *Sesamum indicum* L.
BIOLOGY. The larva feeds as an internal stem-borer; the posterior spiracles are well separated and each bear an ellipse of normally 8 well-defined pores around the black central horn (Fig. 6).
HOLOTYPE ♂, Iraq: Baghdad, Abu Ghraib, ex stem of broad bean, October 1967; paratypes: 1 ♂, same data; 1 ♂, 1 ♀, 16.vi and 6.vii.1970, same locality, ex stems of *Sesamum indicum* (Younis M. Fattah). Holotype and 1 paratype in British Museum (Natural History), 2 paratypes in author's collection.

Discussion

TAXONOMY. This species is clearly not of Palaearctic origin but belongs to the large complex of greenish species occurring in the Ethiopian and Oriental Regions. I have pleasure in naming the species in honour of Dr. ABDULLA AL-AZAWI who first discovered it feeding in broad bean stems and who in recent years has taken an active interest in the leaf-mining Agromyzidae in Iraq.
HOST-RANGE. It is extremely surprising to find the same species feeding on *Vicia faba* (Leguminosae) and *Sesamum indicum* (Pedeliaceae), as these two families are not in any way related and belong to different sub-classes (CRONQUIST, 1968). Further investigations are required to establish the normal host of this species.
ECONOMIC IMPORTANCE. The presence of larvae in the stems of young bean plants in October and November causes them to wither and die (Dr. AL AZAWI,

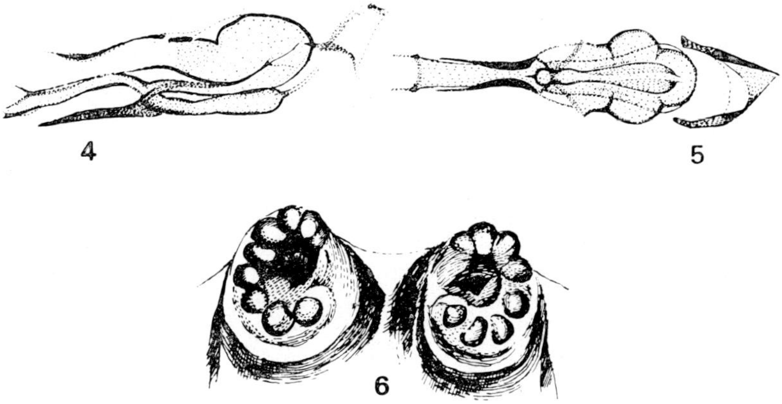

Figs. 4–6. *Melanagromyza azawii* sp.n.: 4, aedeagus, side view; 5, same, ventral view; 6, posterior spiracles of puparium.

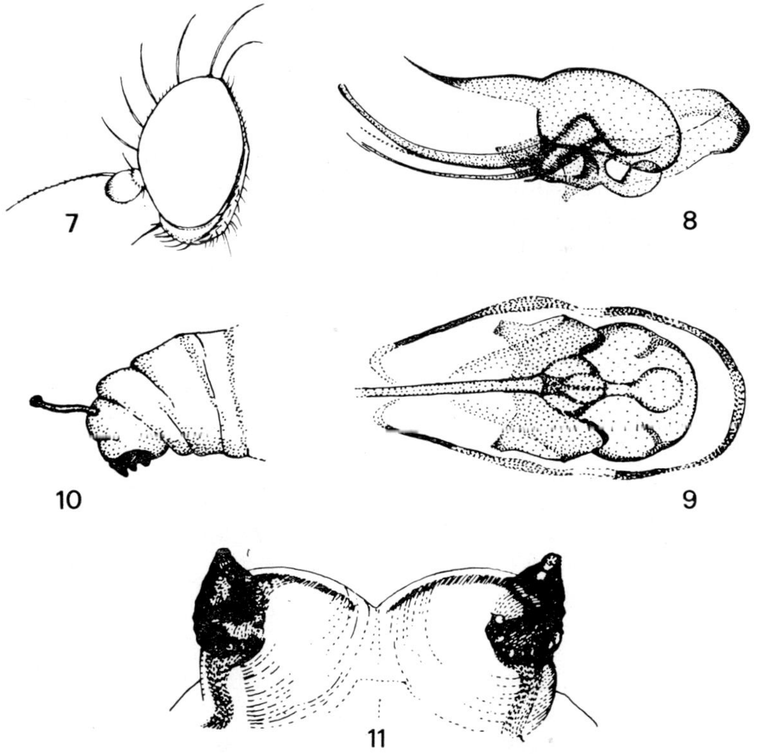

Figs. 7–11. *Melanagromyza bonavistae* Greathead: 7, head; 8, aedeagus, side view; 9, same, ventral view; 10, anterior spiracles of larva; 11, posterior spiracles of puparium.

personal communication). As a new pest of both *Vicia faba* and *Sesamum*, this species deserves further study.

Melanagromyza bonavistae Greathead, 1971 – East Africa

ADULT. Small greenish species.
HEAD (Fig. 7). Frons equal in width to eye, not projecting above eye in profile; 2 reclinate ors, 2 or 3 somewhat incurved ori; orbital setulae sparse, short, reclinate; ocellar triangle well defined, apex extending to lower ors; lunule semicircular; jowls narrow, about 1/10 height of eye; third antennal segment small, round, arista long, distinctly pubescent.
MESONOTUM. 2 strong dc, acrostichals numerous, in some 10 rows.
WING. Length from 1.9 to 2.3 mm, costa extending strongly to apex of vein M_{1+2}, last section of M_{3+4} two-thirds length of penultimate.
COLOUR. Frons mat black, ocellar triangle and orbits weakly shining, faintly greenish; mesonotum distinctly shining greenish, frequently somewhat coppery; abdomen conspicuously shining, greenish or coppery; legs entirely black; squamae and fringe white, margin scarcely differentiated.
MALE GENITALIA. Aedeagus as in Figs. 8 and 9.
EARLY STAGES. Larva: in third instar, mouth hooks with two strong teeth; anterior spiracles unusually long (Fig. 10); posterior spiracles in form of irregular elliptical chitinized plates, each on a conspicuous protuberance and with 12 spiracular bulbs and no trace of a central horn. Puparium: yellowish- to reddish-brown, posterior spiracles as in Fig. 11.
HOST-PLANT. *Lablab niger* L. (Bonavista bean).
BIOLOGY. 'Most eggs are laid in soft young pods in which the seeds are small and the pod walls are still in contact. In these young pods the larva mines in the wall, frequently escaping into the lumen where it feeds on the inner surface of the pod and on the seeds. Larvae developing in older pods tend to remain in the wall of the pod, producing an irregular blotch-mine. As oviposition seems to be concentrated on the side of the pod away from the attachment of the seeds, damage does not affect the ripening of the seeds unless the larvae enter the lumen. When pupation occurs in the mine, the puparium lies with its ventral surface towards the lumen of the pod and a thin "window" is prepared in front of the puparium prior to pupation. The prominent spiracles do not penetrate the wall. When pupation occurs in the lumen of the pod, the puparium adheres by its ventral surface to the side of the pod.' (GREATHEAD, 1971).
DISTRIBUTION. Uganda: Kawanda; Tanzania: Ifakara.
PARASITES. Three species of hymenopterous parasites have been reared from the puparia of this species in Uganda: *Habrocytus* sp. (Pteromalidae), *Eurytoma* sp. (Eurytomidae) and *Eucoilidea* sp. (Cynipidae). In the limited series of rearings of *M. bonavistae* which have been carried out the former species was found to be the most frequent parasite. None of the three, however, are host-specific on *M. bonavistae*, and all have been reared from *M. chalcosoma* Spencer; *Eucoilidea* sp.

has also been reared from stem-mining bean-flies (GREATHEAD, 1969) and *Ophiomyia strigalis* Spencer (GREATHEAD and MILNER, 1971).

ECONOMIC IMPORTANCE. This cannot at the moment be accurately assessed, in view of our limited knowledge of the distribution of *M. bonavistae*. However, its recent discovery at the Kilombara Agricultural and Research Station in Tanzania (E. BOHLEN) suggests that it may be widespread in East Africa. In Uganda *M. bonavistae*, together with the complex of other insects developing in this host which includes Lepidoptera, particularly *Maruca testulalis* (Geyer) (Pyralidae), Cecidomyiids and Eurytomids, together destroy most of the bonavist beans set at Kawanda (GREATHEAD, 1971).

Melanagromyza candidipennis (Lamb, 1912) – Africa

ADULT. Small green species with second cross-vein close to first.

HEAD. Frons narrow, equal to width of eye, not projecting above eye in profile;

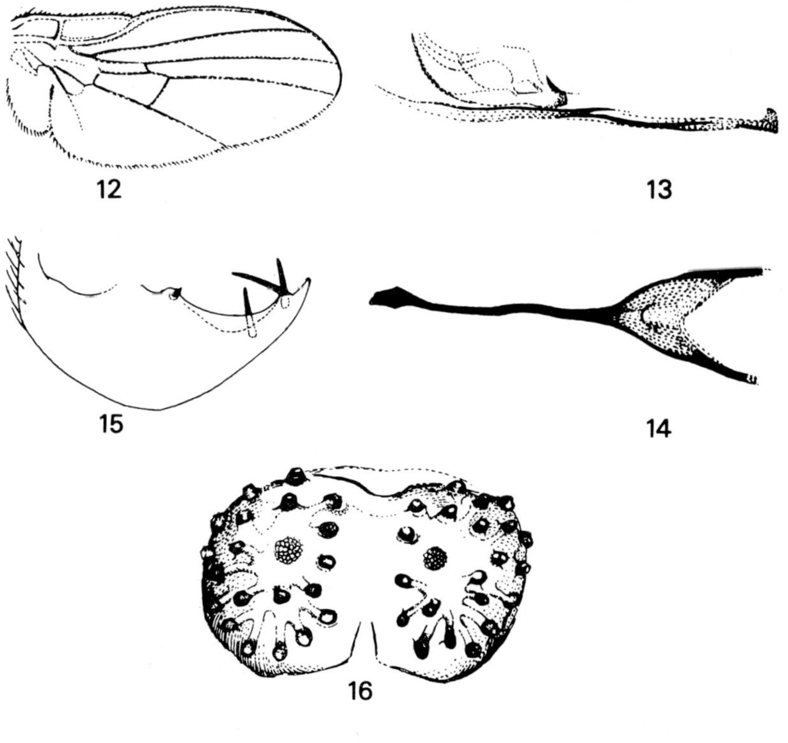

Figs. 12–16. *Melanagromyza candidipennis* (Lamb): 12, wing; 13, aedeagus; 14, ninth sternite; 15, surstylus; 16, posterior spiracles of puparium.

2 equal reclinate ors, 2 incurved ori, orbital setulae sparse, in single row, reclinate; ocellar triangle large, conspicuous, brilliantly shining, apex extending just below level of lower ors; lunule distinctly higher than semicircle; jowls deepest at rear, about 1/6 height of eye, cheeks scarcely differentiated; third antennal segment small, round, arista long, fine, bare.

MESONOTUM. 2 strong dc, acrostichals numerous, in about 8 rows.

WING. Length from 1.8 mm in male to 2.3 mm in female, costa extending strongly to vein M_{1+2}, the two cross-veins closely approximated (Fig. 12), last section of vein M_{3+4} distinctly longer than penultimate.

COLOUR. Frons mat black, ocellar triangle greenish-black; mesonotum greenish, somewhat mat, abdomen more shining green; wings conspicuously milky-white, veins almost colourless; squamae and fringe white.

MALE GENITALIA. Aedeagus as in Fig. 13, with long chitinized section below; ninth sternite with conspicuously elongated hypandrial apodeme (Fig. 14); surstyli with 3 strong spines on inner corner (Fig. 15).

EARLY STAGES. Larva not known; puparium orange-brown, with row of papilli around the centre of each segment; posterior spiracles in form of two adjoining oval plates, each bearing 21 raised, almost spherical bulbs (Fig. 16).

HOST-PLANT. *Vigna unguiculata* (L.) Walp. (cow-pea).

BIOLOGY. Not known in detail, but one specimen has been reared from stems at Ibadan, Nigeria, together with specimens of *Ophiomyia spencerella* (Greathead) and *O. phaseoli* (Tryon).

DISTRIBUTION. Seychelles; Tanganyika; Kenya; South Africa: Barberton; Nigeria: Ibadan.

ECONOMIC IMPORTANCE. Probably slight, as this appears to be an uncommon species. However, it is certainly of interest as an additional species in the complex attacking cowpeas. It is readily distinguishable by the milky-white wings with colourless veins, by the approximated cross-veins and the brilliantly shining greenish-black ocellar triangle.

Melanagromyza chalcosoma Spencer, 1959 – East Africa

Melanagromyza chalcosoma Spencer, 1959: 263. Holotype ♂ from Nyasaland in British Museum (Nat. Hist.).

Melanagromyza kikuyuana Spencer, 1959: 276, **syn. nov.** Holotype ♀ from Kenya ex 'pigeon-pea' in British Museum (Nat. Hist.).

ADULT. Large green species with large ocellar triangle.

HEAD. Frons about $1\frac{1}{2}$ times width of eye, not or scarcely projecting above eye in profile, though orbits well defined; two strong, normally equal ors, reclinate; two equal inclined ori but these weaker than ors; orbital setulae somewhat variable, normally in two rows, the inner row beside eye margin mainly reclinate, the outer row mainly proclinate; inner row occasionally largely lacking and the setulae then appear almost entirely proclinate; ocellar triangle large, pointed at apex, extending to slightly below lower ors; lunule large, slightly higher than a semi-circle, extending to level of upper ori; cheeks linear, jowls narrow, approximately 1/9 vertical height of eye, deepest point at midpoint below eye; third antennal segment large, with moderate pubescence, slightly broader in front, otherwise rounded, arista pubescent, longer than width of eye.

MESONOTUM. Two strong dc, second little more than half length of first, at level of supra-alar; acrostichals thick, extending irregularly in six rows to first dc.

WING. Length 2.8 mm, costa extending strongly to vein M_{1+2}, first cross-vein at mid-point of discal cell, last section of M_{3+4} short, little more than half length of penultimate.

COLOUR. Frons mat black, orbits slightly shining; ocellar triangle almost brilliantly shining black or frequently greenish; lunule grey; mesonotum shining

17

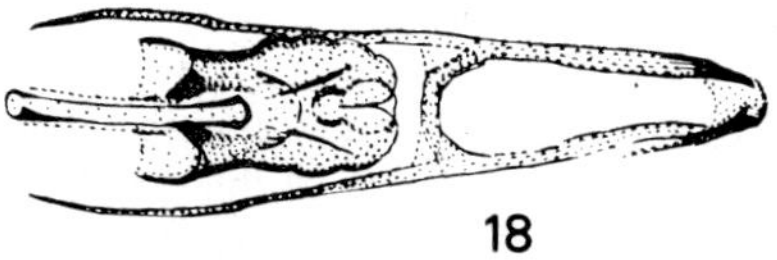

18

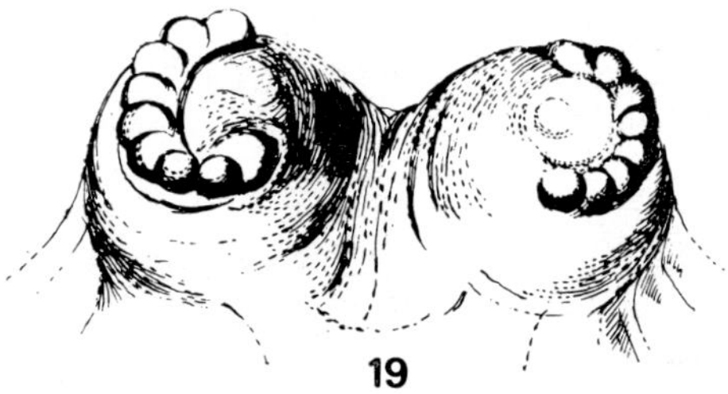

19

Figs. 17–19. *Melanagromyza chalcosoma* Sp.: 17, aedeagus, side view; 18, same, dorsal view; 19, posterior spiracles of puparium.

blackish-green, abdomen bright green, frequently with coppery tinge; wings hyaline, veins pale brown, distinctly yellowish at base; squamae white with yellowish-brown borders and white fringe.

MALE GENITALIA. Aedeagus distinctive as in Figs. 17, 18; surstyli with conspicuous bristles extending into epandrium.

EARLY STAGES. Puparium orange-brown, more rarely darker-brown; posterior spiracles each on a distinct protuberance, bearing an ellipse of 12 irregular bulbs, without a central horn and only lightly chitinized (Fig. 19).

HOST-PLANTS. *Vigna unguiculata* (L.) Walp., *Cajanus indicus* Spreng., *Flemingia rhodocarpa* Bax.

BIOLOGY. The adult fly deposits eggs through the wall of the developing pod. The larva feeds within a single seed, pupates there and emerges as an adult, commonly leaving its puparium within the seed. Attack results in a tiny cylindrical

cavity extending entirely through the long axis of the seed. Emergence of adults occurs within a week or two after the pods mature and begin to dry. No further infestation by this species takes place after the seeds have dried (KOEHLER and METHA, 1971).
DISTRIBUTION. Nyasaland, Kenya, Uganda; apparently not present in West Africa.
PARASITES. Not recorded.

Discussion

TAXONOMY. Before the male genitalia of the Agromyzidae were regularly studied, *Melanagromyza kikuyuana* was differentiated from *chalcosoma* by the slightly different arrangement of orbital setulae which, in the type series, appeared to be entirely proclinate. However, examination of additional material has shown that these setulae can be slightly variable and the genitalia of the male allotype of *kikuyuana* confirms that this species is synonymous with *chalcosoma*.

The male genitalia of *chalcosoma* indicate a close relationship with *Melanagromyza obtusa* (Malloch) which occurs commonly in India, Ceylon and Malaya. However, there are distinct morphological differences and also differences in the biology between the two species. Whereas the larva of *chalcosoma* pupates within individual seeds, in *obtusa* the puparium lies loose within the pod, never in the seed itself.
ECONOMIC IMPORTANCE. Of a sample of 400 cow-pea seeds of the cultivar SVS-47(-a) grown during the short rains which were investigated by KOEHLER and METHA, 13% were infested by *M. chalcosoma*. No important differences in percentage infestation were noted for two other cow-pea cultivars grown during that period. Damage by the fly during the long rains of 1969 however was found to be less than 1%.

M. chalcosoma causes little apparent damage to the seed and KOEHLER and METHA therefore conducted experiments in Kenya in 1969 in order to determine the effect of attack on germination and subsequent plant development, in order to assess whether such seed might be used for normal planting. It was found that the weight of fly-damaged seed averaged 9.95 grams per sample of 100 seeds, while that of undamaged seeds averaged 11.48 grams. For the two cultivars, Saunders Upright and SVS-47(-a), insect damage resulted in germination which was significantly lower than that for undamaged seeds of the same cultivars. For fly-damaged seeds mean radical length was significantly reduced in all cultivars when compared to undamaged seed. Speed of emergence of young seedlings was found to be significantly reduced by seed damage by comparison with emergence of plants produced from undamaged seeds. Final counts of percentage emergence of seedlings in the field, made ten days after planting, revealed that emergence was reduced as a result of planting fly-damaged seeds.

On emergence of the seedlings from the soil a striking difference was noted in the apparent vigour of plants arising from damaged, as compared to undamaged seeds. Damaged seeds produced a significantly higher proportion of plants with crinkled, otherwise deformed, or unusually small cotyledons. On many plants the cotyledons remained confined within the seed coat. Such plants usually died. Plants from damaged seeds weighed only two-thirds as much as those from undamaged seeds thirteen days after planting. However, this difference diminished as the growing season progressed and on the final date of sampling, 27 days after planting, plants from damaged seeds weighed four-fifths as much as those from undamaged seeds.

Most of the performance difference between damaged and undamaged seed in the field was attributable to the reduced emergence of plants from damaged seed; production of abnormal cotyledons unquestionably was ascribable to the planting of fly-damaged seed. When cotyledons could not escape from the seed-coat and the young plant died as a result, some economic loss might be expected. However, for seedlings which survived there was good evidence that normal plant size might be achieved before the flowering period was reached. It was not established whether reduced plant size during the early vegetative period contributes to low yields. (KOEHLER and METHA, 1971.)

It seems clear from the foregoing that *M. chalcosoma* must be treated as a pest of some significance in East Africa. Certainly in seasons during which populations are abnormally large, substantial damage might be expected.

Melanagromyza dolichostigma de Meijere, 1922 – Java, Formosa

ADULT. A small blackish species, with variable metallic colouration.
HEAD. Frons narrow, equal to width of eye, not projecting above eye in profile; 2 strong, equal ors, 2 weaker, incurved ori; orbital setulae sparse, reclinate, apart from 1 or 2 hairs in front; ocellar triangle with apex extending to lower ors; lunule slightly higher than a semicircle; eyes bare in both sexes; jowls narrow, about 1/8 height of eye; third antennal segment small, round, arista long, distinctly pubescent.
MESONOTUM. Two strong dc, acrostichals numerous, in some 8 rows.
LEGS. Mid-tibiae with 2 lateral bristles.
WING. Length from 1.8 mm in male to 2.2 mm in female; costa extending strongly to vein M_{1+2}, first cross-vein at midpoint of discal cell, last section of M_{3+4} about $\frac{2}{3}$ length of penultimate.
COLOUR. Frons mat black, orbits and ocellar triangle only weakly shining; mesonotum blackish but with distinct greenish or purplish tinge, abdomen similar; squamae and fringe white, margin differentiated, pale yellowish-brown.
MALE GENITALIA. Aedeagus as in Figs. 20, 21; ninth sternite with only slightly extended hypandrial apodeme.
EARLY STAGES. Egg yellowish-white, cylindrical, surface marked with distinct longitudinal grooves (VAN DER GOOT, 1930). Larva long and slender, left-hand of the two main teeth of the mouth-hooks substantially longer than the right-hand; anterior spiracles unusually long, with up to 12 minute pores (Fig. 22); posterior spiracles shorter with some 15 pores, central horn lacking (Fig. 23); puparium (Fig. 24) oval, 2.5 mm long, yellowish-brown (DE MEIJERE, 1922).

HOST-PLANTS. *Glycine max* Merr. (=*Soja hispida* Moench.), *Phaseolus calcaratus* Roxb. (Katjang oetji), *P. vulgaris* L. (Katjang boontjes); *Calopogonium mucunoides* Desv., *Crotalaria juncea* L., *Phaseolus radiatus* L., *Pueraria javanica* Benth.; *Phaseolus sublobatus* Roxb. (all Java); 'beans' (Formosa).
BIOLOGY. According to VAN DER GOOT (1930: 71) eggs are laid exclusively on the underside of leaves and are frequently not inserted into the leaf tissue but are merely deposited on the surface. Feeding punctures, however, are made on the

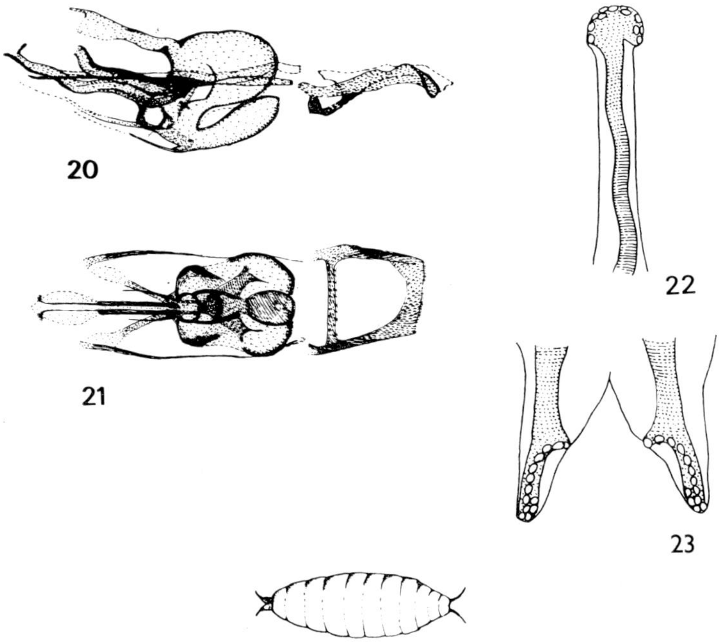

20

21

22

23

24

Figs. 20–24. *Melanagromyza dolichostigma* de Meij.: 20, aedeagus, side view; 21, same, ventral view; 22, anterior spiracle of larva; 23, posterior spiracle (Figs. 22, 23 after VAN DER GOOT); 24, puparium.

upperside of the leaves. Three or four eggs may be laid together on one leaf. Leaves are usually selected which have not fully unfolded and this, together with the hairs on the underside, prevent the eggs falling to the ground before the larva hatches. This takes place within 2 to 3 days of oviposition.

On hatching, the larva immediately eats its way into the leaf tissue and then into the nearest vein and so via the petiole to the stem. Here it feeds initially on the outer layers (bastweefsel) and after feeding down the stem for 2 or 3 cm, then turns and feeds upwards, now eating deeper into the pith of the stem. A substantial hollow is thus eaten out, covered only by the outer layers of the stem tissue and the epidermis.

Frequently two or even as many as six larvae can be found feeding together in *Phaseolus vulgaris* and, particularly in smaller plants, the entire stem tissue is eaten and the shoot dies as a result. In more robust plants the shoot is able to continue its growth and a gall-like swelling develops at the point of larval feeding. This regularly occurs in *Phaseolus calcaratus*, var. *trinervis*.

On one occasion larvae were found in young pods of *Phaseolus calcaratus* but the species normally occurring here is *Melanagromyza obtusa* Mall. (= *Weberi* de Meij).

Duration of larval feeding has not been accurately observed but appears to be from 8 to 10 days.

Pupation takes place in the hollowed-out stem. The puparia are always in the uppermost part of the stem, either in the shoot which has been killed or in the swelling immediately beneath the shoot. On one occasion puparia were found in the swollen petiole of one of the upper leaves.

The time of the entire life-cycle was checked in 18 cases in the laboratory and ranged from 17 to 21 days, averaging 18 days. Females lived in the laboratory for from 11 to 44 days, averaging 22 days.

NATURE OF DAMAGE. This varies considerably (VAN DER GOOT, 1930: 75) according to the specific host attacked.

On *soja hispida* plants are normally attacked when about 3 weeks old and eggs are laid on the third leaf. As long as plants continue their growth prior to blooming and fresh young leaves are available on the main shoot, these are used successively as the site of oviposition by one female after the other. When eggs are laid on young, unfolded leaves, these quickly turn brown and fall. With more mature leaves these withstand the attack longer and generally may fall after a further leaf has developed. The stem itself usually dies for a length of 2 or 3 cm; only in robust plants is the shoot able to continue its growth. In such cases a typical pale yellowish swelling can be seen marking the area where larvae have been feeding; this later turns more greenish. These characteristic features of attack by *M. dolichostigma* can be observed in fields of *Soja hispida* for a period of from 4 to 6 weeks.

Phaseolus calcaratus, *P. calcaratus* var. *trinervis* and *P. mungo* are particularly suitable as hosts owing to their hairy leaves, which provide a rough surface preventing the eggs falling. Leaves on these hosts are better able to withstand the attack and generally survive longer before falling and may even be able to maintain their growth. The entire feeding not infrequently takes place in the petiole. When the larvae reach the shoots, these often die but more frequently than in *Soja hispida* survive and continue normal growth, with a round swelling marking the area of main larval activity.

M. dolichostigma has only once been observed on *Phaseolus vulgaris*, with the shoots showing the characteristic swelling. On *P. radiatus* var. *sepiaria*, although the leaves are virtually without hairs, they nevertheless unfold slowly and thus presumably provide adequate support for the eggs until the larvae hatch. Attacked leaves in due course wither but although the larvae tend to limit their feeding to the petiole, the leaves do not fall until the adults have emerged. Even when feeding does take place in the stem, only a slight thickening results and the stem rarely dies.

On *Calopogonium mucunoides*, *Crotalaria juncea*, *Phaseolus sublobatus* and *Pueraria javanica* attack by *M. dolichostigma* results in the stem dying but the characteristic swelling was not observed (VAN DER GOOT, 1930).

DISTRIBUTION. Java (both at low and higher altitudes); Formosa.

PARASITES. The larval parasite, *Secodella* sp. (Eulophidae), was found in Java to be present in up to 12% of the sample checked on *Phaseolus calcaratus*. Four pupal parasites were also identified: *Cynipoid* sp., *Eurytoma poloni* Girault, *Eurytoma* sp. and *Trigonogastra agromyzae* Dodd. The most important of these was *Eurytoma poloni*. Of 2500 puparia checked in 40 separate observations an average parasitism rate of 10.5% was observed, with a maximum of 47%.

Discussion

TAXONOMY. The significant characters of *M. dolichostigma* are in the adult the unusual metallic reddish-purple colouration, referred to by DE MEIJERE in his description and also noted in specimens seen from Formosa, and in the larva the distinctive long anterior spiracles. The complex aedeagus is remarkably similar to that of *Melanagromyza cordiophoeta* Spencer (1961b: Fig. 43) from Singapore; however, this species is larger and more robust and both mesonotum and abdomen are conspicuously shining green. Male genitalia can appear virtually identical, in certainly distinct species (cf. *M. sojae* and *M. cunctans*, p. 53) and I therefore hesitate to synonymize *cordiophoeta* with *dolichostigma* until additional evidence from larvae and host-plants from the Singapore area is available.

The long larval anterior spiracles and the form of the posterior spiracles indicate a close relationship with *Melanagromyza koizumii* Kato from Japan. However, in the Japanese species the squamal fringe is dark brown and the biology differs, particularly with regard to oviposition. In *Melanagromyza bonavistae* Greathead from East Africa, which has similar larval spiracles, the squamal fringe is also pale and in this species oviposition takes place directly in the young pods.

HOST-PLANTS. Although a number of not closely related Leguminosae serve as hosts, the host range is nevertheless restricted and many other leguminous species and genera are not acceptable. VAN DER GOOT (1930: 74) lists the following: *Arachis hypogaea* L., *Cajanus indicus* L., *Canavalia ensiformis* DC., *Centrosema plumieri* Benth., *Crotalaria usaramoensis* E. G. Baker, *Indigofera suffruticosa* Mill., *Pachyrrhizus erosus* Urb., *Phaseolus lunatus* L., *P. semi-erectus* L., *P. radiatus* L., *P. trilobus* Ait., *Psophocarpus tetragonolobus* DC., *Rhynchosia minima* DC., *Vigna unguiculata* (L.) Walp. (as *sinensis*).

ECONOMIC IMPORTANCE. For various reasons the damage caused by *M. dolichostigma* is generally not serious. Populations are normally small; frequently the first attack takes place shortly before flowering and when the main growth has thus been completed; and finally, despite damage and even death of the shoot for a short length, it frequently survives and is able to resume normal growth. However, when 3- or 4-week-old plants are attacked, stunting of growth occurs and this is rarely compensated for by production of new shoots. The production of pods and the yield is then considerably reduced, as was seen with a crop of *Soja hispida* at Lembang, Java, in 1921. In this case about 50% of plants were attacked. Normally, however, the infestation rate is in the order of 1–2%. In Java it is frequently more serious at higher altitudes.

On *Phaseolus calcaratus* the frequent death of the top shoots inhibits growth and leads to a reduced number of pods. Infestation occurs regularly but generally at a low level but on occasions is severe (VAN DER GOOT, 1930).

On this evidence *M. dolichostigma* cannot be considered a serious pest but nevertheless when populations rise due to favourable circumstances economic loss is clearly caused.

Melanagromyza fabae sp.n. – England

ADULT. Medium-sized species, with proclinate orbital setulae and black squamal fringe.

HEAD (Fig. 25): frons 1½ times width of eye, slightly projecting above eye in profile; 2 strong ors, 3 or 4 largely incurved ori; orbital setulae numerous, entirely proclinate; eye with white pubescence at level of ors; jowls deepest in centre below eye, ¼ height of eye, cheeks linear; third antennal segment small, round, arista virtually bare.

MESONOTUM. 2 strong dc, acrostichals in some 10 rows in front, extending in 4 or 5 rows to level of first dc.

WING. Length from 2.6–2.7 mm in male to 3.2 mm in female, last section of vein M_{3+4} two-thirds length of penultimate, first cross-vein slightly beyond midpoint of discal cell.

COLOUR. Frons mat black, ocellar triangle and orbits scarcely shining; mesonotum mat, almost greyish-black viewed from front, weakly shining black from behind; abdomen weakly shining, greenish; squamae grey, margin and fringe black; halteres black.

MALE GENITALIA. Aedeagus as in Figs. 26, 27, cylindrical bladder below mesophallus curving upwards behind and extending slightly beyond rear of mesophallus.

EARLY STAGES. Mouth-hooks of larva each with single tooth, the left-hand one conspicuously stronger; posterior spiracles separated by own diameter and each with an ellipse of 12–15 elongate pores around the short central horn (Fig. 28); puparium pale, yellowish-white.

HOST-PLANT. *Vicia faba* L.

BIOLOGY. This is not known in detail but oviposition is believed to take place in the stem near ground level in young plants early in June. The larvae feed in the stem below the first node downwards in the centre of the woody root to a depth of about 6 cm and as many as four have been found in a single root. Pupation takes place at the end of July inside the root (Plate 1). There is only a single generation.

HOLOTYPE ♂, Cambs., Chatteris, emerged May, 1971, ex puparium collected

August, 1970; paratypes: 2♂, same data; 1♂, Thornhaugh, Lincs., March, 1972; 3♀, Stamford, Lincs., March, 1972; 4♂, 9♀, Rippingdale Fen, Lincs., emerged May, 1972 ex puparia collected Sept., 1971, all ex roots of *Vicia faba* (A. BIDDLE,

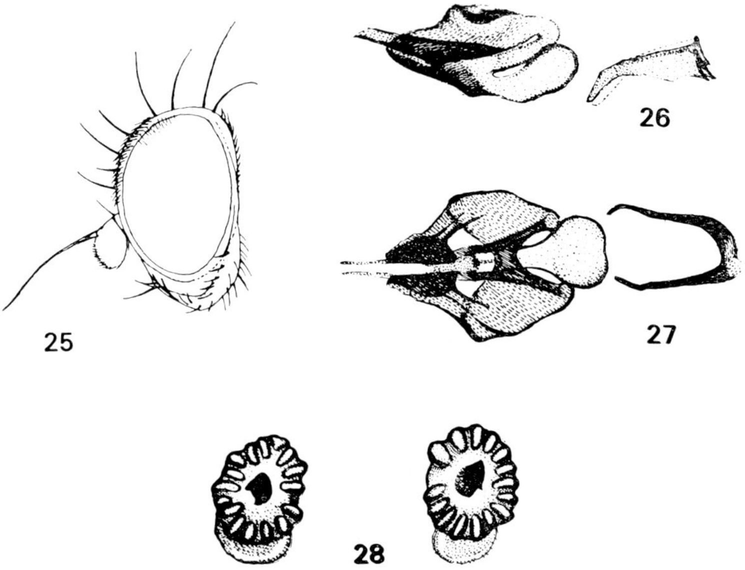

Figs. 25–28. *Melanagromyza fabae* sp.n.: 25, head; 26, aedeagus, side view; 27, ventral view; 28, posterior spiracles of puparium.

Pea Growing Research Organization). Holotype and paratypes in British Museum (Natural History), further paratypes in author's collection and the P.G.R.O.

Discussion

TAXONOMY. This species very closely resembles *M. eupatorii* Spencer, 1957b which feeds in stems of *Chrysanthemum*, *Eupatorium*, *Inula* and *Senecio*, and the only apparent external difference between the adults of the two species is the somewhat larger size in *fabae*, with wing length of 2.6–3.2 mm, as compared to 2.3–2.7 mm in *eupatorii*. Other important characters, such as the proclinate orbital setulae, mat mesonotum and black squamal fringe are identical in the two species.

The general form of aedeagus is also similar but in *fabae* there is greater chitinization and the aedeagus appears conspicuously darker, and also the basal bladder of the mesophallus is larger, with greater upward curvature.

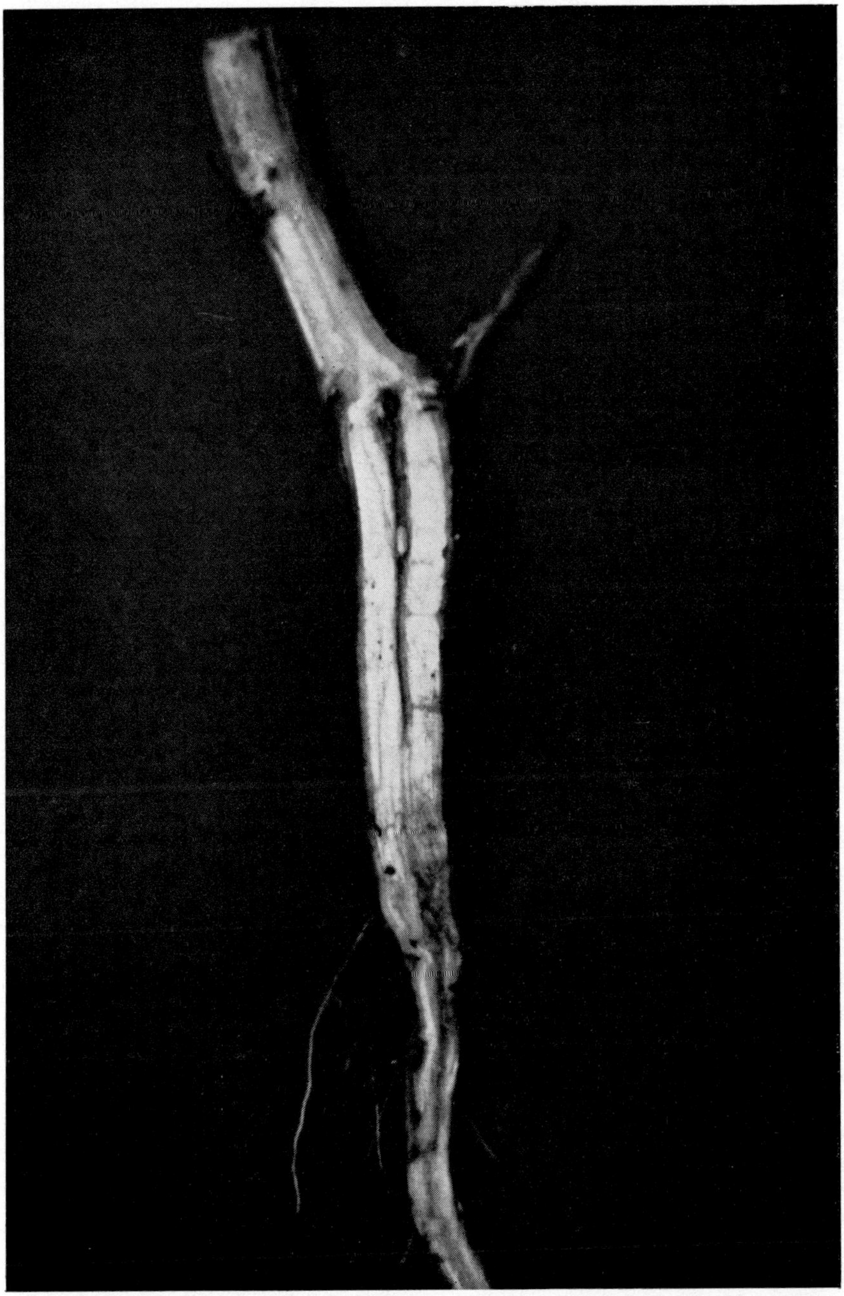

Plate 1: *Melanagromyza fabae* sp.n.: puparium inside root of *Vicia faba* (A. J. Biddle, Pea Growing Research Organization Ltd., England).

The most obvious difference between the two species is in the biology, with the larva of *fabae* feeding largely below ground in the root, while in *eupatorii* it feeds and pupates higher in the stem.

HOST-RANGE AND DISTRIBUTION. The only known host is *Vicia faba*, which has been cultivated in Britain since Roman times. However, it seems probable that native *Vicia* species may also serve as host and that abnormally large populations have built up with the large-scale cultivation of *Vicia faba* in eastern England.

The known distribution extends from the Hull area of S. Yorkshire, north of the Humber estuary, through Lincolnshire to N. Cambs.

PARASITISM. *M. fabae* is heavily parasitized by a Braconid (Dacnusinae), close to but apparently distinct from *Chorebus cybele* (Nixon). No detailed figures are available but more parasites than flies have been reared from several batches of puparia which have been collected. I have also seen a single specimen of an undetermined Pteromalid.

ECONOMIC IMPORTANCE. Samples from a number of fields of *Vicia faba* in Cambs. and Lincs. have confirmed a high level of infestation, with in many instances all plants out of a random sample of 20 containing puparia. Wilting of the plants just prior to harvesting was first found in August, 1970. In August, 1971, top wilting was again reported. 'A field near Holbeach had a broad bean crop growing on very heavy silt on the Wash coastline. Eight out of every ten plants examined were infested with the larvae. The plants belonging to those 80% were almost completely defoliated and in a state of advanced root decay. The fungi isolated from the area of the roots were *Fusarium solani* and in some cases *Fusarium culmorum*, both of which usually attack damaged tissue. The average yield of this field was 25 cwt/acre where the average for the other unaffected fields of similar soil type was in the region of 35 cwt/acre.' (A. BIDDLE, P.G.R.O., personal communication.)

The Pea Growing Research Organization is planning to undertake more detailed investigations of this species but it is already clear that the larval damage, combined with secondary fungal attack, makes this a pest of considerable local significance.

Melanagromyza koizumii Kato, 1961 – Japan

ADULT. Medium-sized blackish-green species, with dark squamal fringe.

HEAD (Fig. 29). Frons not projecting above eye, 2 strong ors, 2 weaker, partially incurved ori; orbital setulae in single row, reclinate; jowls unusually narrow, about $\frac{1}{18}$ height of eye; third antennal segment small, round, arista long, finely pubescent.

MESONOTUM. With 2 strong dc and acrostichals regularly in 8 rows.

WING. Length up to 2.5 mm, costa extending strong to vein M_{1+2}, last section of M_{3+4} about $\frac{2}{3}$ penultimate.

COLOUR. Frons mat, with brownish tinge, ocellar triangle and orbits distinctly shining; mesonotum shining black, with greenish or even purplish reflections, abdomen shining blackish-green, with golden reflections.

MALE GENITALIA. No specimens available for examination.
EARLY STAGES. Larva pale yellowish-white, up to 4.2 mm in length and 1.0 mm in width; mouth-hooks (Fig. 30) with left mandible having single large tooth and right mandible with single, much smaller tooth; anterior spiracles (Figs. 31, 32) long and slender, with 16 to 28 minute pores; posterior spiracles (Fig. 33) on

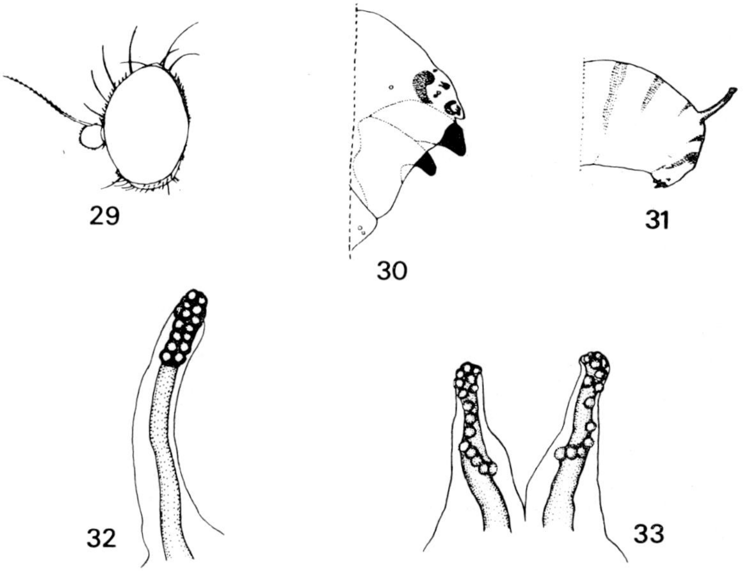

Figs. 29–33. *Melanagromyza koizumii* Kato: 29, head; 30, mouth-hooks of larva; 31, head of larva showing anterior spiracular process; 32, pores of anterior spiracle; 33, posterior spiracles (all after KATO).

conspicuous extensions, each with 13 to 17 pores; puparium brownish-yellow, posterior spiracles dark brown.
HOST-PLANTS. *Glycine max* Merrill, *G. soja* Sieb. and Zucc.; also the wild host *Rhynchosia acuminatifolia* Makino.
BIOLOGY. Oviposition takes place directly in the stem near the top of twigs. Newly hatched larvae mine along the inner side of the cortex layer but from the second instar the larvae penetrate more deeply into the stem. The puparium remains within the stem, with the anterior end protruding through a hole previously prepared by the larva (all from KATO, 1961).
DAMAGE CAUSED. No detailed account is available.
DISTRIBUTION. Japan: Honshu, ? Shikoku.
PARASITES. None recorded.

Discussion

TAXONOMY. The larval characters immediately indicate the close relationship of this species with *Melanagromyza dolichostigma* de Meijere from Java (p. 37). However, the dark squamal fringe and the differing oviposition habits justify considering this species as distinct.
ECONOMIC IMPORTANCE. KATO considers this to be one of the less injurious species attacking *Glycine* in Japan but the damage caused is said on occasions to be severe locally.

Melanagromyza obtusa (Malloch, 1914a) – Tropical Asia

(Bean Pod Fly)

Melanagromyza weberi de Meijere, 1922

ADULT. Medium-sized green species with large green ocellar triangle.
HEAD. Frons relatively narrow, equal to width of eye, not projecting above eye in profile; 2 equal, reclinate ors, 2 largely incurved ori, orbital setulae short, sparse, reclinate, apart from 1 or 2 proclinate hairs in front; ocellar triangle broad, large, apex extending to level of upper ori, almost reaching margin of lunule; jowls unusually narrow, about $\frac{1}{15}$ vertical height of eye; third antennal segment large, rounded, arista long, almost equal to height of eye, distinctly pubescent.
MESONOTUM. 2 strong dc, acrostichals in 12 rows immediately before second dc.
WING (Fig. 34). Length from 2.4 mm in male to 2.8 mm in female, costa extending strongly to vein M_{1+2}, last section of M_4 short, somewhat less than $\frac{2}{3}$ length of penultimate.
ABDOMEN. Basal cone of ovipositor conspicuously elongated (Fig. 35).
COLOUR. Frons mat black, ocellar triangle conspicuously shining greenish-black; orbits moderately shining blackish, lunule grey; mesonotum and abdomen shining green; wing veins pale brown, squamae and fringe white.
MALE GENITALIA. Aedeagus as in Fig. 36, basiphallus with unusually long side-arms; described in great detail by IPE (1967). IPE (1966) has also described the female genitalia.
EARLY STAGES. The egg is white, broad and rounded at the posterior end and narrowed anteriorly into a slightly curved, elongate process, which is an extension of the egg sheath; the egg itself is 0.66 mm long. The white larva has a fine brownish stripe along the entire mid-dorsal line of the body; when full-grown it measures from 3.5–4 mm in length and 1.25–1.5 mm in thickness. The puparium (Fig. 37) is yellowish-white when freshly formed but turns reddish-brown after about 24 hours; the posterior spiracles are each on a short, stout protuberance with an ellipse of some 10–12 pores around a small central scar, in place of the chitinized horn frequently present in this genus.

The larva and puparium have also been discussed and illustrated by SINGH and BERI (1971).

HOST-PLANTS. *Cajanus indicus* Spreng. (tur, dhal, red gram or pigeon pea; Hindi: arhar); in Java *Flemingia* sp.; in Ceylon *Phaseolus radiatus* L.

BIOLOGY. Females copulate 24 hours after emergence and oviposition quickly follows. Eggs are laid within the seed chamber of young pods, projecting through the pericarp inside the pod. AHMAD (1938: 68) found that in six females checked the average number of eggs laid was 38, spread over a period of only four days.

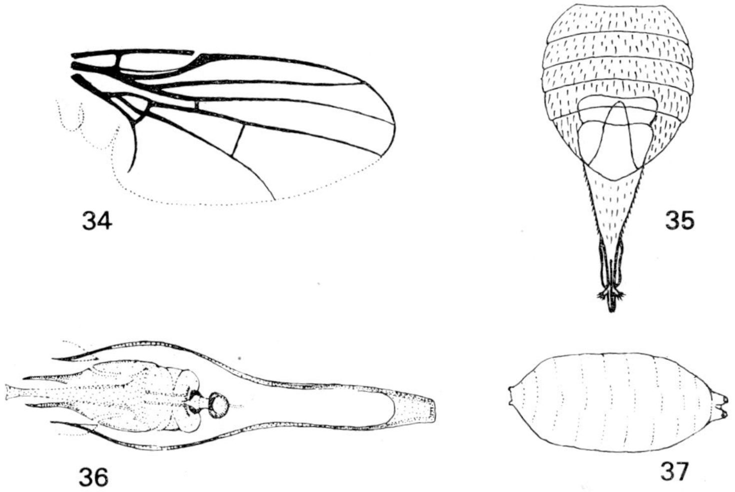

Figs. 34–37. *Melanagromyza obtusa* (Mall.): 34, wing; 35, abdomen (after PANDEY); 36, aedeagus; 37, puparium.

More than one egg is generally found in a single pod, the maximum being seven the average four.

'The average incubation period of eggs, with the temperature fluctuating between 23°C and 28°C is 3.1 days. On hatching the young larvae attach themselves to the soft seeds inside the pods and for the first few hours do not bore into the seed but feed on the surface. Soon after they mine into the seed like leaf-miners, their galleries running just under the epidermis of the seed. It is interesting to find that these galleries, full of faecal matter and some uneaten starch, appear like dirty white bands encircling it and are usually placed at right angles to the filament of the seed. As the larva grows and becomes well developed, it feeds deeper into the seed and even damages the embryo, but so long as it is inside the grain, externally the latter may appear undamaged and complete. On careful examination, however, the burrow of the insect can be distinctly made out and an occasional quivering of the testa locates the place where the larva is feeding. When full grown, the larva comes out of the seed and the hole of exit may be clearly seen. One seed is quite sufficient for the complete development of the larva and though in rare cases a larva may move from one seed to the next, it never leaves the pod in which it is born and thus passes its entire larval life in the same pod. The full-grown larva before pupating comes out of the seed and eats a partial hole into the wall of the pod to enable the future fly to escape. This hole does not communicate with the exterior but a thin circular layer of epidermis is left intact to be forced through by the emerging fly. The larva may either pupate on the groove of the seed itself or very close to the hole mentioned above with the head end directed towards the hole.' (AHMAD, 1938: 69–70).

The average duration of larval feeding at Pusa, India, in March/April was 6.1 days, in December from 18 to 20 days.

Both the eggs and larvae are highly sensitive to desiccation – against which they are fully protected within the enclosed pod. Under natural conditions, the pupal stage coincides with the ripening and drying up of the pods and the viability of the puparia is not impaired in this dry environment. The pupal duration varies according to temperature from 84 days at 27°C to 30.5 days at 18°C.

The main period of activity of *M. obtusa* is March–April and up to 3 generations can be completed in this period. The first flies of the autumn generation appear around mid-October (from AHMAD, 1938).

DAMAGE CAUSED. The young larvae immediately start feeding on the soft seed, initially feeding just below the epidermis, but as they grow they burrow deeper down, consuming the starchy food as well as the embryo, and depositing excreta as they feed. The damaged seeds thus become unfit for human consumption or germination.

In the earliest record of this pest from India at Nagpur, damage to tur-pods was estimated at 12.5% of the whole crop. More recent observations at Pusa and Delhi show that in serious cases damage can be as high as 63%. At Pusa 80% of the pods and 40–60% of the grains were affected during March and April, 1936. At New Delhi the percentage of damaged grains was 42% in April but only 11% in January (AHMAD, 1938). Table III on page 49 gives detailed records of damage observed at Pusa and New Delhi by AHMAD in 1936.

GANGRADE (1963) assessed the damage to different varieties of tur and summed up his findings as follows:

'The early variety Shahdol suffered heavily in the season of 1959–1960, when the damage to pods ranged between 36.0–100.0% and to grains between 30.7–86.8%. During 1960–1961, pods and grains of Shahdol were damaged from 29.0–63.3 and from 11.4–25.3%, respectively. The percentage of damage to pods in No. 148, Hyderabad and Nizamabad varied, respectively, from 45.0–54.0, 40.0–54.0 and 30.0–60.0; whereas in respect of grains of the same varieties, the damaged ranged from 23.3–29.8, 21.3–29.0 and 13.0–32.2. In the case of Local, the infestation of pods ranged between 41.0–57.0% in 1959–1960 and 27.0–58.0% in 1960–1961; while the infestation of grain was 24.5–37.2% and 13.2–31.7% in those years.'

It will be noted that the extent of damage in different varieties ranged from 27–100% of pods and from 11.4–86.8% of individual grains.

SINGH and BERI (1971) report that in the Agra area 'damage to the crop exceeds more than 40% of the yield'.

THEVASAGAYAM and CANAGASINGHAM (1960) investigated pests of dhal in Ceylon. They found that *M. obtusa* was apparently not present as a pest in the Maha season which last from January to March. (However, I found larvae and puparia in pods of *Phaseolus radiatus* L. at Peredeniya on 2 February, 1960.) In the Yala season, on the other hand, with warm, dry weather, *M. obtusa* was the

Table III. Extent of damage to tur pods by *Agromyza obtusa*

Locality	*Date of observation*	*No. of pods*		*Per cent affected pods*	*No. of grains*		*Percentage of damaged grains*
		Examined	*Affected*		*Examined*	*Affected*	
Pusa	27th March 1936	100	83	83	273	142	52.0
	28th March 1936	50	42	84	115	49	42.6
	30th March 1936	100	79	79	233	97	41.6
	1st April 1936	100	86	86	303	192	63.4
New Delhi	16th January 1937	100	28	28	153	18	11.8
	18th January 1937	100	32	32	318	36	11.3
	21st January 1937	100	28	28	152	17	11.2
	15th February 1937	100	12	12	149	10	6.9
	7th March 1937	100	50	50	160	62	38.7
	15th March 1937	100	54	54	158	57	36.1
	1st April 1937	100	59	59	166	71	42.7
	15th April	100	31	31	142	35	24.9

most serious of four pests present. Damage was as high as 50–75%. Partly damaged grains are usually subject to bacterial and fungal attacks and become entirely useless.

DISTRIBUTION. Formosa; India: New Delhi to Madras; Ceylon; Indonesia: Java, Flores.

PARASITES. Two Eulophids are known in India. GANGRADE (1962) described *Euderus agromyzae* and discusses its life history in considerable detail; he records a rate of parasitism of up to 18.7% in Madhya Pradesh. SINGH and BERI (1971) record *Euderus lividus* (Ashm.) in the Agra area; the rate of parasitism increases steadily from February onwards.

Discussion

TAXONOMY. This is a distinctive species, readily recognizable by the greatly enlarged ocellar triangle. It is clearly most closely related to *M. chalcosoma* Spencer in Africa (cf. p. 34) which has a similar biology and generally the same morphology, including the large ocellar triangle.
ECONOMIC IMPORTANCE. *M. obtusa* appears to occur in considerable numbers wherever its host-plants are cultivated in the Oriental Region. Detailed assessment of damage has only been made in India and with records of over 50% both of pods and grains attacked, the economic loss is clearly substantial. With no obvious damage to young, green pods visible externally, the economic importance of this species is probably not always fully appreciated by agriculturists.

Although GANGRADE (1963) reported attacks to tur-pods of above 80% in Hyderabad and Nizamabad (see above), GANGRADE (1964) showed that the actual loss of weight of infested grains was not substantial. In nine samples taken in these two areas it ranged from less than 1–3.3%. However, GANGRADE concludes: 'As a result of the attack of the larva of the fly, the grains become unfit for human consumption in so far as they cannot be turned into "Dal" for which it is used most.'

Melanagromyza phaseolivora sp.n. – Ecuador

HEAD. (Fig. 38). Frons only slightly wider than eye, not projecting above eye in profile; orbits well defined, bearing two (occasionally three) strong ors, the upper reclinate, the lower slightly incurved, and two strong, largely incurved ori; orbital setulae irregularly in two rows, reclinate; ocellar triangle large, broad, apex not extending below level of lower ors; lunule slightly higher than a semicircle; jowls broad, at deepest point at rear about ¼ height of eye; eye in male with conspicuous patch of white hairs at level of ors, in female with only sparse scattered hairs; third antennal segment small, round, arista long, almost equal to vertical height of eye, finely pubescent.
MESONOTUM. Two strong dc, acrostichals numerous, in 12 rows.
WING. Length from 3.1–3.2 mm, costa extending strongly to vein M_{1+2}, first cross-vein at approximately midpoint of discal cell, last section of M_{3+4} relatively long, at least ¾ length of penultimate.
LEGS. Fore-tibia without lateral bristle, mid-tibia with 2 strong bristles.
COLOUR. Frons mat black, ocellar triangle and orbits weakly shining, blackish; lunule somewhat paler, blackish-grey; mesonotum dark blackish-green, moderately shining when viewed from front, conspicuously so from rear; abdomen shining greenish; squamae and fringe white, margin slightly differentiated, pale orange.

Male genitalia. Aedeagus as in Figs. 39, 40; surstyli curving inwards with up to six long slender bristles at end (Fig. 41).

Host-plant. *Phaseolus vulgaris* L. (kidney bean).

Biology. Not recorded but almost certainly internal stem-borer.

Holotype ♂: Ecuador, Pichincha, Sta. Catalina Experiment Station, 1963; paratypes: 2♂, 2♀. Holotype and 2 paratypes in United States National Museum, 2 paratypes in author's collection.

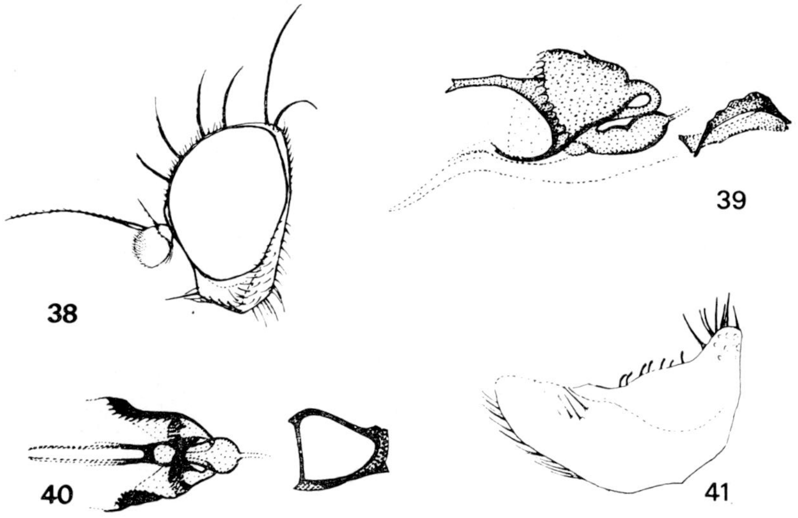

Figs. 38–41. *Melanagromyza phaseolivora* sp.n.: 38, head; 39, aedeagus, side view; 40, same, ventral view; 41, surstylus.

Discussion

This new species can be included in the author's (1973) key to Neotropical *Melanagromyza* species by amending and extending couplet 22 as follows:

Couplet 22, second alternative, read:

—	Larger species, wing length 2.7–3.2 mm	22a
22a	Frons distinctly projecting above eye in profile; wing length 2.7–2.9 mm	*chenopodii* Spencer
—	Frons not projecting above eye; larger species, wing 3.1–3.2 mm	*phaseolivora* sp.n.

Although there is no information available on the detailed biology and damage caused to the host-plant by this species, in view of its substantial size, the larva is clearly in a position to destroy young plants, and the species must be considered of potential economic importance.

Melanagromyza sojae (Zehntner, 1900) – Australia, Tropical Asia to Egypt and Africa

ADULT. Small black and green species.
HEAD. Frons only slightly wider than eye, not projecting above eye in profile; two ors, the upper slightly stronger; 2 slightly incurved ori; orbital setulae in well-defined row, reclinate; ocellar triangle large, broad, apex reaching level of upper ori, distinctly but not brilliantly shining; jowls conspicuous, deepest in centre below eye, about $\frac{1}{7}$ vertical height of eye, cheeks forming narrow ring below eye; third antennal segment round, arista long, virtually bare; eye bare in both sexes.
MESONOTUM. Two strong dc, acrostichals regularly in some 8 rows.
WING. Length from 1.8–2.1 mm, costa extending strongly to vein M_{1+2}, first

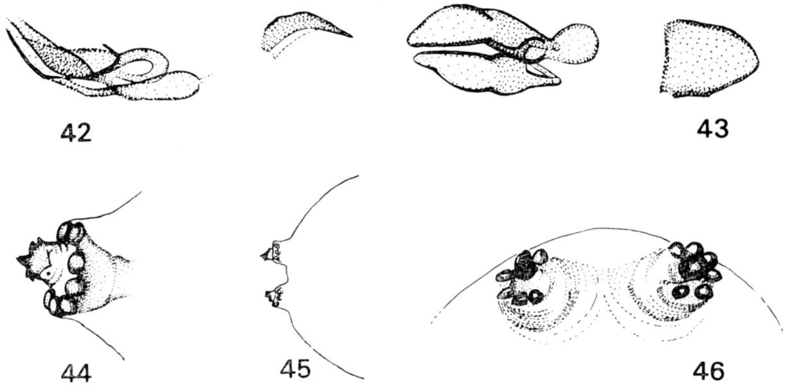

Figs. 42–46. *Melanagromyza sojae* (Zehnt.): 42, aedeagus, side view; 43, same, ventral view: 44–46, posterior spiracles of puparium (Figs. 44, 45 after KATO).

cross-vein slightly before midpoint of discal cell, last section of M_{3+4} slightly more than half length of penultimate.
COLOUR. Frons mat black, orbits and ocellar triangle moderately shining; mesonotum largely black but sometimes with detectable greenish tinge, abdomen conspicuously shining greenish; squamae and fringe entirely white.
MALE GENITALIA. Aedeagus as in Figs. 42, 43; basiphallus separated by distinct gap from distiphallus complex.
EARLY STAGES. Egg whitish, translucent, oval; larva slender, up to 3 mm long, anterior spiracles short, knob-like, with 8 minute pores; posterior spiracles well separated and consisting normally of 6 raised pores around the central truncated horn; puparium yellowish-brown, posterior spiracles as in Figs. 44–46.
HOST-PLANTS. *Glycine soja* Sieb and Zucc. (Java; Formosa); *Cajanus indicus* Spreng. (Katjang hiris), *Indigofera suffruticosa* Mill. (indigo), *I. sumatrana*

Gaertn., *Phaseolus calcaratus* Roxb. (Katjang oetji), *P. radiatus* L. (Katjang idjoe); *Swainsonia galegifolia* (And.) R.Br., Smooth Darling Pea (Australia); *Melilotus* sp. (India); *Medicago sativa* L. (Egypt, Saudi Arabia). Also in Java the wild hosts *Aeschynomene indica* L., *Flemingia* sp. and *Phaseolus sublobatus* Roxb. (VAN DER GOOT, 1930).

BIOLOGY. The life history has been investigated in Java by VAN DER GOOT (1930) and in Japan by KATO (1961). Oviposition takes place exclusively in leaves, never in cotyledons, the petiole or stem, and always on the underside. Many feeding punctures, however, are made on the upperside of the leaves. The eggs are most frequently laid near the base of the leaf but can be found elsewhere in the leaf-blade; they are always near a vein.

The larva hatches after 2 or 3 days and immediately bores into the nearest vein, continuing feeding through the petiole into the stem. Here, it feeds internally in the pith (as opposed to *O. phaseoli* which feeds close beneath the surface), boring down to the root and then again feeding upwards until it is full-grown. It pupates deep within the stem but first eats out a channel to the epidermis, to facilitate the subsequent emergence of the adult. In older plants the pupation site is midway up the stem but in young plants the larva may continue feeding until it again reaches the uppermost part of the stem. Normally only a single larva completes its development in one plant, though on occasions two puparia have been found.

Larval feeding lasts for 9–11 days, the pupal stage 9–10 days. The entire life cycle lasts from 16–26 days and in a sample of 35 averaged 21 days.

Females normally mate within 5 days of emergence. In the laboratory males fed on water and barley lived from 10–46 days, females from 15–36 days.

In Japan in Kyushu Province the larvae first appear in soy bean plants in June and at least four generations may occur until harvest in October (KATO, 1961).

DISTRIBUTION. Indonesia: Java, Flores, Sumbawa; Japan, Formosa, Micronesia, Australia: New South Wales; India: Ludhiana, Pusa; Saudi Arabia, Egypt; South Africa: Natal.

PARASITES. In Java three pupal parasites are recorded: *Cynipoide* sp., *Trigonogastra agromyzae* Dodd and *Eurytoma* sp. (VAN DER GOOT, 1930). A sample of 500 puparia showed a parasitism rate of 30.3%. In addition one larval parasite, *Secodella* sp., was regularly found (also attacking *M. dolichostigma*). In stems of Indigo a rate of parasitism of 24% was observed.

Discussion

TAXONOMY. ZEHNTNER's (1900) description of this species is brief and it has subsequently been discussed and re-described by DE MEIJERE (1922), VAN DER GOOT (1930), SPENCER (1961b, 1963b, 1963d, 1966b). The distinctive characters of this species are the large, broad ocellar triangle, the generally blackish mesonotum and the white squamae and fringe. The arrangement of the larval posterior spiracles is entirely characteristic and will permit immediate identification of the species.

M. sojae is clearly closely related to *M. cunctans* (Meigen) known only from southern and western Europe, where it forms slender stem-galls on *Lotus corniculatus* L. and other less common *Lotus* species. The two species cannot be distinguished either by external characters or male genitalia but the differing biology, supported by distinctive differences in larval morphology, confirm complete but possibly relatively recent, late Tertiary, speciation.

DISTRIBUTION. The wide distribution of *M. sojae* throughout the Pacific and the entire Oriental Region suggests that it originated somewhere within this area. It is particularly abundant on Java, including localities at 500 and 700 m but appeared to be absent from Lembang at 1200 m. In Japan it is believed to be widely distributed throughout Kyushu, Shikoku and the western part of Honshu (KATO, 1961).

In Egypt and Saudi Arabia it is only known in association with cultivated alfalfa (*Medicago*). Its presence in South Africa indicates either a wide distribution in Africa also, or could possibly be explained as an introduction. The species appears to be absent from West Africa.

ECONOMIC IMPORTANCE. Damage to host-plants is limited by the fact that cotyledons are never attacked as in *O. phaseoli* and *O. spencerella*, and plants are usually well established before the larva reaches the stem. Also, with only one or two larvae present, feeding is restricted to the specialized requirement of this species in the stem pith and the damage caused here has little effect on the supply of water and nutrient materials to the upper part of the plant. In exceptional cases, when inadequate food is present in the stem of young plants the larva feeds up into the main shoot which is hollowed out and dies. However, even when this occurs, plants are capable of producing auxiliary shoots and continue normal, though retarded development.

In Java the main host among cultivated plants is *Glycine soja*. VAN DER GOOT (1930: 62) reports that attack on this host varies from 70–100%. Of 21,000 plants checked only 3.5% were unaffected. Among older plants it is rare to find a single one immune. Another important host is *Indigofera suffruticosa* (Indigo). As many as 10 successive larval borings have been found in this host. Attack varies from 28–100%. VAN DER GOOT noted that infested plants tended to droop more rapidly during hot afternoons than healthy plants. Wild *Indigofera sumatrana* was found to be significantly more resistant to attack than *I. suffruticosa*.

On *Cajanus indicus*, *Phaseolus calcaratus* and *P. radiatus* the attack is always limited to very young plants and older plants appear to be avoided. However, even these young plants survive, owing to the habit of only a single larva normally being present in each stem.

In Java the common native plants *Aeschynomene indica*, *Flemingia* sp. and *Phaseolus sublobatus* were invariably infested with *M. sojae* and thus provide a permanent reservoir, from which populations are available for attacking any suitable leguminous hosts which may be cultivated in the vicinity.

In Japan this species mainly attacks soy bean plants during summer and early autumn. After the harvest of soy beans, the larvae are found in garden peas or in *Astragalus sinicus* L. (Chinese milk vetch) in late autumn. In the case of the so-called summer soy beans in Kyushu planted in early June, up to 100% infestation may occur and the damage caused is considered to be serious (KATO, 1961).

The significance of the apparently large population of *M. sojae* which is present in Egypt is not known but samples of the adult flies are regularly found on alfalfa.

Melanagromyza vignalis Spencer, 1959 – Africa

ADULT. Small shining blackish-green species.
HEAD. Frons equal to or slightly wider than eye, not projecting above eye in profile; two equal ors directed upwards, two weaker equal ori directed inwards; orbital setulae regularly reclinate from below lower ori to upper ors, at most 1 or 2 hairs in front more proclinate; ocellar triangle large, extending to mid-way between lower ors and upper ori; lunule slightly higher than a semicircle; orbits widening slightly at midpoint of lunule; jowls narrow, $\frac{1}{12}$ height of eye, slightly deeper behind, cheeks linear; third antennal segment rounded but slightly longer than broad, pubescence short; arista equal in length to width of eye, conspicuously pubescent.
MESONOTUM. Two strong dc, second only slightly more than half length of first, distinctly behind level of supra-alar; acrostichals in 8 rows at front, 4 rows only at first dc.
WING. Length from 2.2–2.4 mm, costa extending strongly to vein M_{1+2}, first cross-vein at midpoint of discal cell, last section of M_{3+4} short, half length of penultimate or slightly longer.
COLOUR. Frons mat black; ocellar triangle and orbits shining black; mesonotum and abdomen distinctly shining blackish-green, occasionally more coppery; squamae and fringe conspicuously silvery-white; wing veins pale brown.
MALE GENITALIA. Aedeagus characteristically asymmetrical, as in Figs. 47, 48, lower tubular process of mesophallus distinctly rotated to right (viewed ventrally from rear); ninth sternite with a finely extended hypandrial apodeme.
EARLY STAGES. Larva unknown, puparium dark orange-brown, 3.5 × 1.3 mm; posterior spiracular processes (Fig. 49) set on a substantial rounded projection on anal segment, each process consisting of a raised pedestal conspicuously darker than the anal projection, each bearing an ellipse of 12 bulbs surrounding a central scar; the two processes separated by more than their own diameter.
HOST-PLANTS. *Vigna catjang* Walp. (black-eyed pea), *Vigna unguiculata* (L.) Walp. (cow-pea), *Glycine* sp.
BIOLOGY. Not known in detail but the larva feeds and pupates within the seed-pods.
DISTRIBUTION. French Sudan; Nigeria; Sierra Leone; Kenya; South Africa: Pietermaritzburg.

Discussion

M. vignalis closely resembles *M. chalcosoma* but is distinctly smaller and can be distinguished by the conspicuously reclinate orbital setulae many of which are proclinate in *chalcosoma*.

As a pod-feeder, the life-history of this species generally resembles that of *chalcosoma* but there may well be essential differences in the feeding pattern which may reduce direct competition between the two species.

With distribution records from East, West and South Africa, *M. vignalis* can be expected to occur in all areas where its host-plants are found.

47

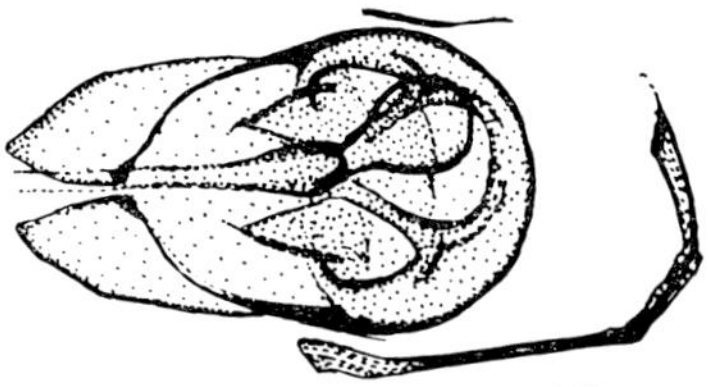

48

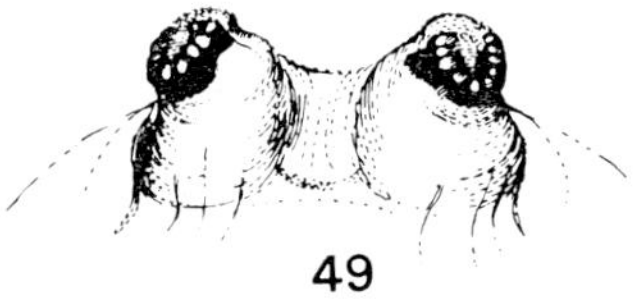

49

Figs. 47–49. *Melanagromyza vignalis* Spencer: 47, aedeagus, side view; 48, same, ventral view; 49, posterior spiracles of puparium.

ECONOMIC IMPORTANCE. The economic importance of *M. vignalis* is still not known but it is regularly present in cow-peas used for human consumption in Nigeria (J. DEEMING, personal communication).

Ophiomyia centrosematis (de Meijere, 1940a) – Australia, Tropical Asia, East Africa

ADULT. Small, shining black species.

HEAD. Frons not projecting above eye; orbits well defined, moderately shining, with 2 reclinate ors, the upper distinctly stronger and 2 partially incurved ori; orbital setulae short, sparse, reclinate; ocellar triangle moderately shining, apex extending to level of lower ors; jowls deepest in centre below eye, $\frac{1}{8}$ vertical height of eye, cheeks linear; third antennal segment small, round, slightly pubescent, arista long, almost bare.

MESONOTUM. Shining black but distinctly more mat viewed from front; two strong dc, acrostichals thick, in some 8 rows.

WING. Length from 1.7–1.9 mm, costa extending strongly to vein M_{1+2}, last section of M_{3+4} about $\frac{2}{3}$ penultimate. Squamae grey, margin and fringe black.
MALE GENITALIA. Aedeagus as in Figs. 50, 51; basiphallus distinctly asymmetrical.
EARLY STAGES. Larva with anterior spiracles on long black stalks, posterior spiracles on short stalks, each with 3 bulbs (Fig. 52); puparium yellowish to reddish-orange.
BIOLOGY. According to GREATHEAD (1969: 546) oviposition takes place not in the leaves but in the hypocotyl and stem. Mining occurs mainly in the stem where pupation takes place, with the long anterior spiracles projecting through the epidermis. SEHGAL (1965: 5) reports the species mining in young pods of *Tephrosia* in India.
HOST-PLANTS. *Glycine soja* Sieb. and Zucc. (Formosa); *Phaseolus lunatus* L., *P. vulgaris* L , *Vigna unguiculata* (L.) Walp. (East Africa). Also the cover crops *Centrosema pubescens* Benth. (Java, Malaya) and *Calopogonium mucunoides* Desv. (Malaya), and the wild hosts *Tephrosia candida* DC. (India) and *Crotalaria mucronata* Desv. (East Africa).
DISTRIBUTION. Java; Malaya; Formosa; Ryuku Is.; Australia; India; East Africa: Kenya, Tanzania, Uganda.
PARASITES. In East Africa *Eucoilidea* sp. (Cynipidae) and *Opius melanagromyzae* Fischer. GREATHEAD (1969) reports that the former is the dominant parasite, sometimes occurring in 50% of the sample investigated; the *Opius* was reared on only two occasions. No parasites have been identified from other areas.

Discussion

TAXONOMY. DE MEIJERE (1940a), considering only external characters, described this species as a *Melanagromyza* but it was transferred to *Ophiomyia* by SPENCER (1966a). Its correct generic position is further confirmed by the habit of the puparial anterior spiracles projecting through the stem epidermis, which is characteristic of many species in *Ophiomyia*.
DISTRIBUTION. The very wide distribution in Asia and the apparently limited distribution in Africa suggests that this is originally an Oriental species, which has later colonized Africa, either naturally or as a result of introduction by man.
ECONOMIC IMPORTANCE. In Java and Malaya *Centrosema pubescens* and in Malaya also *Calopogonium mucunoides* are important cover plants in rubber plantations. DE MEIJERE (1940a: 129) reports that *centrosematis* is a serious pest in Java, and in Malaya the cover plants are sometimes largely destroyed. As with *O. phaseoli* and *O. spencerella*, the stem and root system are seriously weakened and the young plants can only survive if adventitious roots are produced quickly enough and in sufficient numbers.

In East Africa *O. centrosematis* is of significance as one of the complex of species attacking beans and cowpeas. However, it is relatively unimportant by comparison with *O. phaseoli* and *O. spencerella* but was present, even though in small numbers, in all samples investigated by GREATHEAD (see Table IV, p. 72). *Phaseolus* and

Vigna appear to be secondary hosts and the small population present in East Africa supports the suggestion that *O. centrosematis* is not an indigenous species in Africa.

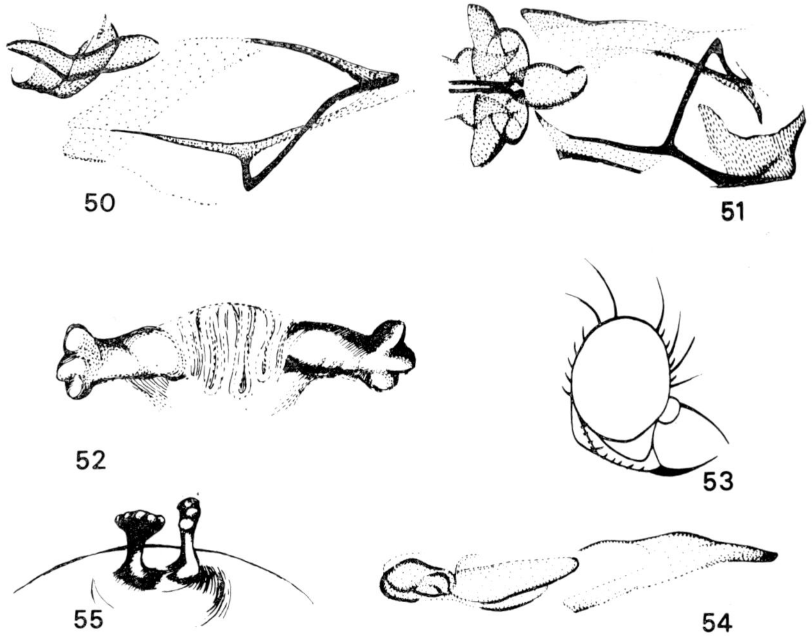

Figs. 50–52. *Ophiomyia centrosematis* (de Meij.): 50, aedeagus, side view; 51, same, ventral view; 52, posterior spiracles of puparium.

Figs. 53–55. *Ophiomyia cicerivora* Spencer: 53, head; 54, aedeagus; 55, posterior spiracles of puparium.

Ophiomyia cicerivora Spencer, 1961b – Pakistan

ADULT. A small shining black species.
HEAD (Fig. 53). Frons wide, almost twice width of eye, not projecting above eye in profile; 2 ors, upper slightly stronger; 2 ori, weaker than lower ors, lower directed inwards; orbital setulae sparse, rather long; orbits not conspicuously differentiated; ocellar triangle large, well defined, apex extending to lower ors; lunule flat, inconspicuous; jowls $\frac{1}{4}$ vertical height of eye, forming angle of 80° in front, almost flat below eye; cheeks broad, $\frac{3}{4}$ width of jowl; vibrissal horn in male conspicuously broad at base; facial keel dividing antennae narrow, scarcely widening, at broadest point narrower than base of vibrissal horn; third antennal segment relatively long.
MESONOTUM. Two strong dc, acrostichals in 6–8 rows.

WING. Length in male 2 mm, costa extending to vein M_{1+2}, first cross-vein only slightly beyond mid-point of discal cell, last section of M_4 shorter than penultimate, in ratio 16:22.
COLOUR. Frons mat, orbits only weakly shining; ocellar triangle conspicuously shining above, less so at apex; mesonotum largely shining but less so when viewed from front; wings hyaline, veins blackish-brown, squamae grey, fringe black, halteres black.
MALE GENITALIA. Aedeagus as in Fig. 54, typical of the genus; ninth sternite with conspicuous hypandrial apodeme, bending ventrally.
EARLY STAGES. Only puparium known, this 2.7 mm long, whitish yellow; anterior spiracles black, 0.08 mm long, projecting through epidermis of stem, surmounted by asymmetrical knob bearing the spiracular pores (number not detectable); posterior spiracles (Fig. 55) on raised stalks, lightly chitinized, pale brownish, each with 6 pores.
HOST-PLANT. *Cicer* sp. (gram).
DISTRIBUTION. Pakistan: Rawalpindi.
BIOLOGY. The larva is an external stem-miner, pupating in the stem, with anterior spiracles projecting through the epidermis.

Discussion

This is a typical *Ophiomyia* species, with the male having a characteristic vibrissal fasciculus and the base of the antennae being separated by a facial keel. It is thus readily distinguishable from *O. phaseoli* which might also be found feeding on this host.

No information is available on the nature of the damage caused. With large populations damage to the stem could be considerable and the plant weakened in consequence. This species deserves further study in view of the importance of the host-plant in Pakistan and India.

Ophiomyia orbiculata (Hendel, 1931) – Europe

Melanagromyza orbiculata Hendel, 1931.
Melanagromyza hexachaeta Hendel, 1931.
Melanagromyza nostradamus Hering, 1933.
Melanagromyza cagliostro Rohdendorf-Holmanová, 1958.
ADULT. Medium-sized, entirely black species.
HEAD. With frons broad, from $1\frac{1}{2}$ to 2 times width of eye, distinctly raised and orbits visible above eye; normally 2 ors and 2 ori but sometimes additional bristle in centre, probably representing a third ori; orbital setulae reclinate; ocellar triangle with apex extending only to lower ors, lunule semicircular; jowls broad, up to $\frac{1}{4}$ height of eye, cheeks forming distinct ring below eye; bases of antennae adjoining, not divided by a raised keel, third antennal segment small, round, arista fine, virtually bare.

MESONOTUM. Two strong, normal dc, with frequently either a strong third one near suture and two or even three smaller presutural ones, which can occur irregularly between the second dc and the suture; acrostichals coarse, in some 8 rows, extending to first dc.
WING. Length 2–2.5 mm, costa extending strongly to vein M_{1+2}, last section of M_{3+4} normally slightly shorter than penultimate.
COLOUR. Frons mat black, ocellar triangle and orbits weakly shining, squamae dark grey, fringe black.

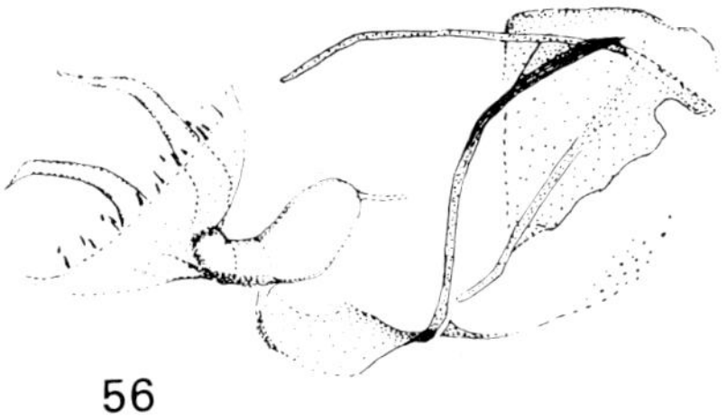

56

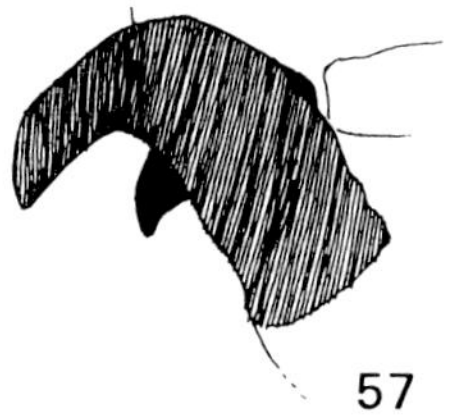

57

58

Figs. 56–58. *Ophiomyia orbiculata* (Hd.): 56, aedeagus; 57, larval mouth-hooks (after GRIFFITHS); 58, posterior spiracles of puparium.

MALE GENITALIA. Aedeagus distinctive, as in Fig. 56.
EARLY STAGES. Larval mouth-hooks each with one tooth, the left substantially stronger (Fig. 57); anterior spiracles on distinct stalks, each with 8–10 pores, posterior spiracles widely separated, irregularly 2-horned, each with 17–22 pores; length of third instar larva 3.1–3.6 mm (GRIFFITHS, 1967a). The puparium is reddish-brown, 3 mm long, with posterior spiracles as in Fig. 58.
HOST-PLANTS. *Pisum sativum* L. and probably wild *Vicia* spp.
BIOLOGY. No detailed study has been made of the life history of this species. It seems probable that oviposition takes place in a young leaf and the larva immediately enters the stem via the petiole. The larva probably feeds down the stem near the outer tissues, but in young stems and as it becomes larger it will certainly feed deeper within the stem. Pupation takes place lower in the stem, frequently below ground level, with the anterior spiracles projecting through the epidermis. Normally only one larva feeds in one plant but as many as three have been found (J. C. DEEMING, personal communication).

DAMAGE CAUSED. Examination of a section of stem containing one puparium shows that a considerable part of the stem tissue has been eaten out and the epidermis is ruptured at the front end of the puparium. Certainly with more than one larva present, young plants would be considerably weakened. J. C. DEEMING has confirmed that attacked plants were definitely stunted.
DISTRIBUTION. Austria; England; Denmark; Sweden; Finland; Germany; Hungary; U.S.S.R.: Moscow area.
PARASITES. None recorded.

Discussion

TAXONOMY. The fact that this species has been separately described four times, twice even by HENDEL in the same paper, is explained by the variable chaetotaxy of the mesonotum. Such variation is not known in any other species in this genus. The presence of two, only two, strong dorso-central bristles is a characteristic feature of this genus (and also of most *Melanagromyza* species). It is therefore not surprising that earlier workers treated specimens with three and with four or five dc, in varying positions, as distinct species. However, the distinctive male genitalia have confirmed that only a single species is involved and the synonymies given above were established by SPENCER (1964a).
RELATIONSHIPS. The male genitalia immediately show that *O. orbiculata* is the sister-species of *O. centrosematis* de Meijere (cf. Figs. 50, 51), which is limited to the Old World tropics. Presumably *O. centrosematis* reached western Europe when this had a sub-tropical climate early in the Tertiary, perhaps at the same time that it reached Africa from its presumed area of origin somewhere in Asia. Speciation might well have occurred in association with its adaptation to the cooler climate which developed later in the Tertiary.
HOST-PLANTS The host of *O. orbiculata* has only once been established – when Mr. DEEMING bred the species from peas growing in his garden at Potter's Bar, Herts. in July, 1962. Specimens have frequently been caught in association with *Vicia* and it seems very probable that this is also a host, possibly even the main one, with transference to *Pisum* occurring only rarely. If *Pisum* was a regular host, it seems inconceivable that the larval feeding would not have been detected previously in such a widely cultivated crop.
ECONOMIC IMPORTANCE. In view of the obvious damage caused to peas by the larvae of *O. orbiculata*, this species must be considered as a potentially serious pest. On the available evidence it can only be assumed that *Pisum* is not the normal host but, given suitable conditions, clearly this crop might be attacked at any time anywhere in western Europe.

Ophiomyia phaseoli (Tryon), **comb. nov.** – Australia, Pacific, Tropical Asia, Africa, Israel

Oscinis phaseoli Tryon, 1895 (Australia).
Agromyza phaseoli Coquillett, 1899.

Agromyza fabalis Jack, 1913 (Rhodesia).
Agromyza destructor Malloch, 1916 (Philippines).
ADULT. A small shining black species.
HEAD (Fig. 59). Frons narrow, equal to width of eye, and only slightly projecting above eye in profile; 2 ors, the upper slightly stronger; 2 weaker, partially incurved ori; orbital setulae sparse, reclinate; ocellar triangle brilliantly shining black, conspicuously elongated, apex extending almost to level of upper ori; orbits only moderately shining; jowls deepest in centre below eye, about $\frac{1}{7}$ vertical height of eye, cheeks little more than linear; third antennal segment small, round, moderately pubescent, arista long, fine, appearing bare.
MESONOTUM. Two strong dc, second midway between first and suture and slightly in front of supra-alar; acrostichals in some six rows.
WING (Fig. 60). Length from 1.8–2.2 mm, costa extending strongly to vein M_{1+2}, first cross-vein slightly beyond midpoint of discal cell, last section of M_{3+4} somewhat variable, from $\frac{2}{3}$ to slightly less than half length of penultimate.
COLOUR. Mesonotum and abdomen brilliantly shining black; squamae pale whitish-grey, margin and fringe black.
MALE GENITALIA. Aedeagus as in Figs. 61, 62; ninth sternite triangular, with short extension at end; surstyli rounded, bearing some weak bristles or hairs.
EARLY STAGES. Egg: pearly-white, oval, measuring about .775 × .35 mm. Larva: relatively long and slender, full-grown 4.25 mm long (Fig. 63); left-hand of the paired mouth-hooks large, with one large but relatively short and two minute teeth, the right-hand mouth-hook half height of the left and only just visible in side view, above mouth-hooks long finger-like process (Fig. 64); cephalopharyngeal skeleton black at front, paler behind; anterior spiracles small, with a circle of 6 minute bulbs; posterior spiracles (Fig. 65) closely adjoining, on conical projections, with normally about 10 bulbs. Puparium pale yellowish, straw-coloured or brownish, up to 3 mm long.
HOST-PLANTS. *Phaseolus calcaratus* Roxb., *P. lunatus* L. (lima bean), *P. mungo* L. (=*P. aureus* Roxb.), *P. vulgaris* L. (kidney bean); *Glycine max* Merrill, *Lablab niger* L. (=*Dolichos lablab* L.), *Vigna catjang* Walp., *V. unguiculata* (L.) Walp. (cowpea); in Java also *Cajanus indicus* Spreng., *Crotalaria juncea* L., *Soja hispida* Moench. Known wild hosts include *Canavalia ensiformis* DC., *Crotalaria mucronata* Desv., *Dolichos uniflorus* Lam., *Phaseolus atropurpureus* DC., *P. lathyroides* L., *P. panduratus* Mart. ex Benth., *P. radiatus* L. and *P. semi-erectus* L. (VAN DER GOOT, 1930; KLEINSCHMIDT, 1970).
BIOLOGY. Oviposition takes place in young leaves, either on the upper surface (GREATHEAD, 1969) or on the lower surface (ALI, 1957). A single female lays between 100 and 300 eggs over a 2 week period (OTANES Y QUESALES, 1918). The eggs hatch in 2–4 days. The larva forms a short leaf-mine but then enters the nearest vein which is followed into the petiole and down into the stem. In young plants the main feeding takes place in the lower, cortical layers of the stem but some larvae can penetrate down into the tap root. With heavy infestations the larvae will feed more deeply inside the stem and also higher up the plant. The larval stage occupies about 10 days and the pupal stage a further 9–10 days. In Rhodesia in warm weather the complete life cycle can take as little as 3 weeks

(TAYLOR, 1958). In Java at higher altitudes VAN DER GOOT (1930) reports that the larval stage can last from 17–22 days and the pupal stage from 13–20 days. On the other hand in Egypt HASSAN (1947) states that the life cycle can be completed in 17 days. The puparium remains beneath the epidermis, head upwards, normally near the base of the stem (GREATHEAD, 1969).

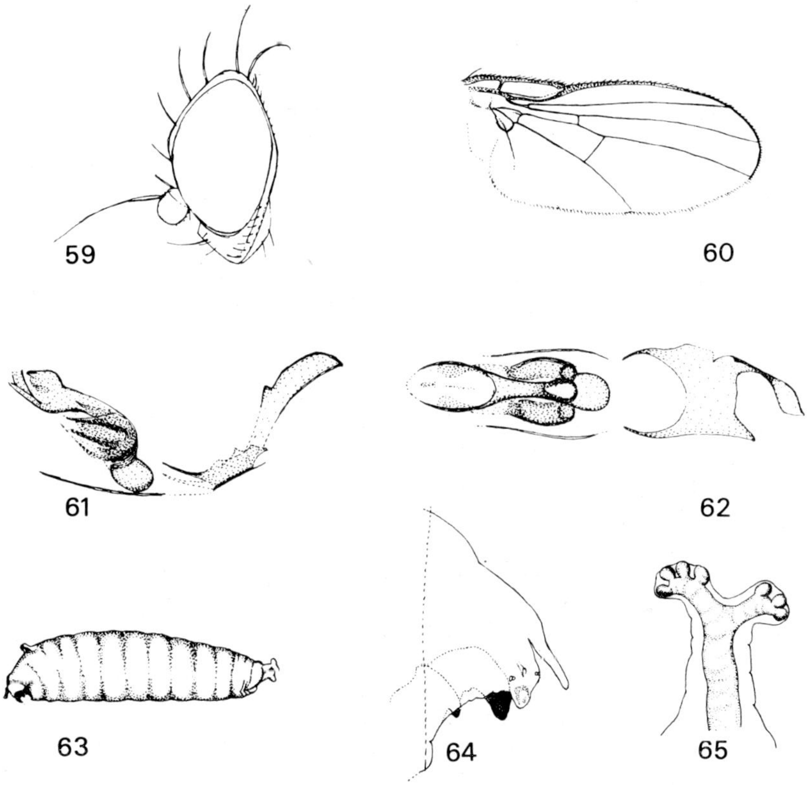

Figs. 59–65. *Ophiomyia phaseoli* (Tryon): 59, head; 60, wing; 61, aedeagus, side view; 62, same, ventral view; 63, larva (after OTANES Y QUESALES); 64, head of larva (after KATO); 65, posterior spiracular process of larva (after SINGH and BERI).

This species has been discussed in some detail by OTANES Y QUESALES (1918) in Java (as *Agromyza destructor* Malloch), WALLACE (1939) in Tanganyika, HELY (1947) and KLEINSCHMIDT (1970) in Australia, KATO (1961) in Japan, GREATHEAD (1969) in East Africa and SINGH and BERI (1971) in India.

DISTRIBUTION. Probably occurring throughout the Old World tropics from Northern Australia and the islands of Micronesia including Okinawa, across Asia to Africa and reaching Egypt and the Jordon Valley in Israel; it is also known in

Hawaii from all the main islands where it is considered a major pest on all types of beans (D. E. HARDY, personal communication).

PARASITES. VAN DER GOOT (1930: 30) records the following parasites in Java:

Species	No. of specimens	Percentage
Trigonogastra agromyzae Dodd	395	59.1
Cynipid sp.	271	40.5
Eurytoma poloni Girault	1	0.2
Eurytoma sp.	1	0.2

In Australia no less than 13 species of parasites from 6 families have been recorded by KLEINSCHMIDT (1970) as follows:

Braconidae: *Opius oleracei* Fischer.
Chalcididae: *Merismorella shakespearei* Girault.
Eulophidae: *Achrysocharis douglassi* Girault, *Hemiptarsenus semialbiclavus* Girault, *Hemiptarsenus* sp., *Neodimmockia agromyzae* Dodd.
Eupelmidae: *Eupelmus gravi* var. *brevicinctus* Girault.
Eurytomidae: *Eurytoma larvicola* Girault, *Eurytoma* sp.
Pteromalidae: *Eurydinotellus viridicoxa* Girault, *Polycystomyia benefica* Dodd, *Pterosema subaenea* Dodd, *Trigonogastra agromyzae* Dodd.

Opius phaseoli Fischer has been recorded from India and the Philippines (FISCHER, 1971).

Opius liogaster Szép. was reported as a parasite in Mauritius by MOUTIA (1932) and in Rhodesia by TAYLOR (1958). In East Africa GREATHEAD (1969) found *O. melanagromyzae* Fischer to be the most important primary parasite. Normal parasitism by this species is usually above 50% and up to 93% has been recorded. A second, undetermined *Opius* sp. is also present but the rate is substantially less at 3–9%. *O. phaseoli* is also attacked by an assemblage of undetermined Chalcidoids but their effect is apparently slight compared with *Opius melanagromyzae*.

In Egypt HASSAN (1947) records *Trigonogastra agromyzae* and *Eurytoma larvicola* Girault as active parasites from September to December.

Discussion

TAXONOMY. Two points require discussion under this heading. Firstly, the correct generic status of *phaseoli* and secondly its relationship to *spencerella* Greathead (p. 70).

On external characters, particularly the form of the head, *phaseoli* appears to be a typical *Melanagromyza*. However, it has been apparent to the author for some years that the larval characters unquestionably place *phaseoli* in *Ophiomyia*. The various relevant facts have now been reviewed, as follows:

1. The biology of *phaseoli* clearly indicates that it correctly belongs in *Ophiomyia*. *Melanagromyza* species feed as internal stem-borers, with oviposition frequently taking place directly in the stem, not in the leaf, and the larva bores in the pith of the stem, pupating deeply imbedded within the stem. Several *Ophiomyia* species, including the type of the genus, *maura* (Meigen), feed exclusively as leaf-miners but the majority are external stem-miners (as opposed to internal stem-borers), feeding close beneath the epidermis of the stem, where pupation also takes place, with the anterior spiracles of the puparium frequently penetrating through the epidermis. It remains to be established whether oviposition normally takes place in the leaf; this may well be so.

The young larva of *phaseoli* reaches the stem after about 2 days' feeding in the leaf and when not subjected to feeding competition probably feeds like typical representatives of *Ophiomyia* beneath the epidermis. However, with the large number of larvae which can occur in a single young plant, clearly to mature completely all available stem tissue must be consumed and this enforced deeper feeding within the stem has in part been responsible for maintaining the belief that *phaseoli* was an internal stem-boring *Melanagromyza*.

2. The external morphology of *phaseoli* is exactly as in typical *Melanagromyza* species. However, the *pulicaria* (Meigen) group of *Ophiomyia* species is now well known both in Europe (SPENCER, 1964a: 802) and North America (SPENCER, 1969a: 93). On external morphology these species were placed in the genus *Melanagromyza*, until genitalia studies showed conclusively their correct generic affiliation in *Ophiomyia*. The *pulicaria* group feed as leaf-miners but are essentially associated with the mid-rib, suggesting a primitive link to stem-boring. There are many parallels between this group and the species pair, *phaseoli*/*spencerella*.

3. The male genitalia are not so obviously typical of *Ophiomyia*, as in the *pulicaria* group. Nevertheless, the somewhat unusual form of the basiphallus (Figs. 61, 62) generally resembles that of *Ophiomyia centrosematis* de Meijere (Fig. 51), which has for some time been accepted as an *Ophiomyia*. The aedeagus of *phaseoli* and *spencerella* suggests that the two species belong to an isolated group; its form is not decisive in indicating generic relationships but it is in any case far removed from that found in typical *Melanagromyza* species.

4. The posterior spiracles of the larva in both *phaseoli* and *spencerella* are entirely typical of *Ophiomyia* and this form, with the spiracular bulbs on raised stalks, is never found in *Melanagromyza*.

The following essential characters separate *Ophiomyia* from the more primitive internal stem-boring genus *Melanagromyza*: vibrissal fasciculus normally present in male; base of antennae normally divided by raised facial keel; in male genitalia, basiphallus with elongate side-arms and distiphallus asymmetrical; posterior spiracles of larva on raised stalks; puparium sometimes black; larva feeds as leaf-miner or external stem-miner.

It is suggested that the *phaseoli* and *pulicaria* groups are primitive representatives of the genus *Ophiomyia*, in which the advanced characters of the genus are clearly shown in the larval morphology, the biology and in part by the male genitalia. Differentiation in external morphology, however, has not taken place and in this respect these species retain the basic forms of head found in most genera, including the most closely related genus, *Melanagromyza*.

On this evidence I transfer *phaseoli* from *Melanagromyza* to *Ophiomyia* herewith.

O. phaseoli and *spencerella* are unique in having a brilliantly shining, unusually elongate ocellar triangle. It has not hitherto been clear whether this was due to convergence or reflects a genuine relationship. Examination of the genitalia of the two species satisfies me that the latter is the case. Not only is there a general similarity in the basiphallus but the close relationship between the two species is particularly apparent from the ventral view of the aedeagus (Figs. 62, 73).

Although the two species are strikingly similar, the ocellar triangle appears to be marginally longer in *phaseoli*, with the apex reaching the level of the uppermost of the lower orbitals, while in *spencerella* the apex lies midway between the lower ors and the upper ori. Also the orbits, particularly above the upper ors, appear more strongly shining in *spencerella*. For a completely reliable identification, however, it will invariably be necessary to examine the male genitalia, which permit the two species to be separated at a glance.

DISTRIBUTION. Although *O. phaseoli* occurs from the islands of Micronesia across Asia to East, West and South Africa, it appears to be less abundant in Africa than in various parts of Asia, such as Java and India. GREATHEAD (see Table IV below, p. 72) has shown that in East Africa *phaseoli* occurs in significantly smaller numbers than *spencerella*. It seems probable that *spencerella* evolved in Africa in isolation from the ancestral stock of *phaseoli* in Asia. Subsequently, after the reproductive isolation of the African and Asian populations was complete and distinct species were established in the two regions, *phaseoli* in its present form again colonized Africa, either in modern times through introduction by man or at least relatively recently, during the past few million years, as opposed to possibly early Tertiary speciation of the original populations. This seems to be the only logical explanation of the present direct competition between the two species in Africa on identical hosts and feeding essentially in the same way on the same parts of the host-plant. It is true that GREATHEAD (1969: 547) indicates that the pattern of emergence of the two species differs somewhat. However, it is inconceivable that this difference would be sufficient to permit sympatric speciation.

Although *O. phaseoli* reaches Iraq, Egypt and Israel, it has not been reported from southern Europe. Presumably it is unable to withstand the cool winters but nevertheless it might at any time appear in favourable localities in such areas as southern Italy or Greece and, in the absence of its normal parasites, could build up large populations, at least for a temporary period, which would be capable of causing severe damage.

O. phaseoli is not known from South America or the southern states of the U.S.A. Great vigilance is called for to prevent any casual introduction which could easily lead to the establishment of this species as a major pest in the Americas.

A map showing the world distribution of this species has been published by the Commonwealth Institute of Entomology, London (1961).

HOST-RANGE. Although *O. phaseoli* attacks most species of *Phaseolus* and many genera of Leguminosae, its host-range is limited. According to VAN DER GOOT (1930: 9), *Phaseolus sublobatus* Roxb. is immune to attack, as also are other common genera of Leguminosae, such as *Cassia*, *Centrosema*, *Desmodium*, *Flemingia*, *Indigofera*, *Lupinus* and *Pisum*. More information is clearly desirable on its exact host-range, as almost certainly populations are maintained on wild hosts, which are entirely unaffected by control measures exercised on cultivated crops.

ECONOMIC IMPORTANCE. All authors who have conducted field investigations into the nature of the damage caused by *O. phaseoli*, whether in Java, Ceylon, East Africa or Egypt, agree that the effect of larval feeding is invariably serious, with young plants frequently being totally destroyed.

Damage in Australia was first recorded by TRYON (1895) and ANON., a (1898). MORGAN (1938) states that in the mild northern half of the New South Wales coastline the main limiting factor in the cultivation of beans is the damage caused by *O. phaseoli*.

WICKRAMASINGHE and FERNANDO (1962) report that damage resulting from oviposition injury to the leaf tissue is of no economic significance but that larval damage is usually very extensive. On seedling plants the larvae pupate below the epidermis of the stem at or immediately above soil level, while on older plants they pupate at the base of the petioles. This concentration of puparia results in the swelling, splitting open and rotting of these parts of the host-plant. It is in the seedling stage that bean plants are most seriously affected when they suffer premature leaf-fall and are either killed or severely stunted. Older plants are similarly affected but are not usually killed by the attack.

DE MEIJERE (1922: 18) reports that on Sumatra young plants were found to contain from 10–20 larvae and were invariably killed by such attack.

HASSAN (1947) writes as follows on the nature of the damage: 'In Egypt, French beans are sown annually, in two crops; the autumn crop in August, September and October, and the spring one in February and March. The autumn crop is the one that is heavily infested, especially in October. The attack begins as soon as the plants appear above the ground. The damage done to the plants is clearly observed in the field in the final stages of the infestation, the plants become stunted and yellow and finally topple over. The stem just beneath the soil becomes swollen, then cracks and gets brown in colour. When such plants are pulled and examined, fully grown larvae and pupae are easily seen in the damaged part of the stem at the surface of the ground or a little above or below it. Replanting in such cases may be necessary in August and September. The death of the infested plants depends upon the number of larvae tunnelling through it. Many plants which appear normal in their growth may harbour a small number of larvae. Dead plants 20 cm long were found to harbour as many as 25 larvae and pupae. Some plants may succumb to 10 or 15 larvae. If the plant is able to form adventitious roots it may recover and in this case cropping is not affected. Old plants examined all through December and January show larvae in the fleshy joints of the branches and the leaf stalks. These larvae usually pupate in the same places'.

GREATHEAD (1969) gives a good summary of the damage caused by *O. phaseoli*: 'The ultimate effect on the crop depends on the powers of recovery of the plants, the ability to produce adventitious roots being the crucial factor. Plants that do not rapidly respond to root damage by developing adventitious roots wilt and die; they are also liable to break off at ground level during windy periods or storms. Plants producing adventitious roots soon recover from the initial heavy infestation and are afterwards sufficiently robust to survive later damage. No attempt to assess the effect of bean-fly on cropping was made but there must be a delay in growth which could be important where rainfall is low. The variety is also an important factor as it has been noticed in trials that varieties showed striking differences in the damage suffered. Again, casual observation suggests that imported varieties from temperate regions suffer more severely than varieties selected from peasant holdings in East Africa. This is to be expected, as beans in temperate climates where (this species) is unknown have not had to evolve the facility for developing adventitious roots to combat infestation, whereas selection for this character must have been practised where the flies occur.'

GREATHEAD considers that most damage is done by the larvae from the first pair of leaves, which when in the third instar destroy the medullary tissue at ground level. Later in the life of the plant, larvae do little damage.

O. phaseoli must be considered at the present time one of the most serious of all known Agromyzid pests. Its range is extensive throughout the Old World tropics and it thrives on many types of beans. Some of the chemical control measures which have been adopted with varying degrees of success are discussed in Chapter 12.

Ophiomyia shibatsuji (Kato, 1961), **comb. nov.** – Japan

Ophiomyia sp., Shibatsuji, 1949.
Melanagromyza shibatsuji Kato, 1961.
ADULT. Small, shining black species.
HEAD (Fig. 66). Frons $1\frac{1}{2}$ times width of eye, projecting narrowly above eye; 2 equal ors, 2 incurved ori, orbital setulae sparse, reclinate; ocellar triangle with apex not quite reaching level of lower ors; jowls of medium width, $\frac{1}{6}$ height of eye, cheeks linear; third antennal segment small, round, arista moderately pubescent.
MESONOTUM. With 2 strong dc, acrostichals in 8 rows.
LEGS. Mid-tibia bearing one weak lateral bristle.
WING. From 2.2–2.4 mm, costa extending strongly to vein M_{1+2}, last section of M_{3+4} only slightly shorter than penultimate, first cross-vein slightly beyond mid-point of discal cell.
COLOUR. Frons mat black, ocellar triangle and orbits moderately shining; mesonotum and abdomen shining black; squamae grey, margin dark brown, fringe black.

MALE GENITALIA. Aedeagus as in Fig. 67, with conspicuous pointed process below; ninth sternite with blunt, broad, only shortly extended hypandrial apodeme.
EARLY STAGES. Third instar larva yellowish-white, 3.5–3.8 mm long; mouthhooks with left-hand tooth very long, long finger-like process above (Fig. 68); cephalo-pharyngeal skeleton as in Fig. 69; anterior spiracles on slender stalks, each with 24–30 pores; posterior spiracles on distinct projections (Fig. 70), each

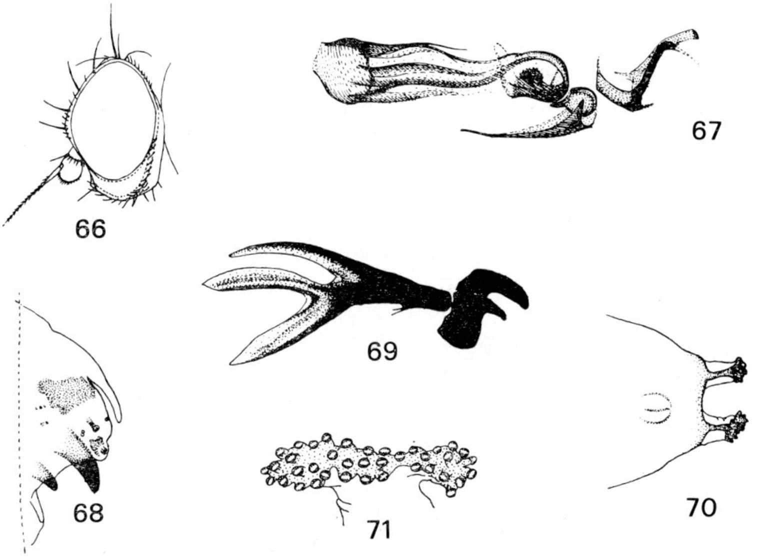

Figs. 66–71. *Ophiomyia shibatsuji* (Kato): 66, head; 67, aedeagus; 68, larval mouthhooks; 69, cephalo-pharyngeal skeleton; 70, posterior spiracles; 71, pores of posterior spiracular process (after KATO).

with 28–41 pores (Fig. 71); puparium entirely black, semi-glossy (from KATO, 1961).
HOST-PLANTS. *Glycine max* Merrill, *G. soja* Sieb. and Zucc. (soy bean).
BIOLOGY. Oviposition takes place in the stem of young plants at or below ground level, never in leaves. The newly hatched larvae mine between the xylem and cortex layers of the stem downwards until they reach the root. Here the main feeding takes place along the outer side of the xylem and the inner side of the cortex but they never bore into the pith. The larvae pupate below ground level, immediately beneath the epidermis. Larval feeding takes place from the beginning of June until the middle of July. The pupal stage lasts throughout the winter, the adults emerging from the end of May till the end of June. There is only a single generation (after KATO, 1961, condensed from SHIBATSUJI).

DAMAGE CAUSED. This is reported as severe throughout Japan, especially in newly cultivated fields in mountainous regions.
DISTRIBUTION. Japan: Hokkaido, Honshu, Kyushu.
PARASITES. None recorded.

Discussion

TAXONOMY. The form of larval spiracles, the black puparium and the biology clearly indicate that this species correctly belongs in the genus *Ophiomyia*, to which it is transferred herewith. It belongs to the group in which the adult lacks the more advanced characters of *Ophiomyia* (see p. 65).

The elongate protuberance above the mouth-hooks is a distinctive character of the larva. From this it can be seen that SASAKAWA (1961: Fig. 24k) was in fact referring to this species when discussing *M. dolichostigma* de Meijere, which, however, does also occur in Japan (p. 37).
ECONOMIC IMPORTANCE. KATO considers that this is the most important of the five Agromyzid pests on soy bean in Japan, causing severe damage on young plants. Its distribution at the present time appears to be limited to Japan and care is clearly required to prevent its introduction into any other of the main areas of cultivation of soy bean in neighbouring countries.

Ophiomyia spencerella (Greathead), **comb. nov.** – East Africa

Melanagromyza spencerella Greathead, 1971.
ADULT. Small, brilliantly shining black, generally resembling *O. phaseoli* (see p. 62).
HEAD. Ocellar triangle conspicuously shining black but invariably apex extending only to midway between lower ors and upper ori and thus appearing distinctly shorter than in *phaseoli*; orbits distinctly shining, particularly between upper ors and vertex.
WING. Length from 1.8–2 mm, thus marginally smaller than in *phaseoli* but in most specimens overlapping with it.
MALE GENITALIA. Aedeagus distinctive, as in Figs. 72, 73.
EARLY STAGES. Posterior spiracles of larva on raised stalks, each normally bearing 10 bulbs; in 86 unparasitized larvae examined, GREATHEAD (1969) found that by far the greatest number had between 9, 10 or 11 bulbs but the number could vary from 2 to 13. Puparium shining blackish, occasionally somewhat paler, more grey, particularly on underside.
HOST-PLANTS. Mainly *Phaseolus vulgaris* L.; less commonly *P. calcaratus* L., *P. lunatus* L., *P. mungo* L., *Lablab niger* L. and *Vigna unguiculata* (L.) Walp.
BIOLOGY. The females scarify the tissue of the leaves in the same way as *M. phaseoli* but rarely oviposit in the leaves. The bulk of oviposition occurs in the hypocotyl at ground level in the first two or three days after the first appearance of

the plant above ground. A few eggs are also deposited in young stem tissue above the level of the cotyledons. The eggs resemble those of *M. phaseoli* and are laid in a similar pocket beneath the epidermis. In the stem this is usually orientated with the opening on the lower side and the egg above the opening. Larvae from eggs in the leaves follow the same life-cycle as those of *M. phaseoli*. Those laid in the hypocotyl or stem mine downwards, feeding extensively in the hypocotyl and taproot, and return to ground level or above to the nearest healthy tissue to pupate. The larva pupates in the same position as that of *M. phaseoli* after preparing a 'window' (GREATHEAD, 1969).

DISTRIBUTION. Kenya: Nairobi. Tanzania: Mbeya, Ilonga, Ukiriguru. Uganda: Kampala area, Serere. Nigeria: Ibadan.

PARASITES. The dominant parasite is *Eucoilidea* sp. (Cynipidae) and at Kawanda in Uganda the rate of parasitism was as high as 57% but is usually less

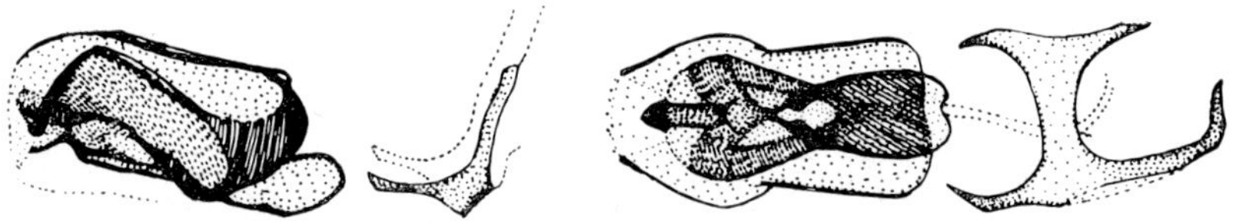

72 73

Figs. 72–73. *Ophiomyia spencerella* (Greathead): 72, aedeagus, side view; 73, same, ventral view.

than 20%. Several undetermined Chalcids were also present in Uganda but occurred in small numbers and were not positively identified. *Opius melanagromyzae*, the main parasite of *O. phaseoli*, apparently does not attack *spencerella* (GREATHEAD, 1969).

Discussion

For the reasons outlined for *phaseoli* (p. 65), *spencerella* certainly correctly belongs in the genus *Ophiomyia*. In this species the black puparium provides additional evidence, as this is a character found in many *Ophiomyia* species but is entirely unknown in the genus *Melanagromyza*.

In East Africa the dominant bean fly is *O. spencerella* and its relative abundance in relation to *phaseoli* and *centrosematis* (de Meijere) is clearly seen in Table IV below, taken from GREATHEAD (1969).

After GREATHEAD had discovered that a second species closely resembling *phaseoli* was present in East Africa, the male genitalia of specimens from all other localities examined appeared to confirm that *spencerella* did not occur elsewhere. Subsequently, however, this species has been confirmed in Nigeria: Ibadan, from stems of *Vigna unguiculata*.

Table. IV. Infestation of *Phaseolus vulgaris* by bean flies and parasitism at three localities over successive plantings

Date planted	*No. plants*	*Plants infested* (%)	Fly puparia			
			Total	*M. phas.* (%)	*M. spenc.* (%)	*O. centr.* (%)
Kawanda						
15.xi.67	100	97	386	16.5	82.1	1.3
13.xii.67	100	92	343	31.4	66.8	1.7
27.ii.68	100	48	48	35.4	58.3	6.3
30.iv.68	100	97	469	17.3	76.9	5.8
28.v.68	50	82	127	10.2	89.0	0.8
17.ix.68	100	68	139	23.7	74.1	2.2
25.ix.68	100	81	230	25.2	71.3	3.5
11.ii.69	100	36	56	28.6	44.6	26.8
5.iii.69	100	38	51	45.1	49.0	5.9
Nairobi						
22.iv.68	50	54.0	27	92.6	—	7.4
23.v.68	102	78.4	139	91.3	6.5	2.2
15.vi.68	127	81.8	174	67.8	28.2	4.0
26.vi.68	99	100.0	453	5.6	93.8	0.02
31.vii.68	100	95.0	230	12.2	86.1	1.7
8.viii.68	100	98.0	284	4.2	92.6	3.2
24.ix.68	55	81.8	147	25.2	57.1	17.7
15.x.68	100	93.0	192	39.1	55.2	5.7
6.xi.68	100	97.0	284	10.9	88.0	1.1
Mtwapa						
25.vii.68	122	26.2	51	92.1	—	7.8
30.vii.68	115	44.3	120	98.3	—	1.7
28.viii.68	100	20.0	32	96.8	—	3.1
26.ix.68	100	47.0	118	85.6	—	14.4
4.xi.68	100	26.0	29	89.6	—	10.3
*ca.*18.xii.68	100	32.0	58	87.9	—	12.1
5.i.69	100	47.0	70	64.3	—	34.3
4.ii.69	66	47.0	41	92.7	—	7.3

As explained on p. 66, it is believed that *O. spencerella* evolved in Africa, in isolation from *phaseoli* but the latter has subsequently colonized many parts of Africa to compete with *spencerella* on the same hosts but effectively isolated from it reproductively.

GREATHEAD (1969: 547) concludes that *O. spencerella* can be a serious pest and when occurring together with *phaseoli*, will cause the main economic damage, since its larvae reach and destroy the root system first. As with *phaseoli*, the ultimate effect of *spencerella* on the crop depends on the ability of the plants to produce adventitious roots and thereby recover.

IV. LEGUMINOUS CROPS 2: LEAF-MINERS

The species discussed below have hosts in the following 11 genera of Leguminosae: *Cicer*, *Desmodium*, *Glycine*, *Lathyrus*, *Medicago*, *Melilotus*, *Pisum*, *Pueraria*, *Trifolium*, *Vicia* and *Vigna*. Most of the species concerned are oligophagous and, as with the internal feeders considered in the previous chapter, are treated systematically, rather than by crops. All belong to the four genera: *Japanagromyza*, *Agromyza*, *Ophiomyia* and *Liriomyza*. Fourteen species are discussed in detail and brief reference is made to three others about which little information is available.

A new *Ophiomyia* sp. is described from Pakistan. Although there is little information about this species, it is believed that the larva feeds as a leaf-miner.

Three of the *Japanagromyza* species, *aequalis* Spencer, *desmodivora* Spencer and *inaequalis* (Malloch), occur only in Florida and the Neotropical region, while the fourth, *tristella* (Thomson), is widespread in the Oriental region. A fifth species, which has not yet been formally described, is known from Peru and is unusual in feeding also on *Brassica*, in addition to its main host, *Phaseolus*. This is possibly identical with a species known as a leaf-miner on *Phaseolus* from Tucuman, Argentina (P. Wygodzinsky). *J. centrosemae* (Frost, 1936) occurs as a leaf-miner on *Centrosema pubescens* Benth. in the Panama Canal Zone. The type specimen is unique and the significance of the species is not yet understood. It was discussed by Spencer (1963a: 301). A leaf-miner on *Phaseolus lunatus* L. has been recorded from Puerto Rico but only a single specimen has been examined. This also has not been described but was briefly discussed by Spencer (1963a: 304).

A recent study of Agromyzidae in Florida and the Caribbean (Spencer and Stegmaier, 1973) has revealed a number of new *Japanagromyza* species which are of wide distribution and probably have hosts in the Leguminosae. Some of these species may well be found on cultivated crops. In this paper a key is given to the 19 species in the genus known in the Neotropical region.

The most important of the four *Agromyza* species discussed below appears to be *A. frontella* Rondani which has appeared in great numbers in recent years on *Medicago sativa* in the eastern United States. This species has no doubt been present in the past in smaller numbers, and the current significant increase in population levels is probably due to disruption of natural control by its parasites, following indiscriminate use of insecticides.

Among five *Liriomyza* species considered, *L. trifoliearum* Spencer was recently described from Florida and New York, also attacking alfalfa, and is now recorded in addition from California. This species, like *Agromyza frontella*, seems to have built up its population to damaging levels in recent years. *L. cicerina* (Rondani) has been known as a pest of *Cicer arietinum* L. since its description in 1875, and there are records of serious outbreaks in southern Europe and the Ukraine.

Liriomyza brassicae (Riley) is normally limited to hosts in the Cruciferae (see Chapter 6, p. 153) but has been recorded on *Pisum* both in Kenya and in India. This unusual transference to an abnormal host occurs in small numbers and no damage to peas by this species has been reported. *Liriomyza crucifericola* (Hering) feeds commonly on *Vicia faba* L. in addition to its normal cruciferous hosts.

Three polyphagous species, *L. huidobrensis* (Blanchard), *L. trifolii* (Burgess) and *Phytomyza horticola* Goureau occur regularly on leguminous crops. Peas are seriously damaged by *L. huidobrensis* in California and by *P. horticola* in India. These species are discussed separately in Chapter 8.

74

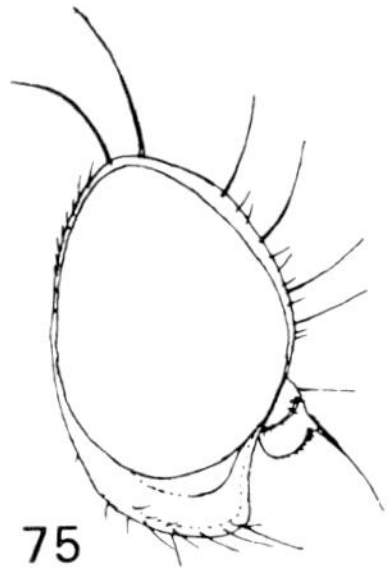

75

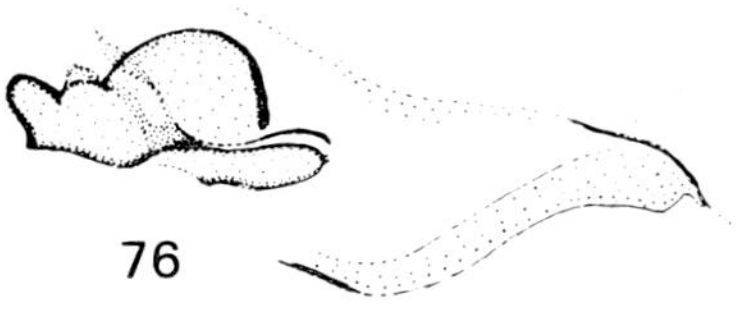

76

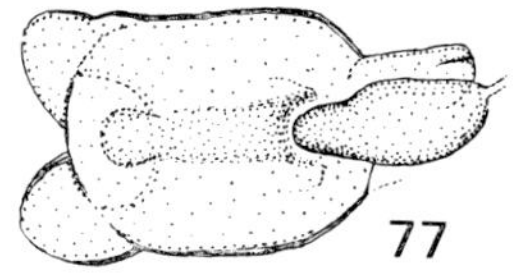

77

Fig. 74. *Liriomyza phaseolunata* (Frost): leaf-mine on lima bean (after FROST).

Figs. 75–77. *Ophiomyia vignivora* sp.n.: 75, head; 76, aedeagus, side view; 77, distiphallus, ventral view.

FROST (1943) described *Liriomyza phaseolunata* (as *Agromyza*) attacking lima beans in New Jersey. The leaf-mine is irregularly blotch-like (Fig. 74). FROST reports that 'This species apparently caused considerable injury during 1942. The infestation covered 2,000 acres and a large percentage of the leaves were infested'. It has unfortunately not been possible to examine types of this species and its identity remains in doubt.

Ophiomyia vignivora sp.n. – Pakistan

HEAD (Fig. 75). Frons narrow, equal to width of eye, slightly projecting above eye in profile; 2 equal reclinate ors, 2 ori, the upper slightly, the lower predominantly incurved; orbital setulae well developed, in single row, reclinate; ocellar triangle slightly raised but scarcely shining, apex extending to level of lower ors; lunule small, semicircular; jowls about $\frac{1}{6}$ vertical height of eye, deepest in front, rounded behind, cheeks conspicuous below eye, about $\frac{1}{2}$ height of jowls; vibrissa slender in male, not developed into fasciculus; third antennal segment small, round, arista fine, bare; facial keel linear, not dividing base of antennae.

MESONOTUM. Two strongly developed dc, second behind level of supra-alar; acrostichals numerous, in some 6 rows.

WING. Length from 1.6–1.8 mm; costa extending strongly to vein M_{1+2}, first cross-vein slightly before midpoint of discal cell, last and penultimate sections of vein M_{3+4} equal.

COLOUR. Head entirely black, frons mat; mesonotum largely mat black, abdomen slightly more shining; legs entirely black; squamae and fringe white, margin scarcely differentiated.

MALE GENITALIA. Aedeagus as in Figs. 76, 77; ninth sternite conspicuously broad at end.

HOST-PLANT/BIOLOGY. *Vigna unguiculata* (L.) Walp. (= *V. sinensis* Endl.).

HOLOTYPE MALE. Pakistan, Rawalpindi, 27.x.1969; paratypes: 3♂, 2♀, same data. Holotype and paratypes in British Museum (Nat. Hist.), 2 male paratypes in author's collection.

DIAGNOSIS. The distinctive characters of this species are the white squamal fringe and relatively broad jowls. It can be included in a revised key to Oriental *Ophiomyia* species as follows:

1	Orbital setulae proclinate above, reclinate below	*beckeri* (Hendel)
—	Orbital setulae all reclinate	2
2	Squamae and fringe white	3
—	Squamae grey, fringe black	5
3	Male with conspicuous vibrissal fasciculus	*cornuta* (de Meijere)
—	Male with slender, normal vibrissa	4
4	Jowls narrow, $\frac{1}{10}$ height of eye; last section of vein M_{3+4} shorter than penultimate	*atralis* (Spencer)
—	Jowls broader, $\frac{1}{6}$ height of eye; last and penultimate sections of M_{3+4} equal	*vignivora* Spencer
5	Jowls broad, in ratio 7:25 with height of eye; very large species, wing length in female 3.8 mm	*ingens* Spencer
—	Jowls narrower, at most $\frac{1}{4}$ height of eye; smaller species	6
6	Facial keel narrow, not protruding	7
—	Facial keel broad, bulbous	8
7	Jowls $\frac{1}{4}$ height of eye; vibrissal horn entirely fused; ocellar triangle conspicuously shining	*cicerivora* Spencer
—	Jowls narrow, $\frac{1}{9}$ height of eye; vibrissal horn differentiated into 2 or 3 distinct bristles; ocellar triangle scarcely shining	*negrosensis* Spencer
8	Acrostichals extending only to second dc	*anguliceps* (Malloch)
—	Acrostichals extending in 3 or 4 rows to first dc	*lantanae* (Froggatt)

Discussion

Details of the biology of this species are not known but it is believed that the larva is a leaf-miner, pupating in the leaf. Large populations could certainly weaken or even destroy young plants and the economic importance of the species deserves further study.

Japanagromyza aequalis Spencer, 1966d – Florida, Caribbean

ADULT. Small species with brown frons.
HEAD. Frons equal to width of eye, not projecting above eye in profile; 2 strong equal ors, 2 equal ori only slightly weaker than ors; orbital setulae fine, sparse, reclinate; jowls narrow, $\frac{1}{12}$ vertical height of eye; third antennal segment covered with short, white pilosity, arista only slightly pubescent.
MESONOTUM. 2 strong dc, frequently a third present, which is short and sometimes developed on one side only; pre-scutellars present, acrostichals sparse, in 6 rows.
WING. Length from 2–2.4 mm, costa extending strongly to vein M_{1+2}, last section of M_{3+4} two-thirds penultimate, first cross-vein before midpoint of discal cell.
LEGS. Fore-tibia with 1, mid-tibia with 2 lateral bristles.
COLOUR. Frons uniformly brown, ocellar triangle and orbits distinctly shining black; mesonotum greenish- or coppery-black, moderately shining, abdomen more distinctly shining; wings somewhat darkened by coarse microtrichia, veins brownish-black; squamae and fringe, also halteres entirely white.
MALE GENITALIA. Aedeagus membranous basally, chitinized at sides towards distiphallus, this short, extending only slightly beyond end of chitinization (Fig. 78); ninth sternite rounded, without apodeme, as in *J. desmodivora* (Fig. 85), side-arms unusually broad, upcurved at end (Fig. 79), surstyli curving inwards, not downwards, without bristles (Fig. 80); cerci without internal bristles but with 5 or 6 short black teeth grouped at end (Fig. 81); epandrium with distinctive narrow extension towards cerci (Fig. 82); ejaculatory apodeme very small, blade linear.
EARLY STAGES. Puparium pale, straw-coloured, posterior spiracles each with 3 pores on widely-separated, well-defined protuberances (Fig. 83).
HOST-PLANT. *Vigna repens* Baker (=*luteola* Jacq.).
BIOLOGY. The larva forms a large primary blotch-mine.
DAMAGE CAUSED. One larva largely destroys the entire leaf surface of a single leaflet.
DISTRIBUTION. Florida: Miami area; Bahamas: Berry Is., Darby Is.; West Indies: Dominica.
PARASITES. None recorded.

Discussion

TAXONOMY. *J. aequalis* generally resembles *J. inaequalis* which occurs on the same food-plant but is readily distinguishable by the stronger dorsal ors, the paler brown frons, the distinctive pilosity on the third antennal segment and the

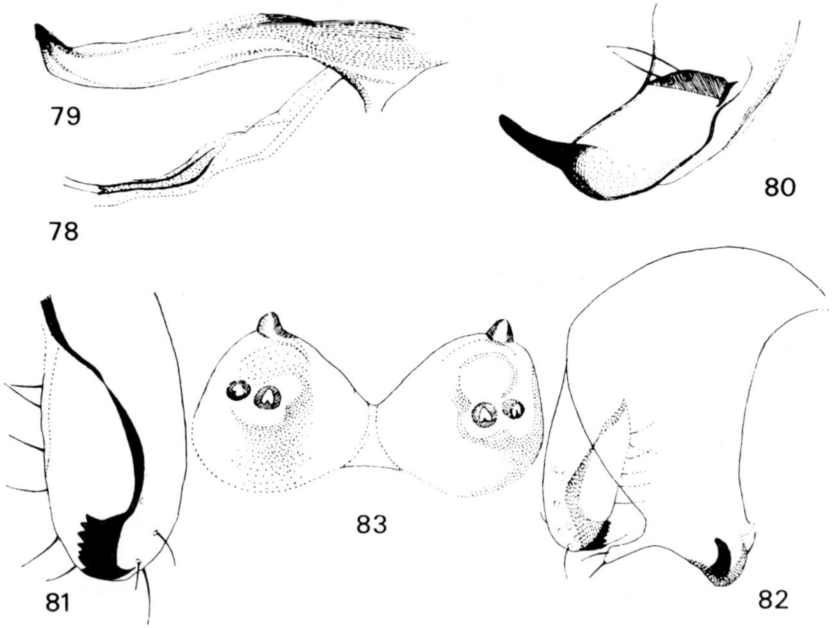

Figs. 78–83. *Japanagromyza aequalis* Spencer: 78, aedeagus; 79, ninth sternite; 80, surstylus; 81, cercus; 82, epandrium; 83, posterior spiracles of puparium.

sparser acrostichals. The larval and puparial characters and also the male genitalia of the two species are entirely distinct.

ECONOMIC IMPORTANCE. Although this species has only been recorded on *Vigna repens*, it very probably feeds also on the important edible crop, *Vigna unguiculata* (= *V. sinensis*), the black-eyed pea, and also on other *Vigna* species which are used as fodder. No significant infestations have been recorded but *J. aequalis* must be considered as a potentially important pest.

Japanagromyza desmodivora Spencer, 1966d – Florida, ?Colombia

ADULT. Small greyish or greenish species, with mat brownish-black frons.
HEAD. Frons narrow, equal to width of eye, not projecting above eye in profile;

2 strong ors, 2 equal weaker ori, the lower incurved; jowls narrow, $\frac{1}{10}$ vertical height of eye; third antennal segment rounded above, flatter below, arista largely bare.
MESONOTUM. 2 dc, pre-scutellars lacking, acrostichals in some 10 rows.
WING. Length from 1.9 mm, in male to 2.2 mm in female, last section of M_{3+4} two-thirds penultimate, first cross-vein well before mid-point of discal cell.
LEGS. Fore-tibia with 1, mid-tibia with 2 lateral bristles.
COLOUR. Frons mat, brownish-black, orbits darker, black, slightly shining; lunule silvery-greyish; mesonotum largely mat, greyish-black, occasionally faintly greenish, abdomen more obviously grey-green; legs entirely black; wings pale, veins yellowish-white; squamae and fringe silvery-white, halteres white.
MALE GENITALIA. Aedeagus as in Fig. 84, basiphallus and base of distiphallus chitinized but area between entirely membranous, distal tubule short, also membranous; ninth sternite as in Fig. 85, without apodeme; surstyli (Fig. 86) short but stout, with some 12 spines; cerci (Fig. 87) with some 20–25 strong spines and unusually long hairs at end; ejaculatory apodeme as in Fig. 88.
EARLY STAGES. Puparium orange-brown, deeply segmented, posterior spiracles each with 3 slightly raised pores (Fig. 89) lying immediately adjacent to the anal depression.
HOST-PLANTS. *Desmodium tortuosum* (Sw.) DC. and almost certainly other *Desmodium* species.
BIOLOGY. The larva produces an upper surface blotch mine, which is green when fresh but later turns brown and, when very old, white (Fig. 90).
DAMAGE CAUSED. An appreciable area of leaf is eaten by a single larva and leaf-mining activity on any scale must weaken young plants, inhibit growth and reduce effective yield.
DISTRIBUTION. Florida: Miami; Colombia: nr. Bogotá (leaf-mines on *Desmodium campylocladus* Hemsl., 10.xii.1958, K.A.S., very probably represent this species).
PARASITES. None recorded.

Discussion

TAXONOMY. This species belongs to a difficult complex and in the adult stage a certain identification will only be possible from the male genitalia.
ECONOMIC IMPORTANCE. Many species of *Desmodium* are cultivated as fodder crops and any substantial increase in existing populations of *Japanagromyza desmodivora* would appreciably affect crop yields.

Japanagromyza inaequalis (Malloch, 1914b) – Florida, Caribbean

ADULT. Medium-sized, greenish-grey species, with uppermost of the two orbital bristles distinctly smaller.

HEAD. Frons little wider than eye, not projecting above eye in profile; 2 ors, the upper conspicuously shorter (Fig. 91), 2 ori, the lower incurved; orbital setulae minute, sparse, reclinate; jowls narrow, about $\frac{1}{10}$ vertical height of eye; third antennal segment large, arista distinctly pubescent.
MESONOTUM. 2 strong dc, pre-scutellars normally well developed but occasionally weak, present on one side only or greatly reduced and similar to acrostichals.
LEGS. Fore-tibia with 1 weak lateral bristle, mid-tibia with 2 stronger ones.

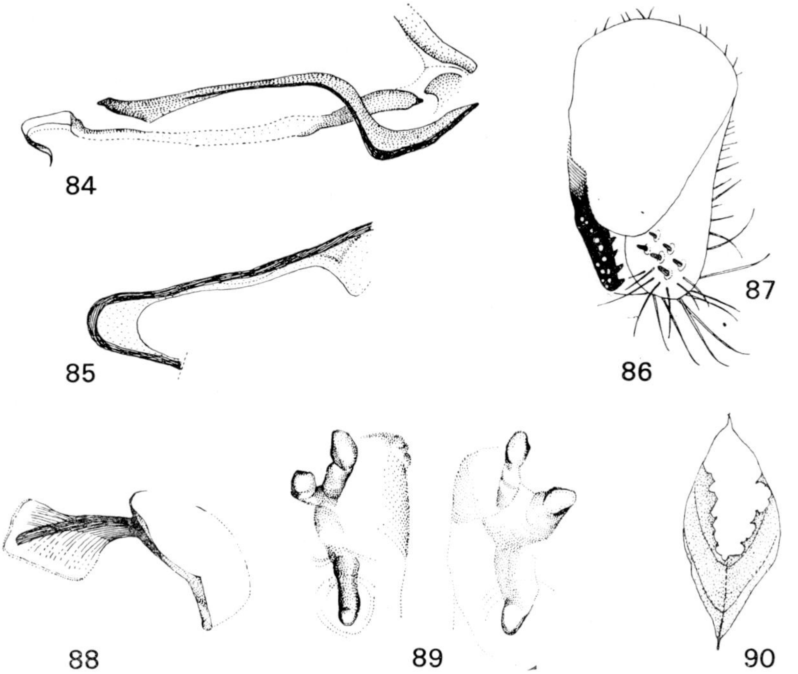

Figs. 84–90. *Japanagromyza desmodivora* Spencer: 84, aedeagus; 85, ninth sternite; 86, epandrium, showing surstylus and cercus; 87, cercus; 88, ejaculatory apodeme; 89, posterior spiracles of puparium; 90, leaf-mine on *Desmodium tortuosum*.

WING. Length normally 2.3–2.4 mm, costa extending to vein M_{1+2}, last section of M_{3+4} two-thirds penultimate, first cross-vein well before midpoint of discal cell.
COLOUR. Frons mat black, more rarely slightly brownish, lunule whitish-grey; mesonotum mat greenish-grey, rarely faintly bluish, more shining from behind; legs entirely black; abdomen shining greenish or faintly bluish; squamae and fringe white.
MALE GENITALIA. Aedeagus as in Fig. 92 with distal tubules pale, coiled; ninth sternite elongate, with slightly extended hypandrial apodeme; surstyli (Fig. 93A)

elongate, directed ventrally, not projecting inwards, bearing from 2–5 strong spines; cerci with 20 or more internally directed spines (Fig. 93B).

Early stages. Puparium oval, dark-brown, posterior spiracles with 3 pores, as in Fig. 94.

Host-plant. *Vigna repens* Baker and probably other *Vigna* species.

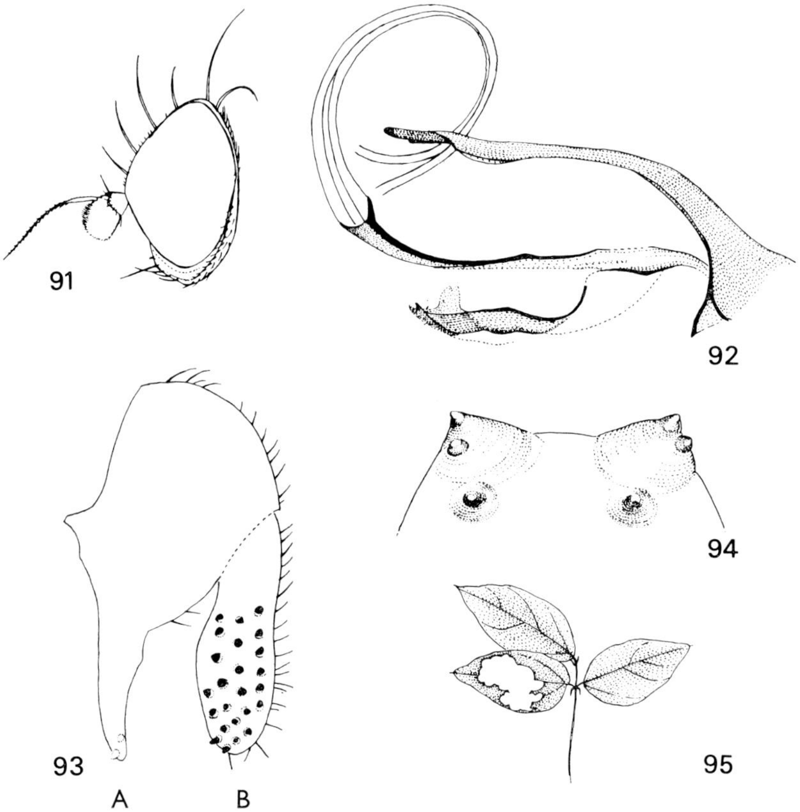

Figs. 91–95. *Japanagromyza inaequalis* (Mall.): 91, head; 92, aedeagus; 93, epandrium, with A, surstylus and B, cercus; 94, posterior spiracles of puparium; 95, leaf-mine on *Vigna luteola*.

Biology. The larva forms a conspicuous greenish-white blotch mine (Fig. 95), pupating externally.

Damage caused. A high proportion of leaflets on a small patch of plants seen at Miami in October, 1968 (K.A.S.) contained mines. If similar populations were present on a large scale where *Vigna* is cultivated the resulting loss of leaf area would seriously reduce crop yield.

Distribution. Florida; Puerto Rico; West Indies: Dominica.

Parasites. In Trinidad *Euopius inaequalis* Fi. (Fischer, 1971).

Discussion

TAXONOMY. The adults of this species are immediately recognizable by the characteristic reduction of the upper orbital bristle. Although both *J. inaequalis* and *J. aequalis* can occur together on the same host-plant, the puparia of the two species are instantly distinguishable, being dark brown in *inaequalis*, pale yellowish in *aequalis*.
ECONOMIC IMPORTANCE. This species is clearly capable of causing substantial damage to *Vigna* as a fodder crop. Further investigations seem desirable to establish its economic importance.

Japanagromyza tristella (Thomson, 1869) – Tropical Asia, Japan, Pacific

Agromyza tristella Thomson, 1869.
Agromyza variihalterata Malloch, 1914a.
Japanagromyza variihalterata, Spencer, 1960a: 17.
Japanagromyza trispina (Thomson), Spencer, 1965d: 25 (in error for *tristella*).
Japanagromyza nawai Kato, 1961, Spencer, 1966a: 3.
ADULT. Medium-sized species with black and white halteres.
HEAD. Frons narrow, equal to width of eye, not projecting above eye in profile; orbits not greatly differentiated, with 2 strong ors and 2 ori, the lower incurved, rarely a third, small ori; orbital setulae minute, sparse, reclinate; ocellar triangle not extended, apex at level of upper ors, lunule small; jowls narrow, $\frac{1}{10}$–$\frac{1}{12}$ height of eye, cheeks linear; eye appearing bare at highest binocular magnification; third antennal segment large, rounded, arista long, fine, virtually bare.
MESONOTUM. 2 strong dc, second just behind level of supra-alar, acrostichals irregularly in 8 or 9 rows, prescutellars well developed.
WING. Length from 2.4 mm in male to nearly 3 mm in female; costa extending strongly to vein M_{1+2}, last section of M_{3+4} about $\frac{3}{4}$ length of penultimate, position of first cross-vein variable but always before midpoint of discal cell.
LEGS. Fore-tibiae with 1 lateral bristle, mid-tibiae with 2.
COLOUR. Frons sooty black, orbits only weakly shining, ocellar triangle entirely mat; lunule grey; antennae entirely black; mesonotum blackish-grey with some subshine but predominantly mat; pleura black; abdomen shining blackish but generally with faint coppery tinge; squamae whitish-grey, margin and fringe black; halteres with stalk black and knob black below, white above.
MALE GENITALIA. Aedeagus as in Fig. 96, with long partially coiled tubule distally and conspicuous forked process above, ninth sternite with elongated hypandrial apodeme, cerci greatly enlarged.
EARLY STAGES. The early stages have been described and illustrated by KATO (1961, as *nawai*), by KURODA (1958, as *Melanagromyza* sp.) and by BERI (1971a, as *variihalterata*). The larva is white, somewhat greenish from 3.5–4.5 mm long; mouth-hooks with 1 large and 5 smaller teeth (Fig. 97); anterior spiracles on short projections, each with 3–5 pores (according to KATO 3 pores, according to

Kuroda 4–5), posterior spiracles characteristically trifurcated, each process with numerous minute pores (Fig. 98), varying from 31–57 (Kato, 1961; Kuroda, 1958, as *Melanagromyza* sp.); puparium reddish-brown.

Host-plants. *Glycine soja* Sieb. and Zucc., *Pueraria hirsuta* Matsum, *P. phaseoloides* Benth., *P. thunbergiana* Benth. (=*triloba* Makino).

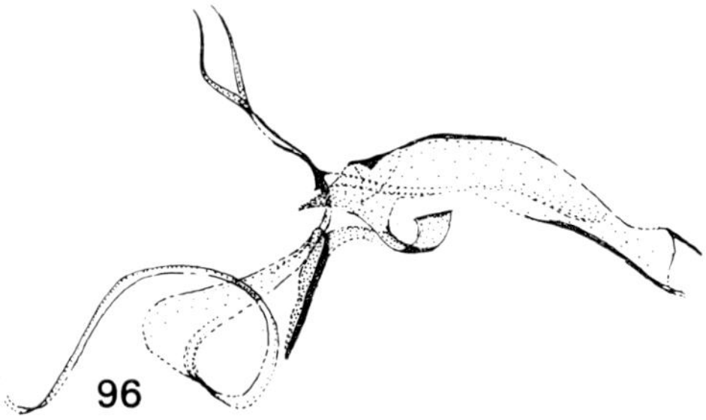
96

97

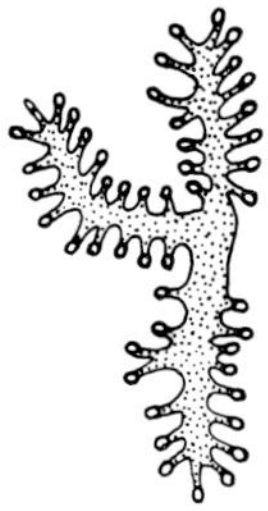
98

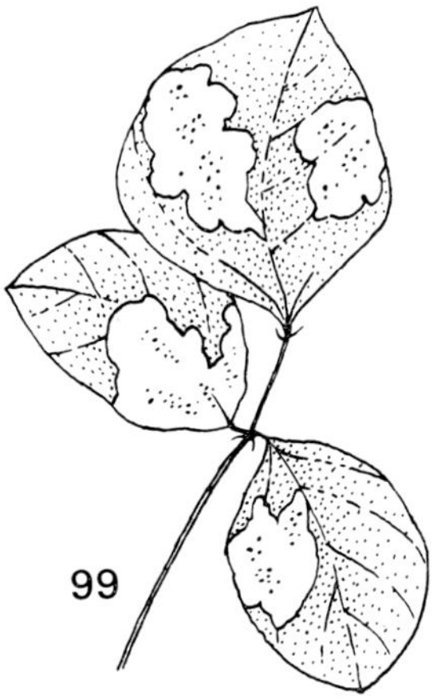
99

Figs. 96–99. *Japanagromyza tristella* (Thoms.): 96, aedeagus; 97, mouth-hooks of larva; 98, posterior spiracles (both after Kuroda); 99, leaf-mine on *Pueraria*.

Biology. The larva feeds exclusively as a primary blotch-miner on the upper surface of young leaves, normally with several larvae in each mine. There are several generations in a year. Two and possibly more mines can be found in a single leaflet (Fig. 99). The larva leaves the mine to pupate on the ground.

Damage caused. A large area of the leaf surface is eaten away by the larvae but from the limited material examined, it appears that growth of the plant is not appreciably affected. Damage on young plants may be more significant.

DISTRIBUTION. S. China; Formosa; Japan: Honshu, Kyushu, Shikoku; Bismarck Archipelago: Manus, New Ireland; Indonesia: Java, Flores; Singapore; Ceylon; Nepal; India.
PARASITES. None recorded.

Discussion

TAXONOMY. THOMSON's species from the S. China seaboard, *Agromyza tristella*, was misnamed in error as *trispina* when establishing the synonymy with *variihalterata* Malloch (SPENCER, 1965d: 25). The same mistake was repeated by SPENCER (1966b: 495) in a paper discussing Agromyzidae from the Bismarck Archipelago and also by SPENCER (1966a: 3) when synonymizing *nawai* Kato with *tristella* Thom. This error was kindly brought to my attention by Dr. D. COLLESS, C.S.I.R.O., Canberra.

The larval posterior spiracles in *J. tristella* are of interest, in that they preserve the primitive arrangement of three pores found in many genera of the Agromyzidae but with a proliferation of each one into a grouping of 12 or more minute subsidiary pores.
HOST-PLANTS. Although the only recorded host-plants are *Glycine* and *Pueraria*, it is possible that other related leguminous genera are attacked. Further collecting will be necessary to establish more accurately the exact host-range.
DISTRIBUTION. This species is widespread in the Pacific area and much of the Oriental region; with records from Nepal and Ceylon, it is surprising that it has not been found in India but this is presumably due to the limited collecting which has been undertaken there.
ECONOMIC IMPORTANCE. On Formosa *J. tristella* was one of a complex of insects attacking *Glycine* and making its effective cultivation impossible without the constant use of insecticides. In Japan the species regularly occurs on *Glycine* and according to KATO may cause rather severe damage locally.

***Japanagromyza** sp. – Peru, Venezuela, ?Argentina

ADULT. Frons brownish, orbits shining black, mesonotum distinctly greyish.
HEAD (Fig. 100). Frons narrow, equal to width of eye, not projecting above eye in profile; 2 equal ors, 2 ori, the upper only slightly weaker than the ors and reclinate, the lower weaker, more incurved; orbital setulae sparse, reclinate; jowls narrow, between $\frac{1}{8}$ and $\frac{1}{12}$ height of eye; third antennal segment large, arista long.
MESONOTUM. 2 strong dc, second only slightly beyond level of supra-alar; acrostichals in 8 rows in front, only isolated hairs at level of first dc; pre-scutellars well developed.
LEGS. Fore-tibia with 1 lateral bristle, mid-tibia with 2.
WING. Length from 2.5 mm in male to 2.75 mm in female; costa extending to

* See also Appendix A.

vein M_{1+2}, last section of M_{3+4} two-thirds penultimate, first cross-vein slightly before midpoint of discal cell.
COLOUR. Frons brownish tinged with grey, orbits distinctly shining black; mesonotum predominantly mat greyish, with only faintest greenish tinge in some specimens, with slight subshine; pleura black, with greyish tinge; legs black;

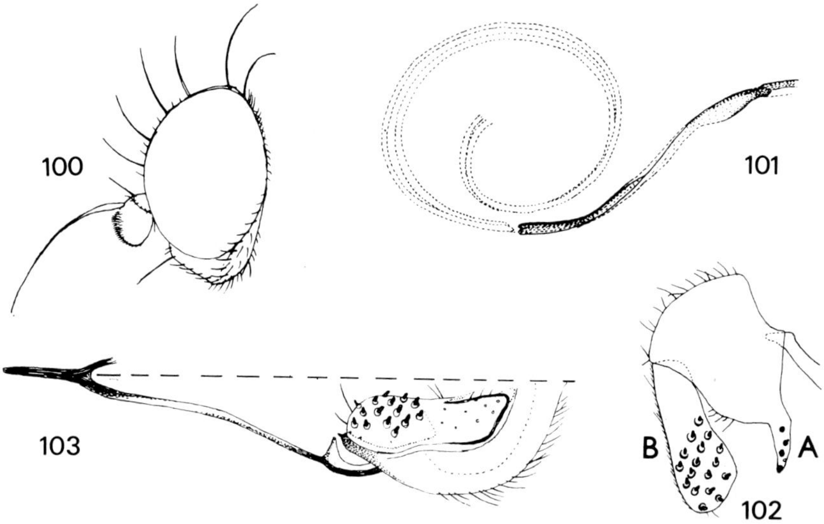

Figs. 100–103. *Japanagromyza* sp. (Peru): 100, head; 101, aedeagus; 102, epandrium with surstylus and cercus; 103, ninth sternite.

abdomen more shining greenish, squamae, including margin and fringe, silvery; halteres white.
MALE GENITALIA. Aedeagus with two pale, coiled tubules (Fig. 101); surstyli distinctly elongate, bearing several strong bristles (Fig. 102A), cerci large, with numerous strong bristles (Fig. 102B); ninth sternite with short hypandrial apodeme (Fig. 103).
HOST-PLANTS. *Phaseolus vulgaris* L. (bean), *Brassica oleracea* L. (cabbage).
DISTRIBUTION. Peru; ?Argentina: Tucuman.
PARASITES. None recorded.

Discussion

TAXONOMY. Distinctive characters of this species are the brownish frons and greyish mesonotum. The formal description of this new species is being prepared by C. A. KORYTKOWSKI, Peru.
ECONOMIC IMPORTANCE. Damage is caused both to beans and cabbage in the coastal areas of central and northern Peru, where the species is considered as an important pest (C. A. KORYTKOWSKI, personal communication).

Agromyza frontella (Rondani, 1875) – Europe, N. America

ADULT. Small species, with reddish frons, shining black mesonotum and costa ending at vein R_{4+5}.
HEAD (Fig. 104). Frons broad, twice width of eye, conspicuously projecting above eye, increasingly so anteriorly; jowls deeply extended at rear, $\frac{1}{2}$ vertical height of eye, cheeks forming broad ring below eye; third antennal segment longer than broad, in male covered with unusually long pubescence, in female with a small tuft of longer hairs at upper corner, arista short.
MESONOTUM. 3 + 1 strong dc, acrostichals in 4 or 5 rows.
WING (Fig. 105). Length from 1.9–2.2 mm, costa ending shortly after vein R_{4+5}, last section of M_{3+4} equal to or slightly longer than penultimate.
COLOUR. Frons reddish-brown, hind-margin of eye and orbits black; third antennal segment black; mesonotum and abdomen blackish-grey, distinctly shining; legs largely black; squamae pale greyish, fringe ochrous to brownish.
MALE GENITALIA. Aedeagus as in Figs. 106, 107.
EARLY STAGES. Anterior spiracles of larva each with 8–12 pores, posterior spiracles with 3; cephalo-pharyngeal skeleton as in Fig. 108; puparium reddish-brown, posterior spiracles as in Fig. 109.
HOST-PLANTS. *Medicago sativa* L. (lucerne); also *Medicago falcata* L., *M. lupulina* L. and *Melilotus* spp.
BIOLOGY. The larva forms a characteristic leaf-mine (Figs. 110, 117A), which is initially narrow and linear, following the margin of the leaf towards the apex, and then turning back and developing into a blotch in the area of the mid-rib. Pupation is on the ground. There are at least two generations between June and September.
DAMAGE CAUSED. In Europe this species is generally uncommon and populations are small. Although leaf-mining damage to lucerne has come to the notice of farmers in England, the effect on the crop has hitherto been negligible.

In the United States, in Massachusetts, it has recently been occurring in large numbers and the leaf-mining damage has been causing concern.
DISTRIBUTION. Widespread in Europe, with confirmed records from Austria, Germany, Netherlands, Italy, Scandinavia, U.S.S.R. and southern England; United States: Massachusetts, New York.
PARASITES. In Europe *Dacnusa dryas* (Nixon); in the United States *Closterocerus cinctipennis* Ashmead and *Diglyphus pulchripes* (Crawford), Eulophidae (det. B. D. BURKS, U.S.N.M.). No information is available on the rate of parasitism encountered.

Discussion

TAXONOMY. This species closely resembles the much commoner European species, *Agromyza nana* Mg., but is readily recognizable by the more blackish mesonotum and abdomen. The characteristic leaf-mine enables the larval stage to be distinguished without difficulty from that of *A. nana*.

It is immediately distinguishable from any other *Agromyza* sp. known in North America by the following combination of characters: presence of strong presutural dc, termination of the costa before vein M_{1+2}, pubescent third antennal segment and ochrous to dark squamal fringe. It can be included in the author's

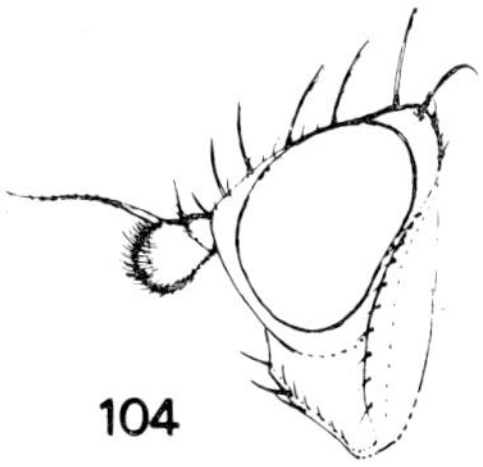

104

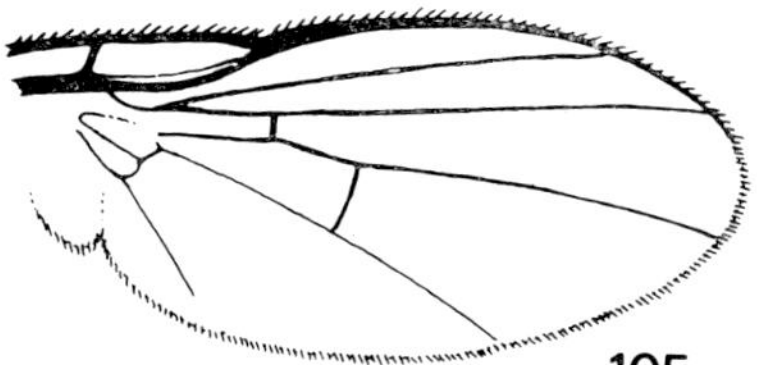

105

106

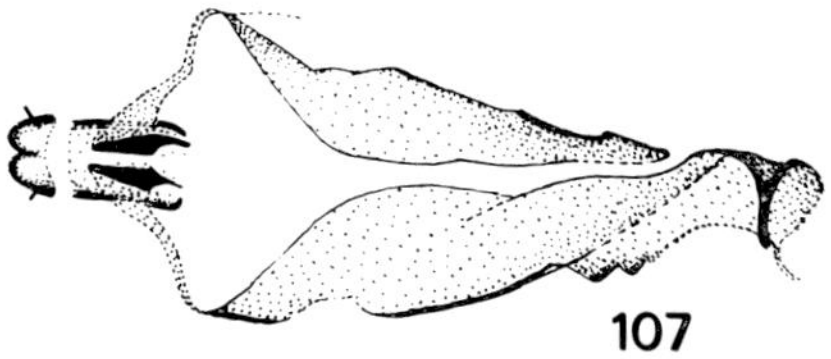

107

108

110

109

Figs. 104–110, 117A. *Agromyza frontella* (Rond.): 104, head; 105, wing; 106, aedeagus, side view; 107, same, ventral view; 108, cephalo-pharyngeal skeleton of larva; 109, posterior spiracles of puparium; 110, 117A, leaf-mine on *Medicago sativa*.

key to Canadian *Agromyza* species (1969a: 31), which can also be used for American species, by extending couplet 29: for sp. indet., read 29a; add new couplet, as follows:

29a Costa ending midway between veins R_{4+5} and M_{1+2}; third antennal segment conspicuously pubescent — *frontella* (Rondani)

— Costa extending to vein M_{1+2}; third antennal segment with normal, short pubescence — sp. indet.

A. frontella generally resembles *A. vicifoliae* Hering known in Europe as a leaf-miner on *Vicia cracca* L. but in this species the antennal pubescence is normal and the squamal fringe is totally black.

In Massachusetts *A. frontella* can occur together on alfalfa with *Liriomyza trifoliearum* Spencer. In the larval stage, although both species have posterior spiracles with only 3 pores, these are elongate (Fig. 109) in *A. frontella* and it is also recognizable by the two conspicuous projections below the posterior spiracles, by the divided anterior spiracles and by the double upper arm of the cephalopharyngeal skeleton (Fig. 108), which is typical of the sub-family Agromyzinae.

ECONOMIC IMPORTANCE. LEHMANN (1934) investigated *A. frontella* in Thuringia in eastern Germany. Although in some cases all three leaflets of lucerne leaves were found to contain mines, he concluded that in most years damage is negligible. VON BOLLOW (1955) reported a serious outbreak in the Würzburg area of southern Germany and noted that growth of infested plants was retarded and considered that the loss in the weight of fodder was considerable. In some years therefore this species can be a pest of some significance.

In the United States the large populations which are now present in Massachusetts and New York are causing some concern.

Agromyza lathyri Hendel, 1923 – Europe

ADULT. Large species with reddish frons.

HEAD (Fig. 111). Frons conspicuously projecting above eye, increasingly so towards base of antennae; 2 strong ors, 2 ori, the lower weaker and incurved; orbital setulae sparse, reclinate; jowls deeply extended at rear, cheeks forming broad ring below eye; third antennal segment longer than broad, somewhat narrowing distally, arista short, bare.

MESONOTUM. 3 + 1 strong dc, the second, third and fourth uniformly decreasing in size; acrostichals coarse, in 6 or 7 rows; pre-scutellars strong.

WING. Length from 2.5–3 mm, costa extending strongly to vein M_{1+2}, last and penultimate sections of M_{3+4} approximately equal, first cross-vein frequently well before midpoint of discal cell.

COLOUR. Frons reddish, orbits variably darkened, from greyish to entirely black; second and third antennal segment black, first slightly paler; face blackish-grey; mesonotum mat-grey, pleura entirely black; legs black, even fore-knees not differentiated; squamae whitish-grey, margin reddish-brown, fringe black; wing base dull reddish.

MALE GENITALIA. Aedeagus as in Fig. 112; ninth sternite triangular, with short hypandrial apodeme; surstyli elongate, with black chitinization and a few strong teeth on inner hind margin.

EARLY STAGES. Larva: anterior spiracles on short stalks, each with up to 16 pores, posterior spiracles each with up to 40 minute pores (Fig. 113); puparium reddish-brown, posterior spiracles with solid black chitinization.

Host-plants. *Pisum sativum* L.; also *Lathyrus latifolius* L. (everlasting pea), *Lath. montanus* L. and probably other *Lathyrus* species.
Biology. Oviposition takes place on the underside of a leaf where a shallow, whitish linear mine is formed in the larva's first instar. After the first moult a large lower surface blotch is formed, generally near the leaf base, which appears like a pale mottled area from the upper surface, due to the variable depth of feeding

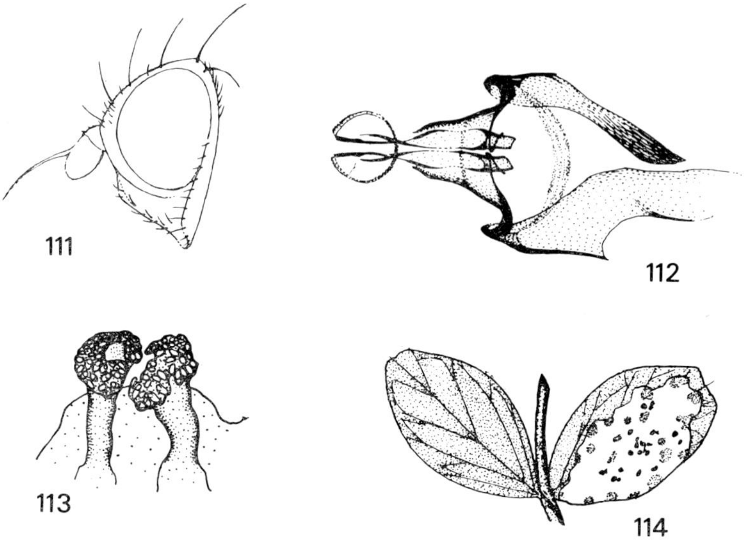

Figs. 111–114. *Agromyza lathyri* Hd.: 111, head; 112, aedeagus; 113, posterior spiracles of larva (after Hering); 114, leaf-mines on *Pisum* (Hering).

through to the upper surface (Fig. 114). The larva occasionally also feeds in the stem or on the upper surface of the leaf. Pupation takes place on the ground.
Damage caused. The numerous large blotch mines can affect the growth of young plants but damage is normally slight.
Distribution. Widespread in western Europe, confirmed in Germany, Holland, S. Sweden, England and Italy (Vesuvius, 20.iv.66, K.A.S.).
Parasites. *Opius turcicus* Fischer.

Discussion

Taxonomy. This is the only *Agromyza* known to feed on *Pisum* and is thus immediately distinguishable from other species on this host. It can occur together with *A. varicornis* Strobl (=*watersi* Spencer) on *Lathyrus latifolius* L., which is frequently cultivated as an ornamental in private gardens and Botanical Gardens but the two species are readily distinguishable both as adults and in the larval stage. In *A. lathyri* the squamal fringe is dark and the posterior spiracles of the larva (puparium) each have up to 40 bulbs; in *A. varicornis* the squamal fringe is pale, yellowish and the larval posterior spiracles each have 3 bulbs.

HOST-RANGE. It is not known whether the primary host of this species is *Lathyrus* or *Pisum*. However, its frequent occurrence on *Lathyrus* certainly provides reservoirs for attacking any peas cultivated in nearby localities.

ECONOMIC IMPORTANCE. HERING (1930) first reported this species attacking *Pisum* in Berlin. At that time he did not consider it to be a serious pest but later

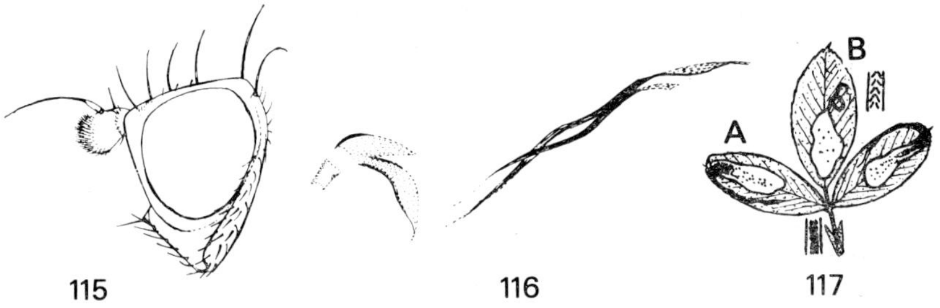

Figs. 115–117B. *Agromyza nana* Mg.: 115, head; 116, aedeagus; 117B, leaf-mine on *Medicago* (Hering).

(1957c: Vol. 2: 783) records it as common and injurious to peas. In view of its considerable size, larval feeding in large numbers would clearly cause substantial damage and the occurrence of the species on cultivated peas could be troublesome.

Agromyza nana Meigen, 1830 – Europe, Iraq, N. India

ADULT. Medium-sized species, with reddish frons, mat grey mesonotum and costa ending at vein R_{4+5}.

HEAD (Fig. 115). Frons broad, twice width of eye, distinctly projecting above eye in profile; jowls broad, up to $\frac{1}{3}$ vertical height of eye, cheeks forming conspicuous ring below eye; third antennal segment small, in male thickly covered in long pubescence, in female with a distinct fringe.

WING. Length from 2–2.5 mm, costa ending shortly beyond vein R_{4+5}, last section of M_{3+4} longer than penultimate, variable, about $1\frac{1}{2}$ times as long.

COLOUR. Frons reddish, hind-margin of eye black, orbits variably darkened; third antennal segment black; mesonotum and abdomen mat grey; legs largely black, fore-knees paler; squamae yellowish-grey, fringe brownish.

MALE GENITALIA. Aedeagus as in Fig. 116.

HOST-PLANTS. *Medicago sativa* L. (lucerne), *Trifolium* spp. (clover); also *Melilotus* spp.

BIOLOGY. Leaf-mine initially irregularly linear, in centre of leaf-blade, not at margin, later developing into a conspicuous white blotch on mid-rib which can entirely fill small leaves (Fig. 117B); pupation externally. There are two or probably more generations from June to September.

DAMAGE CAUSED. Although this is a common species, it does not regularly occur in sufficient numbers for the damage to the leaves to have any appreciable effect on crops of either lucerne or clover. However, PICARD (1914b) refers to considerable damage caused to lucerne in France when mines are numerous (the species is misidentified as *Agromyza nigripes*).

DISTRIBUTION. Widespread throughout Europe, Egypt, Morocco, Tunis; Iraq; India: Delhi.

PARASITES. In Europe *Chrysocharis amyite* (Walk.), *C. naenia* (Walk.), *C. polyzo* (Walk.), *Diglyphus isaea* (Walk.), *Pediobius petiolatus* (Spinola), Eulophidae (BOUČEK and ASKEW, 1968); also *Dacnusa* sp., *Exotela* sp. (Braconidae). In Iraq *Opius filicornis* Thom. (AL-AZAWI, 1971); in Turkey *O. curtifemur* Fischer.

Discussion

TAXONOMY. This species is generally similar to *Agromyza frontella* (Rond.) but is readily distinguishable by the more grey mesonotum and abdomen and the somewhat less extended jowls.

DISTRIBUTION. The discovery of leaf-mines on *Melilotus* sp. at the Agricultural Research Station at Delhi on 19.xi.1962 (K.A.S.) significantly extended the known range of this species. It seems probable that this population entered northern India during some cool stage of the Pleistocene and has been able to adapt to the excessively hot summers well south of its normal range, owing to the cold winters which occur in the Delhi area.

ECONOMIC IMPORTANCE. *Agromyza nana* apparently only rarely causes significant damage to lucerne or clover but its regular occurrence on these two important fodder crops makes it a species of potential importance. It has been discussed by LEHMANN (1934) in eastern Germany.

Liriomyza cicerina (Rondani, 1875) – Europe

ADULT. Very small, dark species, with darkened third antennal segment.

HEAD. Frons $1\frac{1}{2}$ times width of eye, increasingly projecting above eye towards base of antennae (Fig. 118), orbits conspicuously differentiated, raised and paler than frons; 1 reclinate ors, 2 or sometimes 3 incurved ori, orbital setulae sparse, reclinate; jowls deeply extended at rear, there $\frac{1}{3}$ vertical height of eye, cheeks forming broad ring below eye; third antennal segment small, round.

MESONOTUM. 3 + 1 dc, acrostichals sparse, irregularly in 2–4 rows.

WING (Fig. 119). Length from 1.3–1.5 mm, costa extending strongly to vein M_{1+2}, discal cell small, last section of M_{3+4} 3 times length of penultimate.

COLOUR. Frons orange-yellow, orbits distinctly paler yellow, sometimes narrowly darkened beside eye margin and blackish around base of ors; vertex dark, at least outer vertical bristle on black ground; jowls orange, face somewhat greyish, palps black; third antennal segment variably darkened, in darkest specimens almost entirely black, first and second segments yellowish; mesonotum deep black, shining from behind but more mat viewed from front; mesopleura normally black along lower and front margins, paler above, more rarely with only a small black patch on lower margin; legs dark, femora although basically yellow largely darkened by variable black striations.

MALE GENITALIA. Aedeagus as in Figs. 120, 121.

EARLY STAGES. Larval posterior spiracles each with 7–9 pores (Fig. 122), the two processes widely separated, puparium pale brownish-yellow. SHEVTCHENKO's illustration (1937) of the posterior spiracles shows only 3 pores, with the two processes adjoining; this clearly refers to some other species, although it is accepted that the species discussed in his paper does represent *cicerina*. DEL CAÑIZO (1945) reproduces SHEVTCHENKO's illustration without comment.

HOST-PLANTS. *Cicer arietinum* L. (chick-pea); *Ononis* species, including *arvensis* L., *hircina* Jacq., *repens* L., *spinosa* L.; *Hymenocarpus circinnatus* (L.) Savi (Yugoslavia, leg. E. M. HERING).

BIOLOGY. SHEVTCHENKO (1937) made a detailed study of this species in the Ukraine. He found there could be as many as four generations between April and August. The egg stage lasts for 2 or 3 days and, according to his observations, 42% of leaves contain a single egg, 45% two, 9% three, 2% four and 2% five eggs. The larval mine is on the upper or lower surface of the leaf and is linear, shallow, at first greenish, later whitish, winding irregularly and frequently forming a secondary blotch (Fig. 123); pupation takes place externally. The life cycle is completed in between 20 and 30 days, the pupal stage lasting generally from 10–12 days in the early generation. SHEVTCHENKO found up to 59 puparia per sq. decimetre, the equivalent of 1852 per sq. metre. DEL CAÑIZO (1934, 1945) has studied the species in Spain where the main areas of cultivation of *Cicer arietinum* are Castille, Estramadura and Andalucia and he has confirmed the very large populations frequently present in the early generation when fields of chick-pea can be seen with scarcely a single plant unaffected.

DAMAGE CAUSED. In his original description of this species in 1875 RONDANI wrote: 'Larva minatrix foliorum *Ciceri arietini*, cui saepe valde nocens' (Larva mining the leaves of *Cicer arietinum*, frequently causing substantial damage). SHEVTCHENKO records that mined leaves turn yellow, dry up and many fall prematurely. Lower leaves are attacked first and often only three or four healthy leaves remain on each stem. The resulting reduction in assimilation by the plant inevitably reduces its vitality and leads to a significant loss in the harvest. He analysed 10 plants collected on 4 June, 1934, and gives the following results:

Total number of leaflets on the 10 plants	2331
Leaflets destroyed (25.7%)	599
Leaflets with first grade damage (5.3%)	123
Leaflets with second grade damage (3.8%)	88
Total number of leaflets destroyed or damaged (34.8%)	810

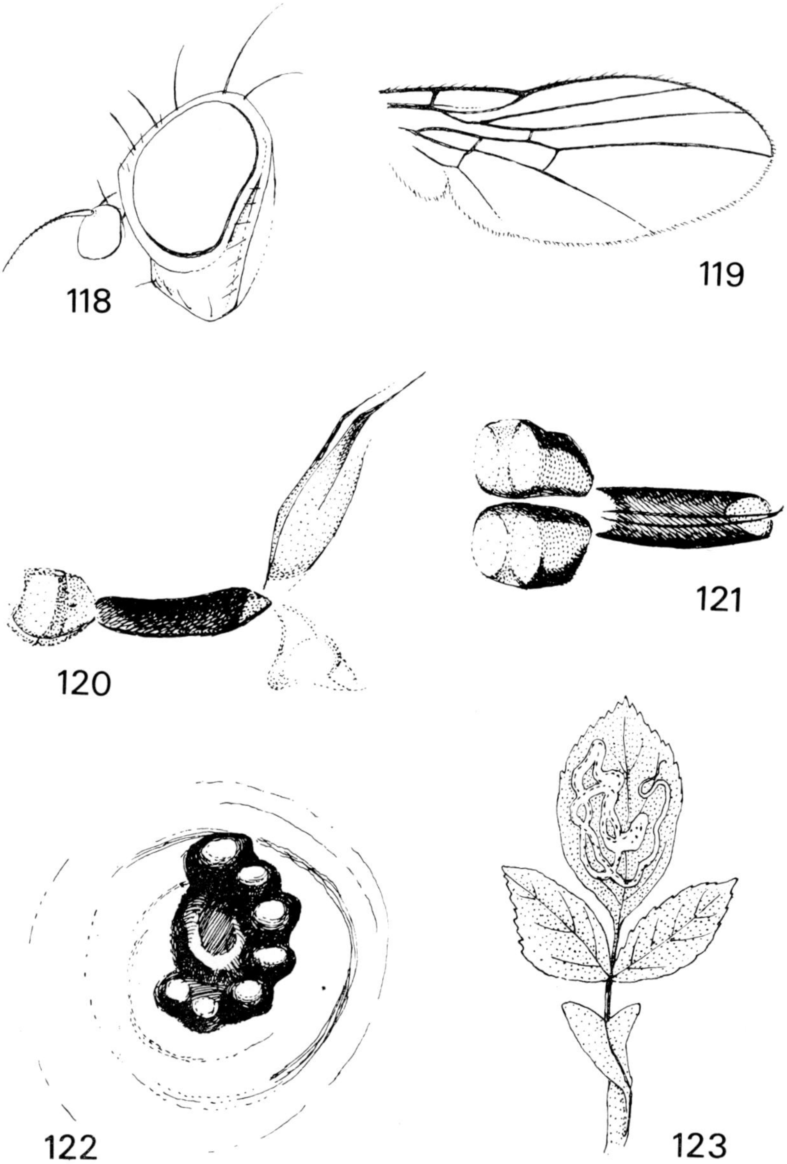

Figs. 118–123. *Liriomyza cicerina* (Rond.): 118, head; 119, wing; 120, aedeagus, side view; 121, same, ventral view; 122, posterior spiracles of larva; 123, leaf-mine on *Ononis* (Hering).

Two weeks later, on 20 June, damage was even more extensive, with an obvious effect on the flowering and fruiting of the plants and thus on the volume and weight of the seed. The following analysis is given of 20 plants:

Total number of leaflets	1033
Leaflets destroyed or damaged (92.4%)	955
Number of flowers	123
Number of pods formed	26

There were no entirely healthy plants available for control but it was clear that on plants suffering partial attack the pods were smaller and less numerous than normal, while in cases of heavy attack, the plants failed to produce any pods. On plants with an attack between 73% and 93%, only 11 flowers or pods were noted; plants with a 100% attack had only 3 flowers. The weight of the peas in these was only 0.98 gr, while it was up to 3.19 gr in the plants suffering less damage (73–93%). The average loss from these figures is 69.3%. With 45% of all plants attacked in this way, the reduction in harvest was 31.18%.

Some variation was observed in susceptibility to attack on plants of different varieties and from different sources. In general local varieties proved to be more resistant to attack but none was immune.

Del Cañizo (1934, 1945) refers to fields of chick-peas in Spain in which virtually all plants show evidence of leaf-mining attack. The plants are not destroyed but substantially weakened, with a consequent reduction of yield. Damage to the leaves actively facilitates subsequent fungal attack, referred to locally as 'la rabia', caused by *Phyllosticta rabiaei* (Pass.) Trotter. Shevtchenko refers to the fungal disease in the Ukraine as '*Ascochyta*'.

Distribution. Widespread in Mediterranean area, confirmed in Italy, Yugoslavia, Spain, Portugal, Morocco, Tunis, Turkey (Lodos, 1961), Ukraine, England, Germany, S. Sweden.

Parasites. A Braconid in Spain was found to parasitize up to 90% of larvae of the first generation, thus effectively reducing populations later in the year (Del Cañizo, 1934: 98). The identity of this species is not certain but it is possibly *Opius pygmaeus* Fischer, which has been confirmed parasitizing *L. cicerina* in Surrey, England (Fischer, 1962: 87).

Discussion

Host-range and distribution. *Liriomyza cicerina* occurs commonly on *Ononis* species in W. Europe and it seems probable that this is the primary host, from which it colonized *Cicer arietinum* when this was introduced into S. Europe from India in classical times. There are very few Oriental *Liriomyza* species and no records of *Liriomyza cicerina* from India or elsewhere in W. Asia where *Cicer arietinum* is native. This is of course not conclusive evidence that the species does not occur there but the balance of evidence suggests that it is native to the Mediterranean area.

ECONOMIC IMPORTANCE. *Cicer arietinum* is important not only as a source of human food but is also a valuable fodder crop. It may well be that in most seasons the populations of *Liriomyza cicerina* are effectively controlled by its parasites, as has been noted by DEL CAÑIZO in Spain. However, some economic loss, both in the pea harvest and in foliage for fodder, undoubtedly occurs wherever this crop is cultivated. It can be assumed that the mass outbreak which occurred in the Ukraine in 1934 was exceptional but nevertheless *Liriomyza cicerina* must be considered as a major pest, liable at any time when a significant build up of population occurs to cause serious damage.

Liriomyza congesta (Becker, 1903) – Europe, N. Africa

ADULT. Very small species, with mat mesonotum and acrostichals in 2 rows.
HEAD. Frons not significantly projecting above eye, 2 equal ors, 2 ori (rarely 3),

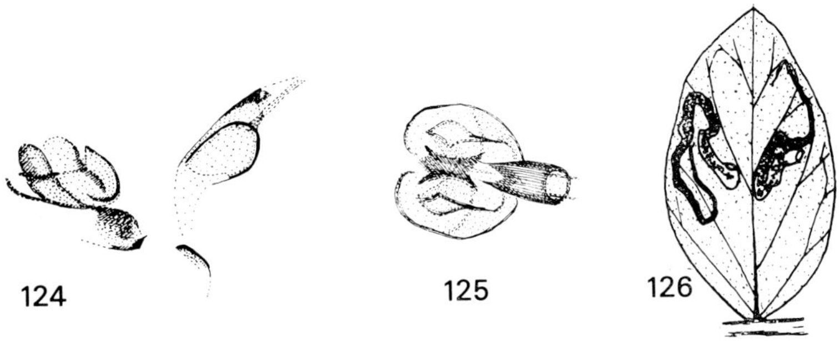

124 125 126

Figs. 124–126. *Liriomyza congesta* (Beck.): 124, aedeagus, side view; 125, distiphallus, ventral view; 126, leaf-mine on *Pisum* (Hering).

the lower weaker, largely incurved; orbital setulae minute, sparse, reclinate, jowls deeply extended at rear, up to $\frac{1}{3}$ height of eye, cheeks forming distinct ring below eye; third antennal segment small, round.
MESONOTUM. 3 + 1 fully developed dc, sometimes one or more small additional differentiated bristles beyond 4th; acrostichals sparse, in 2 rows.
WING. Length from 1.3 mm in male to 1.7 mm in female, discal cell small, last section of vein M_{3+4} thus relatively long, 3–4 times length of penultimate.
COLOUR. Frons bright yellow, upper orbits also yellow and both vertical bristles on yellow ground; all antennal segments bright yellow; mesonotum mat greyish-black, though with some subshine; mesopleura largely yellow, with only small black bar in lower front corner; femora bright yellow, tibiae and tarsi only marginally darker; abdomen with tergites yellow laterally and frequently also hind-margins narrowly yellow; squamae yellow, margin and fringe blackish.
MALE GENITALIA. Aedeagus distinctive, as in Figs. 124, 125.
EARLY STAGES. Larva with posterior spiracles each having 3 pores (contrast *L. pisivora* Hering, Fig. 134); puparium yellowish-brown.

HOST-PLANTS. *Pisum sativum* L., *Vicia faba* L., *Medicago sativa* L.; also *Lathyrus* spp. and many other genera of Papilionaceae, particularly in the tribes Galegeae and Vicieae. BUHR (1953) has discussed the host-range of this and other species on *Vicia faba* in some detail.
BIOLOGY. Leaf-mine short, linear, upper surface (Fig. 126), frass in minute black grains at each side of a greenish central band; pupation externally. At least two generations, from May to October.
DAMAGE CAUSED. The short, individual leaf-mines cause no significant damage to the leaves and it is rare to find more than a single mine per leaf.
DISTRIBUTION. Widespread in western Europe from Scandinavia to the Mediterranean and eastwards to Asiatic areas of the U.S.S.R.
PARASITES. In Europe *Chrysocharis liriomyzae* Delucchi, *C. naenia* (Walk.), *C. pubicornis* (Walk.), *Diglyphus isaea* (Walk.) (Eulophidae, BOUČEK and ASKEW, 1968); *Halticoptera patellana* Dalman (Pteromalidae); *Dacnusa maculipes* Thom., *Chorebus misellus* (Marshall) and *C. pseudomisellus* Griffiths (Braconidae); *Opius aemulus* Hal., *O. ilicis* Nixon, *O. levis* Wesm., *O. lugens* Hal., *O. minor* Fi. and *O. pendulus* Hal. In Iraq *Achrysocharella formosa* West., *Diglyphus crassinervis* Erdös (Eulophidae); *Halticoptera circulus* Walk., *Sphegigaster orobanchiae* Kurdj. and *Thinodytes cyzicus* Walk. (Pteromalidae) (AL-AZAWI, 1967).

Discussion

TAXONOMY. The distinctive characters of this species are the mat greyish mesonotum and the acrostichals in only 2 rows. However, a closely related species, *L. pisivora* Hering, also has these characters and the two cannot be reliably separated on external characters. The male genitalia are entirely distinct, cf. Figs. 124, 125 and Figs. 132, 133. The two species are also readily separable in the larval and pupal stage, with the posterior spiracles each having 3 bulbs in *congesta* and 7 or 8 in *pisivora*.
DISTRIBUTION. Earlier workers confused this species with *Liriomyza trifolii* (Burgess) which is common in the southern United States. However, the distinctive differences in the male genitalia confirm that *congesta* is limited to Europe and parts of Asia, and *trifolii* is exclusively Nearctic and Neotropical.
ECONOMIC IMPORTANCE. The damage caused by this species on two of its main hosts, *Pisum sativum* and *Vicia faba*, is negligible, largely because populations have never yet been known to build up sufficiently to produce a mass outbreak of leaf-mining activity. Effective control is apparently exercised by its various parasites. Nevertheless, the regular occurrence of this leaf-miner on these two important crops is sufficient reason for keeping it under close observation.

Liriomyza crucifericola (Hering), **comb. nov.** – Germany

Dizygomyza (*Cephalomyza*) *crucifericola* Hering, 1951d. Holotype ♀ in Zoologisches Museum, Berlin.

ADULT. Small species, with yellow frons, black third antennal segment and scutellum.
HEAD. Frons distinctly projecting above eye in profile; 1 ors, 2 equal incurved ori, orbital setulae minute (Fig. 127); jowls deeply extended at rear, $\frac{1}{2}$ vertical height of eye; third antennal segment round, arista short.
MESONOTUM. 3 + 1 dc, acrostichals lacking.
WING. Length 1.6 mm, costa extending to vein M_{1+2}, discal cell small, last section of M_{3+4} $2\frac{1}{2}$ times penultimate.
COLOUR. Frons yellow, orbits and entire hind-margin of eye, including base of

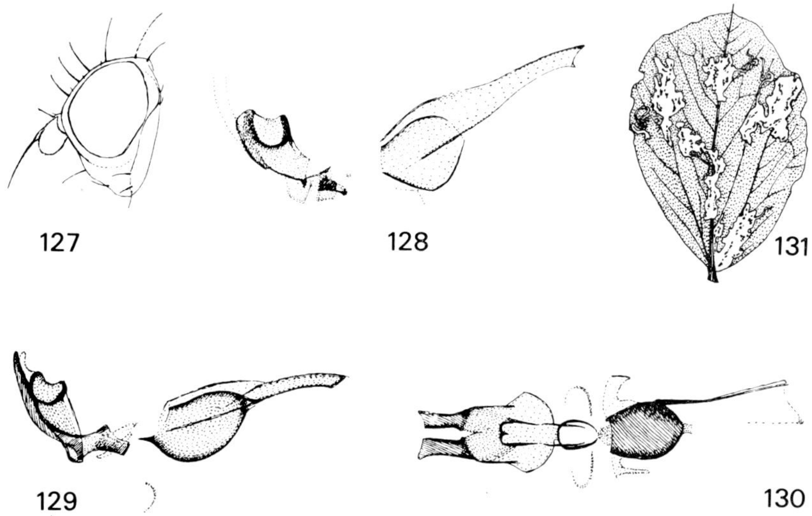

Figs. 127–131. *Liriomyza crucifericola* (Her.): 127, head; 128, aedeagus, side view (*Vicia faba*); 129, same (*Adonis aestivalis*); 130, same, ventral view (*Adonis*); 131, leaf-mines on *Vicia faba*.

vti, black; antennae, face and palps black; mesonotum and scutellum mat black, pleura entirely black; legs largely black but all femora bright yellow.
MALE GENITALIA. Aedeagus as in Figs. 128–130; surstyli as in typical *Liriomyza* species, with single spine at end; ejaculatory apodeme small, base bowl-shaped; ninth sternite with slender side-arms, U-shaped.
EARLY STAGES. Larva with mouth-hooks having two equal, overlapping teeth; posterior spiracles each with 3 pores, anterior spiracles with 7 or 8 minute pores; puparium yellowish-brown.
HOST-PLANTS. *Vicia faba* L.; also 33 genera of Cruciferae and one or more genera in following families: Capparaceae, Caryophyllaceae, Labiatae, Linaceae, Malvaceae, Papaveraceae, Papilionaceae, Polemoniaceae, Resedaceae, Scrophulariaceae, Tropaeolaceae, Umbelliferae and Valerianaceae (BUHR, 1953).
BIOLOGY. The first leaf-mines appear in the second or third week in May,

occurring mainly on Cruciferae, Malvaceae, Valerianaceae and *Linaria*, *Linum* and *Trigonella* species. From the end of May *Vicia faba* becomes a favourite host and mines are always frequent on this host by the end of June. There is apparently only a single generation.

The mine is initially linear, later developing into a large blotch (Fig. 131), which is at first greenish but in due course turns brown.

DAMAGE CAUSED. Many larvae can feed together in a single leaf of *Vicia faba*. The blotch mines coalesce and the leaves dry up, often before all larvae can complete their development.

DISTRIBUTION. Germany: Rostock, Bonn, Neuss, Erfurt.

PARASITES. *Cirrospilus vittatus* Walk., *Diglyphus isaea* (Walk.).

Discussion

TAXONOMY. The systematic position of this species is not wholly clear and further study will be necessary before its affinities can be fully understood. It certainly can no longer be retained in the genus *Dizygomyza* and does not belong in the sub-genus *Cephalomyza* (now in the genus *Amauromyza*, cf. SPENCER, 1971a). It is therefore tentatively transferred to the genus *Liriomyza*, despite the black scutellum, the colour of which can no longer necessarily be considered of generic significance. The form of both ninth sternite and surstyli are typical of *Liriomyza* but the aedeagus and ejaculatory apodeme are somewhat aberrant. It is also not clear whether the variation in the aedeagus noted between specimens on Cruciferae, *Vicia faba* and Ranunculaceae can be considered as of specific significance. The distal tubules are shortest in specimens from *Vicia faba* (Fig. 128), intermediate in specimens from Cruciferae and longest in those from Ranunculaceae (Figs. 129, 130).

DISTRIBUTION. This species was first discovered in the Botanical Gardens at Rostock and later in four other localities in Germany. It is not so far known elsewhere in Europe.

HOST-RANGE. The limited polyphagy found in this species is unusual. Although first discovered feeding on Cruciferae, it has since been found in Germany that the favourite host is *Vicia faba*. It has subsequently been confirmed on a wide range of other families, including *Adonis aestivalis* L. and *Ranunculus arvensis* L. (Ranunculaceae), cf. HERING, 1960: 138 and 1962: 61.

ECONOMIC IMPORTANCE. The leaf-mining damage to young plants of *Vicia faba* can be considerable and HERING (in SPENCER, 1968: 27) records severe damage to broad beans in the Bonn area in 1949. No damage has been reported on Crucifers.

L. crucifericola appears to have a limited distribution and only rarely occurs in sufficient numbers to cause economic damage.

Liriomyza pisivora Hering, 1957c – Europe

ADULT. Very small species, with mat mesonotum and acrostichals in 2 rows. Chaetotaxy, venation and colour almost exactly as in *L. congesta*, p. 94, but abdomen is generally darker, largely black, with only front tergites slightly yellow laterally.

MALE GENITALIA. Aedeagus as in Figs. 132, 133; surstyli distinctive, appearing almost rectangular, with a blunt chitinized area on inner corner representing a rudimentary spine.

EARLY STAGES. Larva with posterior spiracles each having 7–8 pores (Fig. 134, contrast *L. congesta*, p. 94); puparium yellowish-brown.

HOST-PLANTS. *Pisum sativum* L.; also *Pisum abyssinicum* A. Br., *P. elatum* Bieb., *P. jomardii* Schrk. (all in Botanical Gardens, Rostock, Germany) and *Lathyrus odoratus* L., *L. sativus* L., *L. silvester* L., *L. tingitanus* L.

BIOLOGY. Leaf-mine normally almost exclusively on lower surface of leaf, shallow, whitish, frequently adjoining veins and branching; if on upper surface,

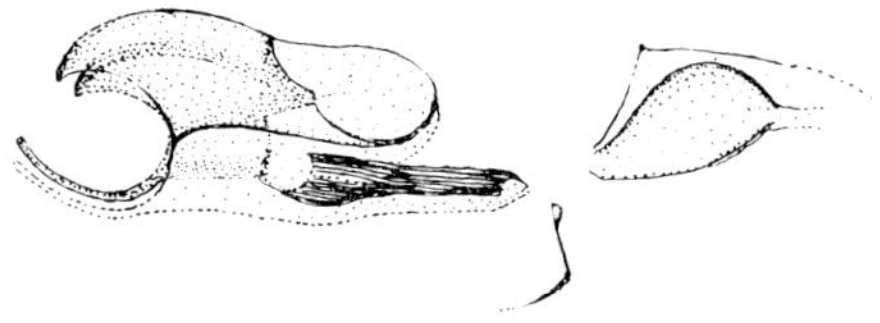

132

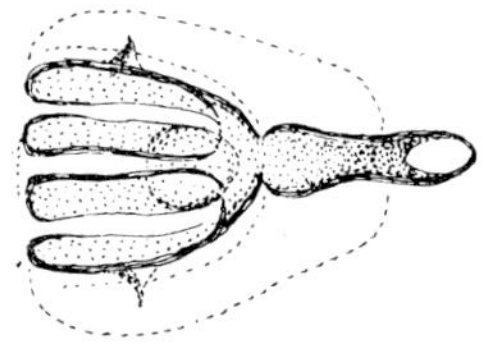

133

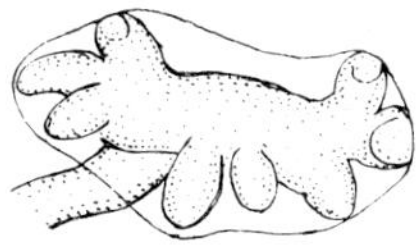

134

Figs. 132–134. *Liriomyza pisivora* Her.: 132, aedeagus, side view; 133, distiphallus, ventral view; 134, posterior spiracles of larva (after DE MEIJERE).

this is usually a short section at the very end. Mines also occur along the midrib and even in the stem. Frass is deposited irregularly in greenish-black strips. The larva leaves the mine to pupate on the ground.

DAMAGE CAUSED. Larval activity is limited to producing shallow mines which cause no significant damage to the leaves, at least with the small populations known hitherto.

DISTRIBUTION. Germany, England, possibly more widespread in W. Europe and overlooked.

PARASITES. *Dacnusa maculipes* Thomson; *Opius tersus* Förster.

Discussion

TAXONOMY. The presence of two species on *Pisum* had been indicated by earlier workers who noted that the larval posterior spiracles normally had only 3 pores but sometimes 7 or 8. HERING (1957c: 12) finally clarified this problem and described *pisivora* as new. He differentiated *L. pisivora* and *L. congesta* on characters of the leaf-mines as well as the larvae but pointed out that on external characters the adults could not be reliably separated, although the abdomen is generally darker in *pisivora*. The male genitalia have since been found to be strikingly different, as can be seen in Figs. 124, 125 and Figs. 132, 133.

ECONOMIC IMPORTANCE. At the present time this can be considered as negligible but *L. pisivora* should be duly noted as occurring regularly on peas in W. Europe.

Liriomyza trifoliearum Spencer, 1973 – N. America

ADULT. Small species with darkened antennae, blackish femora and largely black mesonotum.

HEAD. Frons not projecting above eye in profile, but orbits well differentiated; two equal ors, two ori, the lower slightly weaker, incurved; orbital setulae sparse, in single row, reclinate; jowls deep, extended at rear, up to $\frac{1}{3}$ height of eye; third antennal segment small, round, arista short, only weakly pubescent.

MESONOTUM. 3 + 1 strong dc, third and fourth equal; acrostichals in 4 rows.

WING. Length from 1.8 mm in male to 2.2 mm in female, last section of vein M_{3+4} twice length of penultimate, first cross-vein at midpoint of discal cell.

COLOUR. Frons orange-yellow, orbits normally slightly darkened to lower ors, both vertical bristles on black ground; third antennal segment variable, virtually black in darkest specimens, more normally brownish-yellow; mesonotum moderately shining black; mesopleura largely black, narrowly yellow along upper margin (Fig. 135).

LEGS. Coxae, tibiae and tarsi black, femora basically yellow, particularly below, but variably darkened, possibly appearing largely black; abdomen shining black, all tergites with linear yellow hind-margins; squamae grey, margin and fringe black.

MALE GENITALIA. Aedeagus as in Figs. 136, 137, surstyli with one strong bristle on inner corner.

EARLY STAGES. Puparium dark orange-yellow, posterior spiracles each with 3 pores.

HOST-PLANTS. *Medicago sativa* L., *Pisum sativum* L., *Trifolium incarnatum* L. and *T. repens* L.

BIOLOGY. At Ithaca, New York, adults of the first generation emerged during May. In 1969 mines were observed as early as 25 May but in 1967 and 1968 no mines were found before mid-June. Adults of the second generation emerged during late June to early July. Larvae of a third generation were present during September and October. There is normally a peak of leaf-mining activity early in June but populations varied substantially in the years 1966–1969 (A. G. WHEELER, personal communication). In Florida adults and mines have been found in April.

The larva forms a whitish, upper surface linear mine (Fig. 138), with frass deposited in isolated black lumps. Pupation appears to be variable, either in or outside the mine.

DAMAGE CAUSED. Injury to leaves is caused primarily by larval feeding but secondarily also by feeding punctures of the females. JENSEN and KOEHLER (1970) suggest that the actual feeding damage is secondary to that resulting from the tendency of mined leaflets to drop from the plant before or during harvesting (referring jointly to *L. sativae* and *L. trifoliearum*, given as *L. pictella*).
DISTRIBUTION. United States: California, Florida, New York, Massachusetts; Canada: Prince Edward Is.
PARASITES. *Chrysocharis ainsliei* Cwfd. (Eulophidae) in Massachusetts (A. G. WHEELER). JENSEN and KOEHLER (1970) record 6 parasites in California from a

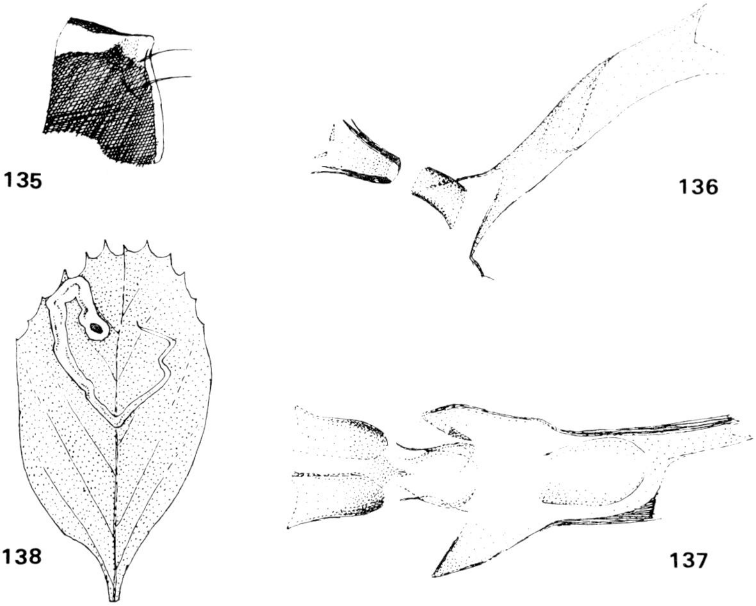

Figs. 135–138. *Liriomyza trifoliearum* Spencer: 135, mesopleuron; 136, aedeagus, side view; 137, same, ventral view; 138, leaf-mine on *Medicago sativa*.

mixed population of *Liriomyza sativae* and *L. trifoliearum* (identified as *pictella*); these species are mentioned in Chapter 8, p. 204.

Discussion

TAXONOMY. This is the darkest of the *Liriomyza* species occurring on crops in the United States. It could most easily be confused with *pictella* (Thomson) but this is a rare species only known from the unique type from California. The main distinguishing characters are the darkened third antennal segment, the largely black mesopleura and the dark legs. In these characters there is some overlapping with *L. huidobrensis* (which is not common on alfalfa) but the two species are immediately separable in the larval and pupal stages, with the posterior spiracles each having 3 pores in *trifoliearum* and 6–9 in *huidobrensis*. Both *L. sativae* and *L. trifolii* are paler species, with bright yellow antennae and femora.

A species which occurs together on alfalfa with *L. trifoliearum* is *Agromyza frontella* (Rondani) (see p. 85). The leaf-mines of the two are readily separable, being irregularly linear in the *Liriomyza* (Fig. 138) and developed into a blotch in the *Agromyza*. The *Agromyza* larvae and puparia are distinguishable by the conspicuous projections below the posterior spiracles and the unusually elongate pores (Fig. 109).

HOST-RANGE. When this species was described the only known hosts were *Medicago* and *Trifolium*, belonging to the sub-division Trifolieae of the Papilionatae. Subsequently mines have been found on *Pisum* on Prince Edward Is., Canada. Further genera in the Leguminosae may therefore now be found as hosts.

ECONOMIC IMPORTANCE. Leaf-mining damage to alfalfa definitely ascribable to this species in New York and to the mixed population of *sativae* and *trifoliearum* in California is considerable. In New York the infestation has only come to notice in recent years, and the increase in population of *trifoliearum* to injurious levels appears to be following the pattern familiar from Florida, California and Texas. It is suggested that this increase might be correlated with the use of insecticides and it would be of interest if this could be confirmed experimentally.

V. MISCELLANEOUS CROPS 1: INTERNAL FEEDERS

Sixteen species are discussed here, attacking 13 genera in 10 different families. Five of these species have not yet been formally described and therefore are referred to only briefly. All species apart from *Melanagromyza azawii* and *M. splendida* are restricted in their choice of host to a single genus. *M. splendida* feeds on both Compositae and Umbelliferae, while a *Melanagromyza* sp. found damaging stems of *Sesamum indicum* L. in Iraq proved to represent *M. azawii* Spencer, described earlier in Chapter 3, with *Vicia faba* L. probably as its main host.

The damage caused by this group of internal feeders is in general less than those attacking leguminous crops but can be locally severe. Of the 10 *Melanagromyza* species, *M. hibisci* can cause substantial damage to *Hibiscus esculentus* L. in India and *M. splendida* has seriously damaged *Apium graveolens* L. in Hawaii. *Ophiomyia lappivora* attacking *Arctium lappa* L. in Japan has caused appreciable local damage in some seasons.

Liriomyza braziliensis causes some damage to potatoes in Brazil and Peru and probably in other areas in South America. *Napomyza carotae* is a recently described species damaging carrots in Europe and it is believed that its economic importance may have been obscured in the past, owing to its confusion with the better known carrot fly, *Psila rosae* L.

Of the five undescribed species, a *Melanagromyza* sp. has been found in stems of *Brassica* (cauliflower) in Rhodesia and Singapore, and two others in the same genus have recently been reported damaging stalks of tomatoes in Colombia and Ecuador.* Their formal description is being prepared by Mr. G. Steyskal, Washington. Two species of uncertain generic affiliation (?*Phytoliriomyza*) apparently cause considerable damage to stems of potatoes in Bolivia and their description is also being prepared by Mr. Steyskal.

Napomyza cichorii may be found feeding in roots but the larva feeds primarily as a leaf- or stem-miner, and it has therefore seemed more convenient to discuss it together with the other serious pest on chicory, *Ophiomyia pinguis*, in Chapter 6. *Phytomyza rufipes* Meigen also feeds both in stems and as a leaf-miner, and is also discussed in the following chapter.

COMPOSITAE: ARCTIUM LAPPA L.

Ophiomyia lappivora (Koizumi), **comb. nov.** – Japan

Melanagromyza lappivora Koizumi, 1953. Types in Entomological Laboratory of Okayama University, Japan.

* See also Appendix A.

ADULT. Small, entirely shining black species.
HEAD (Fig. 139). Frons distinctly projecting above eye in profile, about $1\frac{1}{2}$ times width of eye; orbits conspicuous, with 2 strong ors, 2 ori, the lower incurved; orbital setulae numerous, reclinate; ocellar triangle with apex extending to level of upper ors, lunule semi-circular; jowls deeply extended at rear, $\frac{1}{3}$ height of eye, cheeks forming distinct ring below eye; third antennal segment rounded, arista finely pubescent.
MESONOTUM. 2 strong dc, acrostichals in 8 rows. Legs without differentiated lateral bristles.

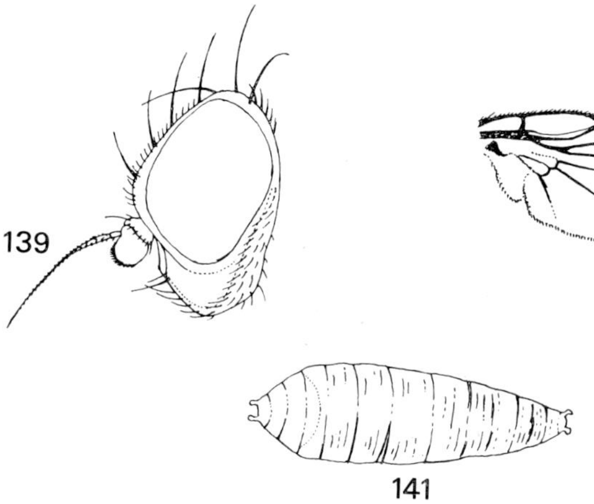

Figs. 139–141. *Ophiomyia lappivora* (Koizumi); 139, head; 140, wing; 141, puparium (all after KOIZUMI).

WING (Fig. 140). Length from 2.3 mm in male to 2.8 mm in female, costa extending to vein M_{1+2}, last section of M_{3+4} two-thirds penultimate, first cross-vein just beyond midpoint of discal cell.
COLOUR. Head black, mesonotum and abdomen entirely shining black, without any greenish tinge; squamae pale grey, fringe deep black.
MALE GENITALIA. Aedeagus typical of *Ophiomyia* species, illustrated by KOIZUMI, 1953: Fig. 3.
EARLY STAGES. Larva creamy-white, up to 8 mm long, slender; anterior spiracles on short stalks, each with 18 pores in 2 rows; posterior spiracles on short raised projections, widely separated, each with a ring of 9 pores. The puparium (Fig. 141) is grey, tapering to rear, 4.2 mm in length.
HOST-PLANT. *Arctium lappa* L. (Burdock; Japanese: *Gobo*).
BIOLOGY. KOIZUMI, quoting MUTO and TAKAHASHI, states that eggs are laid in the cotyledon or at the base of the stalk of a young leaf. The young larva mines

down into the root where the main feeding takes place. When mature it returns above ground and pupates at the base of a leaf stalk.

There are said to be 4–5 generations a year from May to October. Overwintering takes place in the larval state in the cortex of the root.

DAMAGE CAUSED. The larval feeding severely damages the roots of burdock, which are 'dwarfed, with many divided, futile minute roots and no fine, slender good roots, which are of high commercial value, are produced' (KOIZUMI, 1953).

DISTRIBUTION. Japan: Honshu, Gifu district.

PARASITES. None recorded.

Discussion

TAXONOMY. The description, illustrations and biology of this species clearly indicate that it correctly belongs in the genus *Ophiomyia*, to which it is transferred herewith. The tapering puparium, grey in colour, with posterior spiracles on raised projections and no trace of a central horn, the spinulose aedeagus and the form of larval feeding are all typical of *Ophiomyia* species. *O. lappivora* thus belongs to the relatively large group without the conspicuous external adult characters – vibrissal horn in male and raised facial keel – on which HENDEL based his original concept of the genus. SASAKAWA (1961) discusses and illustrates the female genitalia.

ECONOMIC IMPORTANCE. Burdock roots are a valuable, if limited, commercial crop in Japan and the edible roots are seriously malformed and otherwise damaged by the larval feeding of *Ophiomyia lappivora* which is thus an important local pest.

COMPOSITAE: LACTUCA SATIVA L.

Melanagromyza splendida Frick

This species has been bred from lettuce in Hawaii but significant damage has not been reported. It is a more serious pest on *Apium* and is therefore discussed below under Umbelliferae.

CONVOLVULACEAE: IPOMOEA BATATAS Lam.

Melanagromyza caerulea (Malloch, 1913a) – Caribbean, Gulf States

ADULT. Large greenish-blue species.

HEAD. Frons only slightly wider than eye, not projecting above eye in profile; 2 strong, reclinate ors, 2 largely incurved ori; orbital setulae numerous, short, reclinate; ocellar triangle broad, large, apex extending just below level of lower ors; orbits narrow, lunule slightly higher than semicircle; jowls narrow, $\frac{1}{10}$ vertical height of eye; third antennal segment small, rounded, arista with unusually long pubescence.

WING. Length from 2.6 mm in male to 3.4 mm in female, discal cell large, last section of M_{3+4} little more than half length of penultimate.

LEGS. Fore-tibia with one strong lateral bristle.

COLOUR. Frons mat black, ocellar triangle and orbits only weakly shining; mesonotum conspicuously shining bluish-green, sometimes entirely greenish; abdomen green, squamae and fringe white.

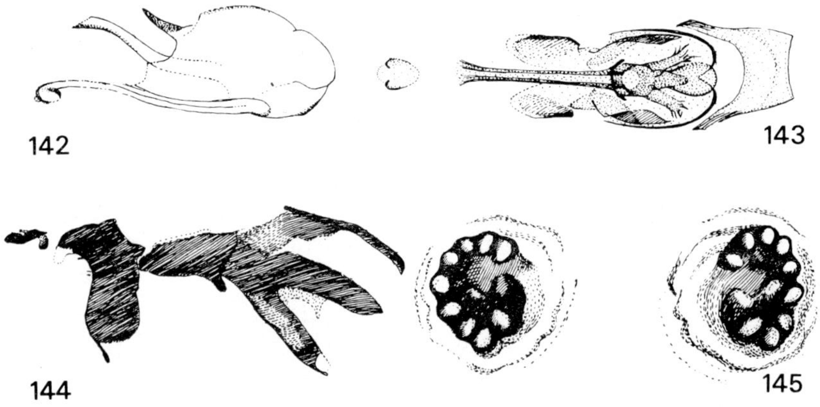

Figs. 142–145. *Melanagromyza caerulea* (Mall.): 142, aedeagus, side view; 143, same, ventral view; 144, cephalo-pharyngeal skeleton of larva; 145, posterior spiracles of puparium.

MALE GENITALIA. Aedeagus as in Figs. 142, 143, distiphallus with conspicuous chitinized process at end.

EARLY STAGES. Cephalo-pharyngeal skeleton of larva as in Fig. 144, mouth-hooks with one strong tooth on left, main right hand one weaker. Puparium reddish-brown, 3 mm in length, posterior spiracles each with an ellipse of 9 pores, central horn lacking (Fig. 145).

HOST-PLANTS. *Ipomoea Batatas* Lam. (sweet-potato); also *Ipomoea lacunosa* L. and *I. sinnata* Orteg., and almost certainly other species in the genus.

BIOLOGY. The only study made of the life history of this species is by BAILEY and PLANK (1940). These authors investigated the habits of *Melanagromyza caerulea* during the course of a sweet-potato breeding programme on Puerto Rico. Sweet-potato seed is borne in small capsules, each containing four ovules. These are attached to a small, short placenta at the base of the capsule. Oviposition takes place probably in the placenta from which the young larva enters a growing seed. Normally not more than one or two mature seeds develop from the four ovules present. The Agromyzid larva can complete its development in a single seed but if two or, exceptionally, three are present, these are usually all attacked – but only by a single larva. Pupation takes place within the eaten out seed or if two have been attacked, it may lie imbedded partly in one and partly in the other.

Ipomoea Batatas, like the Morning Glory, flowers early in the morning and the flowers close in the afternoon of the same day and usually drop the following day. BAILEY and PLANK found that when flowers were protected against insects on the day they opened, the seeds were rarely infested by *M. caerulea* larvae. Of 1039 unprotected flowers the average infestation rate of five varieties was 13.4%, with one as high as 24.5%; of 116 covered flowers the percentage attack was only 1.6. It thus seems clear that the normal time of oviposition is on the day the flower opens.

DAMAGE CAUSED. Seeds infested by *Melanagromyza caerulea* are totally destroyed and with the infestation rate rising to as much as 24% the destruction of seed is considerable.

DISTRIBUTION. Caribbean area, with confirmed records from Costa Rica, Dominica, Grand Cayman, Jamaica, Mexico, Puerto Rico; also Florida, Louisiana and Texas.

PARASITES. None recorded.

Discussion

TAXONOMY. This species belongs to a large complex of greenish species. A distinguishing character is the prominent fore-tibial bristle but this does also occur in a small number of other species. The male genitalia are entirely distinctive.

ECONOMIC IMPORTANCE. Adults of this species occur locally in large numbers and the destruction of seeds under natural conditions must be an important factor limiting propagation of the host-plant. However, there can be no direct effect on the edible root. Where breeding programmes of sweet-potato are undertaken, control measures seem necessary in view of the high incidence of seed destruction which may otherwise be expected, but the simple expedient of covering the flowers each day as they open is an inexpensive and apparently effective method of control.

CRUCIFERAE: BRASSICA OLERACEA L. VAR. BOTRYTIS L.

Melanagromyza sp.

A single female has been reared from stems of cauliflower at Salisbury, Rhodesia, in April, 1961. Similar specimens apparently representing the same species have been seen from stems of *Brassica alboglabra* L. H. BAILEY, Singapore, 5.xi.1960 (SPENCER, 1963c: 107).

The status of these specimens cannot be established until males are available. There is no evidence suggesting that the species has caused any significant damage.

EUPHORBIACEAE: RICINUS COMMUNIS L.

Melanagromyza ricini de Meijere, 1922 – Java

ADULT. Medium-sized, entirely black species.
HEAD. Frons narrow, approximately equal to width of eye, not projecting above eye in profile; orbits scarcely differentiated, with 2 ors and 2 ori, the lower incurved; orbital setulae in single row, reclinate; ocellar triangle with apex not reaching level of lower ors; jowls narrow, about $\frac{1}{12}$ height of eye; third antennal

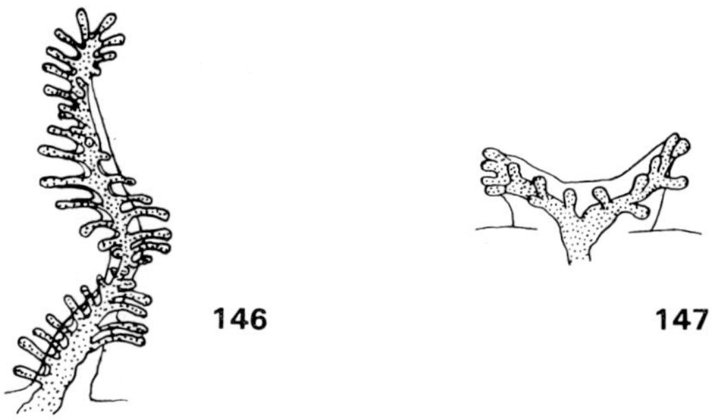

146 147

Figs. 146–147. *Melanagromyza ricini* de Meij.: 146, anterior spiracular process of larva; 147, posterior process (after DE MEIJERE).

segment small, round, arista long, bare. Mesonotum with 2 strong dc and acrostichals in some 10 rows in front, 6 rows extending beyond first dc.
WING. Length in female up to 2.5 mm, costa extending strongly to vein M_{1+2}, last section of M_{3+4} short, little over half length of penultimate, first cross-vein at centre of discal cell.
LEGS. Mid-tibiae with 2 strong lateral bristles.
COLOUR. Frons mat black, ocellar triangle and orbits scarcely shining; mesonotum and abdomen only moderately shining, black; squamae grey, margin and fringe black.
MALE GENITALIA. No males known.
EARLY STAGES. Larva white, up to 4.5 mm long, mouth-hooks each with only single tooth, the left-hand larger; anterior spiracles long, diverging, weakly S-shaped, each with up to 50 pores (Fig. 146); posterior spiracles each with two distinct arms bearing together about 15 pores (Fig. 147); puparium yellow, anterior spiracles black (from DE MEIJERE, 1922).
HOST-PLANT. *Ricinus communis* L. (castor-oil plant).
BIOLOGY. The larvae feed either singly or several together in young and half ripe fruits.

DAMAGE CAUSED. According to the collector of this species, the larvae cause serious damage to *Ricinus* on Java by destroying the fruits.
DISTRIBUTION. Java; tentatively identified from Flores and Sumbawa, and also from Formosa.
PARASITES. None recorded.

Discussion

TAXONOMY. Only the two female type specimens are known. A positive identification of additional material will only be possible after the genitalia of males bred from *Ricinus* can be examined.
DISTRIBUTION. Although *Melanagromyza ricini* is at present only known from Asia, *Ricinus communis* is a native of Africa and it therefore seems probable that it must also occur there.
ECONOMIC IMPORTANCE. Little information is available about this species but, in view of the importance of its host, it must be considered a pest of some significance and deserves further study.

LILIACEAE: ASPARAGUS OFFICINALIS L.

Ophiomyia simplex (Loew, 1896) – Europe, N. America

ADULT. Shining black species with costa ending at vein R_{4+5}.
HEAD (Fig. 148). Frons broad, almost twice width of eye; orbits conspicuously shining and distinctly projecting above eye in profile; normally 5 strong orbital bristles, orbital setulae strong, reclinate behind, more incurved in front; jowls broad, almost $\frac{1}{4}$ height of eye, cheeks forming broad ring below eye; third antennal segment small, round, arista fine, bare.
MESONOTUM. 2 strong dc, smaller third and fourth sometimes present before suture; acrostichals coarse, in some 10 rows.
WING (Fig. 149). Length from 2.2 mm in male to 3 mm in female, costa ending at or shortly behind vein R_{4+5}, first cross-vein well beyond midpoint of discal cell, last and penultimate sections of R_{3+4} equal or last section longer.
COLOUR. Frons sooty black, ocellar triangle conspicuously shining; mesonotum moderately, abdomen brilliantly shining black; squamae grey, margin and fringe black.
MALE GENITALIA. Aedeagus as in Figs. 150, 151.
EARLY STAGES. Larva: creamy-white, up to 5 mm long; anterior spiracles on short stalks each bearing about 10 minute pores, posterior spiracles on broad,

raised processes, each with about 16 irregular pores; puparium pale or darker brown, flattened, about 4 mm long, posterior spiracles as in Fig. 152.

HOST-PLANT. *Asparagus officinalis* L.

BIOLOGY. Adult flies normally emerge in Western Europe from early June to mid-July, with a second generation at the end of August. In the United States emergence in New York State is from mid-May and in California and southern States certainly considerably earlier.

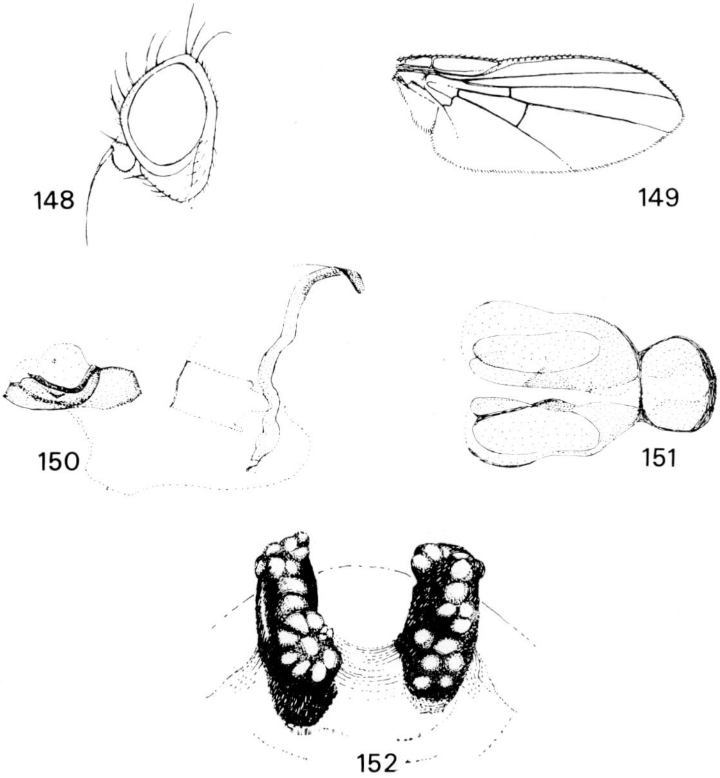

Figs. 148–152. *Ophiomyia simplex* (Loew): 148, head; 149, wing; 150, aedeagus, side view; 151, distiphallus, ventral view; 152, posterior spiracles of puparium.

Oviposition has been described in detail by FINK (1913), as follows: 'Before beginning to oviposit the female is usually seen running up and down an asparagus stalk, searching with her proboscis for a suitable place in which to lay her eggs. After this preliminary test of the entire stalk the female usually begins to oviposit at the base of the stalk at or near the surface of the soil; sometimes, according to the writer's observations, if the soil around the base of the stalk is loose the female even ventures below the surface as far as possible. She then takes up a position with her body parallel to the long axis of the stalk, and, bending her abdomen at almost a right angle to the stalk, begins to work her ovipositor into the epidermis. The ovipositor eventually pierces the epidermis and the egg is inserted just beneath it. The process of probing the stalk is repeated after each egg is deposited.'

On hatching, the larvae initially mine up the stem, according to BARNES (1937) sometimes for as much as 30 cm and then turn and continue feeding downwards. They frequently reach ground level or even slightly below and pupate immediately beneath the epidermis. FINK observed puparia as far as 7 inches below ground level. If full-grown earlier, pupation can occur at any point in the stem. The pupal stage of the first generation lasts from 3–4 weeks but some puparia even from the

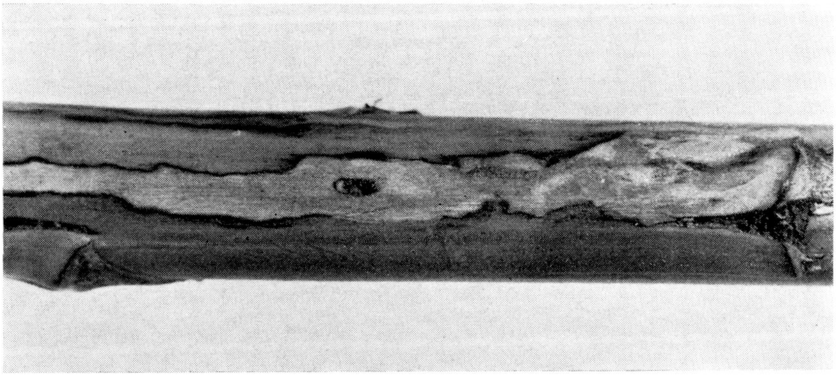

Plate 2: *Ophiomyia simplex:* mine with puparium on Asparagus, Netherlands (Instituut voor Plantenziektenkundig Onderzoek, Wageningen).

first generation do not produce flies until the following spring. A mine with puparium from the Netherlands is shown in Plate 2.

DAMAGE CAUSED. Invariably several larvae are found on a single asparagus plant and in America CHITTENDEN (1907) has reported as many as nine per inch of stem. The zig-zag feeding can result in complete girdling of the stem on which the epidermis cracks and splits. In consequence the plants turn yellow and frequently die. Stalks can rot at the point of maximum infestation and easily break in a strong wind.

DISTRIBUTION. Western Europe: England, France, Germany, Hungary; United States: widespread in Eastern States from Tennessee to New York, also present in California; Canada: Ontario, Quebec.

PARASITES. In Europe *Chorebus rondanii* Giard (Braconidae), *Sphegigaster* sp. (Pteromalidae) and *Pediobius epigonus* (Walker) (Eulophidae). The reference by BARNES (1937) to *Dacnusa bathyzona* Marshall is a misidentification for *Chorebus rondanii* (GRIFFITHS, 1967b: 655).

Discussion

TAXONOMY. For many years this species was treated as a *Melanagromyza*, since the male lacks the vibrissal fasciculus present in most species in *Ophiomyia*. However, larval characters, biology and male genitalia confirm its correct generic placing in *Ophiomyia* and it was transferred to this genus by SPENCER (1966b: 55). The male genitalia indicate that this species occupies an isolated position and it can be assumed that this differentiation has taken place during a long period of restricted association with its only host – asparagus.

DISTRIBUTION. *Ophiomyia simplex* was described from the United States, from caught specimens obtained in the 'Middle States'. After it had been recognized as the asparagus fly, it was found to be widespread in the East, from New England to Tennessee. In Europe it was first recorded from Hungary (SAJÓ, 1896), identified as *Agromyza maura* Meigen, and has since been confirmed from Germany, France and England. It is probably present wherever asparagus is regularly cultivated.

FINK (1913) states that it is a native American species. However, this must be considered as doubtful. *Asparagus officinalis* is native to Europe and Asia and was certainly introduced into the United States by early settlers. The habit of the larvae to pupate in the stem and root would thus make introduction of this insect virtually automatic together with plants intended for subsequent cultivation.

ECONOMIC IMPORTANCE. A number of early American authors record many examples of destructive populations at various localities in the United States. CHITTENDEN (1907), for example, reported that at Concord, Mass., an important area of cultivation of this crop, 'infestation was practically absolute, the insects being as abundant as the common asparagus beetle'. A report from California quoted by FINK, states: 'one noticeable feature of this year's growth has been the premature yellowing and dying of many stalks'.

In England no serious infestation has occurred in recent years; in Germany, in most seasons, yellow wilting plants can be observed in the main areas of asparagus cultivation.

More recently EICHMANN (1943) has re-assessed the significance of *O. simplex* in eastern Washington where asparagus is now an important crop. During the cutting season mines have been found on nearly every stalk. In the fall plants were found to be turning yellow prematurely and it was believed that this was caused by the asparagus miner. Three generations were confirmed between April and September. However, EICHMANN found that damage to asparagus stalks is superficial and only the cortex in the basal part of the stalk is destroyed. 'The food and water transportation system of the asparagus plant is not touched. Observations on the feeding habits of the maggots have shown that only the cortex in the basal part of the stalk is destroyed. The cortex has no vital function in the life process of the asparagus stalk. The cortex is parenchymous tissue which contains chlorophyl. Its sole function is the manufacture of food and the proportion of food produced in the lower part of the stalk is slight indeed compared with that manufactured by the rest of the plant. Heavily mined asparagus stalks may remain green until frost; whereas, a nearby unmined stalk may turn yellow prematurely. The premature yellowing of asparagus in late summer, and the wilting and dying asparagus stalks credited by other writers to the attack of the asparagus miner, has been credited in eastern Washington to infection by *Fusarium oxysporum* Schlecht.' EICHMANN concludes: 'At no time in our observations was it possible to associate asparagus miner injury with dead or dying asparagus stalks. Nor was it possible to associate asparagus miner injury with reduced yields of asparagus.'

It would be of interest to confirm EICHMANN's findings that *O. simplex* is not really a pest in Washington, by more detailed studies of this insect in other asparagus growing areas.

LINACEAE: LINUM USITATISSIMUM L. (FLAX)

Melanagromyza lini Spencer, 1963a – Peru

ADULT. Small greenish species, with broad jowls deepest in centre.
HEAD (Fig. 153). Frons not projecting above eye, about $1\frac{1}{2}$ times width of eye; normally 2 ors and 3 ori (in one male seen 7 orbital bristles present); orbital setulae sparse, reclinate, apart from 1 or 2 hairs in front which are proclinate; ocellar triangle large, with apex extending at least to level of upper ori; eye bare; jowls rounded, deep, slightly more than $\frac{1}{4}$ vertical height of eye, deepest in centre below eye or even forwardly projecting, cheeks forming conspicuous ring below eye; third antennal segment broader than long, arista fine, almost bare.
MESONOTUM. 2 strong dc, acrostichals numerous in front, in some 8 rows, 4 or 5 rows extending to level of first dc.
WING. Length from 2.6 mm in male to 3 mm in female, last section of M_{3+4} short, little over half length of penultimate.
COLOUR. Frons mat black, ocellar triangle and orbits moderately shining; mesonotum shining blackish-green, abdomen greenish-coppery; squamal fringe white.
MALE GENITALIA. Aedeagus as in Figs. 154, 155.
EARLY STAGES. Unknown.
HOST-PLANT. *Linum usitatissimum* L. (Flax).
BIOLOGY. The larva feeds as an internal stem-borer but no further details have been recorded.
DAMAGE CAUSED. Apart from obvious weakening of the plants from larval feeding, no other information is available.
DISTRIBUTION. Peru: Barranca, Cañete.
PARASITES. None recorded.

Discussion

TAXONOMY. This species belongs to a large and difficult complex of greenish stem-boring species, which is particularly well represented in the Neotropical

Region. The most distinctive character of *Mel. lini* is the form of the jowls. In many cases, however, species in this group can only be positively identified by the male genitalia.

ECONOMIC IMPORTANCE. *Linum* is still widely cultivated and, although *Mel. lini* is not known to have caused damage to flax, this is almost certainly its only

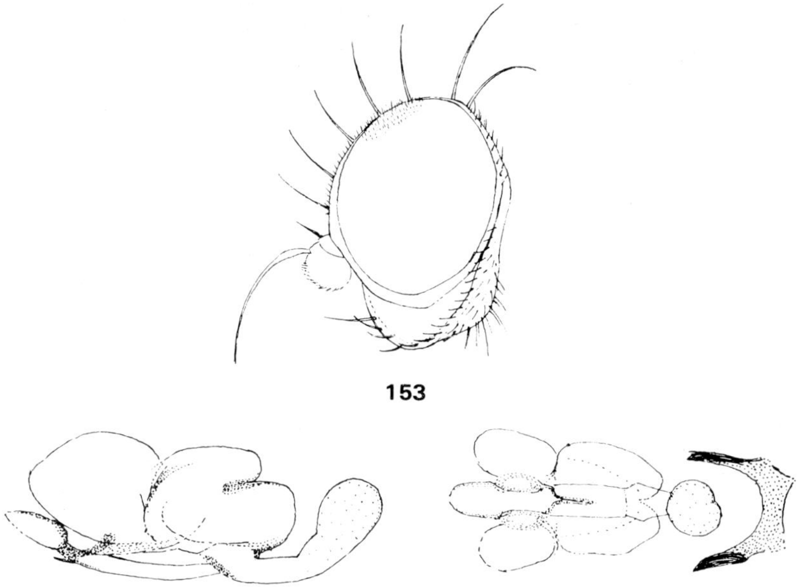

153

154 **155**

Figs. 153–155. *Melanagromyza lini* Spencer: 153, head; 154, aedeagus, side view; 155, same, ventral view.

host and it can therefore be expected to occur in other areas of South America where flax is grown commercially.

MALVACEAE: HIBISCUS ESCULENTUS L.

Melanagromyza hibisci Spencer, 1961b – India

ADULT. Medium-sized greenish species.

HEAD. Frons broad, $1\frac{1}{2}$ times width of eye, not projecting above eye in profile; 2 strong ors, 2 ori; ocellar triangle large, apex extending to level of lower ors;

orbits prominent, widening anteriorly to lower ori, then sharply narrowing to base of antennae; lunule conspicuously high and narrow, upper margin at level of upper ori (Fig. 156); jowls narrow, deepest in centre below eye, about $\frac{1}{8}$ vertical height of eye; third antennal segment rounded, arista distinctly pubescent.
MESONOTUM. 2 strong dc; acrostichals coarse, in 12 rows, only individual hairs extending to first dc.
WING. Length in male 2.4 mm, in female up to 2.8 mm; costa extending strongly to vein M_{1+2}, first cross-vein at midpoint of discal cell, last section of M_{3+4} only slightly more than half length of penultimate.
LEGS. Mid-tibiae with 2 strong bristles.

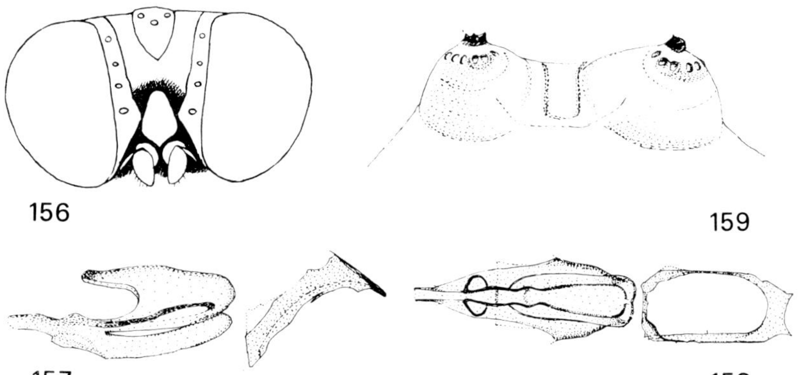
156 159 157 158

Figs. 156–159. *Melanagromyza hibisci* Spencer: 156, head showing lunule; 157, aedeagus, side view; 158, same, ventral view; 159, posterior spiracles of puparium.

COLOUR. Frons and jowls mat black; ocellar triangle and orbits only weakly shining, blackish; mesonotum shining greenish-black, sometimes with faint coppery tinge, abdomen similar; squamae and fringe whitish.
MALE GENITALIA. Aedeagus as in Figs. 157, 158; ninth sternite triangular, with extended hypandrial apodeme; surstyli rounded, bearing fine hairs on inner margin.
EARLY STAGES. Egg white or yellow, broad and rounded at posterior end, narrowed anteriorly into a curved and elongated hollow process; larva white, up to 4.1 mm in length (VENUGOPAL, 1954); puparium creamy yellow, posterior spiracles on separated, rounded protuberances, each with 8 pores around a minute central black horn (Fig. 159).
HOST-PLANT. *Hibiscus esculentus* L. (India: Bhendi; America: Okra).
BIOLOGY. Flies copulate on the day following emergence and oviposition takes place in the upper part of the main stem of seedling plants or side-shoots of more mature plants. Usually one or two eggs are found in each shoot but as many as 4–6 have been found. The larva hatches after 3 or 4 days and feeds internally in the

stem or petiole. Larval feeding lasts for 9–11 days. Usually one or two puparia are found in each infested shoot but in one instance 10 were observed. The pupal stage lasts for 8–9 days (based on VENUGOPAL, 1954).

DAMAGE CAUSED. The larval feeding on stem tissue produces a gall-like swelling and the epidermis frequently splits open longitudinally. Growth of the affected plant is stunted and the shoot above the infested region withers. Young plants wilt and infested leaves die, while in more mature plants with larval feeding in side-shoots, these also can wilt and die.

VENUGOPAL found that eight varieties of 'Bhendi' examined in S. India were all attacked by *M. hibisci* but the three varieties with hairy stems were appreciably less affected than those with smooth stems. Infestation ranged from 15–80%. The results of a further series of nine checks are tabulated by VENUGOPAL in the following table:

Table V. Showing the number of shoots examined, percentage infestation and population counts of the different stages of the insect

No. of shoots examined	*No. of shoots infested*	*% shoots infested*	*No. of maggots*	*No. of live pupae*	*No. of empty pupal cases*
(1) 78	49	62.8	19	63	14
(2) 69	25	36.2	6	59	4
(3) 25	22	88.0	1	25	4
(4) 100	73	73.0	12	115	8
(5) 100	45	45.0	2	54	18
(6) 100	43	43.0	15	28	0
(7) 175	78	44.6	26	71	0
(8) 50	22	44.0	14	35	0
(9) 52	18	34.6	17	28	0

DISTRIBUTION. India: Madras, Coimbatore, New Delhi.

PARASITES. VENUGOPAL reports parasitism by Braconids and Chalcids but no identifications are given.

Discussion

TAXONOMY. VENUGOPAL (1954) confused this species with *Melanagromyza obtusa* (Mall.), which exclusively infests pods of *Cajanus* and some other Leguminosae (see p. 46). SPENCER (1961b) recognized the species on *Hibiscus* to be distinct.

DISTRIBUTION. The only confirmed records are from the three localities in India cited above. Specimens from Flores and Sumbawa are now considered doubtfully referable to *M. hibisci. Hibiscus* is widely cultivated in the southern United States but *M. hibisci* is apparently not present there.

ECONOMIC IMPORTANCE. VENUGOPAL (1954) considers *M. hibisci* to be a serious pest, as 'well over 50% of the plants or branches are infested during a bad attack. The attack is more severe during the period July to November than during the cooler months of December and January. It must be noted, however, that in Madras in 1947 a severe attack of young plants was recorded during the period January to March'.

The immature fruits of *Hibiscus* are an important vegetable crop in many areas of the tropics and the destruction and weakening of plants by *Melanagromyza hibisci* must therefore rank it as a pest of some significance.

SCROPHULARIACEAE: DIGITALIS (FOXGLOVE)

Napomyza scrophulariae Spencer, 1966h – Europe

ADULT. Large species with yellow frons and mat grey mesonotum.

HEAD (Fig. 160). Frons strongly projecting above eye in profile, orbital setulae numerous; third antennal segment small, rounded, virtually bare.

COLOUR. Frons orange-yellow, third antennal segment black; mesonotum mat grey; legs black, all knees bright yellow.

WING (Fig. 161). Length from 2.6 mm in male to 3.9 mm in female, second cross-vein variable, in continuation of, basad or slightly beyond first.

MALE GENITALIA. Aedeagus as in Fig. 162.

EARLY STAGES. The puparium is pale yellowish-brown, somewhat elongate, up to 4 mm long.

HOST-PLANTS. *Digitalis lanata* Ehrh., *D. purpurea* L.

BIOLOGY. There are two generations. Flies emerge in May/June from over-wintering puparia and again in the autumn from larvae which have matured during the summer. It is not known whether oviposition takes place in the leaves or directly in the stem. The larvae can feed at any point in the stem, from immediately above the root system up to the flower-head. Most puparia are found in the stem but it is not uncommon to find them also in the seed-head.

DAMAGE CAUSED. Primary damage is caused to the stems from immediately above the root up to the flower-head; development of the leaves and general growth of the plant are also affected. Severe secondary damage is caused by fungal attack both by *Verticillium alboatrum* Reinke and Berth. and by *Fusarium* sp.

In France, *Digitalis purpurea* is sown in the autumn and planting out takes place towards the end of March. Plants can appear normal until the summer but in 1963 at an experimental plot at Angers growth then ceased and 10% of the plants rotted and were destroyed. *Digitalis lanata* is sown in early spring and symptoms of damage are readily visible by the autumn. Leaves turn brown and rot and stems break off at ground level with the slightest pressure. The neck of the stem appears blackish and necrotic, even though in the early stages the root system itself may still appear healthy. In the spring of 1964 50% of the plants which had been preserved until spring for collection of seed were defective and had to be destroyed (GEOFFRION and HÉRISSET, 1967). In East Germany it was found that even a single larva was able to kill *D. lanata*, while *D. purpurea* is more resistant (NECZYPOR, 1962).

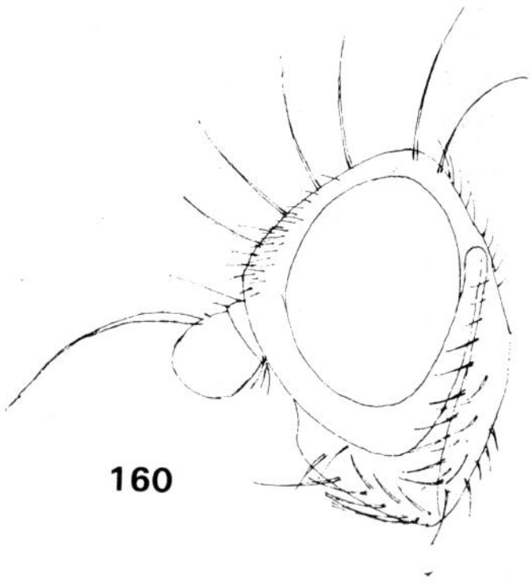

160

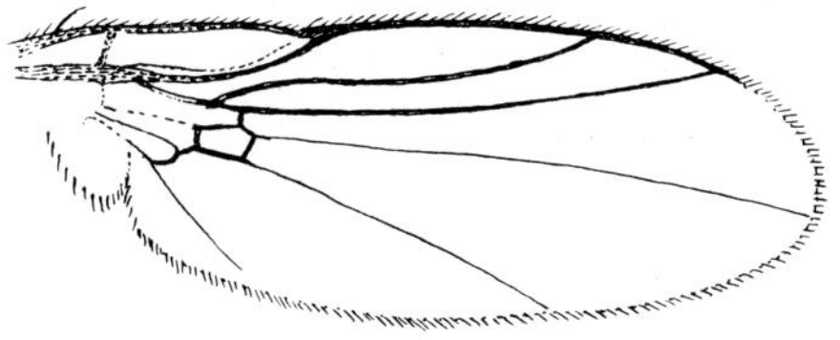

161

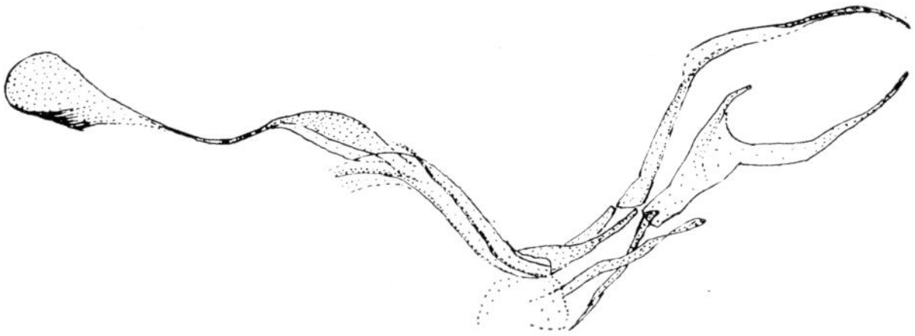

162

Figs. 160–162. *Napomyza scrophulariae* Spencer: 160, head; 161, wing; 162, aedeagus.

DISTRIBUTION. Britain, including Scotland (Edinburgh, Isle of Skye), France, Corsica, Germany (Berlin, Radebeul, Lausitz); apparently widespread in Europe.
PARASITES. None recorded.

Discussion

TAXONOMY. Most species in this small genus show only slight differentiation and a number cannot be reliably separated on external characters alone. The shape and pilosity of the third antennal segment and wing length are useful diagnostic characters but some otherwise similar species can only be satisfactorily identified by the minor but constant differences in the male genitalia. *N. scrophulariae* closely

resembles the common feeder on Composites, *N. lateralis* (Fallén), but the third antennal segment is rounder and the aedeagus is less conspicuously curved.
HOST RANGE. NECZYPOR (1962) reported that while *Digitalis lanata* and *D. purpurea* are heavily attacked in East Germany, no evidence of the fly could be found on *D. ambigua* Murr., *D. ferruginea* L. or *D. lutea* L.
ECONOMIC IMPORTANCE. *Digitalis purpurea* and *D. lanata* are cultivated in a number of countries in Europe for the glucosides extracted from the leaves, such as digitaline, digitoxin, digitonin and digitalein, which are important medicinally (UPHOFF, 1968).

The damage to the stem and root caused by the larval feeding of *N. scrophulariae*, followed by secondary fungal attack of *Verticillium alboatrum* and *Fusarium* sp., has caused concern to growers in France, where between 10 and 50% of plants can be destroyed. NECZYPOR (1962) reports substantial damage by this species (as *N. lateralis* Fallén) to both *D. lanata* and *D. purpurea* in the cultures of the pharmaceutical factory at Radebeul, East Germany. In 1960 in one area of a field containing 5250 plants, only 31 plants survived.

SOLANACEAE: SOLANUM TUBEROSUM L.

*Liriomyza braziliensis (Frost) – Argentina, Brazil, Ecuador, Peru, Venezuela

Agromyza braziliensis Frost, 1939; Frick, 1952a: 402; Spencer, 1963a: 357.
Agromyza ecuadorensis Frost, 1939; Spencer, 1963a: 357.
Liriomyza subandina Blanchard, 1954: 34, **syn. nov.**
ADULT. Exceptionally large species, wing length up to 4.5 mm, with triangular yellow patch on mesonotum adjoining scutellum (Fig. 163).
HEAD. 4 strong orbital bristles, orbital setulae minute, reclinate or upright; jowls deeply extended at rear, up to $\frac{1}{2}$ vertical height of eye; third antennal segment small, rounded, arista finely pubescent, swollen at base.
MESONOTUM. 3 + 1 strong dc, acrostichals irregularly in 4 rows in front, only scattered hairs extending to yellow area behind.
WING. Length from 3.3–3.7 mm in male and up to 4.5 mm in female, last section of M_{3+4} about $1\frac{1}{2}$ times length of penultimate.
COLOUR. Frons orange-yellow, upper orbits black or at least darkened; third antennal segment orange-yellow, brown or black; mesonotum moderately shining black, with yellow patch adjoining scutellum (Fig. 163); mesopleura black at least in lower third or more normally largely black with only narrow upper margin yellow; legs with femora bright yellow, sometimes with variable brownish striations, tibia and tarsi darker, brownish-yellow to almost black; squamae dark grey, margin and fringe black.

* See also Appendix A.

MALE GENITALIA. Aedeagus as in Figs. 164, 165; surstyli elongate (Fig. 166), bearing conspicuously long hairs at end and short, stout bristles basally.
EARLY STAGES. Larva up to 6 mm long, whitish-yellow, posterior spiracles on short projections, each with 3 pores; puparium dark brown, deeply segmented, truncate at posterior end (Fig. 167).

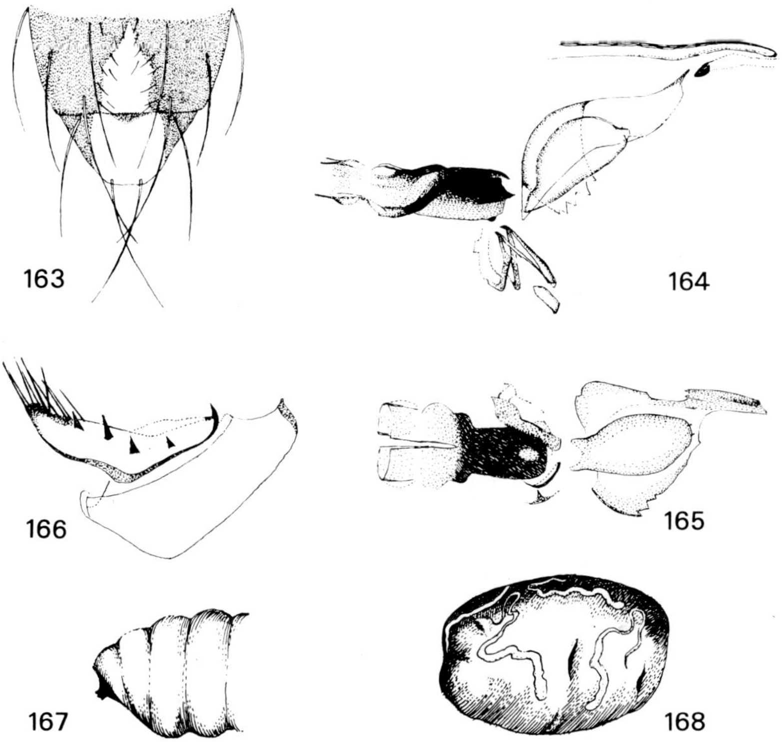

Figs. 163–168. *Liriomyza braziliensis* (Frost): 163, mesonotum; 164, aedeagus, side view; 165, same, ventral view; 166, surstylus; 167, posterior end of puparium; 168, mines on potato tuber (after MENDES).

HOST-PLANTS. *Solanum tuberosum* L. (white potato); *Solanum andigenum* Juz. and Buk.
BIOLOGY. MENDES (1940) studied the life history of this species in Brazil. He was unable to confirm the place of oviposition but believed this to be directly in the potato tuber. It seems more probable that the egg is laid in the lower stem at ground level, from where the larva will mine down into the tuber. The mine always remains near the surface of the tuber (Fig. 168) and never penetrates to any depth. Pupation takes place in a pupal chamber at the end of the mine.

Damage caused. The very visible larval mines considerably reduce the commercial value of infested potatoes. However, from the strictly culinary viewpoint, the damage caused is of no real significance, as the mines would in any case be removed by peeling the potatoes. Damage to seed potatoes is also negligible, as the 'eyes' are not attacked.

A check in a potato field at Cascata (São João da Bõa Vista) made by Mendes on 10 August, 1937, revealed the following degree of larval attack:

Table VI

Variety	*Tubercles examined*				*Percentage of tubercles*			
	Healthy	*Slightly infested*	*Seriously infested*	*Total*	*Healthy*	*Slightly infested*	*Seriously infested*	*Infested*
Irish Cobbler	496	94	188	778	63.8	12.1	24.2	36.3
N. York Rural Do-Dooley	759	110	137	1006	75.4	10.9	13.7	24.6
Golden	1292	133	81	1506	85.8	8.8	5.4	14.2
Green Mountain	756	46	18	820	92.2	5.6	2.2	7.8
Total	3303	383	424	4110	80.4	9.3	10.3	19.6

The average infestation of almost 20% is considerable, particularly as over half this figure is classified as heavy attack.

Distribution. Argentina: Mendoza; Brazil: Cascata, Nova Teutonia; Ecuador; Peru; Venezuela.

Parasites. None recorded.

Discussion

Taxonomy. The size range in this species from 3.1–4.5 mm is considerable and there is also a significant variation in colour, with the darker specimens apparently being limited to higher altitudes in the Andes. However, the male genitalia in all specimens examined are identical.

Liriomyza subandina Blanchard was described from specimens caught on leaves of wild potato in Argentina ('obtenidas de hojas de papas silvestres', Blanchard, 1954). It has hitherto been believed that this species was a leaf-miner and it has therefore been compared with *Liriomyza quadrata*, which feeds as a leaf-miner on a number of genera of Solanaceae. I have now examined two specimens from Mendoza labelled by Blanchard as 'cotipos'. Both were thought by Blanchard to be females but one, in bad condition and with the sex difficult to establish, proved to be a male, and I have now been able to confirm that the species is identical with *braziliensis* Frost. The wing length of this male is 3.1 mm.

L. braziliensis closely resembles *L. quadrata* (p. 176) but the third antennal segment and mesopleura are generally darker. It is also a more robust species and the smallest male seen, from Mendoza, Argentina, with wing length of 3.1 mm, is larger than any male of *quadrata* and also larger than most females. There is therefore little real overlapping in size.

DISTRIBUTION. MENDES found that larvae of *L. braziliensis* were present at all localities examined in the Cascata area of Brazil from altitudes of 1300–1420 m but none could be found at the lower altitudes of 600 m around Campinas. He therefore suggested the species was limited to these higher altitudes. In Peru it is also only known from the mountains in the North. However, I have now seen a confirmed male from southern Brazil at Nova Teutonia, Santa Catarina, at an altitude of 300–500 m and the species therefore has a wide altitudinal range.
ECONOMIC IMPORTANCE. The real significance of this species remains to be established. Although the larval mining damage is superficial, there is no evidence regarding the effect of such infestation on the growth of young potato tubers and clearly this might have some appreciable effect on the total weight of the crop. Recent reports of the species in northern Peru indicate that it is now known as an important pest there.

UMBELLIFERAE: APIUM GRAVEOLENS L.

Melanagromyza apii Hering, 1951a – Australia

ADULT. A medium-sized blackish-green species.
HEAD. Frons slightly wider than eye, orbits conspicuous, just visible above eye, increasingly so towards base of antennae; 2 strong ors, 2 weaker, partially incurved, ori; orbital setulae sparse, reclinate; ocellar triangle with apex extending slightly below level of lower ors, lunule slightly higher than a semicircle; jowls deepest in centre below eye, $\frac{1}{5}$–$\frac{1}{7}$ height of eye, cheeks forming distinct ring below eye; eye bare in both sexes; third antennal segment small, round, arista short, bare.
MESONOTUM. Two strong dc, acrostichals in 8 rows in front but scarcely extending beyond level of first dc.
WING. Length from 2.4–2.7 mm, costa extending strongly to vein M_{1+2}, first cross-vein at midpoint of discal cell, last section of M_{3+4} slightly more than $\frac{2}{3}$ length of penultimate.
COLOUR. Frons mat black, orbits and ocellar triangle only weakly shining; mesonotum predominantly shining black, with only faintest greenish tinge, abdomen brilliantly shining green; squamae and fringe white, margin scarcely differentiated.
MALE GENITALIA. Aedeagus as in Figs. 169, 170; ninth sternite with greatly elongated hypandrial apodeme.
EARLY STAGES. Larval mouth-hooks with one large upper and a smaller lower tooth; puparium pale whitish-yellow to more brownish, posterior spiracles separated by their own diameter, each with an ellipse of 9 or 10 pores on the outer side of the strong central horn (Fig. 171).
HOST-PLANT. *Apium graveolens* L. (celery).
BIOLOGY. The larva feeds as a true miner in the petiole, mining initially in the

upper, later in deeper layers of the parenchyma but avoiding the vascular bundles. Pupation takes place at the base of the petiole. Emergence dates of the adults in the type series from 18 November–2 December and from 26 April–31 May indicate that there are two separate generations (HERING, 1951a).

Larvae have been found near the rootstock and this suggests that feeding occurs also in the stem, at least in young plants.

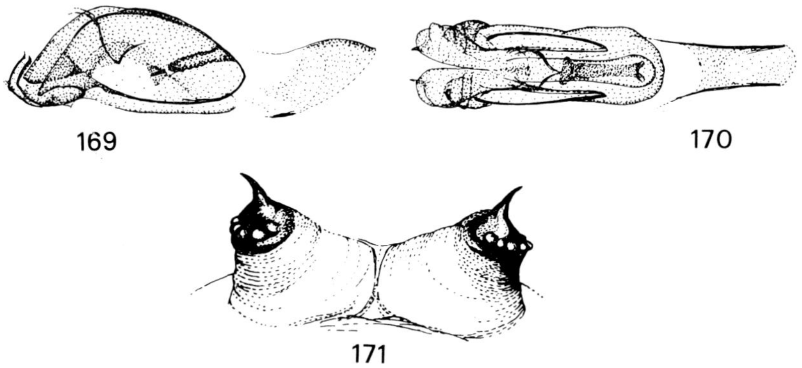

Figs. 169–171. *Melanagromyza apii* Hering: 169, aedeagus, side view; 170, same, ventral view; 171, posterior spiracles of puparium.

DISTRIBUTION. Only known from Australia: N.S.W., Sydney area.
PARASITES. Not recorded.

Discussion

This is a typical *Melanagromyza* but the somewhat unusual form of aedeagus gives no clue to any close relationship with other species. No other species feeding on Umbelliferae are known in Australia.

Apium graveolens was almost certainly introduced by early settlers but the possibility of the simultaneous introduction of *M. apii* can be virtually excluded, as the male genitalia are not in any way representative of the type found in European species. It therefore seems probable that *apii* is an endemic species, with the normal host being *Apium australe* Pet. Thou., which occurs widely in Australia and elsewhere in the southern hemisphere.

ECONOMIC IMPORTANCE. The economic importance of this species is not known. However, larval feeding in the petiole of young leaves would certainly render them unpalatable for human consumption and plants would be weakened if numerous larvae feed together in the stem.

UMBELLIFERAE: APIUM GRAVEOLENS L.; ALSO COMPOSITAE (LACTUCA)

Melanagromyza splendida Frick, 1953 – Caribbean, U.S.A., Hawaii

ADULT. Medium-sized, greenish species, with conspicuously wide orbits.
HEAD. Frons not projecting above eye; 2 equal reclinate ors, 2 or 3 largely incurved ori; orbits broad, conspicuously widening at midpoint, up to $\frac{1}{3}$ width of frons, bearing several rows of setulae, the inner rows erect or partially proclinate, those nearest eye margin reclinate; ocellar triangle with apex extending to level of upper ori; jowls deepest in centre below eye, $\frac{1}{6}$ height of eye; eye in male conspicuously pilose at level of ors; third antennal segment small, round, arista fine, bare.
WING. Length from 2.2 mm in male to 2.7 mm in female, last section of vein M_{3+4} two-thirds penultimate.
COLOUR. Frons mat black, ocellar triangle and orbits weakly shining, blackish; mesonotum shining blackish-green, scarcely mat even viewed from front, abdomen brilliantly shining green; squamae and fringe silvery-white; halteres black.
MALE GENITALIA. Aedeagus as in Figs. 172, 173, tubular process below mesophallus not extending to rear; surstyli curving inwards, fringed with short hairs or weak bristles which lie at right-angles to aedeagal apodeme (Fig. 174).
EARLY STAGES. Puparium pale yellowish, posterior spiracles separated by twice their own diameter, pale, weakly chitinized, each with an ellipse of 7–9 pores around a central scar representing a rudimentary horn (Figs. 175, 176).
HOST-PLANTS. Umbelliferae: *Apium graveolens* L., *Daucus carota* L. (carrot); Compositae: *Lactuca* (lettuce); other Composite hosts known are *Ambrosia, Bidens, Carthamus, Erechtites, Flaveria, Gaillardia, Gnaphalium, Helianthus, Tagetes, Xanthium* and *Zinnia.*
BIOLOGY. The larva feeds as an internal borer in the stem or in the midrib of large leaves. It is not known whether oviposition takes place directly in the stem or in an upper leaf. Pupation occurs within the plant. There is almost certainly only a single generation within each main vegetative period of the host.
DAMAGE CAUSED. No details are available but it has been reported that serious damage has been caused to celery plantings on Hawaii (D. E. HARDY, personal communication).
DISTRIBUTION. United States: Hawaii, also California, Michigan, Florida; Mexico; Bahamas, Jamaica.
PARASITES. None recorded.

Discussion

TAXONOMY. The characteristic feature of *M. splendida* is the unusually broad orbits, bearing numerous rows of setulae. The character also occurs – in an even more conspicuous form – in *Mel. virens* (Loew) and has led to the two species being confused in the past. They were finally separated and clarified by SPENCER (1969a) and SPENCER and STEGMAIER (1973).

HOST-RANGE. All other known species in the genus *Melanagromyza* are host-specific, either monophagous or at most oligophagous. It is also rare for any species in the family to feed both on Umbelliferae and Compositae. Even the

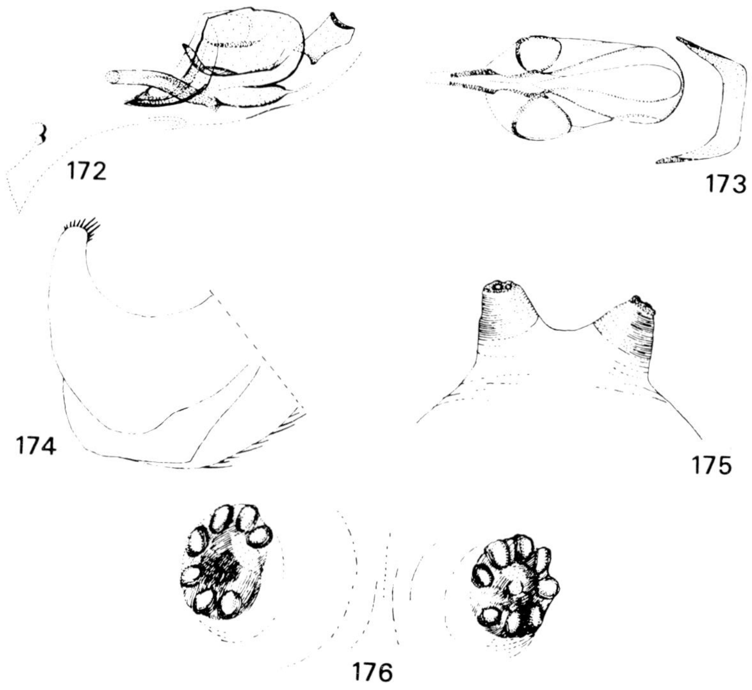

Figs. 172–176. *Melanagromyza splendida* Frick: 172, aedeagus, side view; 173, same, ventral view; 174, surstylus; 175, 176, posterior spiracles of puparium.

highly polyphagous species, *Phytomyza horticola* Goureau, is found only infrequently on hosts in the Umbelliferae. However, careful re-examination of available material, including a paratype of *splendida* ex celery, has failed to give any indication that the populations on Umbelliferae and Compositae might be distinct.

DISTRIBUTION. *M. splendida* appears to be a sub-tropical species. It is common in Florida and the Bahamas and although the only specimens known from further south in the Neotropical Region are from Jamaica, it is no doubt widespread in the Caribbean area. and probably also in Central America. With records from both Mexico and California, it can be expected to occur widely in the Gulf States. The species must be at the northern limit of its range in Michigan. Its presence in Hawaii is certainly due to relatively recent introduction by man.

ECONOMIC IMPORTANCE. The only damage to cultivated plants on record is on celery in Hawaii. However, it seems likely that more extensive damage may be caused by this species, particularly in areas where little information is available, such as Mexico, where it is known to attack carrots. Seedling plants will be the most vulnerable, in which larval feeding in the stem would have the maximum effect on growth. In Hawaii it has also been reared from lettuce.

UMBELLIFERAE: DAUCUS CAROTA L.

Napomyza carotae Spencer, 1966h – Europe

ADULT. Medium-sized species, third antennal segment with conspicuous fringe of white pubescence.
HEAD (Fig. 177). Frons broad, twice width of eye, strongly projecting above eye in profile; 2 equal, reclinate ors, 2 incurved ori; orbital setulae numerous, proclinate; jowls deep, $\frac{1}{2}$ vertical height of eye, cheeks forming broad ring below eye; third antennal segment longer than broad, with distinct fringe of white pubescence.
MESONOTUM. 3 + 1 strong dc, acrostichals sparse, irregularly in 3 or 4 rows.
WING (Fig. 178). Length from 2.7 mm in male to 3.1 mm in female, costa ending at vein R_{4+5}, average costal ratio 39:12:24; position of second cross-vein slightly variable but normally in continuation of first.
COLOUR. Frons orange-yellow, upper orbits black, both vertical bristles on black ground; second and third antennal segments black, first paler, slightly yellowish; palps black; mesonotum mat-grey, pleura blackish-grey, only mesopleura with upper margin narrowly yellow; legs black, with all knees bright yellow; squamae yellow, fringe black.
MALE GENITALIA. Aedeagus as in Figs. 179, 180, distiphallus relatively large, black, with pronounced curvature.
EARLY STAGES. Cephalo-pharyngeal skeleton of larva as in Fig. 181; puparium (Fig. 182) white, slender, up to 5.5 mm long, posterior spiracles blackish-brown, on short projections, each with a double row of about 20 minute pores.
HOST-PLANTS. *Daucus carota* L.; almost certainly other Umbelliferae but exact host-range requires confirmation.
BIOLOGY. This species, as *Phytomyza geniculata* Macquart, was first reported as a pest of carrots in the Berlin area by BÖRNER (1907). A good illustration of the adult was given and it was recorded that the single larva found had been feeding inside the base of the stem, where pupation also took place. The author concluded his paper by expressing the hope that his communication would stimulate the search for this species in other areas, so that its life history and economic importance could be clarified. This was a vain hope, as the paper was entirely overlooked by subsequent workers. More than fifty years were to pass before this pest of carrots was more fully investigated in Switzerland, Germany and the Netherlands.

WIESMANN (1961) has made the most detailed available study of the life history of this species (as *Phytomyza lateralis* Fallén). He has confirmed that nutrient obtained by the females from the feeding punctures made in the leaves of carrot is essential for development of the ovaries. Only after 4 or 5 days of such feeding is the female able to oviposit. Eggs are laid not only in the leaves but also directly

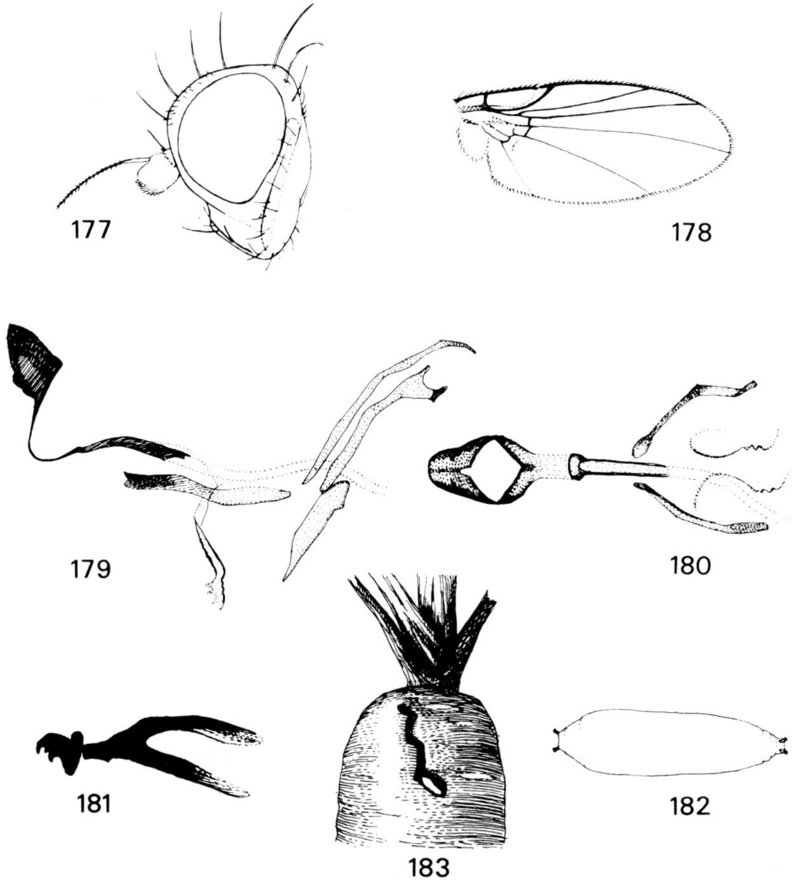

Figs. 177–183. *Napomyza carotae* Spencer: 177, head; 178, wing; 179, aedeagus, side view; 180, same, ventral view; 181, cephalo-pharyngeal skeleton of larva; 182, puparium; 183, mine in carrot.

in the stem and in fact the stem is the preferred site for oviposition. Of 1124 eggs counted, 413 (36%) were laid in leaves and 711 (64%) in stems. The number of eggs laid was investigated and the average was found to be 173, with a maximum of 248 by a single female. The daily average was 49.

From eggs laid in leaves the young larvae form an upper surface mine running towards the main vein of the leaf section and this is then followed to the petiole.

In the first day after hatching a mine 3 or 4 cm long can be formed. After the first moult the larva generally reaches the main root of the carrot, where irregular silvery mines are formed. These later cave in and the mine depressions are readily visible (Fig. 183). Most of the mining actually takes place in the upper half of the carrot and pupation occurs towards the top, near the base of the leaves. When eggs are laid in mature plants, the entire larval development can take place in the stem and pupation follows at the base, without the actual carrot being affected.

Larval development from hatching to pupation lasts at a temperature of 23 24°C from 53–68 days. The duration of the pupal period on the other hand is relatively short in the first generation, lasting only 7–8 days.

Adults of the first generation appear (in Switzerland) from the beginning of May and emergence continues until early in June. The main period of oviposition is from the middle to the end of May. Puparia of the first generation are found towards the end of July. Adults of the second generation appear early in August and continue until mid-September. The last flies are found early in October. These apparently represent a partial third generation and are the offspring of the earliest flies of the second generation which emerged at the beginning of August.

DAMAGE CAUSED. In Switzerland as many as 10 larvae have been found mining in a single carrot. The mine channels soon collapse and lead to serious deformation of the carrot. In damp weather rot sets in in these old mine channels and during storage this can develop further. Early carrot crops suffer most, as the greatest damage is caused by the first generation.

DISTRIBUTION. Switzerland, Germany, Holland; probably widespread in Europe but overlooked.

PARASITES. *Dacnusa pubescens* (Curtis). In samples of over 700 puparia collected in the three years 1941, 1942 and 1943 WIESMANN records a parasitism rate of 50%, 53% and 69% in the first generation. In a sample of 100 puparia of the second generation, however, the rate of parasitism was reduced to only 3%.

Discussion

TAXONOMY. This species was unknown until it was identified as distinct during a revision of the genus *Napomyza* by SPENCER (1966h). All earlier workers had misidentified it either as *Phytomyza geniculata* Macquart or as *Napomyza lateralis* (Fallén). In consequence of this and other misidentifications *Napomyza lateralis* was incorrectly considered to have an unusually wide host-range. Differences between the species in this complex are small but in conjunction with differences in the male genitalia a positive identification is invariably possible. *N. carotae* is readily distinguishable by the slight pubescence of the third antennal segment (not to be confused with the conspicuously long pubescence found in *N. hirticornis* Hendel).

HOST-RANGE. WIESMANN carried out larval transplantation experiments and found that normal development occurred in four other genera of Umbelliferae, namely *Pastinaca sativa* L. (parsnip), *Apium graveolens* L. (celery), *Petroselinum crispum* (Miller) Airy-Shaw (parsley) and *Heracleum sphondylium* L. (hogweed).

In addition, numerous flies were caught on a field containing a strong growth of *Pimpinella saxifraga* L. Puparia have been found in stems of *Anthriscus sylvestris* (L.) Hoffm. in England (K.A.S.), which may represent the same species.

DISTRIBUTION. It is surprising that confirmed records of specimens bred from carrots are only available from Switzerland, Germany and Holland. WIESMANN was able to establish from infested carrots which he examined that the species occurs commonly throughout German-speaking Switzerland. GERSDORF and VON KRAFT (1965) report its occurrence in Bavaria, Lower Saxony, Schleswig-Holstein and Berlin. It was only discovered in Holland as a by-product of the detailed studies on *Napomyza cichorii* Spencer (as *lateralis* Fall.). There seems little reason to doubt that *N. carotae* occurs throughout western Europe on cultivated carrots and still remains largely overlooked. Presumably plant-growers are confusing its damage with that caused by the well-known pest of carrots, *Psila rosae* L. (Psilidae).

ECONOMIC IMPORTANCE. WIESMANN discovered that in Switzerland the proportional infestation of carrots by *Napomyza carotae* was substantially greater than by *Psila rosae*. In the majority of areas from which samples were examined the damage caused was assessed as substantial or considerable. No damage was detectable in only two of the 22 localities concerned.

VAN'T SANT (1961) reported that in the Friesland district of the Netherlands damage has varied from a few percent to 40%. He suggests that as the mine channels are normally on the surface, they can easily be scraped off during preparation for cooking. Furthermore, infested carrots taste as good as healthy ones (!). For these reasons apparently during auctions such infestations have not led to any reductions in price.

It is nevertheless suggested that *Napomyza carotae* is of considerable economic importance and that this pest has not been detected in most countries owing to the superficial similarity in the damage caused to that of *Psila rosae* L.

GERSDORF and VON KRAFT also suggest that *N. carotae*, which has now been found to occur widely in Germany, has in the past been confused with *Psila rosae*. They raise the valid question as to which of the two species is in fact responsible for most of the damage to carrots which is regularly encountered and leads to such an obvious reduction in quality. Further investigations are thus clearly called for in all European countries to establish the real significance of *Napomyza carotae* as a pest of carrots.

VI. MISCELLANEOUS CROPS 2: LEAF-MINERS

Of the 20 species discussed below, 19 are host specific to a single plant family. The only exception is *Liriomyza brassicae* (Riley), which has occasionally been found on *Pisum*. Crops affected belong to the families Chenopodiaceae, Compositae, Cruciferae, Labiatae, Liliaceae, Rosaceae, Solanaceae and Umbelliferae. Two groups deserve special mention – the three chicory flies, *Ophiomyia pinguis* (Fallén), *Napomyza cichorii* and *Phytomyza penicilla* Hendel; and the three onion flies, *Liriomyza cepae* (Hering), *L. chinensis* (Kato) and a new species confirmed during the course of this study, *L. nietzkei* sp.n.

For many years the chicory crop in the Netherlands, France and Switzerland, and more recently also in Italy, has suffered severe damage from one or more of the three chicory flies. *Phytomyza penicilla* Hendel (previously misidentified as *continua* Hendel) has only once been recorded, from the Geneva area of Switzerland, and must be considered as of minor significance. The two other species are serious pests. The relative importance of *Ophiomyia pinguis* in relation to *Napomyza cichorii* has not been clearly established. In Switzerland *O. pinguis* was the dominant species during the initial investigations of Deshusses and Deshusses (1929) and in Italy it accounts for 90% of the infestation of chicory (Süss, 1970). In France *O. pinguis* is the only species which has been recorded; on the other hand in the Netherlands it is apprently entirely absent from cultivated chicory. In Belgium Lounsky (1965a, 1966a) has found that while *N. cichorii* can dominate among puparia at the end of the vegetative season, the larval population is largely of *O. pinguis*. More detailed studies, with these two species clearly separated at all stages – larvae, puparia, adults and also by parasites (the two species do not have parasites in common) – are clearly necessary before a true picture of the relative importance of the complete populations of the two species can be properly worked out. Lounsky (1965b) discusses various possible formulae for establishing a quantitative assessment of damage to the chicory crop. He concludes that the number of leaves infested divided by the number of heads examined (rather than by the number of heads infested) gives the best basis for assessing the damage.

Five species are now known to infest cultivated onions and leeks, three of which have either regularly or sporadically caused serious damage to the crop. In Germany *Liriomyza cepae* (Hering) is considered a major pest.

Nietzke (1943, 1953) clearly differentiated the 'Herxheim race' and the 'Oggersheim race' of *cepae*, and examination of original material has shown beyond doubt that the 'Herxheim race' represents a distinct species which is described below as *nietzkei* sp.n. and the 'Oggersheim race' represents true *cepae*. There is a large degree of overlap in the distribution of these two species in Germany but both have large populations in the Magdeburg area of East Germany. Eichler (1950) reported severe damage to *Allium cepa* in mid-July in the Calbe and Aschersleben areas. NOLTE (1960) reported serious attacks on *Allium porrum* L. in October, in the Halle and Magdeburg areas; in a later paper (1964) he discusses serious damage caused to *Allium cepa* in the Aschersleben area in the summer of 1963. In all these papers the species is discussed as *cepae*. It seems clear that both species were involved. SHANAB and BOGNÁR (1970) discuss aspects of the life history of '*Dizygomyza cepae* Hering' in Hungary. A high rate of infestation in some areas is recorded but it is not clear whether the species referred to is the true *cepae* or *nietzkei*.

Both NIETZKE and NOLTE refer to differences in the type of mines which may be found. These may be entirely linear or more blotch-like. It is suggested that the linear mines are formed by *L. nietzkei* and the blotch mines by the true *cepae*. It is hoped that this can be confirmed by breeding adults from the two types of mines.

The following key is provided which may be helpful to future workers in differentiating these two species:

1	Scutellum entirely dark, pleura generally black, femora blackish at least on upper side; wings and veins pale; medium-sized species, wing length 2–2.5 mm; male genitalia: aedeagus as in Figs. 228, 229.	*Liriomyza cepae* (Hering)
—	Scutellum centrally yellow, pleura more yellowish, femora bright yellow; smaller species; wings and veins darker; wing length 1.6–1.9 mm; male genitalia: aedeagus as in Figs. 235, 236	*Liriomyza nietzkei* sp.n.

Liriomyza chinensis (Kato), which was described as a sub-species of *cepae* from Inner Mongolia, has now been identified also from Japan and Malaya and is raised to full specific status. The degree of damage caused by this species is not known.

In the United States and Venezuela *Liriomyza trifolii* (Burgess) periodically causes significant damage to onions and leeks. *Phytomyza horticola* Goureau is occasionally found mining onions in Western Europe but this is not one of its regular hosts and it has not so far caused any damage.

Phytomyza rufipes Meigen can cause considerable damage to cultivated cruciferous crops in Europe. It feeds both as a leaf-miner, mainly in the midrib, and also in stems.

Ptochomyza asparagi Hering which attacks *Asparagus* is the smallest Agromyzid known, with a wing length in males which is frequently less than 1 mm. The leaf mines on *Asparagus* were originally discovered by BUHR in northern Germany and the adults were bred out by HERING. HERING's excitement on the emergence of the first specimens of this intriguing little species is clearly shown in his letter to BUHR dated 14 February, 1942 (HERING in SPENCER, 1968: 11). The letter even starts with a rhyming couplet:

'Hurra, Hurra, Hurra!
Die Spargelfliege ist da!'

Haplomyza capsici sp.n. is described from Argentina where it has been found heavily mining Red Pepper.

CHENOPODIACEAE: BETA, SPINACIA.

Amauromyza (Trilobomyza) flavifrons (Meigen, 1830) – Europe, Canada

ADULT. Small species, with yellow frons and halteres, and entirely black legs.
HEAD (Fig. 184). Frons not significantly projecting above eye, with 2 equal ors and 3, sometimes 4, incurved ori; orbital setulae sparse, reclinate; jowls deepest in centre below eye, $\frac{1}{6}$ height of eye, cheeks forming narrow ring below eye; third antennal segment small, rounded, arista finely pubescent.
MESONOTUM. 3 + 1 dc, fourth only slightly weaker than third, acrostichals irregularly in 4 or 5 rows.
WING. Length from 1.9 mm in male to 2.7 mm in female; costa extending strongly to vein M_{1+2}, last section of M_{3+4} up to $3\frac{1}{2}$ times length of penultimate, discal cell thus small.
COLOUR. Frons, lunule and jowls yellow; first and second antennal segments yellowish, third largely black but sometimes at least slightly paler on inside; palps black; mesonotum deep black, moderately shining; pleura largely black but upper margin of mesopleura and wing base bright yellow; legs black, only fore-knees at most slightly yellowish; squamae yellowish-grey, margin and fringe black, halteres bright yellow.
MALE GENITALIA. Aedeagus as in Fig. 185, not surrounded in spinulose membrane; ejaculatory apodeme greatly enlarged, with bowl-shaped base (Fig. 186).
EARLY STAGES. Larva yellow, mouth-hooks each with 2 teeth, the 2 mandibles not coinciding and the teeth thus alternating; anterior spiracles each with 7 pores, posterior each with 3, 1 distinctly longer. Puparium reddish-brown, 2 mm long.
HOST-PLANTS. Chenopodiaceae: *Beta vulgaris* L. (Beet), *Spinacia oleracea* L. (Spinach); also Caryophyllaceae: *Dianthus* (Carnation), *Lychnis*, *Saponaria*, *Silene* and other genera.
BIOLOGY. The larva forms an upper surface leaf-mine which is at first linear but after the first moult becomes an oval blotch (Fig. 187); frequently the linear section is not apparent, as it becomes enveloped in the later blotch. Pupation takes place externally. There are normally two generations.
DAMAGE CAUSED. The larva causes normal leaf-mining damage but large numbers rarely occur together and there is little effect on the growth of beet or spinach.
DISTRIBUTION. Widespread in Europe; U.S.S.R.: Kirghiz S.S.R. (DOVNAR-ZAPOLSKI, 1969); Canada: Ontario.
PARASITES. Europe: *Opius ochrogaster* Wesm., *O. tersus* Förster, *O. variegatus* Szépl. (Opiinae); *Dacnusa maculipes* Thomson, *Chorebus trilobomyzae* Griffiths

(Dacnusinae); *Chrysocharis albicans* Delucchi, *C. melaensis* (Walk.), *C. seiugata* Delucchi (Eulophidae).

Discussion

TAXONOMY. This species was included in an enlarged concept of the genus *Amauromyza* but retained in the sub-genus *Trilobomyza* by SPENCER (1971a). The male genitalia clearly indicate the affinity of the partially yellow species, such

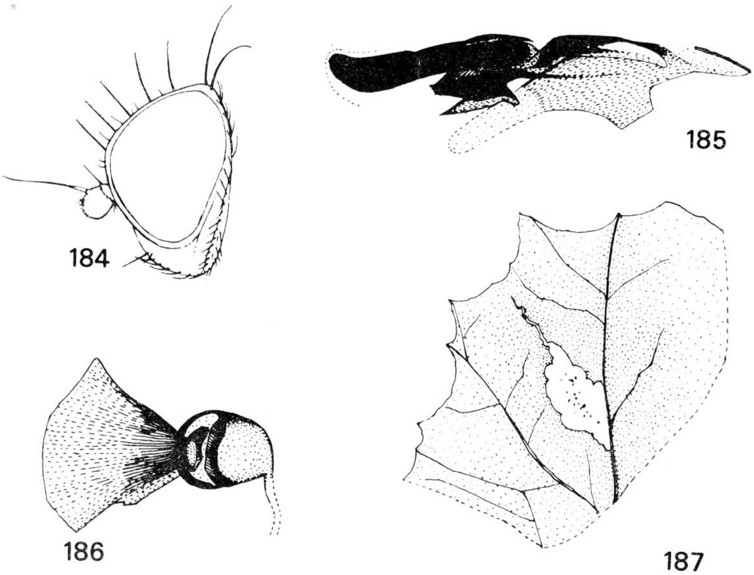

Figs. 184–187. *Amauromyza (Tril.) flavifrons* (Mg.): 184, head; 185, aedeagus; 186, ejaculatory apodeme; 187, leaf-mine on *Beta*.

as *flavifrons* and also *luteiceps* Hendel, with the entirely black species for which the genus *Amauromyza* was originally erected.

DISTRIBUTION. The only known records of *flavifrons* in North America are from Ontario in Canada and it is believed that this represents a case of relatively recent introduction.

ECONOMIC IMPORTANCE. The main hosts of this species are in the Caryophyllaceae and some damage is caused to *Dianthus* species, cultivated as Carnations or Pinks (cf. Chapter 11, p. 320). However, leaf-mines regularly occur on *Beta* and *Spinacia*. HERING (1949a) discusses this species as a miner of *Beta* and suggests that its mines may have been confused in the past with those of the more serious pest, *Pegomya hyoscyami betae* (Curtis).

Although little real damage is caused to these crops, if larvae were present in any numbers in spinach leaves, they would certainly be unpalatable for human consumption. *Amauromyza flavifrons* is a widespread species but population levels have never been known to be other than small.

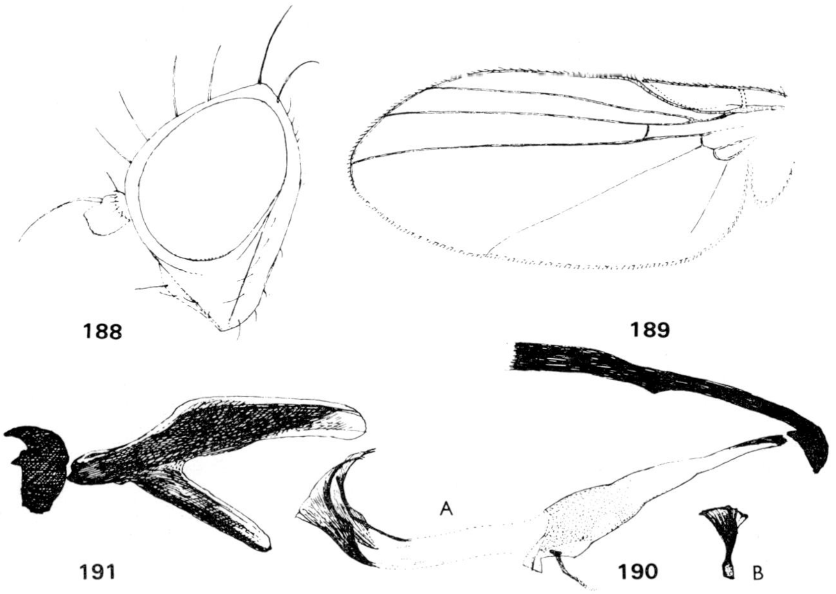

Figs. 188–191. *Liriomyza caulophaga* (Kl.): 188, head; 189, wing; 190A, aedeagus; 190B, ejaculatory apodeme; 191, cephalo-pharyngeal skeleton of larva (after KLEINSCHMIDT).

CHENOPODIACEAE: BETA

Liriomyza caulophaga (Kleinschmidt, 1960) – Australia, Indonesia

ADULT. Head bright yellow, including entire hind-margin of eye, second cross-vein lacking.

HEAD (Fig. 188). Orbits well differentiated, conspicuously projecting above eye in profile; 2 equal reclinate ors, 2 incurved ori; orbital setulae lacking; jowls angular, deeply extended at rear, cheeks forming broad ring below eye; third antennal segment small, with distinct point at upper corner, arista fine, bare.

MESONOTUM. 3 + 1 strong dc; acrostichals sparse, in 2 rows, extending only to level of second dc.

WING (Fig. 189). Length in female from 1.7–1.9 mm, second cross-vein lacking.

COLOUR. Head, including entire hind-margin of eye pale, lemon-yellow; all antennal segments yellow; mesonotum mat ash-grey, almost silvery; pleura predominantly yellow; mesopleura with narrow blackish-grey band along lower and front margins; scutellum broadly yellow centrally; legs: coxae and femora bright yellow, tibiae and tarsi slightly darker, brownish-yellow; abdomen moderately shining black but all tergites with broad yellow hind-margins and broadly yellow at sides; squamae, including margin and fringe, pale yellow.

MALE GENITALIA. Aedeagus as in Fig. 190A; ejaculatory apodeme small (Fig. 190B).

EARLY STAGES. No larvae have been studied but the mouth-parts, extracted from an empty puparium, have been examined by KLEINSCHMIDT, 1970 (Fig. 191). The two halves of the mandibles are equal, each with 2 teeth, with a small, rounded projection above the lower tooth.

The puparium is pale brown, measuring 2.3 × 0.8 mm. Details of the number of pores on the anterior or posterior spiracles have not been recorded.

HOST-PLANT. *Beta vulgaris* L. var. *cicla* L. (Silver Beet).

BIOLOGY. No detailed studies have been made on the life histoıy of this species. KLEINSCHMIDT indicates that larvae have only been found in the midrib or petiole and suggests that the eggs are laid directly into this part of the leaf. 'The larvae tunnel in the soft, white, spongy tissue between the vascular strands of the midribs and stalks. When fully grown, larvae approach the surface, where a "window" or a hole indicates the site of pupation' (KLEINSCHMIDT, 1970). Pupation thus takes place within the mine.

DAMAGE CAUSED. The larval mining can cause leaves to turn brown and die.

DISTRIBUTION. Australia: Queensland, Brisbane; N.S.W., Sydney area: Cobar, Leeton, 40 miles E. of Wilcannia; Indonesia: Lombok.

PARASITES. None recorded.

Discussion

TAXONOMY. This species is significantly different from *Liriomyza chenopodii* and can be distinguished by the angulate third antennal segment, the brighter yellow of the head, the entirely yellow hind-margin of the eye, the projecting frons and the more ash-grey mesonotum.

L. caulophaga is an aberrant species. The form of aedeagus is somewhat unusual for *Liriomyza* and no other case is known of a species in this genus feeding as an internal borer (although the type of the genus, *Liriomyza urophorina* Mik is a seed-feeder). It is thus certain that *L. caulophaga* is not closely related to *L. chenopodii*, although both feed on beet and both lack the second cross-vein.

DISTRIBUTION. In addition to the confirmed records in Australia and New Zealand, the discovery of the single caught specimen on the island of Lombok, east of Bali in Indonesia, suggests that this species is widely distributed in S.E. Asia. It may well be present in Malaya and also New Guinea.

ECONOMIC IMPORTANCE. It seems clear that the Note, ANON., c. in the *Agric. Gaz. N.S.W.*, 1947, discussing *Haplomyza imitans*, in fact refers to *caulophaga*.

Reference to the larvae tunnelling in stalks could not refer to *imitans* (=*chenopodii*), as the larvae of this species are exclusively leaf-miners. *L. caulophaga* must therefore be recognized as a periodical pest of some importance on beet.

CHENOPODIACEAE: BETA, SPINACIA

Liriomyza chenopodii (Watt, 1924) – Australia, New Zealand

ADULT. Very small species, with mat grey mesonotum and second cross-vein lacking.
HEAD. Frons equal to width of eye, not projecting above eye in profile; 2 equal ors, 1 incurved ori; orbital setulae minute, sparse, reclinate; jowls relatively deep, $\frac{2}{5}$ height of eye; third antennal segment small, round, arista only finely pubescent.
MESONOTUM. 3 + 1 dc, first and second strong, third and fourth equal, substantially weaker, distance between first and second twice that between second and third; acrostichals sparse, in 2 rows.
WING. Length 1.6 mm, costa extending strongly to vein M_{1+2}, second cross-vein lacking.
COLOUR. Frons and orbits yellow; hind-margin of eye black, both vertical bristles on dark ground or inner vertical just on yellow; third antennal segment yellow; mesonotum mat grey but with some sub-shine; scutellum bright yellow centrally, black at sides; mesopleura largely yellow, with dark patch at front lower corner which can extend up front margin; abdomen shining black, with tergites narrowly yellow on hind-margin and laterally; legs: coxae yellow, femora bright yellow, with faint, slightly variable brownish striations; tibiae and tarsi darker, brown; squamae yellow, margin and fringe dark.
MALE GENITALIA. Aedeagus as in Figs. 192, 193; ejaculatory apodeme with unusually large blade (Fig. 194).
EARLY STAGES. The larval anterior spiracles each have 4 or 5 minute pores; the posterior spiracles have the primitive arrangement of 3. The mouth-hooks and cephalo-pharyngeal skeleton are shown in Fig. 195. The puparium is brownish-orange, with black spiracles and measures 1.5 × 0.5 mm (KLEINSCHMIDT, 1970).
HOST-PLANTS. *Beta vulgaris* L. (Silver Beet), *Spinacia oleracea* L. (Spinach), *Cheiranthus cheiri* L. (Wall-flower); also *Cerastium glomeratum* Thuil., *C. vulgatum* L., *Chenopodium album* L., *Coronopus didymus* (L.) Sm., *Stellaria media* (L.) Vill.
BIOLOGY. Eggs are laid in the upper surface of the leaf, where the larva forms a narrow, somewhat winding linear mine. This is at first greenish, later becoming white. Mines are mainly upper surface but may be in part or wholly on the lower leaf surface. Larval feeding lasts about 14 days. The larva leaves the mine to pupate on the ground, forming a typical yellowish-brown puparium. Adults can emerge after 12 days.
DAMAGE CAUSED. In New Zealand numerous mines can be found in the leaves of beet and spinach and WISE (1953) gives a photograph of a heavily mined leaf

of beet. In Australia the larvae of this species (as *imitans* Malloch) have been recorded (ANON., c, *Agric. Gaz. N.S.W.*, 1947) as tunnelling in the veins and stalks of Silver Beet and with severe infestations causing the leaves to wither and die. However, it seems virtually certain that the species in this report was misidentified and in fact refers to *Liriomyza caulophaga* (Kleinschmidt), which does not make a true leaf-mine but does feed in the midribs and petioles of leaves.

DISTRIBUTION. Australia: New South Wales, Sydney area; Queensland: Brisbane; New Zealand: Auckland and Christchurch areas.

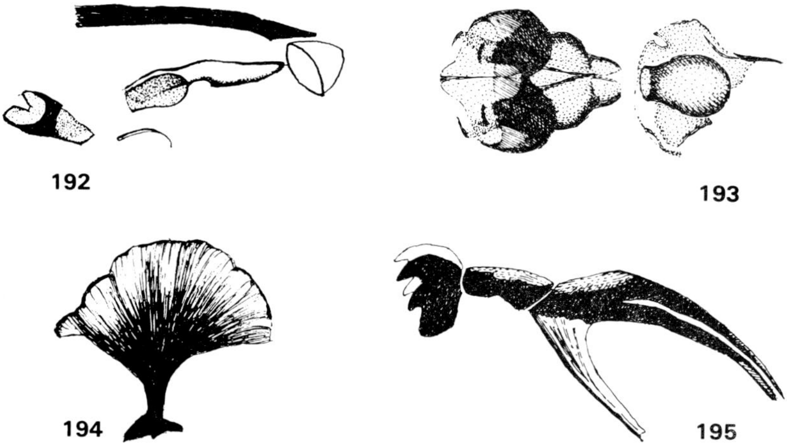

Figs. 192–195. *Liriomyza chenopodii* (Watt): 192, aedeagus, side view; 193, same, ventral view; 194, ejaculatory apodeme; 195, mouth-hooks.

PARASITES. *Opius cinerariae* Fischer (Braconidae); *Chrysocharis* sp., *Hemiptarsenus* sp. (Eulophidae). KLEINSCHMIDT (1970) reports a rate of parasitism at Brisbane of up to 23 %.

Discussion

TAXONOMY. This species was originally described from New Zealand in the genus *Haplomyza* in view of the lack of the second cross-vein. MALLOCH (1934b) described *Haplomyza imitans* from Australia and this was synonymized with *chenopodii* by SPENCER (1963b). The male genitalia confirm this synonymy and also that the species is a true *Liriomyza* despite the lack of the second cross-vein. KLEINSCHMIDT (1970), in her valuable contribution, ignores this earlier work and, using both the wrong generic and specific name, discusses the species as *Haplomyza imitans*.

L. chenopodii is readily distinguishable from *L. caulophaga* by the rounded third antennal segment, the darker colouration of both the yellow areas and the mesonotum, the non-projecting frons and smaller size. There are also significant differences between the mouth-hooks of the two species (cf. Figs. 191, 195).
ECONOMIC IMPORTANCE. If it is accepted that the Note in the *Agric. Gaz. N.S.W.*, 1947, refers to *Liriomyza caulophaga* and not to *chenopodii* (as *imitans*), there are in fact no reports of any significant damage to beet and spinach in Australia. In New Zealand WISE (1953) reports that 'this native species is occasionally a pest in vegetable gardens'.

COMPOSITAE: CICHORIUM

Ophiomyia pinguis (Fallén, 1820) – Europe, Central Asia, Egypt

ADULT. Stout black species, with bulbous facial keel, no upper orbital bristles in male, orbital setulae all proclinate.
HEAD. Frons broad, $1\frac{1}{2}$ times width of eye, distinctly projecting above eye in profile; in female 2 ors and 2 ori, orbital setulae proclinate, sparse (Fig. 196); in male ors lacking, orbital setulae more numerous, thick (Fig. 197); jowls deepest in front, $\frac{1}{4}$ height of eye, no vibrissal fasciculus in male (cf. Fig. 53); base of antennae divided by conspicuously raised, almost spherical protuberance; proboscis greatly elongated.
MESONOTUM. Two strong dc, acrostichals in 8 rows between second dc, in 4 rows between first.
WING. Length from 1.9 mm in male to 2.3 mm in female, costa extending to vein M_{1+2}, last section of M_{3+4} normally slightly shorter than penultimate, first cross-vein at midpoint of discal cell.
COLOUR. Entirely black species, frons mat, ocellar triangle, orbits, mesonotum and abdomen almost brilliantly shining black; squamae grey, margin and fringe black.
MALE GENITALIA. Aedeagus as in Fig. 198.
EARLY STAGES. Larva whitish-yellow, up to 5–6 mm long, anterior spiracles on minute projections, each with 10 pores, posterior spiracles (Fig. 199 on distinct stalks, each with an ellipse of 9–11 pores; mouth-hooks each with one prominent tooth, the left-hand one enlarged. Puparium pale yellow, up to 3.7 mm long.
HOST-PLANTS. *Cichorium endivia* L. (endive), *Cichorium intybus* L. (chicory, 'witloof'), *Lactuca sativa* L. (lettuce); also *Leontodon* sp.
BIOLOGY. In view of the economic importance of this species as one of the two chicory flies which cause such extensive damage to cultivated chicory in Belgium, the Netherlands, France and Switzerland, until recently surprisingly little has been known of its detailed life cycle. During the many years it was investigated in Belgium, it was not successfully bred in the laboratory (LOUNSKY, 1965a: 118)

and much biological information in the literature refers merely to 'the chicory flies', without differentiating between the two species concerned. Recently, however, detailed studies on this species have been made in Italy, where it is a serious pest in the Milan area (SÜSS, 1970).

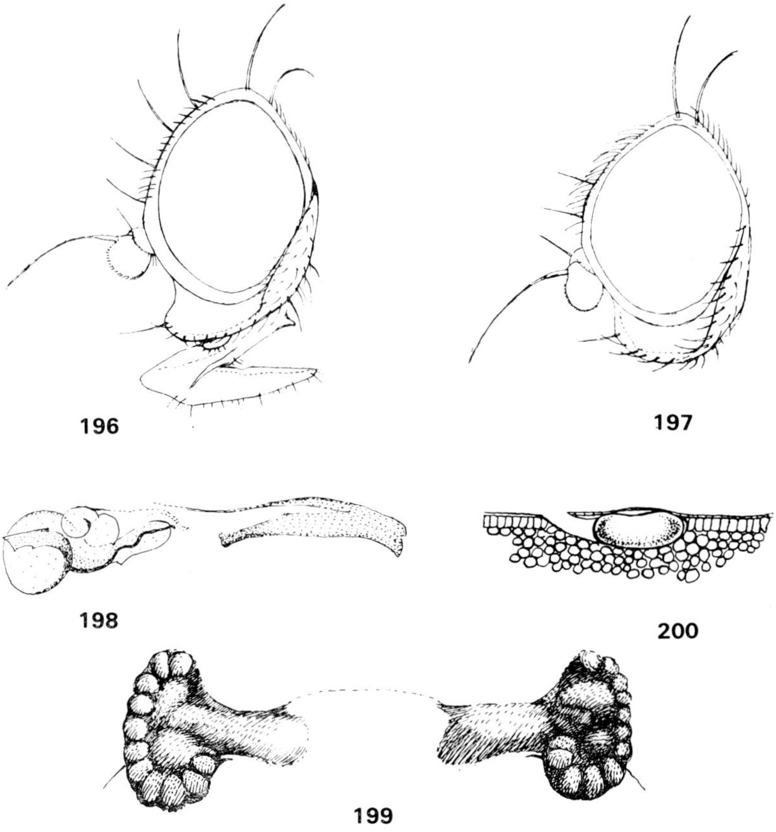

Figs. 196–200. *Ophiomyia pinguis* (Fall.): head, 196, female; 197, male; 198, aedeagus; 199, posterior spiracles of puparium; 200, egg (after SÜSS).

SÜSS points out that duration of the life cycle is strictly temperature-dependent. In Lombardy there are 3 or 4 generations annually. The first adults emerge from the end of March to mid-April from plants of the cultivar 'Bianca a cuore pieno', which have been stored in the open in covered mounds throughout the winter. Oviposition takes place on other cultivars such as 'a grumulo' or 'Spadona' which are already well advanced at this time. The egg (Fig. 200) is laid beneath the epidermis of the upper leaf surface, generally near the midrib. In captivity females live for 8–10 days and lay about 20 eggs. A female will not normally lay more than one egg in the same leaf but other females may choose the same leaf for oviposition. The egg hatches after about 10 days and larval feeding in the first generation lasts for 18–25 days, depending on climatic factors.

The puparium is normally readily visible near the surface of the leaf at the end of the mine. With an ambient temperature of 15°C the pupal stage lasts 2 weeks.

The total life cycle of the first generation takes from 50–60 days, while in the summer generations this is reduced to 40–45 days. Throughout the summer and up to October there is complete overlapping of generations and all stages of the insect can be found together. At the end of the season the cultivar 'Bianca a cuore pieno' is heavily attacked and as many as 18 larvae have been found in a single plant (Süss, 1970).

In Belgium and the Netherlands the first flies emerge towards the end of May. Oviposition takes place in the young leaves early in June. In natural plants larval feeding is believed to be mainly in the leaves, probably in the outer layers of the petiole. However, the majority of species in this genus are external stem-miners and it seems likely that in normally growing plants larval feeding does occur in stems as well as in the petioles.

Larval feeding continues slowly throughout the summer. Normally pupation could be expected by the end of the summer or early in the autumn. However, when cultivated chicory plants are lifted in autumn and prepared for the forcing beds by cutting back the leaves between 2 and 4 cm above the neck of the root, many larvae are always still present in the leaves. The continued infestation during the winter forcing develops from the larvae which have remained in the base of the leaves and, as these die and rot, they migrate via the stem into the young growing shoot and thus into the white chicory head. Under the influence of the increased temperature in the forcing beds the rate of the life cycle naturally increases. While under natural conditions in the field the overwintering larva would remain dormant until shortly before pupation in March (according to van den Bruel, 1939: 225), larvae can be found actively feeding in the blanched heads throughout the winter. I have reared adults from chicory bought in London shops in December and also in March.

There is no evidence to suggest any significant emergence of flies in autumn or of emergence during forcing, with subsequent infestation of the white leaves. However, this has not been positively excluded.

All authors (Deshusses and Deshusses, 1929; Mesnil, 1934; Lounsky, 1965a) agree that the larvae do not penetrate into the root.

Hammad (1956) illustrates the leaf-mine of *O. pinguis* on lettuce in Egypt.

Damage caused. The larval feeding in the white leaves of the forced chicory heads is usually very visible and frequently accompanied by a reddish discolouration. Infestation of the heads thus quickly becomes apparent, either when the chicory is used as a salad or when cooked.

Distribution. Widespread in Europe, with confirmed records from Austria, Belgium, England, Finland, France (Mesnil, 1934), Germany, Hungary, Italy, Netherlands, Spain, U.S.S.R.: Moscow district (Rohdendorf, 1958); Tadzhik

S.S.R., Uzbek S.S.R. (Dovnar-Zapolski, 1969); Sweden, Switzerland, Yugoslavia; most northerly record from Norbotten in northern Sweden; also Egypt (Hammad, 1956).

Parasites. *Chorebus leptogaster* (Haliday). This species is also known as a parasite of *O. cunctata* (Hendel) and *O. pulicaria* (Meigen). Other species which have been reared from mixed infestations of the two chicory flies all have as their host *Napomyza cichorii* Spencer.

Discussion

Taxonomy. *Ophiomyia pinguis* is unusual in being sexually dimorphic, the loss of the upper orbital bristles in the male being compensated for by the more numerous orbital setulae. The male genitalia are entirely typical of the genus.

The type of leaf-mine formed resembles that produced by the common European species, *O. cunctata* (Hendel) and *O. pulicaria* (Meigen), and also by *O. beckeri* (Hendel), which is known from Europe, South Africa and India. It seems probable that *O. pinguis* is closely related to these three leaf-mining species.

Host-range. The discovery of *O. pinguis* feeding in leaves of cultivated lettuce in Egypt (Hammad, 1956) is of considerable interest. Presumably it may in due course also be found infesting this crop in Europe.

Distribution. The species is widely distributed, with *Cichorium intybus* as its normal host. Abnormally large populations have built up in Belgium, France, Switzerland and Italy in association with the cultivation of *Cichorium intybus* and *C. endivia*. Records from the Netherlands, however, suggest that there the species is still restricted to wild *C. intybus* as its host and is present only in small numbers. No cases have been reported of infestation of cultivated chicory.

Early records of the presence of this species in the United States have proved erroneous (Spencer, 1964a: 801).

Parasitism. Most records in the literature give only the combined rate of parasitism of the two chicory flies. However, van den Bruel (1961: 509) gives interesting records of the fluctuation in the population of *O. pinguis* in Belgium which was clearly due in part at least to variations in the rate of parasitism by *Dacnusa leptogaster*. At the end of the 1930–1931 winter ten *Napomyza cichorii* (referred to as *lateralis* or 'mouche grise') were obtained for every one *Ophiomyia pinguis* ('mouche noire'). The following winter *O. pinguis* had disappeared. Simultaneously, the rate of parasitism by *Dacnusa leptogaster* increased from the average 17% to 59%. At the end of the 1933–1934 season two lots of chicory from Tervuren were examined, the one containing larvae exclusively of *N. cichorii*, while the other contained 76% larvae of *cichorii* and 24% of *pinguis*. The population of *O. pinguis* had therefore by 1934 largely re-established itself, following decimation by its parasite the previous year. The next season, 1934–1935, was marked by one of the most serious infestations of chicory flies known and almost entirely due to *O. pinguis* but the year after that *N. cichorii* had again assumed its normal dominance but the overall level of attack was greatly reduced.

Economic importance. In three of the four main areas where chicory is cultivated, namely Belgium, France (Seine-et-Marne) and Switzerland (Geneva), *Ophiomyia pinguis* has been known as a serious pest since 1928. In this year in one production centre near Geneva the entire crop was so badly infested that it was unsalable (DESHUSSES and DESHUSSES, 1929, 1933). First records of damage to chicory in Belgium go back to 1913 but this infestation was apparently due to the other species, *Napomyza cichorii*. In 1931, however, when the problem of the chicory flies was again investigated in Belgium, *O. pinguis* was also identified (VAN DEN BRUEL, 1961: 508). In France, cultivation of chicory became important by 1934, mainly in the Paris area, but between 50 and 80% of the crop was attacked by *O. pinguis*, seriously reducing the sale price and limiting use of the vegetable. However, following intensive research for many years both in Belgium and the Netherlands, satisfactory control measures have now been discovered (see Chapter 12). The most critical point for cutting the leaves and thus removing the maximum possible number of larvae and other mechanical methods of control have also been thoroughly investigated (LOUNSKY, 1963b, 1964). As a result of this work infestations are now normally kept within acceptable limits but infested heads still find their way into shops, as anyone with a critical eye, even in London, can readily see.

The losses caused to growers by *O. pinguis*, the currently increased costs of essential control measures, and the reduction in the market which has certainly occurred following the natural reaction of customers to an unpalatable product containing larvae, combine to make this species an extremely serious pest wherever chicory is cultivated in western Europe.

COMPOSITAE: CICHORIUM

Napomyza cichorii Spencer, 1966h – Europe

ADULT. Large species, with wing length up to 3.5 mm, third antennal segment virtually bare, cut away below (Fig. 201).
HEAD. Frons broad, twice width of eye, conspicuously projecting above eye in profile; normally 2 ors and 2 ori, the latter largely incurved but sometimes 1 or even 2 additional bristles present; orbital setulae numerous, proclinate; jowls deep, up to $\frac{1}{2}$ height of eye, cheeks forming broad ring below eye; third antennal segment distinctly cut away below, appearing virtually bare.
MESONOTUM. 3 + 1 strong dc, acrostichals essentially in 2 rows.
WING. Length from 2.7 mm in male to 3.5 mm in female; average costal ratio 33:15:25, second cross-vein normally in continuation of first.
COLOUR. Frons pale yellow, more orange in front; hind-margin of eye black beyond outer vertical bristle and up to margin of inner, orbits narrowly black adjoining eye margin; all antennal segments black; mesonotum mat ash-grey; pleura blackish-grey, upper margin of mesopleura bright yellow; legs black, all knees bright yellow; squamae yellow, margin and fringe black.

MALE GENITALIA. Aedeagus as in Figs. 202, 203, distiphallus with characteristic angle, rather than curvature present in other species.
EARLY STAGES. The egg is milky white, about 0.3 mm long, partially transparent. The full-grown larva is up to 5.3 mm long. The posterior spiracles of the slender white puparium are blackish-brown each with a double row of 20–24 minute bulbs (Fig. 204).
HOST-PLANTS. *Cichorium endivia* L., *C. intybus* L. (chicory, endive, witloof).
BIOLOGY. Oviposition takes place in the midrib, generally towards the base of

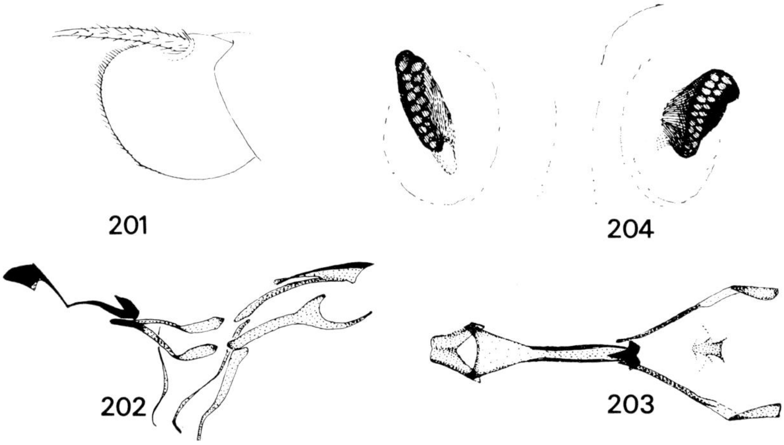

Figs. 201–204. *Napomyza cichorii* Spencer: 201, third antennal segment; 202, aedeagus, side view; 203, same, ventral view; 204, posterior spiracles of puparium.

the leaf. The larva can feed entirely in the leaf but frequently passes via the stem into the root. Pupation takes place internally, either in the leaf or root.

There are two main generations and at least a partial third. According to VAN DEN BRUEL (1939) in the majority of cases overwintering takes place in the larval state, with pupation occurring only at the end of March. By mid-April all larvae have transformed and the first flies begin to emerge during the last ten days of May. VAN'T SANT (1961), however, states that overwintering can take place either as a larva or puparium.

A second generation appears from August, emergences continuing until early October (VAN DEN BRUEL, 1939); VAN'T SANT reports that the first flies of the second generation appear in July. Possibly some of the October flies represent a partial third generation. The life cycle can be completed in 6–8 weeks, the larval stage lasting 23–27 days and the pupal stage 20 days. Adults can live from 20–30 days.

The new growth of cultivated chicory plants does not appear until the end of June. Therefore only the latest flies of the first generation and also the entire second generation are able to infest the plants which will in due course provide the blanched chicory heads during the following winter. It seems probable that any eggs laid in June will have produced adults well before the plants are transferred to the forcing beds in October/November and that it is thus exclusively the second generation which is responsible for the damage caused during winter forcing.

The investigations of LOUNSKY (1965a) suggest that the majority of larvae still in the plants during the winter feed primarily in the roots rather than in the new growth forming the edible heads. However, the data are obscured by the failure to identify the parasites and associate them either with *Napomyza cichorii* or with *Ophiomyia pinguis* (which is not difficult, as only *Chorebus leptogaster* attacks *pinguis* and any other species obtained will represent *cichorii*). However, it is clear that many *cichorii* larvae do feed in the leaves, together with those of *pinguis*.

DAMAGE CAUSED. Damage to the outer leaves of the chicory heads is essentially the same as caused by *O. pinguis*. The irregular mine channels show the same reddish discolouration. However, the larvae of *cichorii* also bore in the centre of the head, destroying much tissue, and this damage may be entirely invisible from a superficial external inspection. In addition, if the plant is infested at all, mine channels will also be present in the root. This will cause not only weakening of the plant but will also frequently lead to stunted or otherwise malformed heads of inferior quality. Damage to leaves and a root in the Netherlands is shown in Plates 3 and 4.

DISTRIBUTION. Belgium, France, Netherlands, Italy.

PARASITES. *Dacnusa pubescens* (Curtis); *Chorebus glaber* (Nixon), *C. senilis* (Nees), *C. parvungulus* (Thomson).

Discussion

TAXONOMY. This species was misidentified as *Napomyza lateralis* (Fall.) until the genus was revised by SPENCER (1966h). The form of the third antennal segment (Fig. 201) is distinctive and the male genitalia also show characteristic differences from other closely related species in this complex.

ECONOMIC IMPORTANCE. Substantial damage to the chicory crop regularly occurs in the main areas of cultivation in Belgium and the Netherlands, although the seriousness of the infestation varies considerably from year to year. The year 1957 was particularly bad in the Amsterdam area, where one farm lost 50% of the crop and others reported losses up to 30% (VAN'T SANT *et al.*, 1961). Comparable damage also occurred in the same year in Belgium (VAN DEN BRUEL, 1961). In the Lombardy area of northern Italy damage to chicory by this species is relatively slight (SÜSS, 1970).

In the Low Countries chemical control measures are now essential for producing a satisfactory crop (Chapter 12, p. 347) but *Napomyza cichorii* remains a major pest in chicory production. The practical results of the infestation by this species can be summarized as follows:

1. Although the external appearance of the heads may be excellent, quality can be diminished by the internal boring of the larvae.

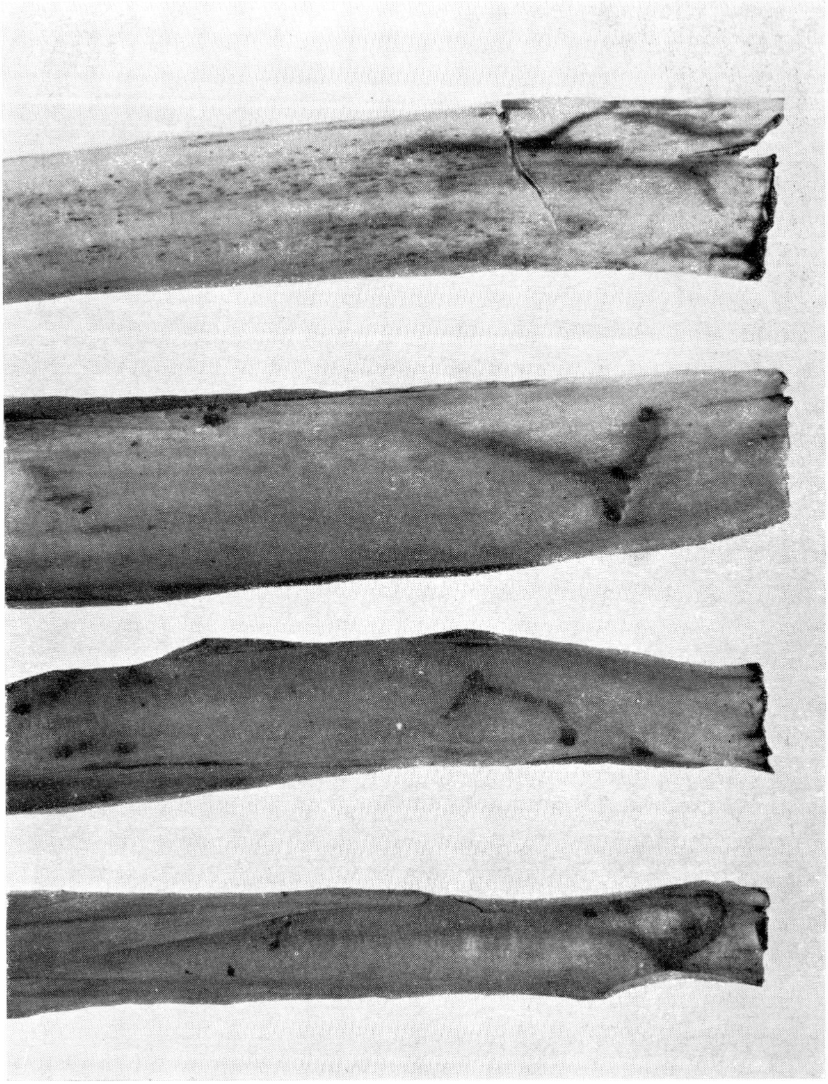

Plate 3: *Napomyza cichorii:* mines in chicory heads, Netherlands (Instituut voor Plantenziektenkundig Onderzoek, Wageningen).

2. Additional labour costs incurred in the individual inspection of chicory heads and removal of infested outer leaves.

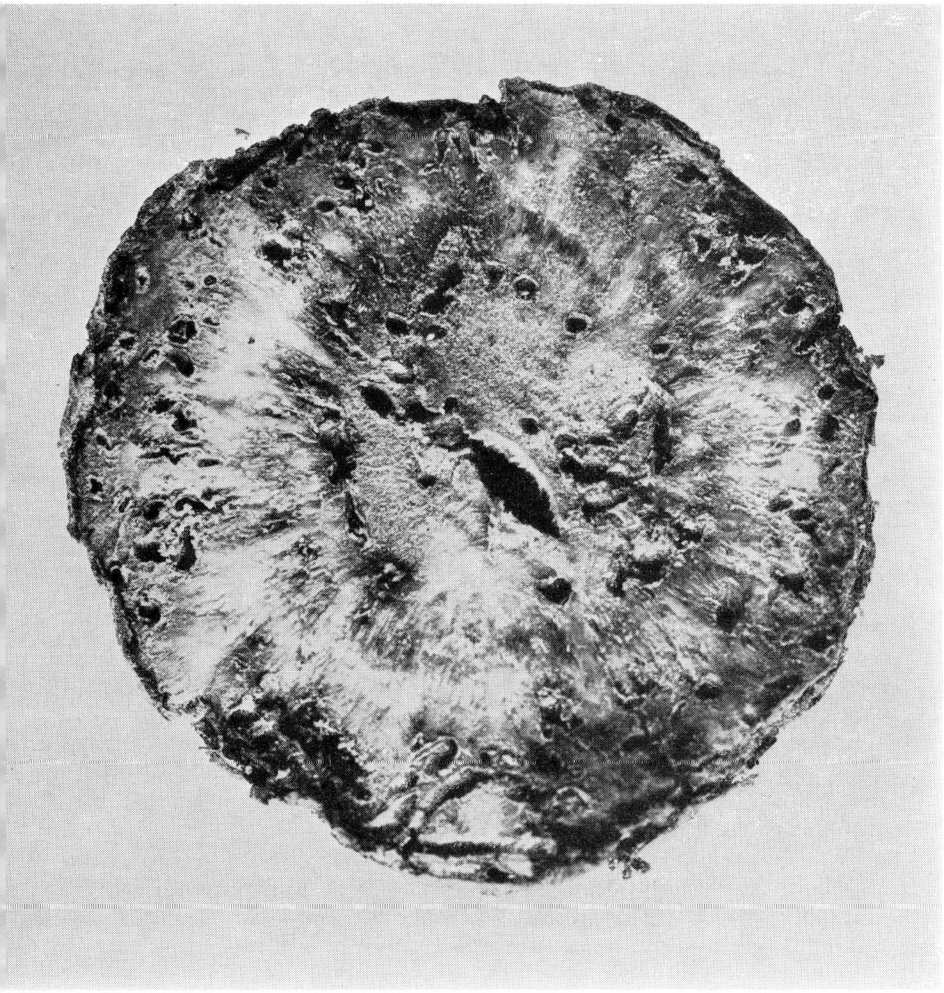

Plate 4: *Napomyza cichorii:* heavily infested root of chicory, Netherlands (Instituut voor Plantenziektenkundig Onderzoek, Wageningen).

3. Loss in weight of the total crop from removal of leaves.
4. Substantial costs of essential insecticide applications.
5. Damage to the market by the sale of unpalatable heads containing larvae or puparia.

COMPOSITAE: CICHORIUM

Phytomyza penicilla Hendel, 1935 – Europe

ADULT. Large species, with yellow frons, mat grey mesonotum and third antennal segment with conspicuously long pubescence.
HEAD. Frons distinctly projecting above eye, 2 strong equal ors, 2 ori, the lower significantly weaker; orbital setulae well developed, proclinate; jowls deep, up to ½ height of eye; third antennal segment with conspicuously long pubescence (Fig. 205).
MESONOTUM. 3 + 1 strong dc, acrostichals basically in 2 rows, normally with 1 or 2 additional hairs in front.
WING. Length from 2.7 mm in male to 3.8 mm in female; costa ending at vein R_{4+5}, second costal section long, about 3½ times length of fourth.
COLOUR. Frons orange-yellow, hind-margin of eye dark, inner vertical bristle generally just on yellow ground; third antennal segment black, first and second usually paler, slightly yellowish; mesonotum mat grey, pleura dark, only mesopleura with narrow yellow upper margin; legs: femora black, with all knees bright yellow.
MALE GENITALIA. Aedeagus as in Fig. 206, with black terminal processes of distiphallus diverging at an angle of 45°.
EARLY STAGES. Puparium white or slightly yellowish, 4 mm long; anterior spiracles with up to 20 minute pores arranged in 2 rows, posterior spiracles only slightly raised, each with from 20–30 pores forming a regular row around the outer margin and scattered irregularly over the whole surface (Fig. 207).
HOST-PLANTS. *Cichorium endivia* L., *C. intybus* L. (Chicory); also *Lactuca serriola* Torn., *Lapsana communis* L., ? *Arctium majus* L.
BIOLOGY. There are two generations. Larvae of the first generation feed in the midrib of the basal leaves in April and May before the flowering stem has appeared. The midrib becomes somewhat thickened as a result of the larval feeding. Pupation takes place at the base of the leaves. The later summer generation feeds in the stem, boring down to and sometimes penetrating the root. These larvae normally overwinter as puparia in the stem or upper part of the root.
DAMAGE CAUSED. Larval feeding in the edible leaves of forced chicory has damaged the crop in Switzerland. It is not known whether the larvae in the stem and root cause any deformation or other damage to the forced heads.
DISTRIBUTION. Switzerland, France, Germany, Austria, (? Rumania).
PARASITES. None recorded.

Discussion

TAXONOMY. *Phytomyza penicilla* belongs to the group of large, internal feeding species which have frequently been referred to as the *robustella* group. These were partially clarified by GRIFFITHS (1964b) and were discussed further by SPENCER

(1971a). Among these species, *penicilla* is readily distinguishable by the conspicuously long pubescence on the third antennal segment.

HERING identified as *continua* Hendel the species damaging cultivated chicory in Switzerland. However, I have examined two specimens from the original material obtained by DESHUSSES in 1928 and am satisfied that this is a misidentification.

BIOLOGY. Although *P. penicilla* was first described in 1935 its biology was not established until 20 years later when BUHR (1955) discussed a number of Agromyzidae forming galls and recorded this species as forming gall-like swellings in

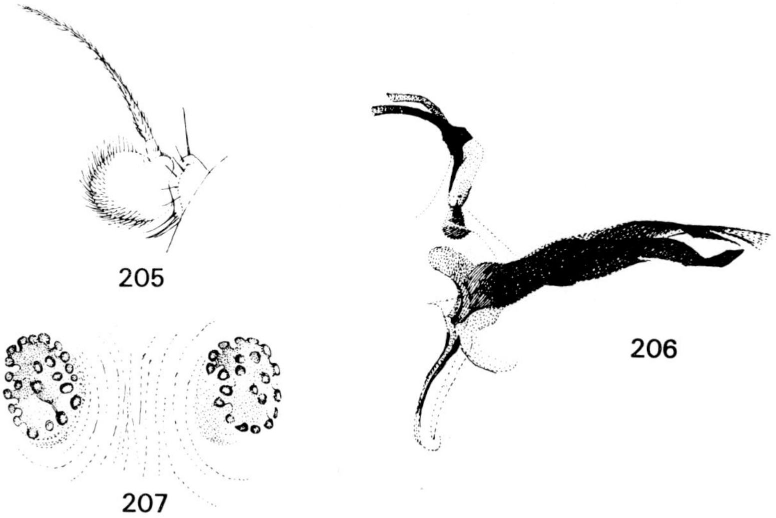

Figs. 205–207. *Phytomyza penicilla* Hend.: 205, third antennal segment; 206, aedeagus; 207, posterior spiracles of puparium.

the leaves of *Lactuca serriola* in Brittany, France. Later he obtained the same species in large numbers from stems of the same host in Thuringia, E. Germany.

DESHUSSES and DESHUSSES (1929, 1933) recorded *P. penicilla* (as *continua* Hendel) in leaves and the upper part of the roots of chicory in the Geneva area of Switzerland.

There is nothing inconsistent in these various accounts of the biology of *P. penicilla* as suggested by HERING (1960). Early in the year or when cultivated chicory is forced, the larva is able to feed in the midrib of the leaves in essentially the same way as later in the stem.

HOST-RANGE. The three confirmed hosts, *Cichorium*, *Lactuca* and *Lapsana* belong to the latex-bearing sub-family of the Compositae, Liguliflorae. A single puparium from Rumania ex *Arctium majus* (leg. Hering), differs only slightly from those of *penicilla* but with this host belonging to the sub-family Tubuliflorae the exact status of this specimen can only be established when adults can also be examined.

ECONOMIC IMPORTANCE. *P. penicilla* has only once occurred in substantial numbers on cultivated chicory, when DESHUSSES and DESHUSSES reported it as causing serious damage, together with *Ophiomyia pinguis*, in the Geneva and Thônex areas of Switzerland. The most important damage was apparently normally caused by *O. pinguis*. One case is cited where 96% of the damage was due to *pinguis* and only 4% to *penicilla*; in another region, however, the relative proportions are given as 55% *pinguis* and 35% *penicilla* (sic). Surprisingly there are no records of *penicilla* on cultivated chicory in France, Belgium or the Netherlands. It thus appears that *Cichorium* is not the most favoured host. However, with one mass outbreak on record, a similar attack in pest proportions might at any time occur again when conditions are particularly favourable for this particular species.

COMPOSITAE: CYNARA

Agromyza apfelbecki Strobl, 1902 – S. Europe, Chile

ADULT. Very large species, wing length up to 4.2 mm, frons and antennae reddish.

HEAD (Fig. 208). Frons exceptionally broad, $2\frac{1}{2}$–3 times width of eye, conspicuously projecting above eye in profile; normally 2 ors and 2 ori but 1 or even 2 additional bristles may be present; orbital setulae short, reclinate; jowls deeply extended at rear, cheeks forming broad ring below eye; third antennal segment ovoid, arista short, only finely pubescent; conspicuously high epistoma above mouth-margin.

MESONOTUM. Dorso-central bristles strong, variable but normally 4 + 2; acrostichals in some 6 rows.

WING (Fig. 209). Length from 3.4 mm in male to 4.2 mm in female; costa extending strongly to vein M_{1+2}, last section of M_{3+4} short, little more than $\frac{1}{2}$ length of penultimate, first cross-vein at or slightly before midpoint of discal cell.

COLOUR. Frons reddish, hind-margin of eye and orbits greyish-black, with silvery pruinosity; all antennal segments reddish, third variably darkened towards apex; mesonotum and abdomen mat, ash-grey; legs black but all knees distinctly reddish; squamae pale grey, fringe black.

MALE GENITALIA. Aedeagus as in Fig. 210, surstyli with a patch of strong bristles on inner face.

EARLY STAGES. Egg white, 0.72 × 0.28 mm (Fig. 211A). The micropile is in the form of two minute protuberances, the larger of which lies towards the dorsal surface of the egg (Fig. 211B). The chorion appears smooth but at high magnification shows a well-defined reticulation (Fig. 211C). The adult larva can attain a length of 7.5 mm; it has a substantial protuberance (Fig. 212A) beyond the posterior spiracles and these each have 3 elongate pores surrounded by 4 sensilli, each bearing 4 long hairs (Fig. 212B). The puparium is reddish-brown, up to 4 mm long (RICCHELLO, 1928 as *andalusiaca* Strobl).

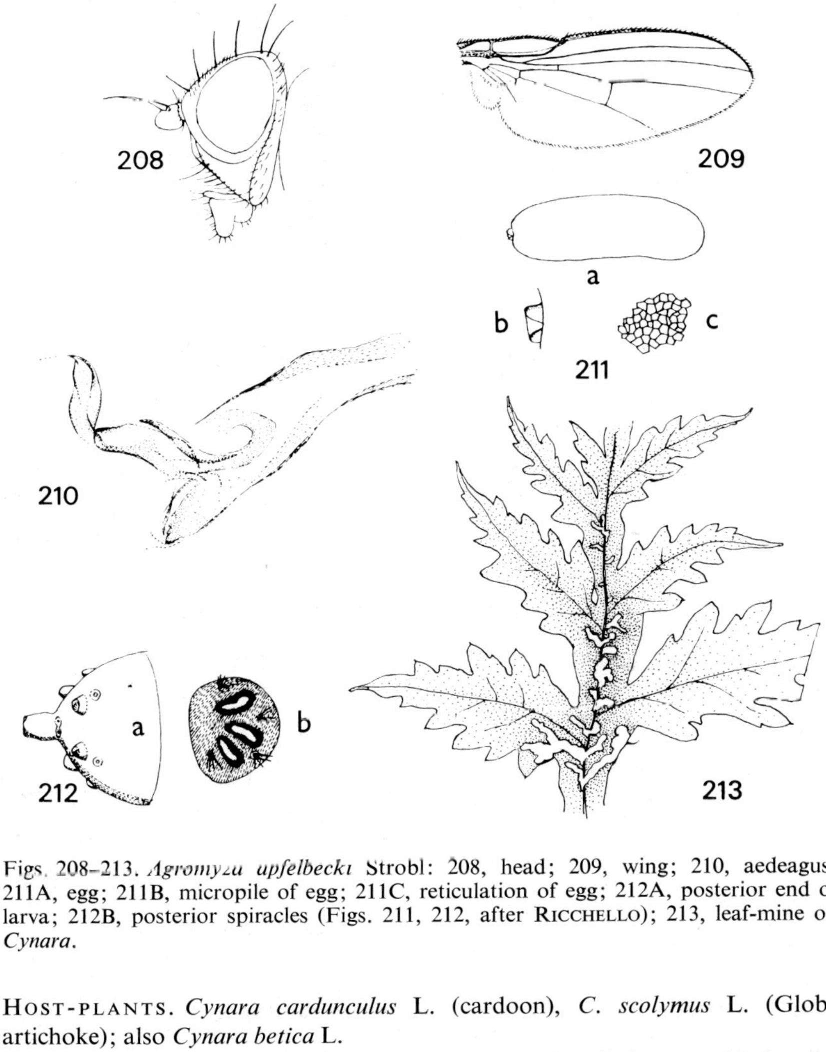

Figs. 208–213. *Agromyza apfelbecki* Strobl: 208, head; 209, wing; 210, aedeagus; 211A, egg; 211B, micropile of egg; 211C, reticulation of egg; 212A, posterior end of larva; 212B, posterior spiracles (Figs. 211, 212, after RICCHELLO); 213, leaf-mine on *Cynara*.

HOST-PLANTS. *Cynara cardunculus* L. (cardoon), *C. scolymus* L. (Globe artichoke); also *Cynara betica* L.

BIOLOGY. RICCHELLO confirmed experimentally that it is essential for the females to imbibe sap from feeding punctures in the host-plant before the eggs will develop. Oviposition normally commences 5 days after emergence and the eggs are invariably laid in the immediate vicinity of the midrib or of a main lateral vein. The larva hatches after 7–8 days.

In the first instar the larva forms a small blotch in the leaf-blade before entering a vein, along which it continues feeding until reaching the midrib where its main development takes place. As it progresses along the midrib, short, relatively broad channels or blotches are eaten into the adjoining parenchyma (Fig. 213).

On completion of feeding the larva generally leaves the leaf by cutting an exit slit through the lower surface near its base. It allows itself to fall to the ground, where pupation takes place.

RICCHELLO states that there are three generations between September and April. No larvae can be found feeding after May. Emergence of the first generation of adults, as with other Mediterranean species which aestivate as puparia throughout the long, hot summers, takes place a few weeks after the onset of the first rains in autumn which induce fresh growth in the host-plants necessary for the ovipositing females. PICARD (1914a) states that in the Montpellier and Perpignan areas in the south of France larvae are particularly active in December and January. Larvae found on 27 May, 1959, at Estartit on the Costa Brava, Spain, produced adults from 10 October until early November (K.A.S.). I have also found the flies in abundance in fields of artichokes south of Barcelona in March.

DAMAGE CAUSED. Larval feeding along the midrib or one or more of the lateral veins seriously affects the growth of young leaves, which turn yellow and die. In old mines the upper epidermis and the thin layer of cells of the lower, spongy parenchyma disintegrate and the leaf then assumes the characteristic appearance indicating earlier larval attack, with holes taking the place of the original mine.

RICCHELLO recounts how the peasant growers in southern Italy ascribed these holes to the action of spring frosts. They were amazed when he demonstrated to them that the holes were not caused by frost but by the feeding of larvae. Others were not at all convinced and countered by saying that the larvae themselves were caused by the frost! The effect of frost is in fact the exact reverse and causes considerable mortality in the larvae of this species. PICARD (1914a) reported that when the temperature dropped to -14.5°C at Montpellier during the middle of January virtually all larvae were found to be dead, with no other obvious cause than the low temperature.

DISTRIBUTION. Dalmatia, Italy, France, Spain, Malta (SPENCER, 1973), Sicily, Greece, certainly present throughout Mediterranean area; Chile.

PARASITES. *Grandia cynaraphila* (Ricchello), Dacnusinae; *Opius apfelbeckianus* Fischer, Opiinae; *Stenomalina muscarum* (L.), Pteromalidae. RICCHELLO found a rate of parasitism as high as 80% at Pompeii by *G. cynaraphila*. Other checks showed rates of 60% and 70%. There is apparently no significant difference in the parasitism between the three generations in November, February and April. Parasitism by the Pteromalid, on the other hand, is significantly less and only isolated cases were observed. No information is available from other countries.

Discussion

TAXONOMY. STROBL originally described *apfelbecki* (named after the Viennese entomologist APFELBECK) as a large, southern variety or race of *Agromyza abiens*

Zett. Later, in 1906, he described *A. andalusiaca* from southern Spain, considering it to be distinct from *apfelbecki* in view of its larger size and minor differences in colour and wing venation. However, examination of the male genitalia has shown the two to be identical and they were correctly synonymized by HENDEL (1931:107).

HOST-PLANTS. MARCHAL (1911) first reported this species (as *abiens* Zett.) feeding on artichokes in the Perpignan area of southern France. He suggested it might also feed on other Compositae, such as thistles. RICCHELLO, despite considerable searching, was unable to find any hosts in Italy other than cultivated artichokes. MARCHAL, however, was correct in his supposition, as I was able to confirm by finding larvae feeding on the wild thistle, *Cynara betica*, on the Costa Brava north of Barcelona. It remains to be established whether other related genera also serve as hosts.

DISTRIBUTION. *Cynara cardunculus* was introduced into South America and has now spread over large areas of the Pampas. The larvae of *Agromyza apfelbecki* were no doubt introduced together with the host-plant and the species is probably present wherever the plant is now cultivated in South America. It has only been confirmed in Chile.

ECONOMIC IMPORTANCE. The damage caused by the larvae of this species can be considerable but varies significantly from year to year and is also largely dependent on the state of growth of the plants attacked and the number of larvae present in a single plant. A single larva can cause yellowing of the leaf but this will have little effect on a vigorous, well-established plant. However, when numerous larvae are present in a plant, particularly in the first year of its growth, it can either be so retarded that no flower-head is produced or it may even die. In a check of 100 leaves RICCHELLO found an average of 5 larvae per leaf, with a maximum of 11. All leaves had at least one larva but of the sample of 100 examined, only 12 had this small number.

Agromyza apfelbecki is thus a pest which cannot be ignored wherever the host-plants are cultivated, either for the flower-heads or for the young stems and foliage as a cooked vegetable.

COMPOSITAE: LACTUCA SATIVA L.

***Amauromyza maculosa** (Malloch, 1913a) – South America, U.S.A., Hawaii

ADULT. Head entirely black, orbital bristles strong; mesonotum brilliantly shining black, with 3 + 1 strong dc; squamae and fringe silvery-white; halteres variegated, primarily white but knob with conspicuous black area above; wing with costa extending to vein M_{1+2}, last section of M_{3+4} two-thirds length of penultimate, length from 2.3–2.5 mm.

* See also Appendix A.

MALE GENITALIA. Aedeagus as in Figs. 214, 215; ninth sternite with broad side-arms, rounded at end but with rudimentary hypandrial apodeme; surstyli divided by complete suture from epandrium (Fig. 216); ejaculatory apodeme with bowl-shaped base (Fig. 217).
EARLY STAGES. Puparium conspicuously reddish, posterior spiracles each with 3 pores.
HOST-PLANTS. Compositae, particularly *Lactuca sativa* L.; also ornamentals (cf. Chapter 11). STEGMAIER (1967b) records the following genera in Florida:

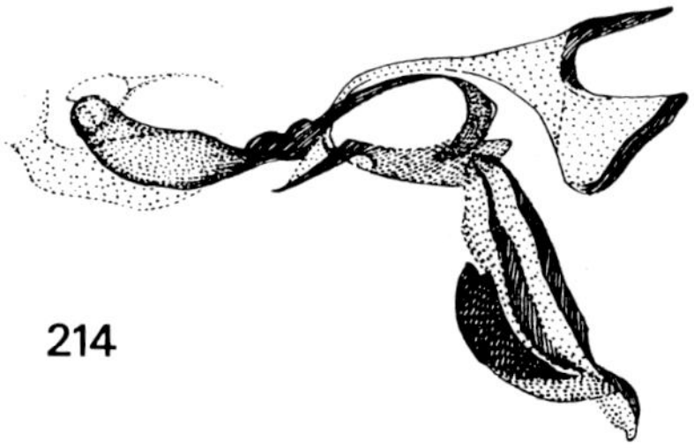

214

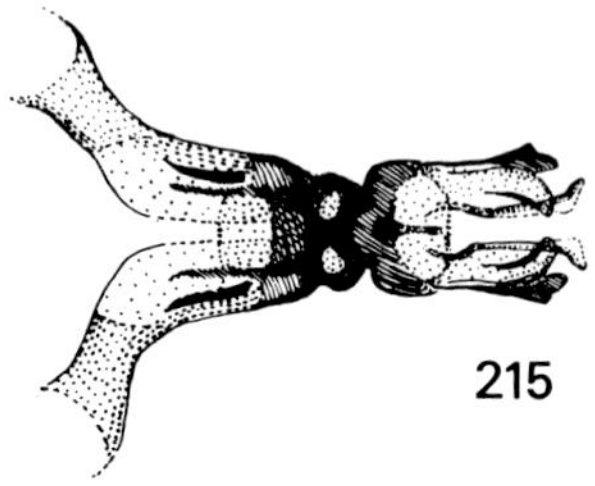

215

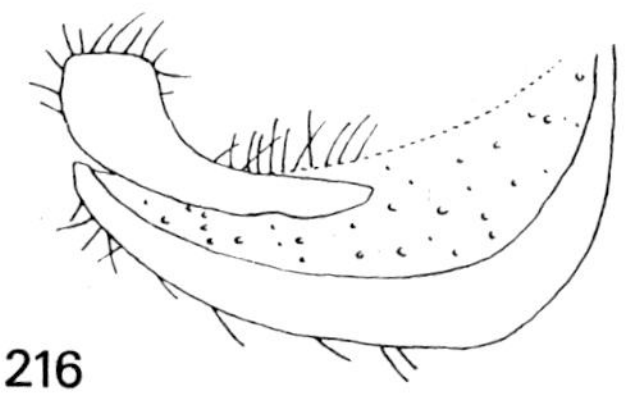

216

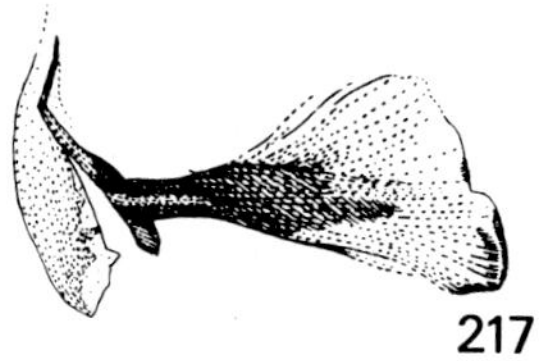

217

Figs. 214–217. *Amauromyza maculosa* (Mall.): 214, aedeagus, side view; 215, same, ventral view; 216, surstylus; 217, ejaculatory apodeme.

Baccharis, *Bidens*, *Chrysanthemum*, *Erechtites*, *Emilia*, *Eupatorium*, *Gnaphalium*, *Gaillardia*, *Helianthus*, *Melanthera*, *Senecio*, *Sonchus* and *Tagetes*.
BIOLOGY. The larvae form a large blotch (Fig. 504), several feeding together; pupation is external. There are several generations in a year but the detailed life history has not been studied.
DISTRIBUTION. South America, confirmed from Argentina, Brazil, Peru, Venezuela; Costa Rica; West Indies, including Barbados, Trinidad, Grand Cayman, Cuba, Bahamas, Bermuda; eastern U.S.A. from Florida to New York; Hawaii.
PARASITES. In Florida 4 species of Eulophidae: *Chrysocharis* sp., *Derostenus variipes* Cwfd., *Derostenus* sp., *Diaulinopsis callichroma* Cwfd. (STEGMAIER, 1967b).

Discussion

TAXONOMY. This is clearly an isolated species and probably deserves separate sub-generic status. It is readily recognizable by the variegated colour of the halteres.

ECONOMIC IMPORTANCE. Lettuce has been damaged in Trinidad (R. CRUTTWELL, personal communication) and the total loss of a small field of the same crop has been reported at Cagua, Venezuela (L. CERMELI, personal communication). The large blotch mines produced by this species will always represent a potential hazard to young plants in the family Compositae.

CRUCIFERAE: MOST CULTIVATED GENERA; ALSO PISUM

Liriomyza brassicae (Riley, 1884) – Cosmopolitan

Oscinis brassicae Riley, 1884: 322. Holotype ♀ from St. Louis, Missouri, in U.S. National Museum.

Liriomyza brassicae Frick, 1952a: 402.

Liriomyza cruciferarum Hering, 1927: 461 (synonymy correctly suggested by HENDEL, 1931: 255).

Phytomyza mitis Curran, 1931: 97 (synonymy established by FRICK, 1957b: 68).

Liriomyza hawaiiensis Frick, 1952b: 513 (synonymy established by SPENCER, 1963a: 356).

Liriomyza bulnesiae Spencer, 1963a: 360 (synonymy established by SPENCER and STEGMAIER, 1973).

ADULT. Small species, with shining black mesonotum and orbits frequently darkened.

HEAD (Fig. 218). Frons not projecting above eye, with 2 equal ors and 2 largely incurved ori; jowls deepest at rear, cheeks forming only narrow ring below eye; third antennal segment small, round.

MESONOTUM. 3 + 1 dc, acrostichals in 4 rows.

WING. Length from 1.25 mm in male to 1.6 mm in female, discal cell small, last section of vein M_{3+4} 3 times length of penultimate.

COLOUR. Frons pale yellow, hind-margin of eye black, with normally both vertical bristles on dark ground; orbits frequently darkened, black or at least brownish, sometimes only very narrowly so adjoining margin of eye, in paler specimens no trace of this darkening detectable; all antennal segments yellow; mesopleura normally with a dark area filling lower third and this may also extend up hind-margin.

LEGS. Coxae and femora bright yellow, tibiae and tarsi darker, brownish; abdomen predominantly shining black, tergites with hind-margins at most narrowly yellow, laterally more broadly so; squamae yellow, margin and fringe dark.

MALE GENITALIA. Aedeagus distinctive, as in Figs. 219A, B; ejaculatory apodeme with unusually large blade (Fig. 220).

EARLY STAGES. Larva bright yellow, up to 4 mm long, posterior spiracles each with 3 pores; puparium pale yellowish-orange.

HOST-PLANTS. Primarily Cruciferae, cultivated genera including *Brassica*, *Cheiranthus*, *Nasturtium*, *Raphanus*; also found commonly on *Tropaeolum* (Tropaeolaceae). Other genera of the Cruciferae recorded by HERING (1957c) are: *Barbarea*, *Cakile*, *Eruca*, *Hirschfeldia*, *Isatis*, *Lepidium*, *Matthiola*, *Moricandia*,

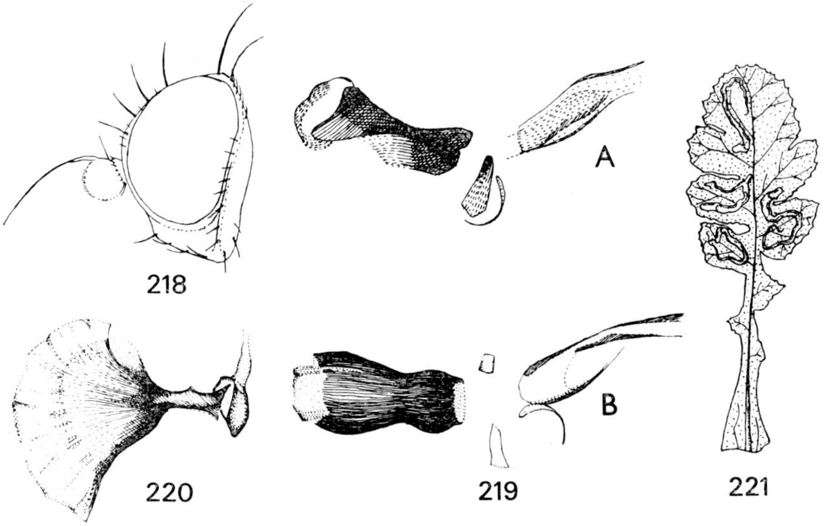

Figs. 218–221. *Liriomyza brassicae* (Riley): 218, head; 219A, aedeagus, side view; B, same, ventral view; 220, ejaculatory apodeme; 221, leaf-mine on *Brassica* (after HERING).

Sinapis and *Sisymbrium*; FRICK (1959) records *Radicula* in the United States and mines were found on *Diplotaxis* in Tasmania (SPENCER, 1963b), HERING found mines at Montreal, Canada, on *Erysimum* and *Lobularia* (SPENCER, 1969a: 171) and SASAKAWA (1964) records *Cardamine* in Hawaii (as *L. hawaiiensis*). *L. brassicae* also feeds commonly on *Capparis*, *Cleome* and *Gynandropsis* (Capparaceae), *Reseda* (Resedaceae) and more rarely on *Pisum* (Leguminosae).

BIOLOGY. The larva feeds as a leaf-miner either on the upper or lower leaf surface. The mine is essentially linear (Fig. 221) but irregular, appearing whitish or greenish; the black frass is in more or less connected threads. Pupation takes place on the ground. In Europe there are two generations but in tropical areas there are certainly more, coinciding with the active growth of the host-plants.

DAMAGE CAUSED. The small, individual leaf-mines are of no significance but larvae can occur in large numbers, with entire leaves being largely eaten out. In mature plants only the outer leaves are affected and this does not influence the growth of the plant. However, similar heavy attacks on seedlings would destroy or certainly weaken them.

DISTRIBUTION. Cosmopolitan. United States: California, Florida, Missouri, Hawaii, probably present in most States; Canada: Ontario, Manitoba; S. America: Guyana, Venezuela. Micronesia: Bonin Is., Chichi Jima; S. Mariana Is., Guam; Philippine Is.: Luzon; Fiji; Australia: Queensland, New South Wales, Tasmania; Cambodia, Thailand, Singapore, Ceylon, India; Africa: Egypt, Ethiopia, Kenya, Mozambique, Rhodesia, Senegal; Europe: Canary Is., Spain, Germany.

PARASITES. Florida: *Chrysocharis* sp., *Diaulinopsis callichroma* Crawford and *Pnigalio* sp. (Eulophidae), STEGMAIER (1967d). California: *Diglyphus begini* (Ashm.), *D. intermedius* (Girault), *Diglyphus* sp., *Derostenus arizonensis* Cwfd., *D. punctipes* Cwfd., *Derostenus* sp., *Chrysocharis petiolata* (Girault), *Chrysocharis* sp. (Eulophidae); *Halticoptera patellana* (Dalman) (Pteromalidae); *Ganaspidium* sp. (Cynipidae); *Opius* sp. near *complicans* (Braconidae), OATMAN and PLATNER, 1969. Australia: *Achrysocharis* sp., *Enaysma* sp., *Hemiptarsenus* sp., *Tetrastichus* sp. (Eulophidae); *Opius cinerariae* Fischer (Braconidae); Gen. near *Anastalus* sp. (Eupelmidae), KLEINSCHMIDT, 1970. Egypt: *Achrysocharella* sp., *Diglyphus* sp. (Eulophidae); *Charips* sp. (Cynipidae); *Diaeretiella rapae* (Curtis) (Braconidae), M. A. ASSEM, 1966 (unpublished thesis).

OATMAN and PLATNER (1969) found that in California 67.5% of the larval population was parasitized, the rate being highest during the October–December quarter at an average of 80.9% and lowest during the January–March quarter at 49.6%. Parasitization during the April–June and July–September quarters averaged 64.6% and 69.6% respectively. The rate was lowest in January at 26.7% and highest in October at 84.1%. No comparative figures are available from other parts of the world.

Discussion

TAXONOMY. As might be expected with a species of world-wide distribution, *Liriomyza brassicae* has been described a number of times under different names from different parts of the world.

A distinctive character is the darkening of the orbits but this is by no means constant. Paler specimens appear to occur more commonly in the Pacific area than elsewhere. FRICK's description of *Liriomyza hawaiiensis* in Hawaii, as distinct from *L. brassicae* occurring on the same food-plants and in the same localities, resulted from the attachment of too great importance to such colour differences. SASAKAWA (1964), when revising Hawaiian *Liriomyza* species, retained *hawaiiensis* as distinct from *brassicae*. However, no attempt was made to record differences between them and the description and illustrations of the genitalia of *hawaiiensis* apply accurately to *brassicae*.

In Florida, California, Venezuela and Hawaii *L. brassicae* occurs together with *L. sativae* Blanchard (cf. p. 219). These species cannot be reliably separated on external characters, unless the characteristic darkening of the orbits of *brassicae* is well developed. However, the male genitalia are entirely distinct (cf. Figs. 219A, B and 323, 324) and the form of the aedeagus in *brassicae* is remarkably constant throughout the world; nothing comparable is known in any other species.

Liriomyza bulnesiae was described from specimens caught on *Bulnesia arborea* Engl. (Zygophyllaceae) in the Botanical Gardens, Caracas. This was an isolated tree, with the leaves heavily infested with silvery epidermal mines and it was assumed that the adults on the leaves belonged to the same species which had produced the leaf-mines. However, it is now clear from the male genitalia that the species was *brassicae* and the adults were merely resting on the foliage and not in any way directly associated with this host. I am now satisfied that the leaf-miner on *Bulnesia* is *Liriomyza schmidti* (ALDRICH, 1929), which occurs commonly on a wide range of hosts in the Neotropical Region producing characteristic epidermal mines.

DISTRIBUTION. *Liriomyza brassicae* is the most truly cosmopolitan species known in the Agromyzidae. It is not common in Europe and is absent from Scandinavia and Britain. It occurs only locally in Canada but is abundant in California and Florida and no doubt in other southern States. There is only a single record from South America – from Guyana. Its wide distribution throughout the Pacific area, including Australia, and also many parts of the Oriental and Ethiopian regions, may in part be explained by recent introductions by man. Its wide occurrence in the tropics and only limited distribution in northern temperate areas is unusual in this genus.

HOST-RANGE. Although the primary hosts are in the Cruciferae, the occurrence of *L. brassicae* on Capparaceae and Tropaeolaceae can be logically explained by the presence of myrosin in all three families, which is presumably one of the attractive substances determining the females' choice of host. It is more difficult to explain why *Pisum* is also occasionally used as a host. This has been observed in India (SEHGAL, 1965), in S. Rhodesia (SPENCER, 1964b), in Kenya (K.A.S.) and in Australia (KLEINSCHMIDT, 1970).

ECONOMIC IMPORTANCE. A study of pests of cabbage in S. California in 1963–1965 suggested that the damage caused by *L. brassicae* was not significant. Even on young plants the leaf-mining injury was overcome by the rapid growth of the plants and was limited by the increasing rate of parasitism during the period of growth. On older plants the damage was confined to the outer leaves which dry up and drop off as the heads mature (OATMAN and PLATNER, 1969).

Large populations were observed in California, with peaks averaging 178 mines per plant on 6 August, 1963, and 47 per plant on 25 August, 1964. Very heavily infested mature cabbages were seen by the author at Dakar, Senegal, early in August, 1957 (SPENCER, 1959).

In Hawaii *L. brassicae* is considered to be a serious pest on cabbage, broccoli cauliflower and other cruciferous crops on all the main islands (D. E. HARDY, personal communication). This may be accounted for by abnormally high populations due to a lower rate of parasitism, in the absence of some of the parasite species present in mainland America.

In Egypt where the populations of *L. brassicae* are small, it is not considered a serious pest but it does occur regularly on cabbage and cauliflower.

It seems clear that in all areas where it occurs *L. brassicae* must be considered as a potentially serious pest both of cruciferous plants and also peas.

CRUCIFERAE: BRASSICA

Phytomyza rufipes Meigen, 1830 – Europe; Canada, U.S.A.

ADULT. Large, yellow and grey species, with broad frons and deep jowls.
HEAD (Fig. 222). Frons exceptionally broad, up to 3 times width of eye, conspicuously projecting above eye in profile; 2 equal ors, 2 or 3 ori; orbital setulae sparse,

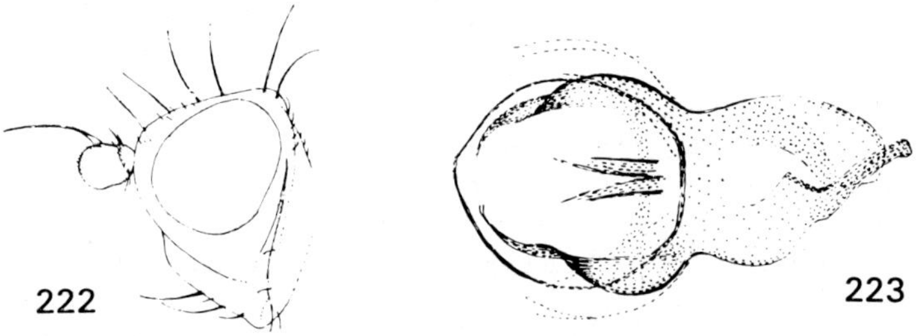

222 223

Figs. 222–223. *Phytomyza rufipes* Mg.: 222, head; 223, aedeagus, ventral view.

proclinate; jowls unusually deeply extended at rear, up to $\frac{3}{4}$ height of eye, cheeks almost $\frac{1}{2}$ depth of jowls; third antennal segment slightly longer than broad.
MESONOTUM. 3 + 1 strong dc, acrostichals sparse, limited to 1 or 2 hairs in front or even entirely lacking.
WING. Length from 2.5 mm in male to 3.5 mm in female; costa ending at vein R_{4+5}, second costal section short, just less than twice length of fourth; second cross-vein lacking.
COLOUR. Frons yellowish or somewhat darker, orange-brown, orbits and lunule yellow; hind-margin of eye partially grey, outer vertical bristle on dark ground, inner on yellow; third antennal segment varying from almost entirely black to largely yellow; face and palps yellow; mesonotum mat, ash-grey; humerus, notopleural triangle and upper margin of mesopleura yellow, mesopleura mat grey at least in lower three-quarters.
LEGS. Coxae yellow, femora largely yellow but with variable brownish striations, tibiae and tarsi more brownish; abdomen basically black but tergites yellow laterally and on hind-margins; squamae and fringe pale.
MALE GENITALIA. Aedeagus as in Fig. 223.

EARLY STAGES. Egg shining white, without sculpturing; larva up to 6 mm long, posterior spiracles each with 25–30 minute pores. The puparium is somewhat elongate, mat yellowish, 3 mm long (SMITH, 1927; FREY, 1951).

HOST-PLANTS. Exclusively Cruciferae, common on cultivated *Brassica* species, particularly *B. oleracea* var. *acephala* DC (Kale), var. *botrytis* L. (Cauliflower), var. *capitata* L. (Cabbage), var. *gemmifera* Zeker (Brussels Sprouts), var. *italica* L. (Sprouting Broccoli, Calabrese), *B. juncea* Coss. (Leaf Mustard); also *Brassica Napus* L. (Rape), var. *napobrassica* (L.) Rebh. (Swede) and *B. Rapa* L. (Turnip); other genera recorded include *Alliaria, Armoracia, Conringia, Diplotaxis, Moricandia, Peltaria, Raphanus, Rorippa* and *Sinapis* (HERING, 1957c).

BIOLOGY. Oviposition takes place in the leaf-blade, usually near a vein towards the margin of the leaf. FREY (1951) found that females laid up to a total of 81 eggs, with on average 5 eggs laid per day. In 94 observations the average time of hatching was 3.7 days, larval feeding continued for 11.9 days and the pupal period lasted 18.7 days.

The young larva makes a true mine towards the nearest vein and then feeds inside this downwards towards the midrib and petiole where the main feeding takes place. At the end of the petiole, if it is not by now fully grown, it can either turn and continue feeding in the reverse direction or, particularly in young plants, it may enter the stem. Rarely, larvae have been observed to eat their way out of one stalk and bore into another. Very few Agromyzid larvae are able to do this.

Not infrequently 10 larvae can be found in a single stalk and as many as 25 were observed in individual plants in Silesia in 1935 (HOCHAPFEL, 1937).

In England in 1949 soil populations of puparia of from 50,000 to 1,650,000 per acre were reported in fields of swedes in Norfolk (M.A.F.F., unpublished report).

The first flies appear in May or June and at least three or four generations follow, with larvae still actively feeding as late as November.

DAMAGE CAUSED. The damage caused by larval feeding varies considerably with the time of year and the type of crop being attacked.

In the seedling stage infested plants become malformed with conspicuous yellowing of the heart leaves (German: Gelbherzigkeit). ROESLER (1937) considers that given favourable conditions such young plants may still develop into strong mature plants but they are inevitably somewhat weakened, with reduced resistance and are less able to survive unfavourable conditions or attacks by other pests.

Infestation in older plants is almost entirely restricted to the outer leaves, which will normally in any case be discarded. FREY (1951) found that in Schleswig-Holstein at the end of October 89% of all outer leaves contained larvae, while a month later this percentage had declined to 62%. The percentage of inner leaves infested was 1.4% in October and 2.7% in November.

In England damage has been on occasions severe in the eastern counties of Norfolk and Suffolk on the crop of Italian sprouting broccoli (*Brassica oleracea* var. *italica*) which is subsequently processed by quick freezing. The young shoots about 4 inches in length contain larvae throughout the cutting season from August to December and in 1957 the infestation was so severe that for a short period the processing firm refused to accept further supplies.

'Examination of at least 100 shoots on each of four calabrese crops in Norfolk on successive dates from 12 August to 12 November, 1958, showed that young larvae were present in each crop on each occasion. The mean figures for percentage infestation of shoots at the stage at which they are normally cut (about 4 in. long) were:

	12 Aug.	27 Aug.	9 Sept.	9 Oct.	12 Nov.
Shoots infested, per cent	2	11.5	14	18.5	19

(CARDEN, 1962).

DISTRIBUTION. Probably present throughout Europe, also confirmed in U.S.S.R. (Leningrad), Egypt, Canary Islands and Madeira (HENDEL, 1931–1936), in Finland reported as far north as Ostrobothnia (VAPPULA, 1965); Canada (SPENCER, 1969a); United States (FRICK, 1959).

PARASITES. In Europe: *Dacnusa pubescens* Curtis, *Chorebus thusa* (Nixon) (Braconidae); *Pleurotropis flaviscapus* Thoms. (Eulophidae); *Stenomalina muscarum* L. (Pteromalidae). HOCHAPFEL (1937) quotes a rate of parasitism in Silesia of 8%, while CARDEN (1962) in England found a substantially higher rate of 27%.

Discussion

TAXONOMY. *Phytomyza rufipes* is a distinctive species and no other comparable Agromyzid is known to occur on Cruciferae. It somewhat resembles *Phytomyza flavicornis* Fallén, an internal stem-borer in *Urtica*, and has been misidentified as this species by several workers. However, *flavicornis* is even more yellow, with the third antennal segment and mesopleura being entirely yellow and it has only a single upper orbital bristle.

ECONOMIC IMPORTANCE. Damage to cultivated cruciferous crops has occurred in many parts of Europe. In Italy DEL VECCHIO (1917) reported severe damage to the cauliflower crop in Lombardy. BERG (1956) gave details of widespread attacks on cauliflower in the Lower Rhine vegetable growing area but concluded that in most cases commercial damage was not substantial. Severe damage has been reported in Finland, in the worse cases with up to 70% of leaves being destroyed (VAPPULA, 1965).

As reported above the presence of larvae in late Broccoli shoots used in England for quick freezing has on several occasions proved locally serious. CARDEN (1962) comments as follows: '. . . the larvae and mines were usually very small and were not likely to be noticed by the consumer unless they were numerous. However, on cooking the shoots some of the larvae might be freed from the mines and float on the water where they would be more conspicuous'. Frozen broccoli has become an increasingly important commercial crop but housewives may well be put off if they realize to what extent the shoots are infested.

Many additional reports of damage to a variety of crops have been recorded in England (M.A.F.F., unpublished reports). In 1925 appreciable damage was caused to turnips in the S.E. Counties. In the same year cauliflower, kale, rape and other crops were rejected by a merchant in Boston, Lincs., owing to the rapid yellowing of the leaves after harvesting in August. In 1949 there was a report from Norfolk of feeding punctures causing much distortion to seedlings. In 1956 the kale crop in Cardigan, Wales, suffered stem damage, followed by root rot. Mining in the stems of cabbage just below the skin causing cracking and browning associated with poor growth of the plants was reported from the Ely area in 1951. In 1958 10 acres of Brussels Sprouts were severely damaged in Kent and had to be replanted.

Phytomyza rufipes must thus be considered as a major pest of cultivated Crucifers in Europe. Few areas of cultivation will be found not to contain larvae but these will not in all cases necessarily either physically damage the crop or result in commercial loss. However, both risks will always be present and the exact significance of the infestation will not be predictable.

Young plants can suffer the most obvious damage. As the infestation will in most cases be detectable at the planting out stage, even though the majority of such plants may recover, the general practice is for these to be destroyed and replaced. Later in the year, larvae can be expected in large numbers in outer leaves but as these in any case will not be used for human consumption, they are not of real significance.

There are no reports of damage having been caused in Canada or the United States.

LABIATAE: MENTHA (MINT)

Phytomyza petoei Hering, 1924 – Europe, Asia

ADULT. Small, dark species, with only single upper orbital bristle.
HEAD (Fig. 224). Frons broad, twice width of eye, not projecting above eye in profile; orbits broad, differentiated but not raised; only 1 strong ors, 2 ori, the lower frequently greatly reduced; orbital setulae proclinate, in single row; jowls deep, $\frac{2}{5}$ height of eye, cheeks forming distinct ring below eye; third antennal segment large, somewhat longer than broad.
MESONOTUM. 3 + 1 strong dc, acrostichals sparse, in 2 rows.
WING. Length from 1.6–2 mm, second costal section little more than $1\frac{1}{2}$ times length of fourth.
COLOUR. Frons dark brown, orbits mat grey, third antennal segment jet black; mesonotum mat, greyish, legs black, fore-knees at most inconspicuously paler; squamae grey, margin and fringe black; abdomen shining black.
MALE GENITALIA. Aedeagus as in Fig. 225.
EARLY STAGES. Anterior spiracles of larva oval, each with about 14 pores, posterior spiracles on conspicuous conical projections, each with about 25 pores; puparium greyish-brown.

Host-plants. *Mentha* species, particularly *M. longifolia* (L.) Huds. and *M. spicata* L., emend. Huds.
Biology. The larva forms an upper surface leaf-mine (Fig. 226), which is essentially linear but irregular and sometimes appearing to form a secondary blotch. Pupation is normally externally but not infrequently the puparium remains in the leaf at the end of the mine. There are two to three generations from May to October.

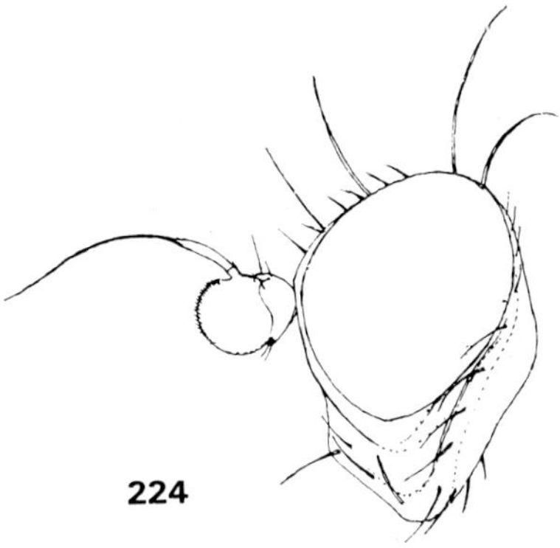
224

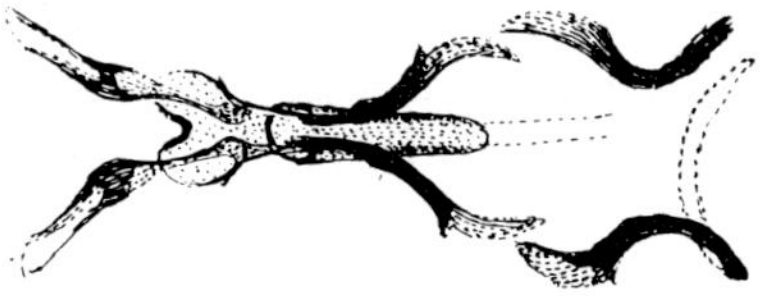
225

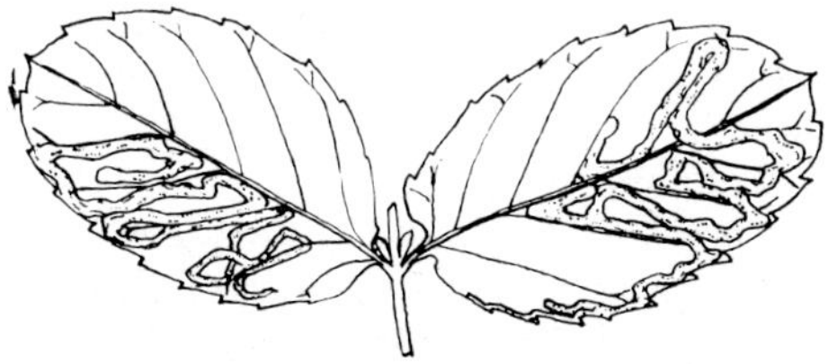
226

Figs. 224–226. *Phytomyza petoei* Her.: 224, head; 225, aedeagus, ventral view; 226, leaf-mine on *Mentha* (Hering).

Damage caused. Larvae frequently occur in considerable numbers on cultivated mint and a significant number of leaves can be rendered useless for culinary purposes.
Distribution. Germany, Czechoslovakia, Rumania; Yugoslavia (K.A.S.); U.S.S.R.: Kirghiz S.S.R., Frunze; Uzbek S.S.R., Tashkent (Dovnar-Zapolski, 1969).
Parasites. *Dacnusa maculipes* Thomson, *Chorebus nana* (Nixon); *Chrysocharis pubicornis* (Walker).

Discussion

Economic importance. *Phytomyza petoei* has not been known to occur in sufficient numbers to affect the growth of cultivated mint. However, it occurs commonly on garden-mint in southern England where it has at least an appreciable nuisance value.

LILIACEAE: ALLIUM CEPA L.

Liriomyza cepae (Hering, 1927) – Germany

ADULT. Medium-sized species, with conspicuously angulate third antennal segment, black scutellum, largely dark coxae and femora, and milky wings.
HEAD (Fig. 227). Frons and orbits very strongly projecting above eye; orbits well differentiated, with 1 reclinate ors and 3 largely incurved ori; orbital setulae sparse, mainly upright; jowls broad, deeply extended at rear, $\frac{2}{5}$ height of eye, cheeks forming broad ring below eye; third antennal segment conspicuously angulate, with distinct point at upper corner; broad epistoma present above mouth margin.
MESONOTUM. 3 + 1 dc, third and fourth equal, acrostichals sparse in 2 rows.
WING. Length from 2 mm in male to 2.5 mm in female, discal cell small, last section of vein M_{3+4} just less than 4 times length of penultimate.
COLOUR. Frons, jowls, antennae and palps entirely yellow; face only slightly greyish; upper orbits and hind-margin of eye variably darkened, outer vertical bristle on dark ground, inner sometimes at margin with yellow; mesopleura and sternopleura almost entirely black, notopleura yellowish but often indistinctly so; mesonotum mat grey, scutellum similar, without any trace of yellow; legs: coxae black, femora appearing largely black on upperside, more yellow below, all knees bright yellow, tibiae and tarsi dark yellowish-brown; abdomen black; wing normal above but becoming milky-white below vein M_{1+2}, R veins brown, M_{1+2}, both cross-veins and all veins at base white; wing base yellow, squamae and fringe yellowish-white.
MALE GENITALIA. Aedeagus as in Figs. 228, 229, paired distal tubules long, projecting far beyond bowl-shaped mesophallus; one strong spine within epandrium at upper hind-corner.
EARLY STAGES. NIETZKE (1943) made a detailed study of the early stages and biology of this species, which he treated as the Oggersheim race of *cepae*. The whitish-yellow larva when full-grown is about 2 mm long or possibly more; the anterior spiracles each have 7 pores (Fig. 230A), the posterior spiracles 9 (Fig. 230B). The average measurements of the reddish-brown puparium are 3.05 × 1.16 mm; the average weight was found to be 0.0015 gr.
HOST-PLANT. *Allium cepa* L. (Onion).
BIOLOGY. No detailed studies have been made specifically on this species, although in general the life history is similar to that of *Liriomyza nietzkei* described below. However, emergence dates of the adults are somewhat later and the larval feeding time somewhat longer. Specimens of the type series were caught on fields of onions at Calbe, near Magdeburg in August, 1926.

DAMAGE CAUSED. Larval feeding in the leaves can seriously interfere with the transport of nutrient materials to the developing onion. However, in this species the later appearance of the mines allows the onion more time to grow and significant damage to the leaves apparently occurs in most cases only after the onion has fully developed.
DISTRIBUTION. Germany: Calbe, near Magdeburg and Oggersheim, Pfalz.

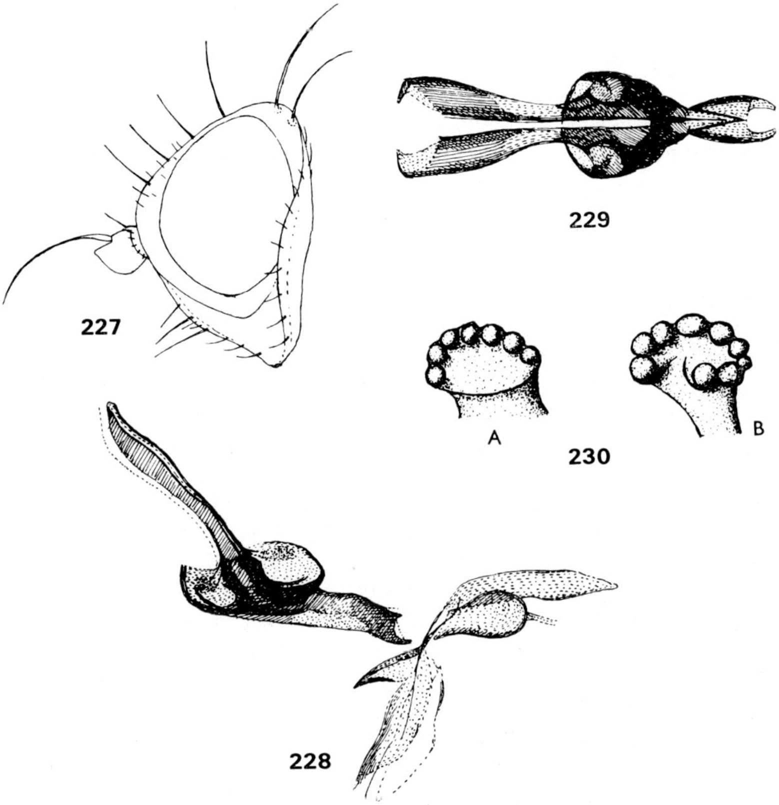

Figs. 227–230. *Liriomyza cepae* (Hering): 227, head; 228, aedeagus, side view; 229, same, ventral view; 230A, larval anterior spiracles; 230B, posterior spiracles (HERING).

PARASITES. None recorded but possibly *L. cepae* is parasitized by the same species which attack *L. nietzkei* (see below).

Discussion

TAXONOMY. In view of its black scutellum, this species was described by HERING in the genus *Dizygomyza*. NOWAKOWSKI (1962: 96) suggested that it nevertheless correctly belongs in *Liriomyza* (although it is not clear whether he examined *cepae* s. str. or possibly *nietzkei*).

NIETZKE (1943) noticed distinct differences between populations at Herxheim and Oggersheim, Pfalz, in West Germany but, in view of their overall similarity and with the same foodplant, he understandably hesitated to describe a second species as distinct from *cepae*. Instead he differentiated the two forms as races – the 'Herxheim race' and the 'Oggersheim race' – and he rightly noticed that the Oggersheim race was identical with specimens from the type locality at Calbe.

NIETZKE found that the Oggersheim race was larger, with consistent differences in the male genitalia, and emerged later. However, he apparently did not notice the significantly paler wings in the true *cepae* (Calbe and Oggersheim) nor the more acutely angulate third antennal segment (Figs. 227, 234). HERING (1956, 1957b) also noted differences in both the anterior and posterior spiracles (Figs. 230, 237).

DISTRIBUTION AND EVOLUTION. The true *cepae* is so far only known from Calbe and Oggersheim. HERING (1956) discusses two alternative theories to explain the presence in close affinity in Germany of what he refers to as subspecies. One possibility is rapid sub-speciation within Germany. This theory is, however, rejected in favour of the probable importation into Germany of subspecies, which have evolved elsewhere, probably in east Asia (Japan), together with young onions still retaining their leaves. Later, HERING (1957b) also discusses the population from Peking, described as ssp. *chinensis* Kato, 1949.

I think it is far more probable that the present distribution of *cepae* and *nietzkei* is the result of natural dispersal and subsequent evolution, from some centre of origin which probably was in eastern Asia. HERING found that ssp. *chinensis* had the same form of larval spiracles as true *cepae*. The male genitalia also show greater affinity with *cepae* than with *nietzkei* and both have the characteristically pale wings and entirely black scutellum. All these characters are derived, departing significantly from the typical form of *Liriomyza*, represented by *nietzkei*. It is therefore suggested that *nietzkei* is, or is most closely related to, the original population. Subsequent speciation has produced two further species, *chinensis* now occurring in China, Japan and Malaya and closely related to it, *cepae* which has spread into western Europe and is now reproductively isolated from *nietzkei*. The three species form a compact group, as indicated by the distinctively angulate third antennal segment.

ECONOMIC IMPORTANCE. As indicated above, the damage to the onion crop is apparently substantially less with this species than with *Liriomyza nietzkei* owing to the later appearance of both adults and larvae. More detailed studies will be necessary, however, to establish whether this is invariably the case.

Liriomyza chinensis (Kato, 1949), **comb.** and **stat. nov.** – China, Japan, Malaya

Dizygomyza cepae ssp. *chinensis* Kato, 1949: 12.

ADULT. Small species, with angulate third antennal segment, black scutellum,

bright yellow coxae and femora and pale wings.
HEAD AND MESONOTUM. As in *L. cepae*.
WING. Length from 1.6 mm in male to 1.9 mm in female, otherwise as in *L. cepae*.
COLOUR. Head, including face, yellow; hind-margin of eye either entirely yellow or small black area around base of outer vertical bristle but inner vertical clearly on yellow ground; humerus, notopleura and upper half of mesopleura diagonally yellow, with hind- and lower-margins black, sternopleura mainly black but broadly yellow along upper margin; mesonotum and scutellum greyish-black; legs: coxae and femora bright yellow, tibiae and tarsi darker, brownish;

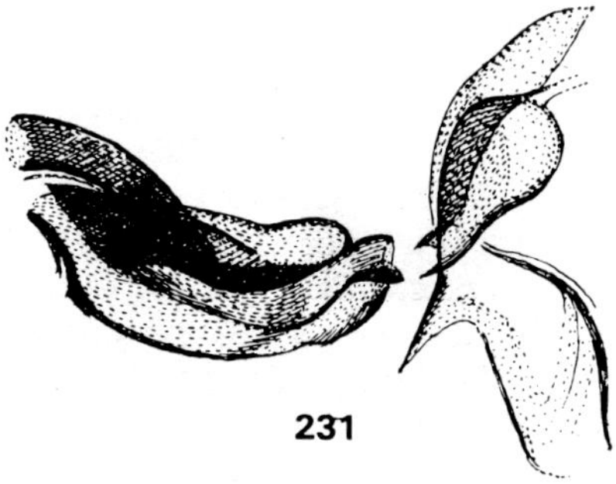
231

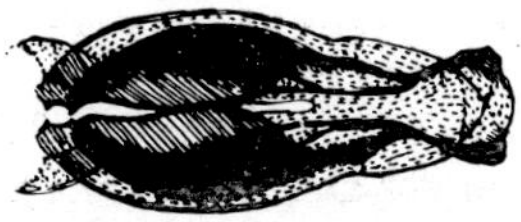
232

233

Figs. 231–233. *Liriomyza chinensis* (Kato): 231, aedeagus, side view; 232, distiphallus, ventral view; 233, larval mouth-hooks (after SASAKAWA).

abdomen black; wing as in *cepae* with lower half and vein M_{1+2} conspicuously pale; squamae and fringe yellow.
MALE GENITALIA. Aedeagus enlarged, as in Figs. 231, 232; also illustrated by SASAKAWA, 1961: Fig. 46d, as *Phytobia* (*Cephalomyza*) *cepae*.
EARLY STAGES. Larval mouth-hooks according to SASAKAWA (Fig. 233) each with 2 teeth, not alternating. HERING (1957b) examined larvae collected by BUHR on Allium at Peking on 12 August, 1956, and found the same spiracular arrangement as in *cepae*, with the anterior spiracles having 7–8 pores and the posterior 8–10. SASAKAWA's illustrations of material from Japan largely confirm this, showing only a single additional pore on both the anterior and posterior spiracles. The puparia of specimens from Singapore and Johore in Malaya are yellowish-brown.
HOST-PLANTS. *Allium cepa* L.; in Japan also *A. bakeri* Regal, *A. fistulosum* L. and *A. odorum* L. (SASAKAWA, 1961); in China 'Spring Onion' (KATO, 1949).

BIOLOGY. No details have been recorded but the life history of this species as a leaf-miner on onions is no doubt very similar to that of *Liriomyza cepae* and *L. nietzkei*. The apparently intermittent feeding by the larva is discussed on p. 168 (see under *nietzkei* below). SASAKAWA (1961) gives dates for adults and also immature stages as 4–6 July at Shimogamo and Kyoto and 20 October at Fukuoka.
DAMAGE CAUSED. KATO (1949) treats this species as a pest of onions in Inner Mongolia and HERING (1957b) states that it causes damage in Japan and Singapore but no detailed information is available.
DISTRIBUTION. China: Inner Mongolia, Peking; Japan: Honshu, Kyoto, Fukuoka, Shimogamo; Malaya: Johore, Singapore.

Discussion

TAXONOMY. With the existing state of knowledge KATO was entirely justified in treating this species merely as a sub-species of *cepae* Hering. He particularly noted the paler colour of the coxae and femora. However, the hind-margin of the eye is also paler and the species is substantially smaller. The male genitalia are entirely distinct but clearly indicate a closer relationship with *cepae* than with *nietzkei*. SASAKAWA's (1961) illustration of the aedeagus of a specimen from Japan agrees exactly with that of a male from Singapore.
DISTRIBUTION. All known records of this species are from eastern Asia. DOVNAR-ZAPOLSKI (1969) records *cepae* mining Allium in the Kirghiz Republic of the U.S.S.R. but unfortunately no adults are available. For the time being it must remain in doubt whether the species is *chinensis* or, more probably, one of the two species present in western Europe.
ECONOMIC IMPORTANCE. No assessment of the importance of this species is possible on present information (see above under Damage caused).

LILIACEAE: ALLIUM CEPA L., A. PORRUM L.

Liriomyza nietzkei sp.n. – Germany

Small species, with angulate third antennal segment, partially yellow scutellum and yellow femora and pleura.
DESCRIPTION OF ADULT.
HEAD. Generally as in *L. cepae*; frons conspicuously projecting above eye in profile; normally 1 reclinate ors and 3, 4 or sometimes more incurved ori; orbital setulae sparse, mainly upright; jowls deeply extended at rear; third antennal segment angulate but frequently more rounded at upper corner, less obviously pointed (Fig. 234).
WING. Length from 1.6 mm in male to 1.9 mm in female, costa extending strongly to vein M_{1+2}, discal cell small, last section of M_{3+4} $2\frac{1}{2}$–3 times penultimate.

COLOUR. Frons, antennae, palps yellow, face predominantly yellow, sometimes slightly greyish; hind-margin of eye generally dark at vertex but in palest specimens at least inner vertical bristle may be on yellow ground; mesonotum mat blackish-grey, scutellum either broadly yellow centrally or at least with a yellow line; pleura: humerus, notopleura and upper third of mesopleura bright yellow; legs: coxae, femora bright yellow, latter sometimes with faint brownish striations, tibiae and tarsi darker brown, at least on hind-legs; wing hyaline, all veins dark; abdomen black; squamae and fringe yellow.

MALE GENITALIA. Aedeagus as in Figs. 235, 236.

HOLOTYPE ♂, Germany, Pfalz: Landau, 16.vii.31, ex *Allium cepa*; paratypes 1♂, 2♀♀, 5–16.vii.31, otherwise same data; Thuringia, Aschersleben, 3♂♂, 1.vi., 2.vii.60 and 1.iv.1961, 1♀, 5.iv.1961, ex *Allium porrum* (H. W. NOLTE); 1♂, 1♀, 4. and 6.viii.60, otherwise same data (H. J. MÜLLER); Neuss, 1♀, 11.vii.57 (W. BERG); Halle (Saale), 1♂, 31.iii.1961, ex *A. porrum*; Thuringia, Mühlhausen, 1♂, 3♀, 30 and 31.viii.60, ex *A. porrum* (all H. BUHR). Holotype and paratypes in author's collection, paratypes presented to British Museum (Natural History).

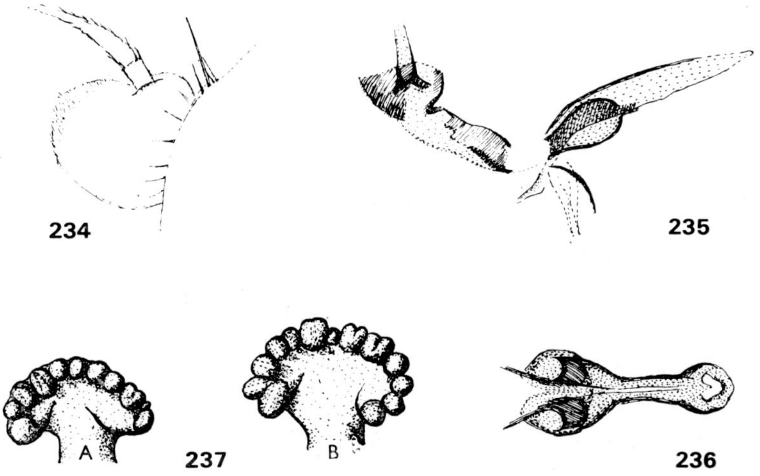

Figs. 234–237. *Liriomyza nietzkei* sp.n.: 234, third antennal segment; 235, aedeagus, side view; 236, distiphallus, ventral view; 237A, larval anterior spiracles; 237B, posterior spiracles.

EARLY STAGES. The egg is yellowish-white, without any sculpturing of the chorion; its average length is 0.31 mm. According to HERING (1956) the anterior spiracles of the larva (Fig. 237A) each have between 9–13 pores, the posterior spiracles 14–15 (Fig. 237B). The reddish-brown puparium is about 2 mm long.

HOST-PLANTS. *Allium cepa* L. (Onion); also *A. porrum* L. (Leek), possibly *A. montanum* F. W. Schmidt (=*A. senescens* auct.).

BIOLOGY. This species is a leaf-miner on *Allium*. In West Germany where NIETZKE conducted detailed investigations into its life history in 1938, adults began to emerge at the beginning of May, building up to a peak early in June, with a subsidiary peak towards the end of June. It was found that peak emergences correlated closely with periods of high temperature and low humidity. From checks in two different localities it was found that males made up only slightly more than 8% of the population. Life of the adult flies varied between 1 and 2 weeks. Prior to oviposition the females make numerous feeding punctures, mainly in the upper part of the leaves. Oviposition can take place in any part of the leaf but the majority of eggs are laid towards the base. During the period of peak emergence as many as 1390 eggs were found in a sample of 100 leaves.

There is still some uncertainty about the exact way the larva feeds. Initially a fine linear mine is formed and later the mine often shows a series of breaks. NIETZKE states that the larva leaves the mine, moves internally within the hollow, cylindrical leaf and then bores its way into the leaf tissue at another place to continue feeding. HERING (1957c), however, states that the breaks in the mine visible from the outside of the leaf merely indicate where the larva has changed its feeding path to the inner surface of the leaf, thus making a continuous mine but alternating the leaf surface where feeding takes place.

The larvae feed downwards to the base of the leaves and, when full-grown, cut their way out and fall to the ground. Larvae of this species, like many in this genus, can 'jump' by coiling themselves up and by quickly releasing the tension throw themselves several centimetres through the air. This procedure takes place several times prior to pupation. They they burrow into the ground and NIETZKE found the optimum depth for pupation to be 5–9 cm, with a few even as deep as 12 cm below the surface.

NIETZKE (1953) reports that there is only a single generation and in some cases it was even found that diapause was not broken until the end of the second winter.

DAMAGE CAUSED. Some damage to young seedlings can be caused when numerous feeding punctures are made by the females. The assimilation process can be interrupted and there is excessive loss of water, resulting in wilting which is particularly noticeable in hot weather. However, in damp weather additional young leaves can be produced in as little as 3 or 4 days. If further feeding punctures are made in these leaves, the seedlings' power of recovery is exhausted and they then invariably die.

The really serious damage is caused when numerous larvae are present in a leaf. Their combined feeding destroys a large quantity of parenchyma and vascular bundles are cut, impeding the interchange of nutrients and products of assimilation. Premature wilting of the leaves occurs and growth of the onion is affected. With a heavy infestation the onion harvested may be little larger than when it was originally planted out.

DISTRIBUTION. West Germany: Pfalz, Neuss; East Germany: Thuringia, Mühlhausen, Aschersleben, Halle.

PARASITES. *Halticoptera patellana* Dalm. (Chalcididae); *Microstilba bidentata* Först. (Cynipoidea).

Discussion

TAXONOMY. NIETZKE (1943) recorded a number of differences between the two populations he studied, treating them as distinct races. I am satisfied that his 'Herxheim race' represents a separate species and I have pleasure in naming it in his honour.

Liriomyza nietzkei is not only smaller than *cepae* but is substantially paler, with yellow femora and a partially yellow scutellum; however, it lacks the pale wings and veins of *cepae*. The male genitalia are significantly different from those of *cepae*, and there are also consistent differences in the form of the larval spiracles.

HOST-RANGE. This species occurs not uncommonly on both *A. cepa* and *A. porrum*. Larvae have also been found on wild *Allium montanum* (=*A. senescens*) in Thuringia (H. BUHR). HERING suggests that this choice of host is probably secondary; he also found possibly significant differences in the larval spiracles (1956: 260 and Fig. 3). The correct status of this population thus remains to be established.

DISTRIBUTION. The only localities where this species is known are in West and East Germany. It is almost certainly more widely distributed in western Europe and it will be of considerable interest to work out its exact range, particularly in relation to *L. cepae*.

ECONOMIC IMPORTANCE. The main area where *Liriomyza nietzkei* has caused serious losses to cultivated onions is in the Pfalz, west of the Rhine. NIETZKE (1953) found that in 1938 of 51 fields examined there was either a serious reduction in or a total loss of the crop in 53% of the area. This species was therefore at that time a pest of the utmost importance. It is not clear to what extent more recent outbreaks in East Germany refer to this species rather than to the true *cepae*.

LILIACEAE: ASPARAGUS OFFICINALIS L.

Ptochomyza asparagi Hering, 1942 – Europe, China

ADULT. Minute yellow species, wing length little more than 1 mm.

HEAD (Fig. 238). Frons conspicuously projecting above eye towards base of antennae; orbital bristles short, 1 ors, 4 or 5 ori, orbital setulae minute, proclinate; jowls deeply extended at rear, $\frac{2}{3}$ height of eye, eye slanting; third antennal segment small, round, arista short.

MESONOTUM. Up to 8 dc, becoming very short towards suture, no stronger than acrostichals, these sparse in 2 rows.

WING. (Fig. 239). Length from slightly less than 1 mm in male to 1.3 mm in largest females, veins R_{2+3} and R_{4+5} curving, second costal section unusually short, only slightly longer than fourth.

COLOUR. Frons and entire hind-margin of eye yellow, third antennal segment somewhat darker, orange-brown to greyish; mesonotum mat greyish-black but

broadly yellow centrally adjoining scutellum, the dark areas normally divided into 3 distinct bands; scutellum largely yellow, with small dark patches laterally; mesopleura entirely yellow; legs: coxae and femora bright yellow, tibiae and tarsi more brownish; squamae yellow, with dark margin and fringe.
MALE GENITALIA. Aedeagus as in Fig. 240, a simplified structure ending in a single slender tubule.
EARLY STAGES. The larvae are yellowish and from 1.5–2.5 mm in length. The mouth-hooks each have 2 teeth which do not alternate in side view. The puparium is slender, basically yellow but variably darkened, up to 2 mm long; the anterior spiracles are on short, diverging projections, each with about 10 minute pores,

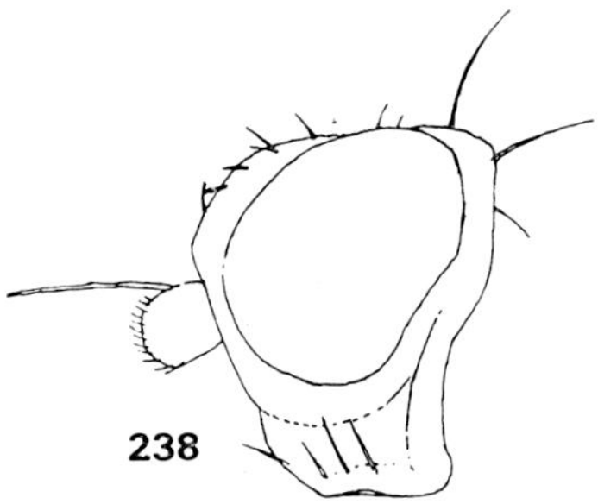

238

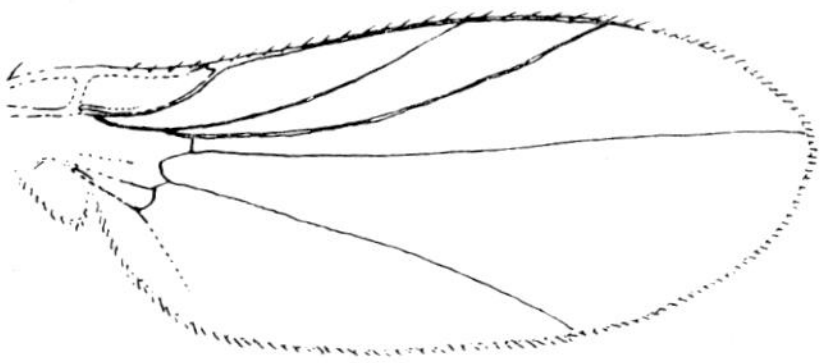

239

240

Figs. 238–240. *Ptochomyza asparagi* Hering: 238, head; 239, wing; 240, aedeagus.

heavily chitinized at end; the posterior spiracles are conspicuously diverging, distinctly pointed, each with 6–10 minute pores (from DE MEIJERE, 1943).
HOST-PLANT. *Asparagus officinalis* L.
BIOLOGY. The larvae mine predominantly in the fine leaves, which are completely hollowed out. Pupation takes place within the leaf, with the anterior spiracles projecting through the epidermis. More rarely mines are found below the epidermis in the main stem.
DAMAGE CAUSED. The entire assimilation parenchyma in the mined area is destroyed by the larval feeding. Wherever the species has been found, populations have been large and, despite the small size of the individual larva, the combined effect of very large numbers must have some effect on the growth of young plants.
DISTRIBUTION. Germany: Brandenburg, Mecklenburg, Saxony, Thuringia; China: Peking (BUHR, 1960).
PARASITES. None recorded.

Discussion

TAXONOMY. The main character on which HERING based his decision to erect a new genus for this species was the presence of only a single notopleural bristle at the centre of the lower margin of the notopleural triangle, while in the closely related genus *Phytomyza* there are 2 bristles, one at each corner. However, this is not a valid generic character, as two further species subsequently placed in this genus, *Pt. asparagivora* Spencer, 1964b and *Pt. czernyi* (STROBL, 1909), cf. SPENCER, 1966c, each have the normal arrangement of 2 notopleural bristles found in *Phytomyza*. Nevertheless, it is considered justified to retain *Ptochomyza* as a separate genus, in view of the distinctive characters of the larva, puparium, male genitalia and the characteristic form of the head and wing.
DISTRIBUTION. BUHR (1960) found the species to occur in large numbers in many localities in northern and eastern Germany and it seems probable that it has merely been overlooked elsewhere in Europe. BUHR visited China to attend the 8th International Plant Protection Conference in Peking in 1956 and among the flowers presented to the East German Delegation at the airport were leaves of asparagus which to his surprise contained mines of *Pt. asparagi*. *Asparagus* decorating the hotel rooms contained similar mines (BUHR, 1960). It thus seems certain that this species occurs throughout the Palaearctic Region.

The related species, *Pt. asparagivora*, also has a wide distribution and has been recorded from southern Europe, Ethiopia and Rawalpindi in Pakistan (SPENCER, 1964b and 1969b).
ECONOMIC IMPORTANCE. HERING (1942) considered that the damage caused by *Pt. asparagi* should not be underestimated in view of the effect on the assimilation of the plant of the destruction of parenchyma by large numbers of larvae. However, this cannot be considered a significant pest and it is obviously less important than the other widespread species attacking cultivated asparagus, *Ophiomyia simplex* (Loew).

ROSACEAE: FRAGARIA

Agromyza fragariae Malloch, 1913a Canada, United States

ADULT. Small species, with brownish frons and 3 + 1 dorso-centrals. This species agrees closely with *Agromyza spiraeae* (see below). Apart from the more brownish frons, the mesonotum is more obviously grey and the legs are almost entirely black, less yellowish.
MALE GENITALIA. Aedeagus as in Figs. 241, 242; ejaculatory apodeme greatly enlarged (Fig. 243).
EARLY STAGES. No detailed information is available on specimens feeding on *Fragaria*. The puparium is reddish-brown and the posterior spiracles each have 3 well-defined pores.

Host-plants. *Fragaria vesca* L., *F. virginiana* Duchesne (Strawberry); possibly also *Rubus occidentalis* L. (Black-Raspberry, Thimbleberry), *R. odoratus* L. (Flowering Raspberry).
Biology. The leaf-mine on *Fragaria* has not been described. The only mine I have seen is on *Fragaria vesca*, collected in the Canadian Rockies at Blairmore, Alta., 26 June, 1966 (K.A.S.). This is broad, almost blotch-like (Fig. 244).

Frost (1924) identified as *fragariae* the species which is common in Ontario, Canada and at Ithaca, New York, forming entirely linear mines on *Rubus* species

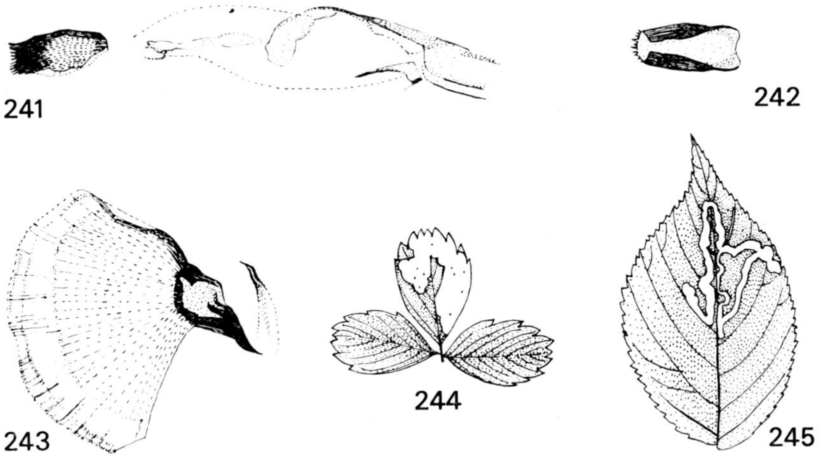

Figs. 241–245. *Agromyza fragariae* Mall.: 241, aedeagus, side view; 242, distiphallus, ventral view; 243, ejaculatory apodeme ($\frac{1}{2}$ scale of Fig. 249); 244, leaf-mine on *Fragaria*; 245, leaf-mine on *Rubus occidentalis* L. (Canada).

and described both the larva and leaf-mine. However, it is by no means certain that the species concerned is identical with *fragariae*. A mine of this type on *Rubus occidentalis* L. from Toronto, Rosedale, 11 October, 1965 (K.A.S.), is shown in Fig. 245.
Damage caused. Normal leaf-mining damage is caused by larval feeding. There is no information to suggest that populations have ever been other than small and there are no records of strawberries in North America having suffered damage.
Distribution. United States: California; Canada: Alberta, Quebec.
Parasites. None recorded.

Discussion

Taxonomy. A number of species in this group can only be satisfactorily identified by the male genitalia (Spencer, 1969a). In *Agromyza fragariae* the frons is definitely brown rather than black and in the limited material which has been seen this appears to be a reliable character for differentiating this species.

The only mine seen is distinctly blotch-like rather than linear and therefore without confirmation from male genitalia it cannot be assumed that the species forming linear mines on *Rubus* species in Canada and the United States is *fragariae*.

ECONOMIC IMPORTANCE. It is believed that insignificant damage is caused to cultivated strawberries. As with the Holarctic species, *Agromyza spiraeae*, population levels are apparently consistently small.

ROSACEAE: FRAGARIA, RUBUS IDAEUS L.

Agromyza spiraeae Kaltenbach, 1867 – Europe, Western U.S.S.R., Canada

ADULT. Small dark species, with black frons, mat greyish-black mesonotum and 3 + 1 dorso-centrals.

HEAD (Fig. 246). Frons usually slightly projecting above eye in profile; 2 equal ors, 2 ori, the lower weaker and incurved; orbital setulae sparse, reclinate; jowls $\frac{1}{4}$–$\frac{1}{5}$ vertical height of eye, deepest at rear; third antennal segment small, rounded, with conspicuous fringe of pubescence at upper corner.

MESONOTUM. 3 + 1 strong dc, the fourth only slightly shorter than the third; acrostichals coarse, in some 5 or 6 rows in front, a few hairs extending almost to level of first dc; pre-scutellars well-developed.

LEGS. Femora normally with developed lateral bristle but rarely a weak one present.

WING. Length from 2 mm in male to 2.6 mm in female, costa extending to vein M_{1+2}, last section of M_{3+4} variable, from $\frac{2}{3}$–$\frac{4}{5}$ penultimate, first cross-vein at mid-point of discal cell, rarely more basad.

COLOUR. Frons sooty-black in front, invariably more brownish behind, all antennal segments black; mesonotum mat, greyish-black; femora black but knees and also tibiae and tarsi more yellowish; squamae pale or darker grey, fringe brownish to black.

MALE GENITALIA. Aedeagus as in Figs. 247, 248; ejaculatory apodeme small (Fig. 249).

EARLY STAGES. The two larval mandibles are equal and exactly coincide, and each have two strong, blunt teeth (Fig. 250); the anterior spiracles each have about 9 pores, the posterior spiracles 3. The puparium is yellowish-brown.

HOST-PLANTS. *Fragaria ananassa* Duchesne, *F. vesca* L. and certainly other *Fragaria* species, *Rubus idaeus* L. and other *Rubus* species; among ornamentals *Geum* and *Potentilla*; also many other wild rosaceous genera, such as *Agrimonia*, *Alchemilla*, *Comarum*, *Filipendula* and *Sanguisorba*.

BIOLOGY. The larva initially forms a narrow linear mine, which after the first moult immediately develops into a somewhat elongate, primary blotch (Figs. 251, 252). Mines appear from the end of June into July and there is a second generation in the autumn. The larva leaves the mine to pupate on the ground.

DAMAGE CAUSED. Each larva consumes a considerable amount of leaf tissue and if numerous larvae are feeding together the growth of strawberry plants can be affected.

DISTRIBUTION. A widespread species occurring virtually throughout Europe, including Estonia (ELBERG, 1964) and the Moscow area of the U.S.S.R.

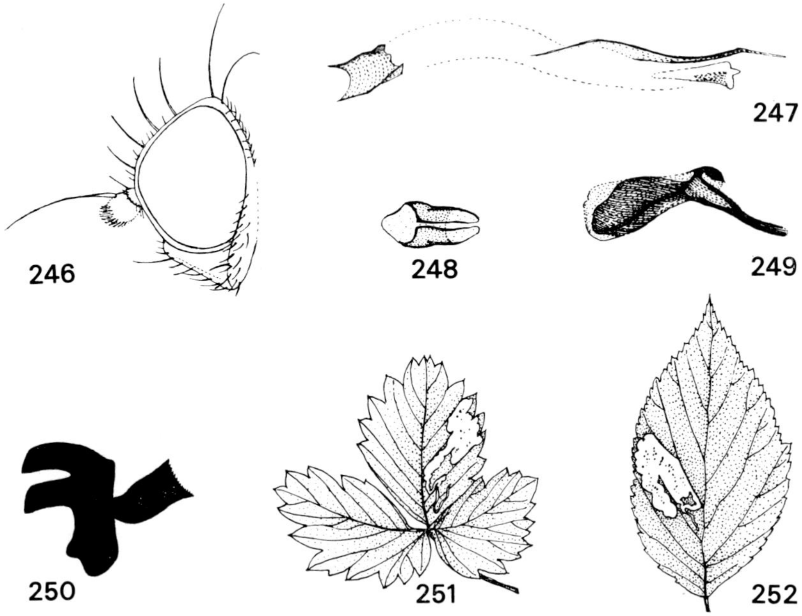

Figs. 246–252. *Agromyza spiraeae* Kalt.: 246, head; 247, aedeagus, side view; 248, distiphallus, ventral view; 249, ejaculatory apodeme (2× scale of *fragariae*, Fig. 243); 250, larval mouth-hooks (Hering); 251, leaf-mine on *Fragaria*; 252, mine on *Rubus idaeus*.

(ROHDENDORF-HOLMANOVÁ, 1958); Canada: British Columbia, Ontario, Quebec; not confirmed in United States.

PARASITES. Europe: *Exotela nowakowskii* Griffiths, *Dacnusa evadne* Nixon, *D. laeta* (Nixon), *Chorebus bres* (Nixon), *C. credne* (Nixon) (Dacnusinae); *Opius agromyzicola* Fischer, *O. cingulatus* Wesmael, *O. instabilis* Wesm., *O. pallipes* Wesm., *O. similis* Szépl., *O. singularis* Wesm.; *Stictomischus groschkei* Delucchi (Petromalidae); *Pnigalio pectinicornis* L. (Eulophidae).

Discussion

TAXONOMY. The adult of *Agromyza spiraeae* is not readily distinguishable from *A. alnibetulae* Hendel feeding on *Betula* but the characteristic leaf-mines of

spiraeae on genera of Rosaceae permit immediate identification in the larval stage.
ECONOMIC IMPORTANCE. Although *A. spiraeae* has a wide distribution, populations are invariably small, possibly in consequence of the high rate of parasitism to which it is subject. However, in 1962 damage was caused to the strawberry crop at Gloucester in the West of England, when it was reported that 'the leaves were ruined' (M.A.F.F., unpublished report).

The leaf-mines on *Rubus* can scarcely have any noticeable effect on such a robust plant.

Agromyza spiraeae is thus not an important pest but minor damage to strawberries may occur whenever circumstances result in a significant build-up in local population levels.

SOLANACEAE: CAPSICUM ANNUUM L.

Haplomyza capsici sp.n. – Argentina

HEAD (Fig. 253). Orbits well differentiated but not projecting above frons; 1 reclinate ors, 3 incurved ori; orbital setulae sparse, reclinate behind, a few in front incurved and even proclinate; eye large, only slightly slanting, jowls deepest at rear; third antennal segment small, round, arista fine, short.
MESONOTUM. 3 + 1, acr irregularly in 3 or 4 rows.
WING. Length 2 mm, second cross-vein lacking.
COLOUR. Frons and orbits yellow, the latter somewhat paler; hind-margin of eye entirely yellow; jowls, face and third antennal segment bright yellow; mesonotum mat, blackish but variable, either entirely dark or largely yellowish behind first dc to margin of scutellum, there yellow, with only small black triangles at sides; pleura yellow, mesopleura only faintly greyish below; abdomen black centrally but in paler specimens broadly yellow laterally.
MALE GENITALIA. Ninth sternite unusually long and narrow, equal in length to aedeagal apodeme; surstyli broadly divided from epandrium, narrow, bearing 2 short spines (Fig. 254); ejaculatory apodeme small, linear (Fig. 255); aedeagus essentially as in *H. tigrensis* Spencer, 1963a: Fig. 96B.
EARLY STAGES. Puparium orange-yellow, posterior spiracles each with an ellipse of some 8 minute pores.
HOST-PLANT/BIOLOGY. *Capsicum annuum* L.; oviposition takes place on the upper leaf surface and the larva initially forms a linear mine, which later develops into an irregular greenish blotch (Fig. 256); several larvae often occur in the same leaf and appear to feed together, due to their mines coalescing (P. WYGODZINSKY, personal communication).
HOLOTYPE ♂, Argentina, Tucuman, between Choromora and La Higuera, 2.v.1954; paratypes: 1♂, 5♀, same data; 1♂, 1♀, Tucuman City, 30.iii.1953 (all P. WYGODZINSKY). Holotype and paratypes in American Museum of Natural History, New York, 2 paratypes in author's collection.

Discussion

TAXONOMY. This species differs from *H. tigrensis* Spencer from Buenos Aires by the blacker mesonotum. Although the aedeagus of the two species appears similar, this is relatively undeveloped in many species in this genus. It seems probable that *H. capsici* is more closely related to two species from Brazil, *H. bullati* Spencer, 1963a and *H. verbascifolii* Spencer, 1963a, feeding on two different

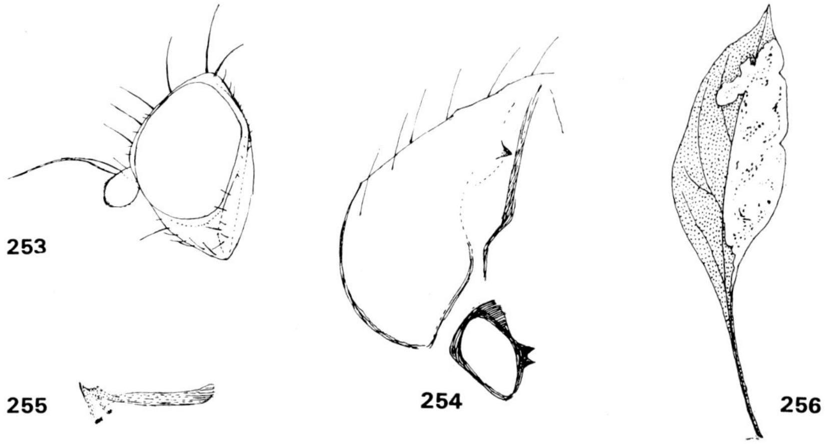

Figs. 253–256. *Haplomyza capsici* sp.n.: 253, head; 254, surstylus; 255, ejaculatory apodeme; 256, leaf-mine on *Capsicum*.

species of *Solanum*. A further species on *Solanum umbelliferum* Eschk. in California is being described by Mr. G. STEYSKAL, Washington, who is preparing a revision of the genus.

ECONOMIC IMPORTANCE. Red pepper is a plant very widely grown in the Tucuman area of Argentina and is economically quite important. The leaf-mining infestation was in some cases quite heavy, with 80% of all plants in some fields attacked and with each plant having an average of 10% of leaves mined (P. WYGODZINSKY, personal communication).

SOLANACEAE: SOLANUM TUBEROSUM L.

*Liriomyza quadrata (Malloch, 1934a) – South America

ADULT. Large, conspicuously yellow species, with yellow mesonotal patch adjoining scutellum.

* See also Appendix A.

HEAD. 4 strong orbital bristles, lower ori somewhat incurved; orbital setulae minute, sparse, reclinate; jowls deep, between $\frac{1}{3}$ and $\frac{1}{4}$ vertical height of eye; third antennal segment small, rounded, arista gradually tapering, only finely pubescent.
MESONOTUM. 3 + 1 strong dc, acrostichals in 4 rows in front, reduced to isolated hairs on yellow patch behind, these frequently incurved rather than reclinate.
WING. Length from 2.3 mm in male to 3.2 mm in female, discal cell large, last section of M_{3+4} only $1\frac{1}{2}$ times length of penultimate.
COLOUR. Frons, orbits, antennae, palps bright yellow; hind-margin of eye black but in palest specimens both vertical bristles on yellow ground, in darker specimens on dark ground; mesonotum moderately shining black with variable yellow patch centrally adjoining scutellum (Fig. 257); mesopleura bright yellow with only small black patch on lower margin but in darker specimens lower half largely black; legs with coxae and femora bright yellow, latter at most with faint brownish striations, tibia and tarsi yellowish or darker brown; squamae yellow, margin and fringe black.
MALE GENITALIA. Aedeagus as in Figs. 258, 259; surstyli with 2 or 3 stout bristles and some longer hairs (Fig. 260).
EARLY STAGES. Puparium pale yellowish, posterior spiracles each with 3 bulbs.
HOST-PLANTS. *Solanum tuberosum* L. (potato), *Lycopersicum esculentum* L. (tomato), *Nicotiana tabacum* L.; also *Solanum nigrum* L.
BIOLOGY. No detailed studies have been made of the life history of this species. The larva forms an upper surface leaf-mine (Fig. 261) which is linear in the first instar and then develops into a whitish blotch, with frass in conspicuous black grains.
DAMAGE CAUSED. The substantial leaf-mine produced by this large species causes appreciable damage to leaves and results in their almost total destruction when several larvae mine in a single leaf. However, this species has not so far been identified as causing significant damage to potatoes or tomatoes.
DISTRIBUTION. Chile, Argentina, Colombia, Ecuador, Peru, Venezuela.
PARASITES. *Opius chilenis* de Santis, *O. eurytenoides* de Santis.

Discussion

TAXONOMY. There has recently been uncertainty regarding the status of this species, in view of the small size – 2–2.25 mm – given by MALLOCH in his original description. I have now checked the four type specimens, all males, in the British Museum (Natural History) and found that the wing length is 2.8 mm in two specimens, 2.5 mm in one and 2.3 mm in the smallest. I have seen specimens from Chile bred from potato and from Argentina bred from tomato with a wing length of 2.3 mm; a female from Ecuador from potato has a wing length of 3 mm and in a series I obtained from *Solanum nigrum* at Bogotá, Colombia, the largest female has a wing length of 3.2 mm.

The variation in colour of *Liriomyza quadrata* affects not only the upper orbits and mesopleura but also the yellow patch on the mesonotum can be larger or smaller than that shown in Fig. 257. However, in all specimens examined the male genitalia are identical. With a species having such a wide range and occurring at such varying altitudes, some local variation is not surprising.

ECONOMIC IMPORTANCE. This species must be considered as a potentially serious pest. It has been recorded in Chile on tomatoes (PORTER, 1936) and on potatoes (PORTER, 1938). I have seen specimens from Tucumán, Argentina, leg. WYGODZINSKY, bred from tomato, from Valparaiso, Chile, and Pichincha, Ecuador, bred from potato. It is reported to be a serious pest in the Peruvian Andes on potato and secondarily on tomato (KORYTKOWSKI, personal communication). Its occurrence on *Solanum nigrum* and probably other wild species of

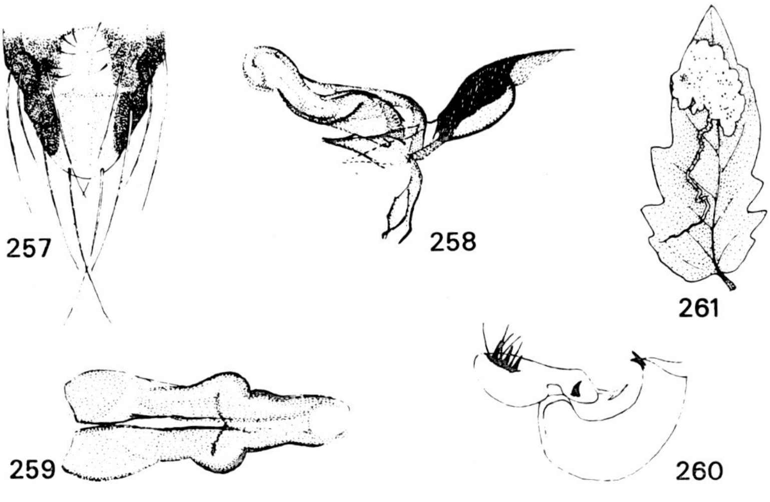

Figs. 257–261. *Liriomyza quadrata* (Mall.): 257, mesonotum; 258, aedeagus, side view; 259, distiphallus, ventral view (both holotype); 260, surstylus; 261, leaf-mine on tomato.

Solanaceae provides reservoirs of population from which it can transfer to cultivated potatoes and tomatoes. Some of the damage to tomatoes in Venezuela ascribed to *Liriomyza munda* may in fact be caused by *quadrata*. Further detailed records of distribution and damage caused are therefore obviously desirable.

UMBELLIFERAE: DAUCUS CAROTA L.

Phytomyza ferulae Hering, 1927c—S. Europe

ADULT. Small species, wing length 2 mm; frons broad, twice width of eye, bright yellow; 2 ors, the upper short or lacking; third antennal segment black, small,

round. Mesonotum mat grey, mesopleura similar but with trace of yellow at upper margin; humerus and notopleura greyish-yellow. Legs grey, knees yellowish.
MALE GENITALIA. Aedeagus as in Fig. 262.
EARLY STAGES. Puparium shining black, posterior spiracles each with an ellipse of some 15 minute pores.
HOST-PLANTS. Umbelliferae: *Daucus carota* L., also *Ferula communis* L., and probably other genera.
BIOLOGY. The larva forms an upper surface linear mine (Fig. 263), which may

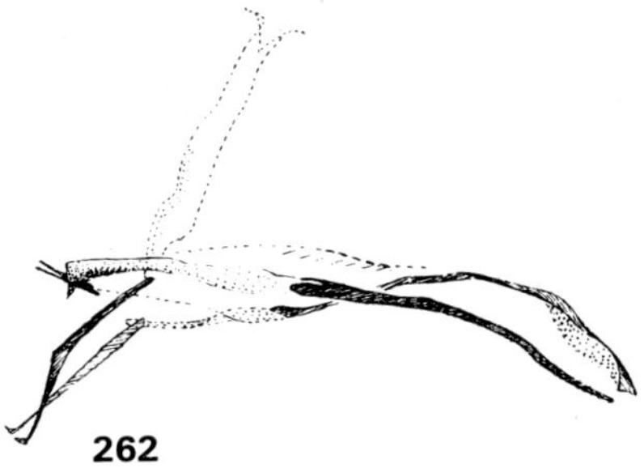

262

263

Figs. 262–263. *Phytomyza ferulae* Her.: 262, aedeagus; 263, leaf-mine on cultivated *Daucus carota*.

develop into a secondary blotch in small leaf sections; there are several generations during the winter and spring from November to May.
DISTRIBUTION. A Mediterranean species, known from Spain: Canary Is., Algeciras, Menorca; Yugoslavia, Sicily, Malta.
PARASITES. *Exotela cyclogaster umbellina* (Nixon); *Opius minor* Fischer.

Discussion

TAXONOMY. This species was originally described from the Canary Islands. Specimens bred from *Daucus carota* L. in southern Spain were described as *Phytomyza dauci* by SPENCER, 1957c and this species was later synonymized with *ferulae* by SPENCER, 1966g.
ECONOMIC IMPORTANCE. Leaf-mines can be very abundant on both wild and cultivated *Daucus carota* and damage might be caused to seedlings. A small patch of carrots in Malta was heavily attacked in March, 1971 (K.A.S.).

VII. A NEW GENUS EMBRACING LEAF-MINERS ON TEA, COFFEE AND OTHER CROPS BASED IN THE OLD WORLD TROPICS AND NORTHERN AUSTRALIA

The group of diminutive leaf-miners occurring widely in the Old World tropics and reaching Australia and southern Japan has hitherto been placed in the genus *Melanagromyza*. However, *Melanagromyza* species are robust and frequently large, and are feeders in stems, roots or seed-heads; the male genitalia and larval spiracles are characteristic and the genus was re-defined when the palaearctic species were revised by SPENCER (1966b). A new genus is therefore now proposed for this tropical and sub-tropical group of leaf-miners.

Tropicomyia gen. nov. (Greek: *tropikos* + *myia*)

ADULT. Small to minute, wing length from 1.2–2.3 mm; costa ending at vein R_{4+5} or extending to M_{1+2}; second cross-vein always present, position of first variable, near midpoint or towards distal end of discal cell (Fig. 265).
HEAD (Fig. 264). Frons not projecting above eye, normally 2 equal ors and 2 ori, the lower weaker and incurved; jowls frequently very narrow, never deep; third antennal segment small, round, arista generally long, somewhat pubescent.
MESONOTUM. 2 strong dorso-centrals, acrostichals numerous in 6–8 rows.
LEGS. Mid-tibia normally without lateral bristle but one weak bristle sometimes present in larger specimens.
COLOUR. Entirely black, mesonotum brilliantly shining or more mat, greyish- or brownish-black, abdomen black without any greenish colouration; squamae dark grey, fringe black in all known species but some with a white fringe may correctly belong here.
MALE GENITALIA. Aedeagus with basiphallus normally broadly U-shaped (Fig. 284), with the two side-arms extending each side of the distiphallus complex, rarely shorter (Fig. 267), well-defined ventral bladder leading to the distiphallus, with two symmetrical lateral processes of varying size and degree of chitinization; in most species a single chitinized process lies below the midpoint of the distiphallus at an angle of approximately 45° (Figs. 267, 276); surstyli projecting inwards, with some hairs but no bristles (Fig. 298); ninth sternite with broad side-arms, slightly extended apically.
EARLY STAGES. Mouth-parts of larva appearing 'serrated', with at least 5 teeth on upper mandible and up to 10 on lower (Figs. 269, 288); cephalo-pharyngeal skeleton with 2 fully developed upper arms (Fig. 269); anterior spiracles on long

stalks, each with up to 8 minute pores (Fig. 270), posterior spiracles with 3 larger pores. Puparium flattened, yellowish brown, with variable markings above or mat black; posterior spiracles arising from an elongate extension (Fig. 271).
BIOLOGY. All known species are leaf-miners. The main group (*Tropicomyia* s.s.) consists exclusively of upper surface epidermal feeders, producing a characteristic irregular silvery mine (Plates 6, 12), with the deeper parenchymatous tissues of the leaf not being eaten; pupation takes place in the leaf at the end of the mine, with the anterior spiracles projecting through the epidermis.

An intermediate group forms a shallow epidermal mine only in the first instar, subsequently modifying the feeding habit to form a normal upper surface linear mine.

In a further group (*Tropicomyia* s. lat.) the larva in all stages forms a deeper upper surface mine with well-defined frass deposits (Plate 5).

Type of genus: *Melanagromyza flacourtiae* Séguy, 1951 – Madagascar, Africa.

It would go beyond the scope of this paper to include here a complete revision of all species now transferred to this genus. However, all material available to me has been re-examined and, in the light of the additional specimens studied, the status of all species has been re-assessed.

The male genitalia and also the larval characters of the *pisi* (Kleinschmidt) group from Australia (Figs. 280, 281) and of *capeneri* (Hering) and some other species in Africa (Figs. 272, 273) diverge considerably from those of the epidermal miners which form the primary group in this new genus. When further material becomes available these groups may be properly given at least separate subgeneric status. At the present time, however, it is proposed to retain them in *Tropicomyia* s. lat. Twenty-four species are now transferred to *Tropicomyia*, as listed below. Of these, ten feed regularly on cultivated plants and can be considered as of potential economic significance.

It is now accepted that the five highly polyphagous species, *T. atomella* (Malloch), *flacourtiae* (Séguy), *polyphaga* (Spencer), *polyphyta* (Kleinschmidt) and *styricicola* (Sasakawa) are distinct. External differences are in some cases slight but differences in the male genitalia appear to be constant. The species feeding on tea in Ceylon, referable to *theae* (Cotes), has been found to be distinct from that on tea in northern India and Japan. The status of *coffeae* (Koningsberger) cannot be clarified in the absence of material from the type locality, Java.

Species of potential economic importance

Tropicomyia atomella (Malloch), **comb. nov.** – Formosa, India, Ceylon.

Agromyza atomella Malloch, 1914. Holotype ♀ from Formosa in Hungarian Natural History Museum, Budapest.
Melanagromyza atomella, Hennig, 1941; Spencer, 1961a: 67; 1966a: 7 (as *theae* Green); Singh and Beri, 1971: 245.

ADULT. Minute black species, mesonotum only moderately shining; head (Fig. 264) with frons not projecting and narrow jowls; wing length from 1.2–1.5 mm, costa extending to vein M_{1+2}, last section of M_{3+4} somewhat shorter than penultimate (Fig. 265).
MALE GENITALIA. Aedeagus as in Figs. 266, 267.
EARLY STAGES. Larva (Fig. 268) with serrated mouth-hooks (Fig. 269); anterior spiracles on long stalks, each with 7 or 8 minute pores (Fig. 270); posterior spiracles with 3; cephalo-pharyngeal skeleton with 2 developed upper arms (Fig. 269) puparium yellowish-brown (Fig. 270), measuring 1.8 × 0.8 mm (SINGH and BERI, 1971).
HOST-PLANTS. The main cultivated plant attacked in India is *Camellia sinensis* (L.) Kuntze. The following genera and families are also known as hosts: *Nerium* (Apocynaceae), *Daemium* (Asclepiadaceae), *Combretum*, *Quisqualis* (Combretaceae), *Flacourtia* (Flacourtiaceae), *Albizzia* (Leguminosae), *Gloriosa* (Liliaceae), *Sida* (Malvaceae), *Barringtonia* (Myrtaceae), *Bougainvillea* (Nyctaginaceae), *Passiflora* (Passifloraceae), *Murraya* (Rutaceae), *Wightia* (Scrophulariaceae).
BIOLOGY. Oviposition takes place on the upper leaf surface and the larva forms an irregular, epidermal mine, pupating in the leaf.
DISTRIBUTION. Formosa, India, Ceylon; probably Thailand.
PARASITES. No species have been identified but SINGH and BERI report that 'in the late breeding season the incidence of parasitism increases considerably and Eurytomid (Chalcidoidea) parasites emerge in large numbers'.

Discussion

TAXONOMY. It has previously been thought that this species was identical with *polyphyta* Kleinschmidt in Australia and with *theae* (Cotes), described from Ceylon. However, differences in male genitalia confirm that all three represent distinct species. *T. atomella* is a particularly small species; the wing length of the female holotype is 1.5 mm.

The holotype is from Pilam, Formosa, August, 1912 (not a male as stated by MALLOCH). Paratypes are mentioned from Taiwan, Takao, Pilam and Anping. Part of the type series was deposited in the Hungarian National Museum, Budapest and others in the D.E.I., Berlin. HENNIG (1941) listed the Berlin specimens and refers to a male from the locality of the holotype, Pilam, but unfortunately this cannot now be traced. Three other specimens represent *Mel. conspicua* Spencer, 1961b. It is hoped that the Pilam male will in due course be found, as examination of the male genitalia is essential to establish a positive identification in this group. In the meantime the species which is common in northern India is accepted as *atomella*. This has been discussed and well illustrated by SINGH and BERI (1971).

SINGH and BERI (1971) record slight differences between *T. atomella* and a species from Assam identified as *theae* Green. The species on tea in Assam is clearly distinct from the true *theae* from Ceylon. The large population of *atomella* present in Bihar feeds equally frequently on tea as on numerous other hosts and it

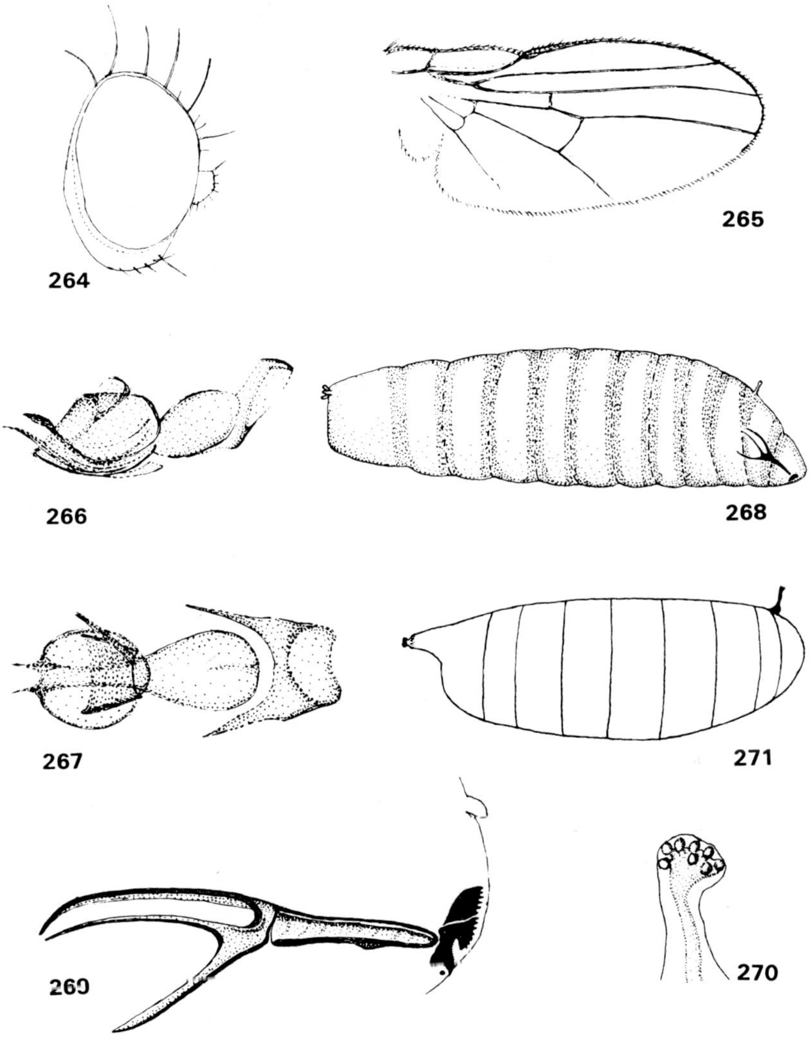

Figs. 264–271. *Tropicomyia atomella* (Mall.): 264, head (holotype, third antennal segment missing); 265, wing (holotype); 266, aedeagus, side view, Namkum, India ex tea; 267, same, ventral view, Colombo, Ceylon ex *Passiflora*; 268, larva; 269, mouth-hooks and cephalo-pharyngeal skeleton; 270, anterior spiracle of larva (Figs. 268–270 after Singh and Beri); 271, puparium.

seems possible that the minor differences detected in the Assam population on tea are not of specific significance.
ECONOMIC IMPORTANCE. Tea is the only host among cultivated plants which is regularly attacked by this species but no significant damage is known to have been caused.

Tropicomyia capeneri (Hering), **comb. nov.** and **stat. rev.** – Africa

Melanagromyza capeneri Hering, 1957a. Holotype ♂ in Transvaal Museum, Pretoria.
ADULT. Small shining black species; wing length from 1.3–1.6 mm, costa extending strongly to vein M_{1+2}.
MALE GENITALIA. Aedeagus as in Figs. 272, 273.

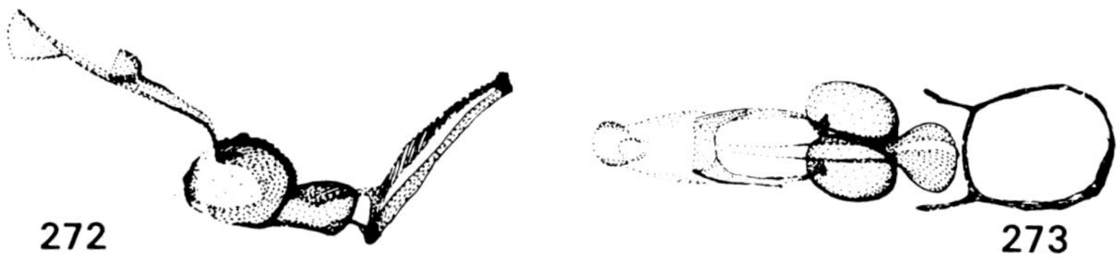

Figs. 272–273. *Tropicomyia capeneri* (Her.): 272, aedeagus, side view; 273, same, ventral view.

BIOLOGY. The larva forms an upper surface linear mine, beginning with a reddish spiral (HERING, 1957a: Fig. 9); frass is deposited in large black grains. Pupation takes place in the leaf at the end of the mine.
HOST-PLANT. The host from which the type was bred in Natal was not identified. However, the aedeagus of the holotype is identical with that of specimens bred from *Vigna unguiculata* (L.) Walp. in Senegal. This is the only known host.
DISTRIBUTION. Natal, Senegal.

Discussion

TAXONOMY. The series from Senegal bred from *Vigna* was originally misidentified as *vigneae* Séguy (SPENCER, 1959). After it was discovered that this series was identical with *capeneri* Hering, *capeneri* was synonymized with *vigneae* (Spencer, 1964b: 29). Now that the true identity of *vigneae* has been established (see p.198 below), this synonymy can be withdrawn.
ECONOMIC IMPORTANCE. The species has been present in large numbers on cultivated *Vigna* in Senegal but there is no information regarding damage caused to the plants by the larval leaf-mines.

Tropicomyia flacourtiae (Séguy), **comb. nov.** – Africa, Madagascar

Melanagromyza flacourthiae Séguy, 1951: 320. Holotype ♂ in Museum National d'Histoire Naturelle, Paris.

Melanagromyza dioscoreae Séguy, 1951: 319, **syn. nov.** Holotype ♂ in Museum National d'Histoire Naturelle, Paris.

Melanagromyza coffeae Hering, 1940 (nec *coffeae* Koningsberger, 1897), **syn. nov.** Holotype ♂ from Tanganyika in B.M.

Melanagromyza passiflorae Hering, 1957a: 72, **syn. nov.** Holotype ♀ from South Africa in Transvaal Museum, Pretoria.

Melanagromyza citri Spencer, 1959: 264, **syn. nov.** Holotype ♀ from Sierra Leone in B.M.

Melanagromyza gossypii Spencer, 1959: 273, **syn. nov.** Holotype ♀ from Nigeria in B.M.

ADULT. Minute, shining black species.

WING (Fig. 274). Length from 1.9 mm in male to 2.1 mm in female; costa ending shortly after vein R_{4+5}.

MALE GENITALIA. Aedeagus as in Figs. 275, 276.

EARLY STAGES. Larva: mandibles each with many minute teeth; anterior spiracles elongate, each with 8 pores, posterior each with 3, closely adjoining at end of elongate extension. Puparium yellowish-brown, darker above.

HOST-PLANTS. *Camellia thea* Link (Madagascar), *Coffea arabica* L. (Madagascar, East Africa); *Citrus* sp., *Gossypium* sp. (West Africa).

A complete list of genera recorded as hosts includes: Apocynaceae (*Mascarenhasia*, *Pachypodium*), Asclepiadaceae (*Cryptostegia*, *Tylophora*), Crassulaceae (*Kalanchoe*), Dioscoreaceae (*Dioscorea*), Flacourtiaceae (*Doryalis*, *Flacourtia*), Leguminosae (*Cassia*, *Clitoria*), Malvaceae (*Gossypium*), Myrsinaceae (*Maesa*), Myrtaceae (*Eugenia*), Nyctaginaceae (*Bougainvillea*), Passifloraceae (*Adenia*, *Passiflora*), Pittosporaceae (*Pittosporum*), Rhamnaceae (*Ziziphus*), Rubiaceae (*Coffea*, *Mussaenda*), Rutaceae (*Citrus*, *Thevetia*), Theaceae (*Camellia*), Verbenaceae (*Duranta*).

BIOLOGY. The larva forms an upper surface, epidermal mine, rarely penetrating into the parenchymatous tissues of the leaf. The mine (Fig. 277) frequently follows the leaf margin but may be quite irregular; it has a characteristic silvery appearance. Pupation takes place in the leaf at the end of the mine, with the anterior spiracles projecting through the epidermis.

Details of the life cycle have not been studied. The pupal period lasts for about 11 days. There are certainly many generations, the number probably only limited by the duration of rainy seasons and the availability of fresh growth in the host-plants.

DAMAGE CAUSED. The exceptionally shallow mine causes only superficial damage to mature leaves and can have no effect on fully grown plants. However, oviposition frequently takes place in very young leaves and in these the epidermis frequently breaks open along the mined area. Subsequent fungal attack is not uncommon, the entire leaf then being destroyed. Young nursery plants of coffee were seen at the Coffee Research Station, Ruiru, Kenya in October, 1971, with several such leaves damaged in this way.

DISTRIBUTION. Madagascar; South, East and West Africa.

PARASITES. In Kenya and Tanganyika: *Achrysocharis purpurissata* Kerrich (as *A. purpurissatus* Kerrich, 1970).

CROWE and GREATHEAD (1970) give a comprehensive list of parasites of coffee leaf-miners but in many cases it is not clear whether the species were parasitizing

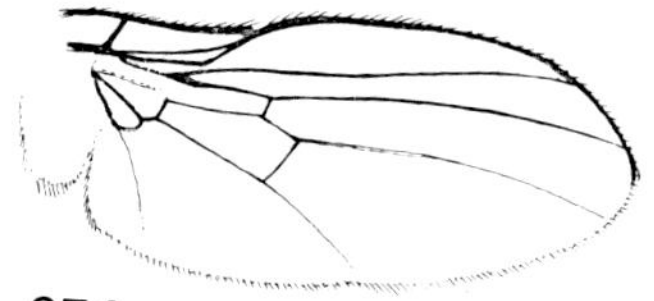
274

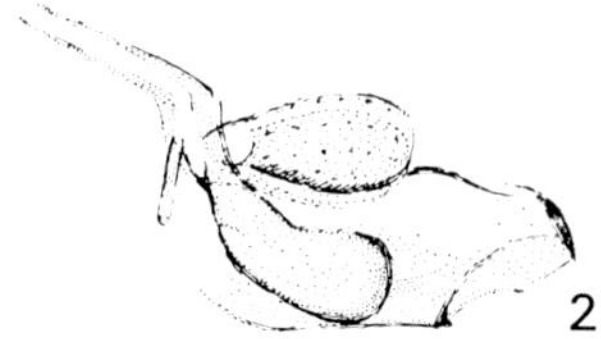
275

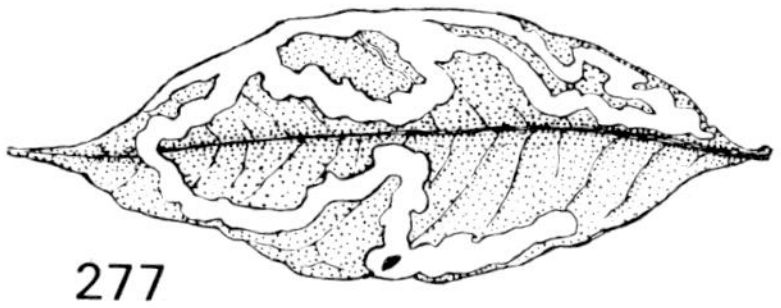
277

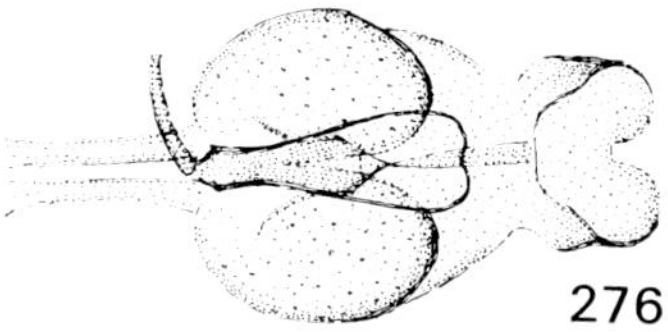
276

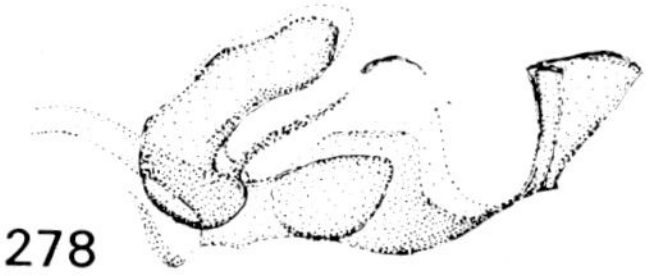
278

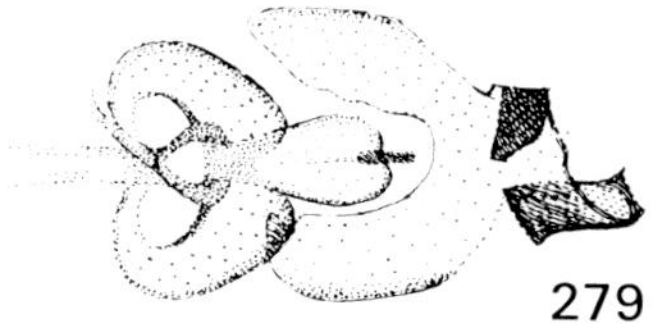
279

Figs. 274–277. *Tropicomyia flacourtiae* (Séguy): 274, wing; 275, aedeagus, side view; 276, same, ventral view; 277, leaf-mine on coffee, Kenya.

Figs. 278, 279. *Tropicomyia nigriclava* (Bezzi): 278, aedeagus, side view; 279, same, ventral view.

lepidopterous *Leucoptera* spp. or *T. flacourtiae*. More accurate rearing records will thus be necessary before the complete range of parasites attacking *T. flacourtiae* can be established.

Discussion

TAXONOMY. SÉGUY described this species as *flacourthiae*, although the host was correctly listed as *Flacourtia Ramontchi*. I consider that the introduction of an h can only be understood as an 'inadvertent error' and under the terms of Article 32 of the 1961 Code of Zoological Nomenclature emend the name to *flacourtiae*.

The relative position of the two cross-veins is somewhat variable in this species. This led Séguy to describe as distinct *Mel. dioscoreae* with the cross-veins approximated (1951: Fig. 11) and *Mel. flacourthiae* with the distance between them greater (1951: Fig. 12). Examination of the genitalia of the holotype of *M. dioscoreae* and of a male cotype of *flacourtiae* confirms that the two are identical and *dioscoreae* is therefore synonymized with *flacourtiae* herewith. Although *dioscoreae* (type bred from *Dioscorea*, a monocot) has page priority in Séguy's paper, it seems preferable to accept the name *flacourtiae* (bred from a wide range of dicots), as clearly this species feeds primarily on dicots and it would be misleading for the valid name to refer to the only monocot on which it is so far known to feed.

Hering's description (1940) of *Mel. coffeae* represented a homonym of *coffeae* Koningsberger (1897) from Java, and was synonymized with *M. passiflorae* Hering (1957a) by Spencer (1963c). Both species can now be synonymized with *flacourtiae* Séguy.

M. citri Spencer, 1959 from Sierra Leone and *M. gossypii* Spencer, 1959 from Nigeria were described when this group was very imperfectly understood. I am now satisfied that both species represent *flacourtiae* and the formal synonymies are established herewith.

Host-range. Genera in 14 families of dicots are known as hosts of this species. An analysis of the distribution of these families among Cronquist's (1968) sub-classes shows 13 to belong to the more advanced sub-classes as follows: Dilleniidae (3), Rosidae (6), Asteridae (4). Only one belongs to the more primitive sub-class Caryophyllidae, none to the two most primitive sub-classes Magnoliidae and Hammamelidae. A single genus, *Dioscorea*, is known in the monocots.

Economic importance. The main crops on which this species has been recorded are coffee, tea, citrus and cotton. Of these the preferred host appears to be coffee. At existing population levels damage to the plants is negligible. However, any substantial increase in populations could be a cause of concern. The growth of young plants could certainly be affected, as a result mainly of secondary fungal damage to mined leaves.

At the present time the species appears to be effectively controlled by parasites. At both Nairobi and Ruiru in Kenya a sample of 40 mined coffee leaves collected by the author in October, 1971, produced only parasites. This 100% parasitism must be exceptional but the average rate is obviously high.

Tropicomyia nigriclava (Bezzi), **comb. nov.** – Rodrigues

Domomyza nigriclava Bezzi, 1926. Holotype ♂ in B.M.
Melanagromyza nigriclava (Bezzi), Spencer, 1959: 281.

Adult. Small, shining black species, wing length in male 1.6 mm, agreeing exactly on external characters with *T. flacourtiae*.

MALE GENITALIA. Aedeagus as in Figs. 278, 279.
HOST-PLANTS. Not known.

Discussion

TAXONOMY. This species clearly belongs to the same original population as *T. flacourtiae* but differences in the aedeagus, although the general form is similar in the two species, indicate some degree of speciation in the population now isolated on Rodrigues. It is therefore proposed to retain *nigriclava* as distinct.
ECONOMIC IMPORTANCE. No hosts are known but presumably this species has the potential to feed on the same range of hosts as *T. flacourtiae*. It is not impossible that some specialization in the preferred host(s) may have developed on Rodrigues.

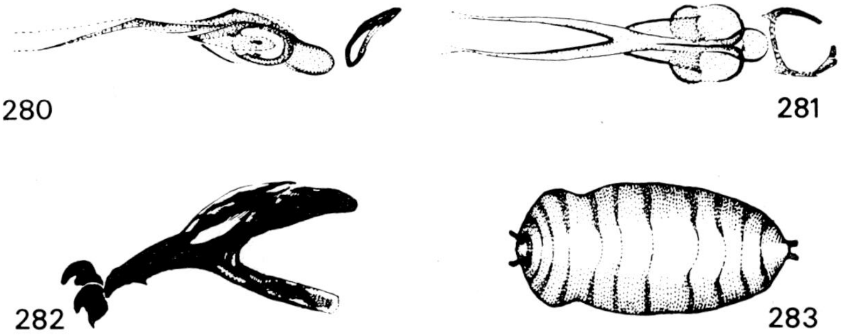

Figs. 280–283. *Tropicomyia pisi* (Kl.): 280, aedeagus, side view; 281, same, ventral view; 282, larval mouth-parts and cephalo-pharyngeal skeleton; 283, puparium.

Whether or not cultivated plants are included in such hosts remains to be established.

Tropicomyia pisi (Kleinschmidt), **comb. nov.** – Australia

Melanagromyza pisi Kleinschmidt, 1960. Holotype ♀ in Queensland Museum, Brisbane.
ADULT. Minute, shining black species; wing length 1.2 mm, costa extending strongly to vein M_{1+2}, last section of M_{3+4} equal to or slightly longer than penultimate.
MALE GENITALIA. Aedeagus as in Figs. 280, 281; ninth sternite with broad, strong side-arms and greatly elongated hypandrial apodeme (cf. Fig. 301).
EARLY STAGES. Larva pale yellow, with minute, scarcely detectable pores on the anterior spiracles and posterior spiracles each having three well-defined pores; mouth-parts and cephalo-pharyngeal skeleton as in Fig. 282; puparium (Fig. 283) shining black, with distinct constriction at anterior end (KLEINSCHMIDT, 1970).

HOST-PLANTS. *Pisum sativum* L., *Trifolium* sp. (white clover); probably other related genera.
BIOLOGY. The larva forms an irregular, whitish upper surface mine, pupating in the leaf (Plate 5) with the anterior spiracles projecting through the epidermis. Frass is deposited in three distinct pellets, associated with each moult (KLEINSCHMIDT, 1970).
DISTRIBUTION. Australia: Queensland.
PARASITES. *Opius cinerariae* Fischer (Braconidae); *Hemiptarsenus* sp., *Tetrastichus* sp. (Eulophidae); sp. indet. (Pteromalidae).
The rate of parasitism reaches 50% (KLEINSCHMIDT, 1970).

Discussion

TAXONOMY. This species forms a compact group with the other Australian leaf-mining species now transferred to *Tropicomyia*, *dianellae* Kl., *indigoferae* Kl. and *wikstroemiae* Kl.
ECONOMIC IMPORTANCE. Mines have only been found on mature plants in which leaf-mining damage is not significant. However, occurrence in large numbers in seedlings could cause significant weakening of the plants and the species must be considered as of potential economic importance. Effective control appears to be exercised by its parasite complex.

Tropicomyia polyphaga (Spencer), **comb. nov.** – Ceylon, India

Melanagromyza polyphaga Spencer, 1961b: 75. Holotype ♂ from Ceylon in author's collection.
ADULT. This species is distinctly larger and more robust than *atomella*. The wing length is consistently from 1.6–1.8 mm and the last section of vein M_{3+4} is significantly shorter than the penultimate. The puparium is always black. The aedeagus is of the same general form as in *atomella* but the distiphallus appears more triangular in ventral view, less obviously rounded (Fig. 284).
HOST-PLANTS. Confirmed hosts are *Tylophora flava* Trimen, Ceylon; *Cryptolepis buchmanani* Roem. and Schult., India (both Asclepiadaceae); *Symplocos spicata* Roxb. (Symplocaceae), Ceylon; *Smilax* sp. (Liliaceae), India; probably *Thevetia neriifolia* Juss. (Apocynaceae), Ceylon, India. Other apparent hosts are *Allamanda cathartica* L. (Apocynaceae) and *Dregia volubilis* Benth. (Loganiaceae), on which mines were found at Peredeniya, Ceylon (K.A.S.).

Discussion

DISCUSSION. The species occurring commonly in India in Mysore State on coffee possibly represents this species, although a positive identification is not

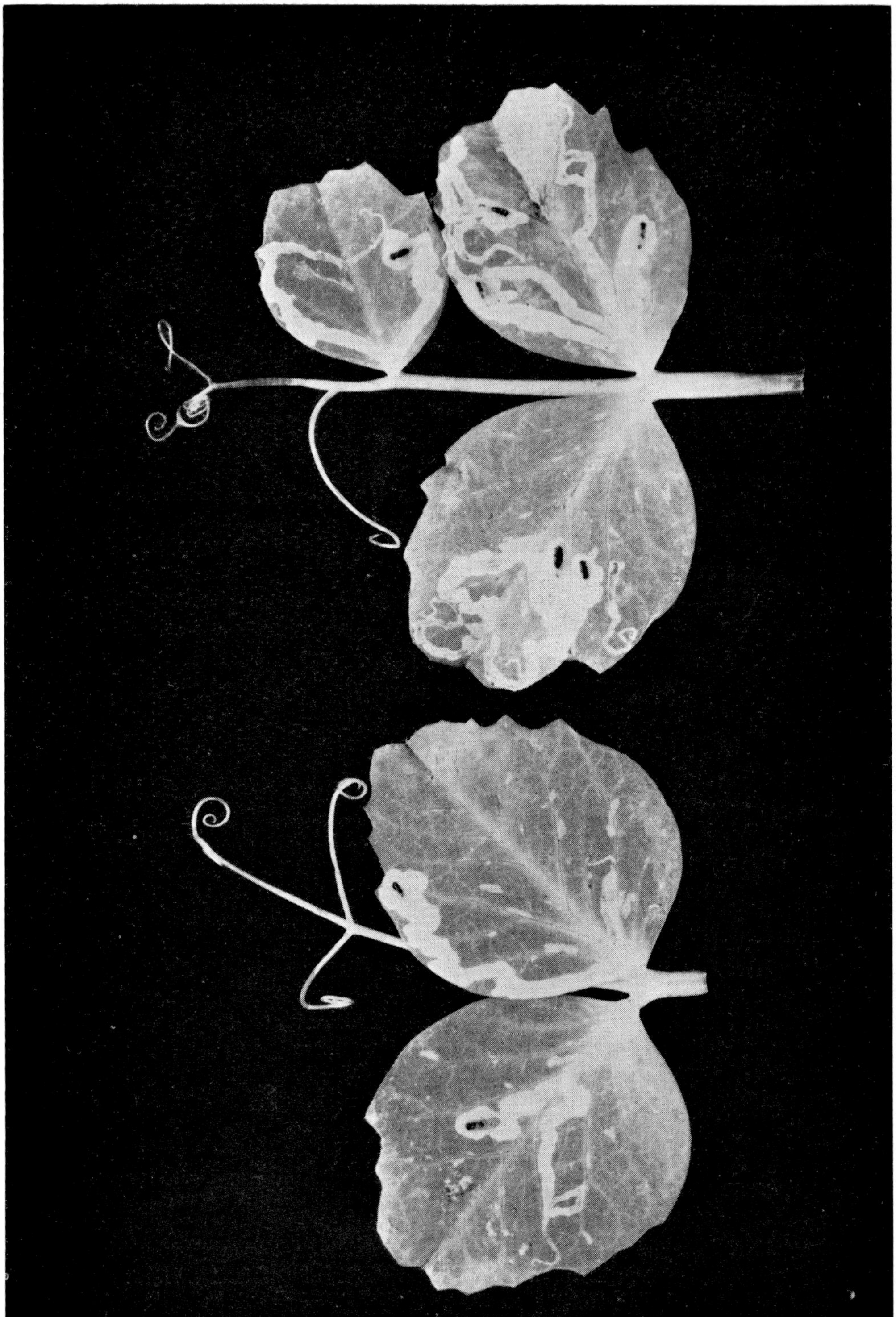

Industries, Brisbane, Queensland)

possible from the single female I have examined. The wing of this specimen is shown in Fig. 285.

LE PELLEY (1968) refers to a recent report from India suggesting that the coffee leaf-miner 'appears to be gaining in importance as a major pest in many coffee areas'. The species is referred to as *Melanagromyza coffeae* (Koningsberger). The identity of this species, described from Java, remains to be established (see p. 181).

Tropicomyia polyphyta (Kleinschmidt), **comb. nov.** – Australia

Melanagromyza polyphyta Kleinschmidt, 1960. Holotype ♀ in Queensland Museum, Brisbane.
Melanagromyza murrayae Spencer, 1963b. Holotype ♀ in author's collection.
ADULT. Minute, shining black species; wing length 1.4–1.6 mm, costa extending to vein M_{1+2}, last section of M_{3+4} distinctly shorter than penultimate.
MALE GENITALIA. Aedeagus as in Figs. 286, 287.
EARLY STAGES. Larva with serrated mouth-hooks (Fig. 288); anterior spiracles on long stalks, each with 6 minute pores, posterior spiracles each with 3. Puparium essentially as in *atomella*, frequently black but also sometimes yellowish-brown.
HOST-PLANTS. Adults have been reared from mines found on 20 different families. Hosts among cultivated plants include *Phaseolus*, *Vigna* (Leguminosae); *Coffea arabica* (Rubiaceae); *Citrus* (Rutaceae); *Camellia* (Theaceae). Other known hosts are as follows: *Plumeria* (Apocynaceae), *Tylophora* (Asclepiadaceae), *Kigelia* (Bignoniaceae), *Lonicera* (Caprifoliaceae), *Celastrus* (Celastraceae), *Quisqualis* (Combretaceae), *Acalypha*, *Breynia*, *Euphorbia* (Euphorbiaceae); *Barringtonia* (Lecythidaceae); *Cassia*, *Erythrina* (Leguminosae); *Eustrephus* (Liliaceae); *Melia* (Meliaceae); *Cudrania* (Moraceae); *Angophora* (Myrtaceae); *Passiflora* (Passifloraceae); *Alphitonia* (Rhamnaceae); *Gardenia* (Rubiaceae); *Murraya* (Rutaceae), *Salix* (Salicaceae), *Hydrangea* (Saxifragaceae), *Cestrum* (Solanaceae), *Brachychiton* (Sterculiaceae).
BIOLOGY. The larval mine is irregular, upper surface, epidermal, appearing conspicuously silvery. A photograph showing mines on *Vigna sesquipedalis* Fruwirth is shown in Plate 6. Pupation takes place on the upper surface of the leaf at the end of the mine. Several larvae may be found in a single leaf and frequently the entire surface of a leaf can be mined. There are numerous generations. According to KLEINSCHMIDT (1970) the pupal period lasts from 18–20 days in late autumn but is shorter in the spring.
DISTRIBUTION. Australia: Queensland, New South Wales, Northern Terr. (Darwin).
PARASITES. KLEINSCHMIDT (1970) records three species, *Opius atricornis* Fischer, *Achrysocharis* sp. and *Enaysma* sp. The rate of parasitism was often high on all hosts.

Discussion

TAXONOMY. SPENCER (1963b) considered this species to be synonymous with *atomella* (Malloch). However, this synonymy was not accepted by KLEINSCHMIDT

(1970) and the further review of all available Australian and Oriental material which has now been undertaken confirms that *polyphyta* can justifiably be treated as a distinct species. The last section of vein M_{3+4} is consistently shorter than the

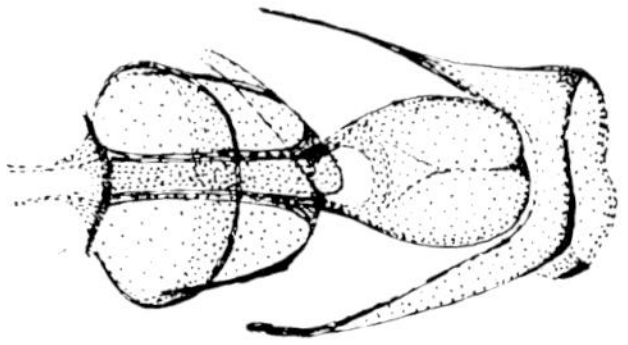

284

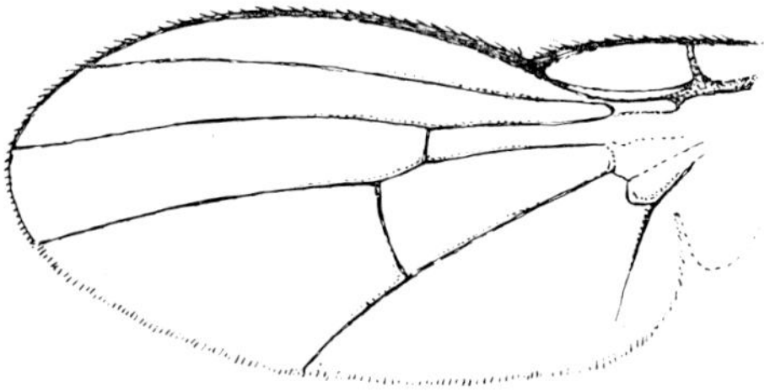

285

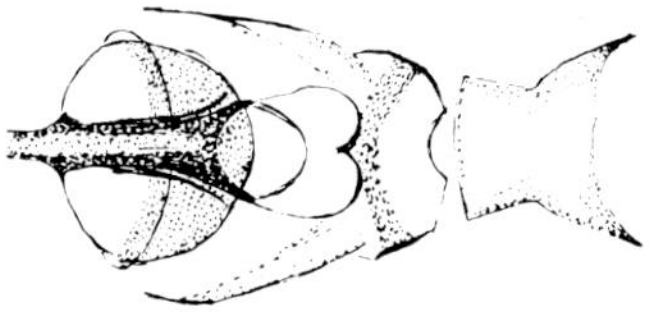

287

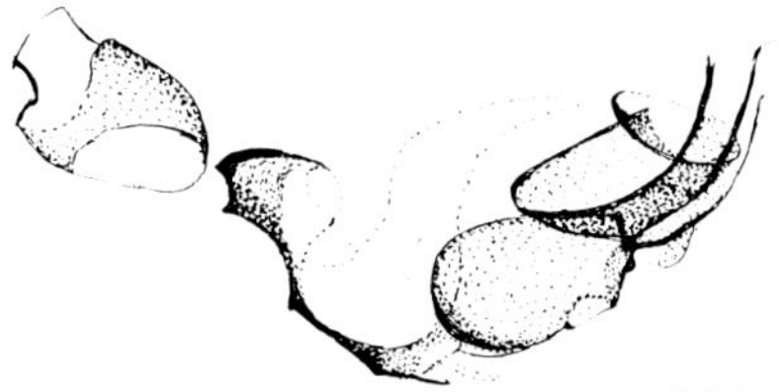

286

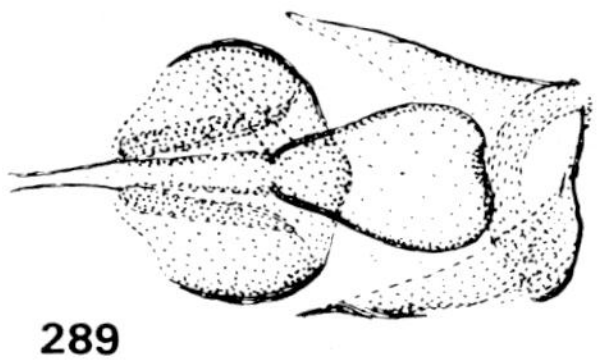

289

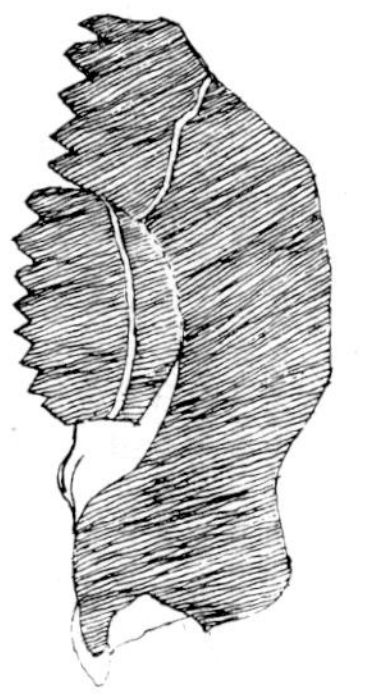

288

Figs. 284–285. *Tropicomyia polyphaga* (Sp.): 284, aedeagus; 285, wing of specimen ex coffee, India.

Figs. 286–288. *Tropicomyia polyphyta* (Kl.): 286, aedeagus, side view; 287, same, ventral view; 288, larval mouth-hooks (after KLEINSCHMIDT).

Fig. 289. *Tropicomyia styricicola* (Sas.): aedeagus.

penultimate and there is a constant, although slight, difference in the aedeagus (Figs. 286, 287) in all Australian specimens examined.

ECONOMIC IMPORTANCE. No significant damage has yet been reported to cultivated plants but the very wide host range suggests that this species could occur on many crops. If the parasite complex was ever seriously affected by insecticides, an outbreak of leaf mining activity in pest proportions could result.

Tropicomyia styricicola (Sasakawa), **comb. nov.** and **stat. rev.** – Japan

Melanagromyza styricicola Sasakawa, 1954b. Holotype ♂ ex *Styrax* in collection of Kyoto Prefectural University.

Plate 6: *Tropicomyia polyphyta:* mines and puparia in *Vigna sesquipedalis*, Toowong, Queensland (Dept. of Primary Industries, Brisbane, Queensland).

ADULT. This is a relatively large species, with wing length from 2–2.3 mm; the last section of vein M_{3+4} is substantially shorter than the penultimate (cf. Fig. 265).
MALE GENITALIA. The aedeagus in ventral view is shown in Fig. 289 (see also SASAKAWA, 1961: Fig. 32D).

HOST-PLANTS. *Actinidia* (Actinidiaceae), *Marlea* (Alangeaceae), *Ilex* (Aquifoliaceae), *Celastrus*, *Euonymus* (Celastraceae), *Fraxinus* (Oleaceae), *Pittosporum* (Pittosporaceae).

Discussion

TAXONOMY. SPENCER (1966a) synonymized *styricicola* with *theae* Green. This synonymy is now withdrawn. It can be seen from differences in the aedeagus of the additional specimens which have been examined that *styricicola* deserves separate status.

SASAKAWA (1961: 352) treated a species mining leaves of tea in Japan as *theae* Green. The illustration given of the aedeagus shows this identification to be incorrect and the exact status of this species remains to be clarified.

Tropicomyia theae (Cotes), **comb. nov.** – Ceylon

Oscinis theae Cotes, 1896: 28. Types from Ceylon lost.
ADULT. Relatively large species within this group, wing length 2 mm, last section of vein M_{3+4} distinctly shorter than penultimate, costa extending weakly to vein M_{1+2}; mesonotum moderately shining black.
MALE GENITALIA. Aedeagus as in Figs. 290, 291.
EARLY STAGES. Details not recorded.
HOST-PLANTS. *Camellia thea* Link.
BIOLOGY. Oviposition takes place on the upper surface of the leaf, the egg being laid immediately beneath the epidermis. As many as five larvae have been observed in a single leaf but this is unusual and normally not more than two occur together. The second and third leaves from the bud seem to be preferred; older leaves are sometimes used but the youngest leaves are avoided. The larva forms an irregular, upper surface, silvery epidermal mine (Fig. 292), pupating after about 11 days at the end of the mine, with the anterior spiracles projecting through the epidermis. Adults emerge after about 15 days (C. H. GADD, 1945).
DAMAGE CAUSED. When fresh, the mine is superficial and subsequent growth is scarcely affected by the larval feeding. However, older leaves are subject to considerable distortion and, ultimately, destruction by fungal attack. Many such leaves were found by the author at Peredeniya, Ceylon in January, 1960.
DISTRIBUTION. Ceylon.
PARASITES. *Closterocerus insignis* Waterston, *Trigonogastra* sp. and an undetermined Braconid.

From 93 puparia the number of these species obtained were 45, 5 and 5 respectively; from 41 larvae 27 specimens of the Braconid were obtained (C. H. GADD, 1945).

Discussion

TAXONOMY. The aedeagus of specimens from Talawakella (Figs. 290, 291) is entirely different from that of other specimens bred from tea in northern India and in Japan (SASAKAWA, 1961: Fig. 33); KURODA (1960) has also discussed the larva of a species noted in Japan, as *theae*. The name *theae*, however, must now be restricted to specimens from tea from Ceylon, as these clearly represent a distinct species within this group.

The name *theae* was first used by COTES (1896) as '*Oscinis theae*, Bigot, MS.'. No formal description was ever published by BIGOT. The specimens referred to

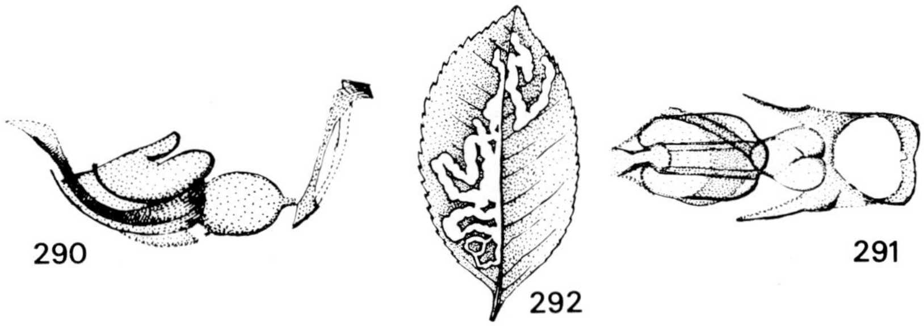

Figs. 290–292. *Tropicomyia theae* (Cotes): 290, aedeagus, side view; 291, same, ventral view; 292, leaf-mine on tea, Ceylon.

had been bred by E. E. GREEN and the name *theae* has hitherto been ascribed to him. This, however, is formally incorrect and the species should in future be known as *theae* Cotes.

DE MEIJERE (1922) gave a detailed description of '*theae*' and later (1940a) described the larva, based on specimens from Java. However, the exact identity of the species present in Java is uncertain and, without examination of males, it cannot be assumed that it is conspecific with the true *theae* from Ceylon.

SPENCER (1966a) treated *theae* as senior synonym of both *atomella* Malloch and *styricicola* Sasakawa. These synonymies are now withdrawn, following clarification of specimens from the type locality, Ceylon.

ECONOMIC IMPORTANCE. GADD (1945) writes: 'Although leaves injured by tea leaf miner are frequently observed, the insect has never assumed the importance of a major pest. In August it became unusually prominent on St. Coombs but had completely disappeared by October'. The rate of parasitism found by GADD was 70% and it appears that at the present time effective control of *theae* is being exercised by its parasite complex.

Tropicomyia vigneae (Séguy), **comb. nov.** – Cape Verde Islands to Indonesia

Melanagromyza vigneae Séguy, 1951. Lectotype ♂ from Madagascar in Musée National d'Histoire Naturelle, Paris.

Melanagromyza alustrata Spencer, 1959, **syn. nov.** Holotype ♂ from Cape Verde Islands in University Zoological Museum, Helsinki.

Melanagromyza dolichi Sehgal, 1965, **syn. nov.** Holotype ♂ from India in collection of Zoological Survey of India, Calcutta.

Melanagromyza jonesi Spencer, 1965b, **syn. nov.** Holotype ♂ from Ethiopia in author's collection.

ADULT. Minute black species, wing length from 1.5–1.75 mm, costa extending to vein M_{1+2}, first cross-vein around midpoint of discal cell.

HEAD. Jowls narrow, rounded, deepest in front, about $\frac{1}{8}$ height of eye; third antennal segment small, round, arista long, little shorter than height of eye, pubescent.

COLOUR. Frons mat brown, orbits scarcely differentiated, mesonotum conspicuously mat, black, almost brownish, abdomen shining black, squamal margin and fringe black.

MALE GENITALIA. Aedeagus as in Figs. 293, 294, with a distinct semicircle of chitinization beyond end of distiphallus.

EARLY STAGES. The larval mouth-hooks (Fig. 295) have a number of small teeth and distinctively diverge from the characteristic serration of true epidermal miners (Figs. 269, 288). The puparium is whitish-grey, with the posterior spiracles each on a separate conical protuberance, strongly chitinized at the end, with 8–10 pores (Fig. 296).

HOST-PLANTS. *Vigna angivensis* Baker, *Dolichos sesquipedalis* L., *Glycine javanica* L., *Lablab niger* L. (= *Dolichos lablab* L.), *Phaseolus mungo* L.; also *Dolichos gibbosus* Thbg.

BIOLOGY. The larva feeds in part as an epidermal miner, probably in the first instar and mainly on the lower side of the leaf, later forming a shallow, irregular linear mine on the upper surface (Fig. 297), with frass in widely-spaced black grains; pupation takes place in the mine, with the anterior spiracles projecting through the leaf epidermis.

DISTRIBUTION. Madagascar: Tsimbazaza; Sudan: Khartoum; Cape Verde Islands: S. Tiago, Maio, Nicolau, S. Antão, Boavista; India: Namkum; Indonesia: Sumbawa.

PARASITES. None known.

Discussion

TAXONOMY. The important characters of this species are the entirely mat mesonotum, the costa extending to vein M_{1+2} and the black squamal fringe. The male genitalia are distinctive.

Examination of the aedeagus of males from *Vigna* sp. and from *Phaseolus mungo* from the type series shows this to be identical to that of *M. alustrata* from the Cape Verde Is., illustrated by SPENCER, 1964b: Figs. 7, 8, and *alustrata* is therefore synonymized with *vigneae* herewith. A male bred from *Vigna*, after which SÉGUY named the species, is now designated as lectotype.

Re-examination of the type series of *M. jonesi* Spencer, 1964b bred from *Glycine javanica* L. at Addis Ababa (E. M. HERING) shows that this species also represents *vigneae*. *M. dolichi* Sehgal, 1965, bred from *Dolichos lablab* L. at

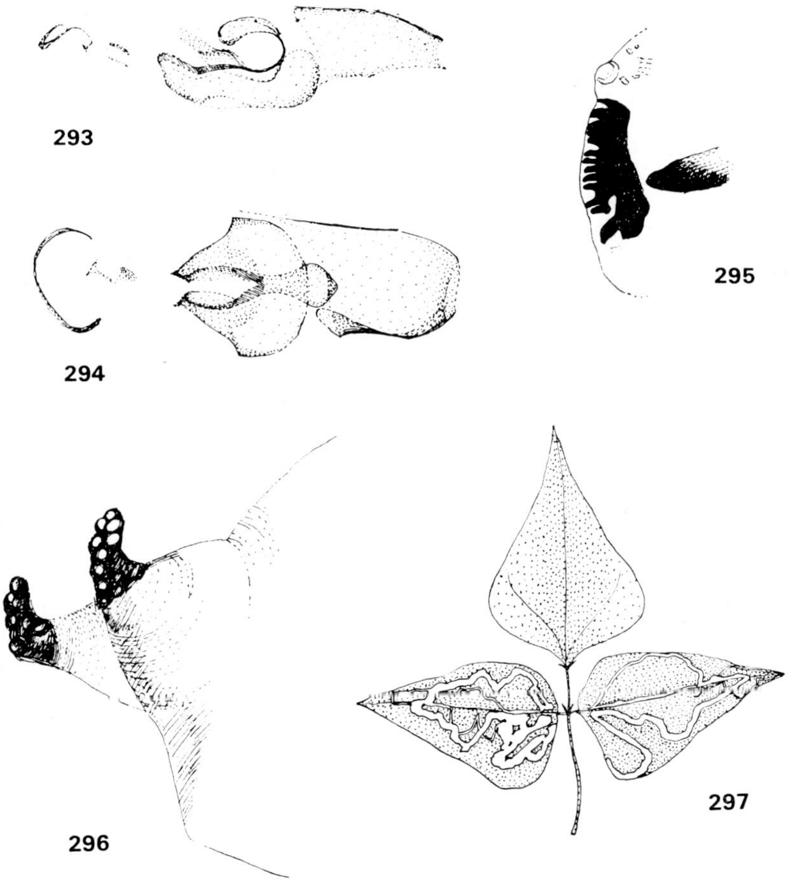

Figs. 293–297. *Tropicomyia vigneae* (Séguy): 293, aedeagus, side view; 294, same, ventral view (both paratypes, ex *Phaseolus*, Madagascar); 295, larval mouth-hooks (after SEHGAL); 296, posterior spiracles of puparium; 297, leaf-mine on *Dolichos*, Cape Town.

Namkum, India (V. K. SEHGAL) was synonymized with *jonesi* by SPENCER, 1965b: 249. Both *jonesi* and *dolichi* are therefore now also synonymized with *vigneae*.

Previously (SPENCER, 1959), a series bred from *Vigna unguiculata* (L.) Walp. in Sénégal was identified as *vigneae*. It was also found that the aedeagus of the Sénégal specimens was identical to that of *M. capeneri* Hering, 1957a, bred from an unidentified plant in South Africa, and *M. capeneri* was therefore synonymized with *vigneae* Séguy (SPENCER, 1964b). This synonymy can now be seen to be incorrect (cf. p. 184).

HOST-RANGE. In his description of *M. vigneae* Séguy recorded as hosts *Vigna angivensis* and *Phaseolus mungo*, and in the collection in Paris he also included under *vigneae* four specimens from Madagascar bred from *Crotalaria incana* L. Examination of these shows the species to be *M. crotalariae* Hering, 1957a, described from South Africa; the aedeagus was illustrated by SPENCER (1963c).

Further confirmed hosts are *Dolichos lablab* in India and *D. sesquepedalis* ('Lubia') in the Sudan (leg. A. BODAWY), *Glycine javanica* in Ethiopia and also the wild plant *Dolichos gibbosus* Thbg. in South Africa (K.A.S.).

It seems probable that *M. vigneae* will also be found to occur on other genera of the Leguminosae.

ECONOMIC IMPORTANCE. *M. vigneae* is clearly a widespread species in Africa and its presence both in India and Indonesia indicates that it also has an extensive distribution in the Oriental Region. It is now confirmed on four cultivated genera of the Leguminosae and, although there are no records of the leaf-mines having occurred in sufficient numbers to cause damage, the species must be recognized as a potential pest. Its life history deserves further study.

Additional species now transferred to Tropicomyia

Africa

Tropicomyia cassinis (Hering), **comb. nov.**

Melanagromyza cassinis Hering, 1957a: 64. Holotype ♀ ex *Cassine sphaerophylla* O. Ktze. from Cape Province in Transvaal Museum, Pretoria.

Tropicomyia ceratiosicyi (Hering), **comb. nov.**

Melanagromyza ceratiosicyi Hering, 1957a: 65. Holotype ♀ ex *Ceratiosicyos ecklonii* Nees from Cape Province in Transvaal Museum, Pretoria.

Tropicomyia clutiae (Spencer), **comb. nov.**

Melanagromyza clutiae Spencer, 1963c: 100. Holotype ♂ ex *Clutia pulchella* L. from Cape Province in author's collection.

Tropicomyia crotonella (Spencer), **comb. nov.**

Melanagromyza crotonella Spencer, 1964b: 21. Ex *Croton macrostachys* Hochst. ex A. Rich. Holotype ♂ from Addis Ababa in Zoologisches Museum, Berlin.

Tropicomyia dicksoni (Hering), **comb. nov.**

Melanagromyza dicksoni Hering, 1957a 70. Holotype ♀ ex unidentified host from Natal in Transvaal Museum, Pretoria.

Tropicomyia gymnosporiae (Hering), **comb. nov.**

Melanagromyza gymnosporiae Hering, 1957a: 71. Holotype ♂ ex *Gymnosporia buxifolia* (L.) Szyszl. from Cape Province in Transvaal Museum, Pretoria.

Tropicomyia haemanthi (Spencer), **comb. nov.**

Melanagromyza haemanthi Spencer, 1963c: 104. Holotype ♂ ex *Haemanthus*

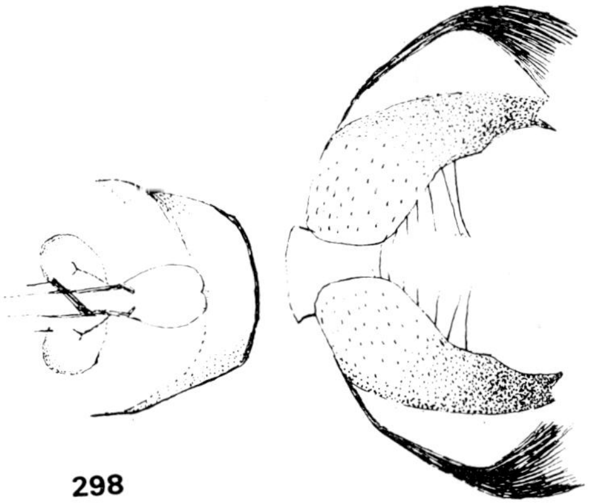

298

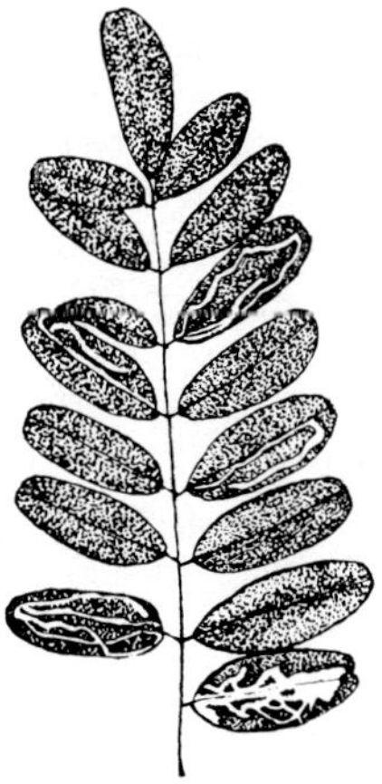

300

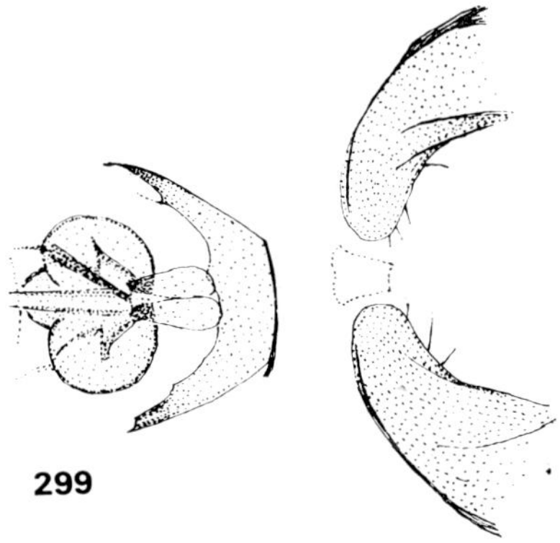

299

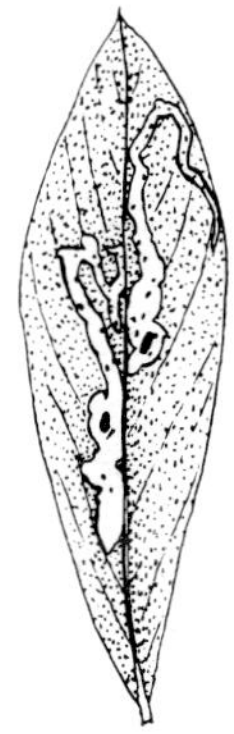

302

301

Fig. 298. *Tropicomyia haemanthi* (Sp.): aedeagus and surstyli.

Fig. 299. *Tropicomyia thunbergivora* (Sp.): aedeagus and surstyli.

Fig. 300. *Tropicomyia indigoferae* (Kl.): leaf-mines on *Indigofera*, N.S.W.

Figs. 301, 302. *Tropicomyia wikstroemiae* (Kl.): 301, aedeagus and ninth sternite; 302, leaf-mines on *Pimelea ligustrina*.

magnificus Herbert from Natal in author's collection. The aedeagus and surstyli are shown in Fig. 298.

Tropicomyia laburnifoliae (Spencer), **comb. nov.**
Melanagromyza laburnifoliae Spencer, 1964b: 25. Holotype ♀ ex *Crotalaria laburnifolia* L. in Zoologisches Museum, Berlin.

Tropicomyia philocroton (Hering), **comb. nov.**
Melanagromyza philocroton Hering, 1957a: 67. Holotype ♀ ex *Croton subgratissimus* Prain from Cape Province in Transvaal Museum, Pretoria.

Tropicomyia thunbergivora (Spencer), **comb. nov.**
Melanagromyza thunbergivora Spencer, 1963c: 106. Holotype ♂ ex *Thunbergia natalensis* Hk. from Natal in author's collection. The aedeagus and surstyli are shown in Fig. 299.

Asia

Tropicomyia coffeae (Koningsberger), **comb. nov.**
Oscinis coffeae Koningsberger, 1897. Types ex coffee from Java lost.

Australia

Tropicomyia dianellae (Kleinschmidt), **comb. nov.**
Melanagromyza dianellae Kleinschmidt, 1960: 11; 1970: 344. Holotype ♀ ex *Dianella caerulea* Sims in Queensland Museum, Brisbane.

Tropicomyia indigoferae (Kleinschmidt), **comb. nov.**
Melanagromyza indigoferae Kleinschmidt, 1960: 9; 1970: 347. Holotype ♂ ex *Indigofera suffruticosa* Mill. in Queensland Museum, Brisbane. The leaf-mine is shown in Fig. 300.

Tropicomyia wikstroemiae (Kleinschmidt), **comb. nov.**
Melanagromyza wikstroemiae Kleinschmidt, 1960: 1; 1970: 359. Holotype ♂ ex *Wikstroemia indica* Endl. in Queensland Museum, Brisbane.

The aedeagus and ninth sternite are shown in Fig. 301 and the leaf-mine on *Pimelea ligustrina* Labill. in Fig. 302.

Many further species certainly await description in this genus. Specimens from Guam in the Pacific, previously identified as *atomella* (Malloch), clearly represent a distinct species; and specimens recently bred from *Eulophia porphyroglossum* (Reich. f.) Bolus (Orchidaceae) and *Cassia floribunda* Cav. (Leguminosae) in Kenya (K.A.S.) are also new species, which will be described later. It seems probable that mines found on *Lasia heterophylla* Schott (Araceae) at Bangkok, 16.i.60 and on *Clerodendron* sp. (Verbenaceae) at Kawanda, Uganda, 24.x.71 (both K.A.S.) are also referable to distinct species.

VIII. POLYPHAGOUS LIRIOMYZA AND PHYTOMYZA SPECIES

It has seemed convenient to treat together in this chapter seven of the most polyphagous species known in the Agromyzidae, five in the genus *Liriomyza* and two in the genus *Phytomyza*. True polyphagy, as found in these seven species, is only known elsewhere in the family in the genus *Tropicomyia* (Chapter 7) and has been briefly discussed in Chapter 1.

Of the five *Liriomyza* species considered below, two – *L. strigata* (Meigen) and *L. bryoniae* (Kaltenbach) – occur exclusively in the Palaearctic Region and are themselves closely related; the other three – *L. trifolii* (Burgess), *L. huidobrensis* (Blanchard) and *L. sativae* Blanchard – occur widely, though with somewhat differing distributions, in North and South America, the two latter also being present on islands in the Pacific. The European species pair, *strigata/bryoniae*, is closely related to the New World species, *huidobrensis*, and it is now obvious that the three species are derived from a single ancestral population, probably in eastern Asia. *L. trifolii* and *L. sativae* are also related, but not closely so with the other three. In the genus *Liriomyza* the polyphagous feeding habit thus appears to have arisen independently only on two occasions, with the two original populations of these five species. With some degree of isolation and restricted association with different hosts, the process is probably now being reversed and further speciation can be expected, with new species evolving having a more typically monophagous or oligophagous feeding habit.

It is not surprising that species feeding on a wide range of plants and having an extensive distribution should have been described by different authors with different names. This is particularly true of the American *Liriomyza* species, and nine new synonyms have been discovered in this group which are formally established below. Owing to their general similarity these species have also been frequently misidentified and it is unfortunate that inaccurate names have appeared in many important papers in American economic literature. With no original material having been preserved, it is now not possible to be certain which species are being discussed in several detailed and otherwise valuable papers.

In America damage to a number of crops, ascribed to '*Agromyza pusilla*', was discussed even before World War I by WEBSTER and PARKS (1913). They were clearly confusing many different species but one of the species involved was certainly *Liriomyza sativae*. McGREGOR (1914) discusses a *Liriomyza* sp. (misidentified as the European species *scutellata* Fall.) damaging cotton in South Carolina. The illustrations given suggest that the species might be *L. trifolii* (Burg.) but this requires confirmation. In June, 1913, the degree of infestation was as high as 98% of plants attacked in some fields. The leaf-mine on cotton is shown in Fig. 334.

Immediately following World War II leaf-miner damage, particularly in Florida, California and Texas increased significantly and a number of papers were published discussing various aspects of the problem, with emphasis on control measures. The papers by WOLFENBARGER (1947 and others) discussing *Liriomyza pusilla* (Meigen) in Florida certainly refer to *L. sativae*. LANGE (1949) published the first of several papers discussing leaf miners and their control in California. LANGE referred to the main species causing damage on tomatoes at that time as *Agromyza* (*Liriomyza*) *subpusilla* Frost. MICHELBACHER et al. (1951) use the same name in discussing control measures against a species damaging melons, while WILCOX and HOWLAND (1952) refer mainly to *Liriomyza* sp. which was causing damage to many different hosts in California between 1939 and 1951. They report that 'in 1947 some growers estimated that their crops (of tomatoes) were reduced 25% to 50%'.

OATMAN and MICHELBACHER (1958, 1959) and OATMAN (1959a, 1959b, 1960, 1961) discuss the life history, ecology, and parasitism of 'the melon leaf-miner', as *L. pictella* (Thomson), in a series of excellent papers. Unfortunately none of the original material used in these detailed studies can now be traced and re-examination of specimens to determine their exact status is thus not possible. The species was certainly not *L. pictella* (see p. 224). It was probably *sativae* but OATMAN (1961) found that his species on melon would not satisfactorily interbreed with the 'tomato leaf-miner', referred to as *munda* Frick. OATMAN concludes that 'the melon and tomato leaf miners appear to be sibling species from sympatric natural populations' (OATMAN, 1961: 57).

Any firm judgment at this stage on the status of the 'melon leaf-miner' is impossible. If there really is no consistent difference between the populations on tomato and on melon, either in the external characters of the adults, in the male genitalia, in larval characters or leaf-mines, no useful purpose would be served by describing the melon leaf-miner as a distinct species. With speciation at such an early stage, the melon leaf-miner can best be treated merely as a local 'strain', which may have become reproductively isolated on melon in one particular area of California. It was apparently established that specimens from tomato would also readily feed on melon in California, and both in Florida and Venezuela apparently one and the same species attacks both hosts. Any attempt to treat the melon leaf-miner as distinct, even as a strain, thus encounters almost insurmountable practical difficulties, since specimens on melon may either be the 'melon strain' or part of the main population which will feed equally well on tomato or presumably also the other hosts acceptable to this species. Complex cross-breeding experiments would thus always be necessary to discover whether one is dealing with the 'melon strain' or typical *sativae*.

In this context further cross-breeding experiments between populations from other hosts would be of considerable interest and could indicate whether strains are also developing in association with other hosts apart from melon. It would also be of interest to repeat the experiments with populations from melon and tomato in other areas where the species commonly occurs, such as Florida, Hawaii, Venezuela or Argentina.

SMITH, BOSWELL and WAVE (1962) and WEBB and SMITH (1969, 1970) discuss the life history and damage caused to chrysanthemums in greenhouses in Delaware, New Jersey, Ohio and Pennsylvania by a species identified originally by FRICK as *Liriomyza* sp.n. and later by STEGMAIER and STEYSKAL as *Liriomyza munda* Frick. Examination of specimens from New Jersey kindly sent to me by Mr. STEYSKAL has shown that this species is in fact *Liriomyza trifolii* (Burgess), see p. 226 and Chapter 11, p. 325.

Detailed studies of the parasites of the *Liriomyza* species attacking commercial crops have been made in the United States by HILLS and TAYLOR (1951) in Arizona, MICHELBACHER et al. (1951), OATMAN (1959a), JENSEN and KOEHLER (1970) in California and by HARDING (1965) in Texas. Unfortunately the exact identity of the leaf-miners considered is uncertain in all these valuable papers. Nevertheless, the identifications given in these papers are important and they are therefore reproduced below. It is to be hoped that a more accurate association with the species of leaf-miner can be established by future workers.

HILLS and TAYLOR (1951) identified 15 species of parasites obtained from leaf-miners attacking watermelons, cantaloups and lettuce in the Salt River Valley of Arizona. The leaf-miner may have been '*munda*' or two species may have been involved. They list the following species, with an interesting breakdown of specimens obtained from cantaloup and lettuce:

Species	*Numbers of parasites from* *Cantaloup*	*Lettuce*
Halticoptera aenea (Walk.)	103	1,501
Derostenus arizonensis Cwfd.	93	294
Solenotus begini (Ashm.)	68	244
Closterocerus utahensis Cwfd.	12	16
Solenotus websteri Cwfd.	11	2
Diaulinopsis callichroma Cwfd.	4	58
Derostenus variipes Cwfd.	2	4
Chrysocharis parksi Cwfd.	3	41
Anagrus giraulti Cwfd.[1]	—	—
Trichogramma minutum (Riley)	10	8
Opius dimidiatus Ashm. *Opius aridis* Gahan *Opius suturalis* Gahan *Opius* sp. A *Opius* sp. B	29	48
Total	335	2,216

[1] Not obtained in 1950

In California MICHELBACHER *et al.* (1951) report *Solenotus intermedius* Gir. (Eulophidae) as the most effective parasite against the melon leaf-miner in the Patterson area; in addition small numbers of *Chrysocharis ainsliei* Cwfd., of an *Opius* sp. and an undetermined Pteromalid were also reared. OATMAN (1959a) lists 19 species parasitizing the melon leaf miner, as follows:

Pteromalidae: *Halticoptera aenea* (Walker).

Eulophidae: *Solenotus begini* (Ashm.), *S. pulchripes* (Cwfd.); *S. Websteri* (Cwfd.); *Derostenus arizonensis* Cwfd., *D. variipes* Cwfd.; *Diaulinopsis callichroma* Cwfd.; *Chrysocharis ainsliei* Cwfd., *C. parksi* Cwfd., *Chrysocharis* spp.; *Zagrammosoma mirum* Grlt., *Zagrammosoma* spp.; *Closterocerus trifasciatus* Westwood; *Astichus* sp.

Braconidae: *Opius suturalis* Gahan; *Opius* spp.

Encyrtidae: *Mirini.*

Myrmaridae: *Polynema* sp.

Scelionidae: *Telonomus* spp.

JENSEN and KOEHLER (1970) record six parasites attacking a mixed population of *L. sativae* (referred to as *munda*) and *L. trifoliearum* (referred to as *pictella*) in the Sacramento and San Joaquin Valleys of California. No attempt was made to associate these parasites with their actual hosts but it may nevertheless be helpful to record the six species, as follows: Pteromalidae: *Halticoptera patellana* (Dalman); Eulophidae: *Chrysocharis ainsliei* Cwfd., *C. parksi* Cwfd. and *Diglyphus begini* (Ashmead); Braconidae: *Opius aridis* Gahan and *O. suturalis* Gahan.

In Texas HARDING (1965) lists in his Table III 21 species reared from a leafminer identified as *L. munda*. However, there is no certainty that all larvae collected from the wide range of hosts on which mines were found represented *L. munda*. Parasites were obtained from larvae on the following hosts: 'Bean (green), Bean (lima), Cantaloup, Cucumber, Eggplant, Mustard, Okra, Pepper, Potato (Irish), Southern pea, Squash, Tomato, Watermelon'; also 'borage, castor beans, cotton, dahlia, gailardia, marigold, sunflower, sweet basil, sweet potatoes and verbena'. HARDING's list shows the percentage of each species obtained from pupae and mined leaves:

Parasite species	*Percent of total from* *Pupae*	*Mined leaves*
Halticoptera patellana	30.9	2.1
Ganaspidium sp.	16.2	
Opius dimidiatus	13.6	1.1
Cothonaspis (*Erisphagia*) sp.	10.1	3.2
Opius sp.	9.2	
Eucoilidea sp.	5.5	
Chrysocharis ainsliei (Crawford)	3.2	0.5
Chrysocharis parksi (Crawford)	2.9	
Chrysocharis sp.	2.0	2.7
Pseudeucoila sp.	2.0	
Opius bruneipes (Gahan)	1.7	
Ganaspidium pusillae	1.4	
Opius suturalis	0.9	
Opius aridis (Gahan)	0.3	
Closterocerus cinctipennis		38.8
Derostenus fullawayi (Crawford)		23.9
Derostenus arizonensis		13.3
Diglyphus websteri (Crawford)		8.5
Derostenus varipes		4.3
Achrysocharis sp.		1.1
Derostenus agromyzae (Crawford)		0.5

The two *Phytomyza* species, *P. syngenesiae* (Hardy) and *P. horticola* Goureau, were treated as the single species, *atricornis* Meigen, until the group was revised by GRIFFITHS (1967c). GRIFFITHS, in my opinion correctly, discarded the name *atricornis* Meigen, 1838 as a nomen dubium. It had been revived somewhat arbitrarily by HENDEL, 1920 but was not used by 19th century authors and there is no real evidence to support HENDEL's concept. *P. horticola* and *P. syngenesiae* are virtually identical morphologically but there is a consistent difference in the male genitalia. This has been confirmed by STEYSKAL (1969) and by VON TSCHIRNHAUS (1969). However, GRIFFITHS (1972) rejects other suggestions made by STEYSKAL regarding differentiating characters of these two species. *P. syngenesiae* seems to favour Compositae and occurs only rarely on hosts in other families; its distribution is mainly Palaearctic but it is also found in North America, Australia and New Zealand, probably as a result of introduction by man. *P. horticola* is more truly polyphagous, having been recorded on 34 plant families; it is common in Europe but is also widespread in Africa, India and other parts of Asia. It is entirely absent from North America. Neither species is known from South America.

Without making genitalia preparations individual adults of *horticola* cannot be separated on external characters from *syngenesiae*. However, the geographical distribution and host range of these two species only partially overlap and in many cases the species can be confidently deduced from the area of origin and host. For example, specimens from North America (on the basis of existing knowledge!) will represent *syngenesiae*, while those from India will be *horticola*; and in Europe specimens from non-Composite hosts will almost certainly be *horticola*.

HUSSEY and GURNEY (1962) discuss two possible races or strains of *atricornis* and also indicate a useful method of assessing host preference by establishing the ratio of female feeding punctures to actual eggs laid (F/E ratio). Race A was obtained from chrysanthemum but the original host of Race B is not given and it is therefore not possible to determine whether these two races, in which minor morphological and biological differences were apparently detected, might perhaps represent *syngenesiae* and *horticola*.

Few records of parasites of *P. horticola* and *P. syngenesiae* have been published since they were segregated by GRIFFITHS in 1967. In a number of cases in earlier papers referring to '*P. atricornis*', it is now not possible to be certain which of the two species were concerned. GRIFFITHS (1964a) records *Dacnusa areolaris* (Nees), *D. laevipictus* Thomson, *D. maculipes* Thomson, *D. pubescens* Curtis (Braconidae), *Chorebus canariensis* Griff. and *C. sativi* (Nixon). ASKEW (1968a) records four Eulophids: *Diglyphus poppoea* Walker, *D. minoeus* (Walker); *Pnigalio soemius* (Walker); and *Hemiptarsenus dropion* (Walker). In addition, BOUČEK and ASKEW (1970) record 10 further Eulophids from Europe: *Achrysocharella formosa* (Westw.), *Chrysocharis appendigaster* (Masi), *C. elongata* (Thom.), *C. liriomyzae* Del., *C. melaensis* (Walk.), *C. pubicornis* (Zett.), *Cirrospilus vittatus* Walk., *Diglyphus isaea* (Walk.), *D. phytomyzae* (Ruschka), *Pediobius acantha* (Walk.); and three from Japan, *Cirrospilus hytomyzae* (Ishi), *Hemiptarsenus mitsukurii* (Ashmead) and *Pnigalio katonis* (Ishi). GRAHAM (1969) records three Pteromalids: *Cyrtogaster vulgaris* Walk., *Halticoptera circulus* (Walker) and *Seladermi berani*

Table VII

Host	*Liriomyza*					*Phytomyza*	
Sub-class/Order/Family	*bryoniae*	*strigata*	*huidobrensis*	*sativae*	*trifolii*	*horticola*	*syngenesiae*
DICOTYLEDONES							
I MAGNOLIIDAE							
5 Ranunculales							
1 Ranunculaceae						×	
6 Papaverales							
1 Papaveraceae	×	×				×	
II HAMMAMELIDAE							
4 Urticales							
3 Moraceae		×					
4 Cannabaceae						×	
III CARYOPHYLLIDAE							
1 Caryophalles							
7 Caryophyllaceae	×	×			×	×	
9 Basellaceae	×						
10 Chenopodiaceae		×	×		×	×	
11 Amaranthaceae	×	×				×	
3 Polygonales							
1 Polygonaceae						×	
IV DILLENIIDAE							
3 Malvales							
6 Malvaceae	×	×		×	×	×	×
6 Violales							
3 Violaceae	×	×	×			×	
6 Passifloraceae				×			
18 Loasaceae	×						
21 Cucurbitaceae	×	×	×	×	×	×	
8 Capparales							
2 Capparaceae	×	×				×	
3 Cruciferae	×	×		×		×	
4 Resedaceae	×	×				×	
5 Moringaceae				×			
12 Primulales							
3 Primulaceae	×	×					
V ROSIDAE							
1 Rosales							
17 Leguminosae	×	×	×	×	×	×	×
4 Myrtales							
10 Onagraceae						×	
9 Celastrales							
3 Celastraceae		×					
10 Euphorbiales							
2 Euphorbiacae	×	×		×		×	

Host	*Liriomyza*					*Phytomyza*	
Sub-class/Order/Family	*bryoniae*	*strigata*	*huidobrensis*	*sativae*	*trifolii*	*horticola*	*syngenesiae*
12 Sapindales							
7 Sapindaceae	×						
11 Anacardiaceae						×	
15 Rutaceae						×	
17 Zygophyllaceae					×		
13 Geraniales							
4 Tropaeolaceae	×	×				×	
14 Linales							
3 Linaceae	×	×				×	
16 Umbellales							
2 Umbelliferae	×	×	×	×	×	×	
VI ASTERIDAE							
1 Gentianales							
2 Gentianaceae	×						
2 Polemoniales							
1 Nolanaceae	×						
2 Solanaceae	×	×	×	×	×	×	
3 Convolvulaceae	×					×	
6 Polemoniaceae	×	×				×	
7 Hydrophyllaceae	×	×				×	
3 Lamiales							
1 Boraginaceae		×				×	
3 Verbenaceae	×	×				×	
5 Labiatae	×	×				×	×
4 Plantaginales							
1 Plantaginaceae		×		×		×	×
5 Scrophulariales							
3 Scrophulariaceae	×	×				×	
8 Bignoniaceae	×						
9 Acanthaceae	×						
10 Pedeliaceae	×						
6 Campanulales							
3 Campanulaceae	×	×				×	
8 Dipsacales							
3 Valerianaceae	×	×				×	
4 Dipsacaceae	×	×				×	
9 Asterales							
1 Compositae	×	×	×	×	×	×	×
MONOCOTYLEDONES							
IV LILIIDAE							
1 Liliales							
3 Liliaceae			×		×	×	

(Delucchi). Finally FISCHER (1971) records four *Opius* species in Europe *O. exiguus* Wesm., *O. gracilis* Fi., *O. lugens* Hal. and *O. pulicariae* Fi.

An analysis and comparison of the host ranges of the seven polyphagous species considered in this chapter has been of considerable interest. All confirmed hosts are shown in Table VII. This has followed CRONQUIST's (1968) botanical classification, in which all Orders are grouped into six sub-classes. All six sub-classes are represented among the hosts but there are substantial gaps both in orders and families. CRONQUIST's numbering of these groups has been used, as this more readily indicates these gaps and shows that even the most polyphagous species of Agromyzidae are in fact significantly restricted in their choice of host, with many important orders and families never being attacked.

The polyphagy shown by the two closely-related species, *Liriomyza bryoniae* and *L. strigata*, is substantially similar and would probably be found to overlap closely with that of the related New World species, *L. huidobrensis*, when this has been more intensively collected. The host range of this group is remarkably similar to that of *Phytomyza horticola*, which is of course in no way related. The ranges of the two other *Liriomyza* species, *sativae* and *trifolii* and also of *Phytomyza syngenesiae*, appear to be appreciably smaller.

BUHR (1953), in a discussion of leaf-miners attacking *Vicia faba* L., includes an interesting comparison of the hosts ranges of *Liriomyza bryoniae*, *L. strigata* and *Phytomyza horticola* (as *atricornis*).

FRICK (1952a, 1959) made a valuable start in clarifying American Agromyzidae in two major papers and also attempted to clarify the difficult group of leaf-mining species which were causing concern in Florida and California and which are considered here. Although it can now be seen that mistakes were made in his interpretation of species, his work nevertheless represented a major step forward, which has greatly facilitated all studies by subsequent authors.

The following key is provided in order to facilitate differentiation of the three polyphagous *Liriomyza* species which are constantly causing damage in the United States and several countries in South America

1 Discal cell large, last section of vein M_{3+4} at most $2\frac{1}{4}$ times length of penultimate (Fig. 314); third antennal segment slightly enlarged, frequently darkened, from pale to dark brown; legs: femora with conspicuous brownish-black striations; mesonotum black but somewhat mat; wing length from 1.7–2.25 mm; male genitalia as in Figs. 316, 317; posterior spiracles of larva and puparium each with an ellipse of 6–9 pores ... *huidobrensis* (Blanchard)

— Discal cell smaller, last section of vein M_{3+4} 3–4 times length of fourth (Fig. 322); third antennal segment small, bright yellow; legs: femora normally bright yellow, less frequently slightly darkened; mesonotum brilliantly shining black or obviously mat grey; wing length from 1.3–1.7

mm; posterior spiracles of larva each with 3 pores — 2

2 Hind-margin of eye largely yellow, both vertical bristles on yellow ground; mesonotum mat grey; male genitalia: aedeagus as in Figs. 329, 330 — *trifolii* (Burgess)

— Hind-margin of eye partially black, outer vertical always on black ground, inner normally on dark ground or at margin with yellow; mesonotum brilliantly shining black; male genitalia: aedeagus as in Figs. 323, 324 — *sativae* Blanchard

GENUS LIRIOMYZA MIK: PALAEARCTIC SPECIES

Liriomyza bryoniae (Kaltenbach, 1858)

Agromyza bryoniae Kaltenbach, 1858.
Liriomyza solani Hering, 1927a.
Liriomyza citrulli Rohdendorf, 1950.

ADULT. Small species with shining black mesonotum, largely yellow femora and both vertical bristles on yellow ground.

HEAD (Fig. 303). Orbits distinctly projecting above eye in profile, with 2 equal ors and 2 ori (rarely 3), orbital setulae minute, sparse, reclinate or almost entirely lacking; jowls deep, up to $\frac{1}{3}$ vertical height of eye, cheeks forming broad ring below eye; third antennal segment small, round, arista gradually tapering.

MESONOTUM. 3 + 1 dc, acrostichals irregularly in 4 rows.

WING. Length from 1.75–2.1 mm, costa extending strongly to vein M_{1+2}, last section of M_{3+4} approximately twice length of penultimate.

COLOUR. Frons bright yellow, orbits slightly differentiated, paler; both vertical bristles on yellow ground but black at hind margin of eye sometimes reaching base of outer vertical; third antennal segment normally bright yellow, rarely slightly darkened; mesonotum black, largely shining but with distinct mat undertone; mesopleura largely yellow, normally with small black bar on lower margin which sometimes also extends up front margin; legs: femora mainly bright yellow but with variable brownish striations, tibiae and tarsi more brownish; abdomen with tergites yellow laterally and narrow yellow hind-margins; squamae yellowish, margin dark, fringe ochrous to brown.

MALE GENITALIA. Aedeagus as in Figs. 304, 305; surstyli with single spine at end, epiphallus broad (cf. Fig. 318).

EARLY STAGES. Egg white, oval, length 0.25 mm; larva when newly hatched 0.5 mm, when full-grown 3 mm long, mouth-hooks each with 2 teeth which are strongly alternating (Fig. 306); posterior spiracles each with an ellipse of 7–12 pores (Fig. 307); puparium: pale yellowish-brown.

HOST-PLANTS. A highly polyphagous species, of economic importance on

Lycopersicum esculentum L. (tomato), *Citrullus vulgaris* Schrad. (watermelon), *Cucumis melo* L. (melon), *Cucumis sativus* L. (cucumber) and *Lactuca* sp. (lettuce). The species has also been reared from many genera in the following families:

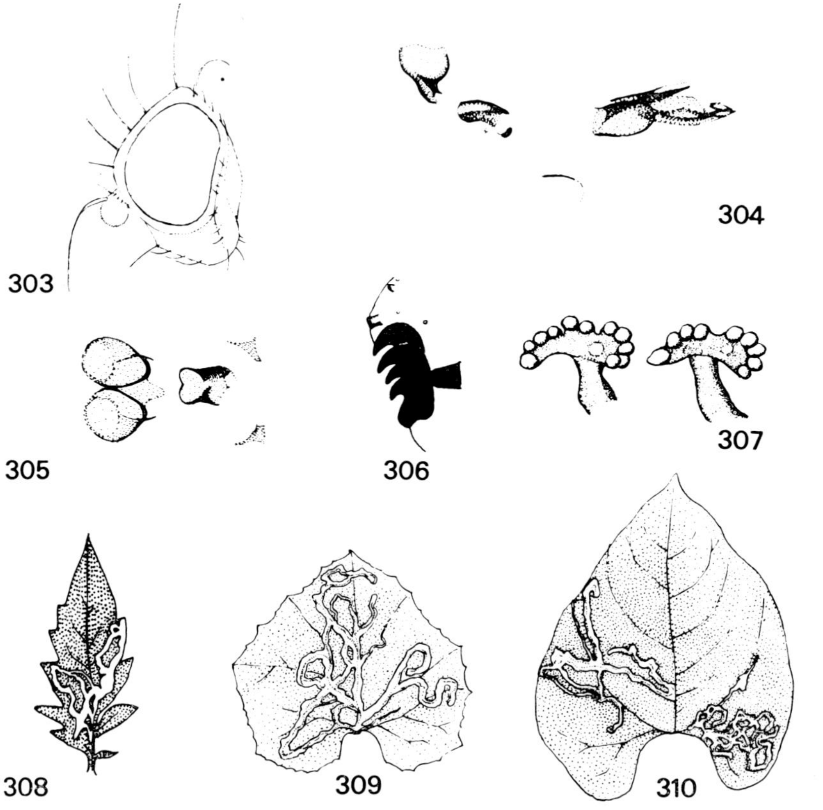

Figs. 303–310. *Liriomyza bryoniae* (Kalt.): 303, head; 304, aedeagus, side view; 305, distiphallus, ventral view; 306, larval mouth-hooks; 307, posterior spiracles (306, 307, HERING); 308, leaf-mine on tomato; 309, mines on melon; 310, mines on bean.

Amaranthaceae, Basellaceae, Capparaceae, Caryophyllaceae, Cruciferae, Dipsacaceae, Euphorbiaceae, Gentianaceae, Leguminosae, Malvaceae, Nolanaceae, Papilionaceae, Polemoniaceae, Pedeliaceae (= Martyniaceae), Scrophulariaceae, Tropaeolaceae, Umbelliferae, Valerianaceae and Verbenaceae; in addition, leaf-mines almost certainly referable to this species have been found on Acanthaceae, Bignoniaceae, Campanulaceae (= Lobeliaceae), Convolvulaceae, Hydrophyllaceae, Labiatae, Linaceae, Loasaceae, Papaveraceae, Polygonaceae, Primulaceae, Resedaceae, Sapindaceae and Violaceae (BUHR, 1954: 278).

BIOLOGY. The information given below is condensed from the detailed studies made by SPEYER (1936, 1938) and SPEYER and PARR (1944, 1946, 1948, 1949, 1951) on tomato plants in glasshouses at Cheshunt, Herts., England. Prior to oviposition females make numerous feeding punctures, either in cotyledons or young leaves. Eggs may be inserted in the lower surface but in the great majority of cases in the upper surface of the leaves. Duration of the egg stage varies at normal temperatures from 4–8 days, as indicated in Table VIII.

Table VIII. Period of Incubation

Eggs			*Period in days*		*Temperature °F.*			
No.	*Laid*	*Hatched*	*Mean*	*Range*	*Mean*	*Mean min.*	*Mean max.*	*Range*
57	4–14.II	9–21.II	5.6	4–6	69.7	64.3	81.6	62–89
40	8–18.IV	12–21.IV	6.2	5–8	67.1	52.5	94.5	43–104
100	8–18.IV	14–25.IV	6.3	5–8	66.8	51.9	95.1	
62	12–21.IV	19–27.IV	5.9	4–8	68.7	53.0	95.7	

A single female was observed to lay as many as 104 eggs, at an average rate of 7 per day, with a maximum of 15 in one day.

Larval feeding is rapid and a distance within the leaf of 2 inches can be covered in a 3 day period. The average duration of the larval stage is 9 days, with a range from 7–13 days. If a cotyledon or young leaf is not large enough to provide sufficient nourishment, larvae can move up the stem via a petiole into a second or even a third leaf. The larva is unable to penetrate leaves from the outside.

Prior to pupation the larva cuts a characteristic semi-circular exit slit in the epidermis of the upper leaf surface, remains within the mine for a further hour or so and then normally drops to the ground, penetrating just below the surface before actually pupating. Some cases were observed of puparia remaining attached to the upper or even lower surface of the leaves but this is exceptional.

The duration of the pupal instar varies considerably according to the generation. In normal spring and summer generations this is about 3 weeks. In a very early generation in February/March adults emerged only after 5–9 weeks. With the latest autumn generation diapause will normally last until the following spring. The typical life cycle of four generations is indicated by SPEYER's log (1938: 68):

May 24th	Flies emerge from pupae.	
May 31st	Pit-marks on foliage.	
June 2nd	Beginning of larval mines ..	Emergence to hatching of eggs, 9 days.
June 7th–12th	Larvae pupate	Larval stages, 5–6 days.
June 23rd	First pupae emerge	Pupal stage about 15 days.
June 30th	Beginning of larval mines ..	Emergence to hatching of eggs, 7 days.

July 4th–10th	Larvae pupate	Larval stages, 4–6 days.
July 22nd–25th	Flies emerge from pupae ..	Pupal stage, 14–18 days.
July 27th	Beginning of larval mines ..	Emergence to hatching of eggs, 5 days.
July 30th–Aug. 3rd ..	Larvae pupate	Larval stages, 3–5 days.
Aug. 14th	Flies begin to emerge from pupae	Pupal stage, 13–16 days.

The irregular linear mine produced by the larva on tomato, melon and bean is shown in Figs. 308–310.

DAMAGE CAUSED. Larval feeding in cotyledons and the leaves of young tomato plants can seriously weaken and even destroy plants. Destruction of the tissue in the cotyledons prevents normal development of the plants and may bring about their total collapse. Through removal of the collenchyma the plant is unable to avail itself of phosphate taken up by the roots and additional loss of the greater portion of the parenchyma in the rough leaves brings assimilation practically to a standstill (SPEYER and PARR, 1949).

In France in addition to tomatoes, leaves of melon, cucumber and lettuce have been found containing up to ten larvae and with such a degree of infestation they turn yellow and dry up. Attacks on young plants can result in the complete destruction of cotyledons and death of the plants.

According to AUDEMAR and D'AGUILAR (1969) 'the tissues of young leaves of melon react to the attacks by becoming thicker and hardening in an area more or less encircling the mines. This reaction causes damage differing from that on other plants and not directly related to larval damage. The drying up, when affecting numerous leaves, may result in the fruit dropping'.

In Egypt damage is caused to cucurbit plants particularly in the seedling stage. Infested plants are mined at first on the underside of the leaves and with a heavy attack affected leaves appear dwarfed, droopy and may dry up. Such plants are weaker than normal and the fruit will inevitably be affected. ABUL-NASR and ASSEM (1961) noted that 'seedlings and young plants are very sensitive to the attack and may perish'.

DISTRIBUTION. Albania (UTECH, 1962), Denmark, Egypt, England (also Channel Islands: Guernsey), France, Germany, Morocco, Spain, including Canary Islands, Netherlands; U.S.S.R.: Ukraine (ROHDENDORF, 1950), Caucasus (UTECH, 1962).

PARASITES. Europe: *Dacnusa hospita* (Förster), *D. maculipes* Thomson, *D. sibirica* Telenga; *Chorebus daimenes* (Nixon); *Aphidius ervi* Hal., *Opius pallipes* Wesm. (Braconidae); *Chrysocharis pubicornis* (Zett.), *Hemiptarsenus zilahisebessi* Erdös, *Pediobius acantha* (Walk.) (Eulophidae); Egypt: *Chrysocharis* sp., *Diglyphus* sp.; *Charips* sp. (Cynipoidea).

Discussion

TAXONOMY. This species is readily recognizable from *Liriomyza strigata* (Mg.), with which it has sometimes been confused in the past, by the generally paler coloration and in particular by both vertical bristles being on yellow ground. Specimens from Egypt occasionally have the third antennal segment distinctly infuscated.

The male genitalia indicate the close relationship between *L. bryoniae* and *L. strigata* and there seems little doubt that the two are sister-species. Larval characters are very similar, although there are generally slightly more pores on the posterior spiracles in *strigata*.

The similarity in the genitalia between *L. bryoniae* and *L. huidobrensis* Blanchard (=*langei* Frick) and also the occasional darkening of the third antennal segment in *bryoniae* which is a characteristic feature of *langei*, led me earlier to synonymize *langei* and also *decora* Blanchard with *bryoniae* (SPENCER, 1963a: 359). However, I am now satisfied that the European species is distinct and these synonymies are formally withdrawn.

DISTRIBUTION. *L. bryoniae* occurs most commonly in the wild in southern Europe, whereas in England, Germany and the Netherlands it is found most frequently in glasshouses and in Botanical Gardens (Kew, Berlin, Rostock). Its distribution in eastern Europe is inadequately known and there are no records east of the Ukraine and the Caucasus.

ECONOMIC IMPORTANCE. This polyphagous species causes significant damage to two entirely different crops – tomatoes and cucurbits.

It first came to notice as a pest of tomatoes in glasshouses in N.W. Europe (Denmark, England, Netherlands) but is also recognized as a major problem in tomato cultivation in Spain. In England the main area of infestation was in the Lea Valley, Hertfordshire where the species has been known since 1936. Some 15 years ago a new area of glasshouse tomato cultivation was developed in West Sussex. According to Dr. N. W. HUSSEY of the Glasshouse Crops Research Institute (personal communication 18 May, 1971) 'the species was unknown there between 1956 and the late autumn of 1968. At that time a nursery near Littlehampton imported a large number of tomato plants from Holland. The following spring a severe outbreak of leaf-mining occurred on these plants during the fruiting stage. In the tomato propagating season of 1969–1970, *L. bryoniae* appeared on three or four nurseries within a mile or so of the original site of infestation. From a practical commercial point of view the outbreak is now under control but the pest is now endemic on all these nurseries and constant control measures have to be taken'.

Dr. HUSSEY further states that *L. bryoniae* is particularly damaging to young plants, as seedlings are often killed. Control is difficult, as many insecticides have proved phytotoxic or create residue problems, from which the puparia are virtually protected in the soil.

In the Netherlands damage to tomatoes was serious locally in 1965 and 1966. A massive increase in population levels has been periodically noted in the glasshouses of the Province South-Holland and in general the leaf-mining activity of *L. bryoniae* has been increasing in recent years (DE BROUWER and VAN OFFEREN, 1966).

Cucurbits, including water melon, sweet melon, cucumber, marrow and snake cucumber, are regularly attacked in early spring in Egypt. Infested crops were observed at Ismailia, Salhia, Dokki and Giza. The water-melon, Egyptian melon (*Cucumis melo* var. *aegyptiacus* Sick.) and squash can suffer severe damage. (F. A. HERAKLY, personal communication; ABUL-NASR and ASSEM, 1961). In the South of France, in the Rhone Valley, melons and cucumbers have also suffered damage. *L. bryoniae* was recorded in England, near Reading, attacking glasshouse-grown cucumbers in 1954 (unpublished M.A.F.F. Report).

Apart from regular damage to tomatoes and cucurbits, in view of its exceptional polyphagy *Liriomyza bryoniae* must be considered as a dangerous potential pest on a number of other crops.

Liriomyza strigata (Meigen, 1830)

ADULT. Small species with shining black mesonotum, largely yellow femora and at least outer vertical bristle on dark ground. Mesopleura normally with black

311

312

313

Figs. 311–313. *Liriomyza strigata* (Mg.): 311, aedeagus, side view; 312, distiphallus, ventral view; 313, leaf-mine on *Pisum*.

patch along lower margin but in darkest specimens entire lower half black. In all other characters agreeing closely with *L. bryoniae* above.

MALE GENITALIA. Aedeagus as in Figs. 311, 312; surstyli normally with 2 spines at end, sometimes even with 3, epiphallus broad (cf. Fig. 318).

EARLY STAGES. Larva having posterior spiracles each with 10–12 pores.

HOST-PLANTS. A highly polyphagous species, main vegetable crops attacked are *Beta vulgaris* L. (Chenopodiaceae), *Lactuca* sp. (Compositae) and *Pisum sativum* L. (Leguminosae). Ornamentals regularly serving as hosts are discussed in Chapter 11. HERING (1957c) records 187 genera as hosts, in following 31 families: Amaranthaceae, Boraginaceae, Campanulaceae, Capparaceae, Caryophyllaceae, Celastraceae, Chenopodiaceae, Compositae, Cruciferae, Cucurbitaceae, Dipsacaceae, Euphorbiaceae, Hydrocotylaceae, Hydrophyllaceae, Labiatae, Leguminosae, Linaceae, Malvaceae, Moraceae, Papaveraceae, Plantaginaceae, Polemoniaceae, Primulaceae, Resedaceae, Scrophulariaceae, Solanaceae, Tropaeolaceae, Umbelliferae, Valerianaceae, Verbenaceae, Violaceae. The additional genus, *Phlox* (Polemoniaceae) has also been recorded as a host.

BIOLOGY. Surprisingly no detailed studies have been made of the life history of this species. According to DE MEIJERE (1924: 279) the egg is laid near the margin of the leaf and the young larva mines initially on the lower surface to the midrib above which, now on the upper surface, the main mine is formed, with lateral offshoots into the leaf-blade, frequently following one of the subsidiary veins (Figs. 313, 502, 514). HENDEL (1931: 250) confirms DE MEIJERE's report that the early mine is normally on the lower surface of the leaf. He states further that on reaching the midrib the larva first mines towards the leaf-base and then turns and feeds along the midrib apically, forming offshoots of varying length into the leaf-blade but always returning again to the midrib. Pupation takes place externally.
DAMAGE CAUSED. Young plants are not normally attacked and, although the species is common and widespread, large populations are rarely found in a single plant. The larval damage to mature plants is thus negligible.
DISTRIBUTION. Certainly present in all countries in western Europe, including U.S.S.R. Eastern limit of its range is not known.
PARASITES. In Western Europe: *Dacnusa discolor* (Förster), *D. maculipes* Thomson; *Chorebus daimenes* (Nixon); *Opius exiguus* Wesm., *O. levis* Wesm., *O. pallipes* Wesm., and *O. propodealis* Fischer; *Halticoptera smaragdina* (Curtis); three Eulophids are recorded by BOUČEK and ASKEW (1970): *Chrysocharis albicans* Del., *C. pubicornis* (Zett.) and *Pediobius acantha* (Walk.). In Iraq AL-AZAWI (1971) records *Achrysocharella formosa* West.

Discussion

TAXONOMY. *Liriomyza strigata* belongs to a group in which the adults are difficult to distinguish on external characters. The male genitalia, however, are distinctive and in Europe only directly comparable with *L. bryoniae*. The leaf-mines are entirely characteristic and this form is not found in any other European *Liriomyza* sp.
ECONOMIC IMPORTANCE. No cases have been recorded of *L. strigata* causing appreciable damage to crops. Although it must be considered a common species, its numerous parasites apparently exercise effective control on population levels. However, with such an exceptionally wide host range, its potential to build up populations of pest proportions should be recognized. Its occurrence on ornamentals is discussed in Chapter 11, p. 327.

NEARCTIC, NEOTROPICAL SPECIES

***Liriomyza huidobrensis** (Blanchard, 1926)

Agromyza huidobrensis Blanchard, 1926.
Liriomyza huidobrensis Blanchard, 1938: 356; 1954: 32.
Liriomyza cucumifoliae Blanchard, 1938: 352, **syn. nov.**

* See also Appendix A.

Liriomyza langei Frick, 1951, **syn. nov.**
Liriomyza dianthi Frick, 1958, **syn. nov.**
ADULT. Medium-sized species, with largely black mesopleura, blackish femora and, normally, darkened third antennal segment.
HEAD. Orbits distinctly projecting above eye; 2 ors, 2 ori, the lower weaker and largely incurved (rarely an additional third present), orbital setulae sparse, reclinate; eye somewhat slanting, jowls deep, up to $\frac{1}{3}$ height of eye, cheeks forming distinct ring below eye; third antennal segment small, round, arista tapering gradually from base.
MESONOTUM. 3 + 1 dc, third and fourth equidistant each side of suture, fourth only slightly smaller, acrostichals irregularly in 4 rows.
WING (Fig. 314). Length from 1.7–2.25 mm, costa extending strongly to vein M_{1+2}, last section of M_{3+4} from slightly less than 2 to $2\frac{1}{2}$ times length of penultimate, position of first cross-vein variable, before, at or beyond midpoint of discal cell.
COLOUR. Frons yellow, generally more orange rather than pale lemon-yellow; upper orbits slightly darkened at least to upper ors, both vertical bristles on dark ground; third antennal segment normally brownish yellow, in palest specimens only faintly infuscated; mesonotum shining black but not brilliantly so, somewha mat, particularly when viewed from front, hind-corners with yellow patch adjoin ing scutellum; mesopleural variable but generally black on lower three-quarter (Fig. 315), mesopleural bristle always on black ground, in darkest specimens only narrow yellow band along upper margin; legs: coxae yellowish-black, femor; basically yellow but variably darkened with black striations, in darkest specimens appearing almost completely black but yellow always detectable on inner sides tibiae and tarsi frequently appearing black, sometimes paler, more brownish-squamae yellow, margin and fringe black.
MALE GENITALIA. Aedeagus as in Figs. 316, 317, with mesophallus separated by membranous gap from paired distiphallus; surstyli with single curving spine on inner corner; epiphallus broad (Fig. 318).
EARLY STAGES. Larva up to 3.25 mm long, posterior spiracles each with a regular ellipse of about 6–9 minute pores (Fig. 319); puparium yellowish- or reddish-brown.
HOST-PLANTS. A highly polyphagous species, with confirmed records on nine families: Chenopodiaceae, Compositae, Cucurbitaceae, Leguminosae, Liliaceae, Linaceae, Solanaceae, Tropaeolaceae, Umbelliferae and Violaceae. Crop plants attacked include *Beta vulgaris* L., *Spinacia oleracea* L.; *Lactuca sativa* L.; *Cucumis melo* L.; *Pisum sativum* L., *Vicia faba* L.; *Allium cepa* L.; *Linum usitatissimum* L.; *Lycopersicum esculentum* L., *Solanum tuberosum* L.; *Capsicum annuum* L.; *Apium graveolens* L.

Many ornamentals also serve as hosts (see Chapter 11).
BIOLOGY. The larva feeds primarily as a leaf-miner but on peas may also feed on the outer surface of young pods (cf. LANGE *et al.*, 1957: 3). The leaf-mine is frequently associated with the midrib and lateral veins (Figs. 320, 505) but can also be irregularly serpentine and, particularly when several larvae are present in a single leaf, can form a secondary blotch.

No detailed studies have been made on the life history but generations obviously follow in quick succession as long as growing conditions of the host-plants provide suitable food. LANGE *et al.* (1957) suggest there may be as many as five or six generations in a year.

DAMAGE CAUSED. Plants can be significantly damaged not only by larval feeding but also by the feeding punctures of females. Many larvae frequently occur together in the same leaf and leaves attacked in this way suffer premature wilting

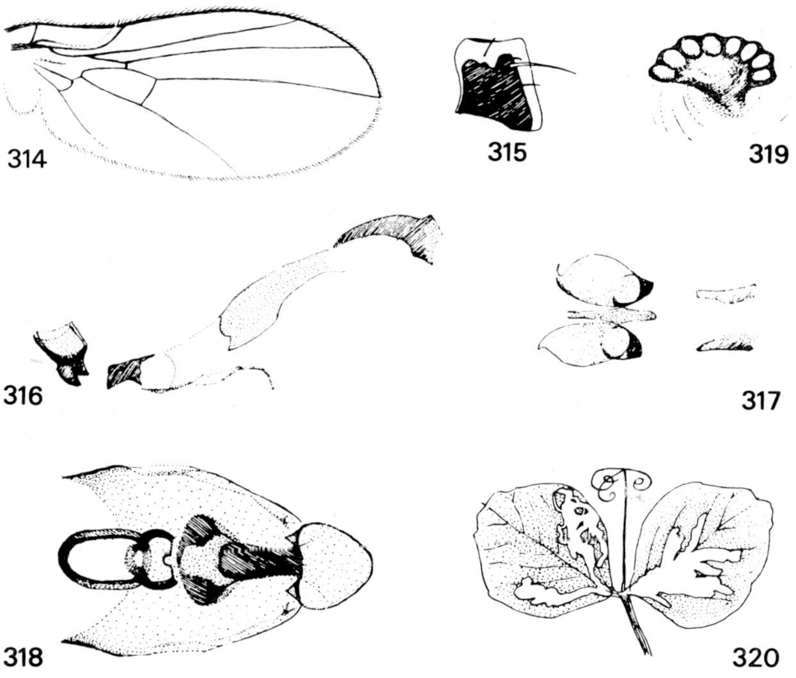

Figs. 314–320. *Liriomyza huidobrensis* (Blanchard): 314, wing; 315, mesopleuron; 316, aedeagus, side view; 317, distiphallus, ventral view; 318, aedeagal hood and epiphallus; 319, posterior spiracles of puparium (one side); 320, leaf-mine on *Pisum*.

and even die (WILCOX and HOWLAND, 1954). LANGE *et al.* (1957) also report that leaves turn white and wither. WILCOX counted as many as 100 feeding punctures on a single spinach leaf. General stunting of the plants results from the feeding both of the larvae and of the females.

DISTRIBUTION. Argentina, Brazil, Chile (including Juan Fernandez Is.), Peru, Colombia, Venezuela; California, Hawaii.

PARASITES. California: *Solenotus intermedius* (Girault), *S. begini* (Ashmead), *Halticoptera aenea* (Walker), Chalcididae; *Mesora* sp., *Opius* sp., *Chrysocharis ainsliei* Crawford, *C. parksi* Crawford (LANGE, 1957). LANGE states that a combined rate of parasitism of from 50–90% is not unusual.

No parasites have been identified in South America or in Hawaii.

Discussion

TAXONOMY. This species was originally described in Argentina by BLANCHARD from specimens from *Cineraria*. Subsequently, *L. cucumifoliae* was described from cultivated melon and *L. decora* from *Vicia faba*, both with only minor colour differences from *huidobrensis*. I have examined specimens of *cucumifoliae* and *decora* and am satisfied that they are identical with *huidobrensis*, with which they are now formally synonymized.

FRICK (1951) described *langei* from California, which at that time was known there primarily as a pest of peas. Later, *L. dianthi* Frick (1958) was described also from California but FRICK himself in 1964 synonymized *dianthi* with *langei*. These two species are clearly identical with *huidobrensis* and are also synonymized with it.

The variation in colour led both BLANCHARD and FRICK to re-describe their own species and FRICK of course had little reason to suspect that his *langei* from California could merely represent *huidobrensis* Blanchard from Argentina.

The larger discal cell and thus relatively shorter last section of vein M_{3+4} immediately separates *huidobrensis* from *sativae*. It is also larger and significantly darker, with frequently a characteristic darkening of the third antennal segment. Both the larval and puparial stages are readily recognizable by the more numerous pores on the posterior spiracles.

DISTRIBUTION. In South America *L. huidobrensis* appears to have used the Andes as a migration route to the south and is widespread in Argentina and has also reached the Juan Fernandez Is. (SPENCER, 1964c). The only record in Brazil is from Campinas. The Amazon Basin has apparently proved an effective barrier on the East of the Continent. In North America it is known only from California and is apparently entirely lacking from the southern States, Florida and the East. Its presence in Hawaii can certainly be explained by a relatively recent introduction and it is not present in other Pacific islands.

HOST-RANGE. The host range, although showing a wide degree of polyphagy, is nevertheless restricted, with many large and widespread families never attacked. The host range overlaps remarkably closely with that of *L. trifolii* (p. 226) and *L. sativae* (p. 221) but, perhaps not surprisingly with a species having such a wide distribution, some local preferences can be noted. *Beta* and *Pisum* are two of the most favoured hosts, both in Argentina and California. Tomato serves as a host sparingly in Argentina and has been confirmed experimentally in California (LANGE, 1957). *Allium* is known as a host both in Venezuela and California. The only record on Cucurbitaceae is from Argentina. Composites are regularly attacked but damage has been recorded mainly on ornamentals (see Chapter 11).

ECONOMIC IMPORTANCE. *L. huidobrensis* is a serious pest and substantial damage has been recorded on differing hosts from most parts of its range.

In Argentina serious damage has been caused to *Vicia faba* (BLANCHARD,

1954: 32) and at Tucuman in 1953 many plants died, in consequence of the large numbers of leaflets per plant which were destroyed by the larvae. From 5–13 larvae per leaflet were counted. *Beta* has also been heavily attacked, with 80% of all plants in one field infested and with up to 10 mines per leaf (WYGODZINSKY, personal communication). In Peru PORTER (1939) records damage to beans.

In Venezuela the most severe damage is caused to garlic (*Allium sativum* L.) in the highlands of the States of Lara and Miranda (1,000 m). *L. huidobrensis* is considered a permanent pest in these areas but the intensity of the damage varies from year to year (CERMELI, personal communication).

In California localized attacks occur in most years but major outbreaks appear to be cyclic in nature. Damage was particularly serious in the years 1938, 1945, 1948 and 1956. For example, in 1945 a 10 acre field of spinach was completely destroyed, while in 1957 a 50% loss of the same crop occurred in the Salinas Valley (LANGE *et al.*, 1957). Extensive damage has been caused to seedling pepper plants and in 1953 'five hundred acres of chili peppers were affected, stunting of the plants being severe in part of the acreage and mild in other parts. Economic loss was due to this stunting and to the cost of replacing missing plants. Damaged plants were about half as large as undamaged ones'. Experimentally a reduction of 54% in plant growth due to leaf miner attack was observed (ELMORE and RANNEY, 1954). This species has also been responsible for appreciable damage to lettuce.

A number of ornamentals are also heavily attacked in California (see Chapter 11).

***Liriomyza sativae** Blanchard – South and North America, Pacific

Liriomyza sativae Blanchard, 1938 (Argentina). Host of holotype, *Medicago sativa* L.

Liriomyza pullata Frick, 1952b, **syn. nov.** (Hawaii). Host of holotype, *Datura*.

Liriomyza canomarginis Frick, 1952b, **syn. nov.** (Hawaii). Host of holotype, *Indigofera*.

Liriomyza minutiseta Frick, 1952b, **syn. nov.** (Hawaii). Host of holotype, 'tomato'.

Liriomyza munda Frick, 1957a, **syn. nov.** (California). Host of holotype, *Lycopersicum*.

Liriomyza guytona Freeman, 1958, **syn. nov.** (Alabama). Host of holotype, 'beans'.

ADULT. Very small species, with brilliantly shining black mesonotum.

HEAD (Fig. 321). Frons very slightly projecting above eye, just less than $1\frac{1}{2}$ times width of eye; 2 equal ors, 2 ori, the lower weaker, orbital setulae sparse, reclinate; jowls deep, almost $\frac{1}{3}$ height of eye at rear, cheeks forming distinct ring below eye; third antennal segment small, round, noticeably pubescent but not exceptionally so.

MESONOTUM. 3 + 1 dc, third and fourth substantially weaker; distance between first and second up to twice that between second and third; second, third and fourth almost equidistant; acr irregularly in 4 rows.

* See also Appendix A.

WING (Fig. 322). Length 1.3–1.7 mm, discal cell small, last section of M_{3+4} from 3–4 times length of penultimate.

COLOUR. Head, including antennae and face, bright yellow; hind-margin of

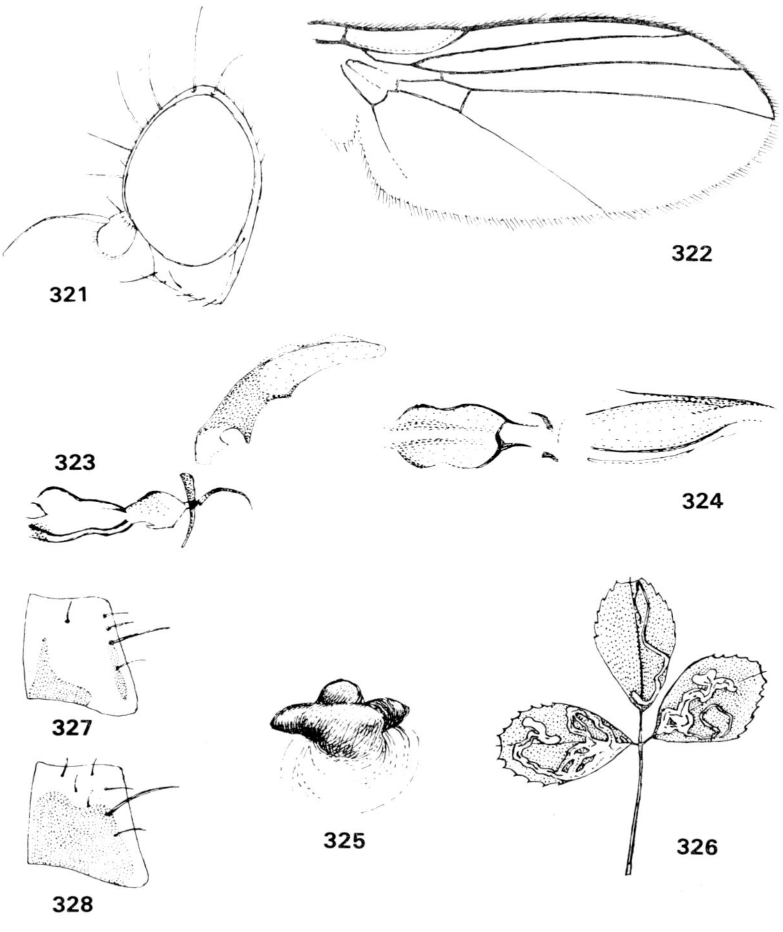

Figs. 321–328. *Liriomyza sativae* Blanchard: 321, head; 322, wing; 323, aedeagus, side view; 324, same, ventral view; 325, posterior spiracles of puparium (one side); 326, leaf-mine on alfalfa, Argentina; 327, mesopleuron of *munda* Frick (holotype); 328, same of '*pictella*', California.

eye black, vte always on black ground, the dark colour diminishes towards the upper orbits and becomes paler, brownish, towards base of vti, which may be just on dark ground or on the yellow; mesopleura predominantly yellow, with a variable dark area, which in the palest specimens is limited to a small grey bar along the lower margin and in the darkest extends along the entire lower margin, well up the front margin and also narrowly up the hind-margin; sternopleura largely filled by large black triangle but always with broad yellow margin above; legs: femora and coxae bright yellow, tibiae and tarsi darker, brownish-yellow on fore-legs, brownish-black on hind-legs; abdomen largely black but tergites variably yellow at sides; squamae yellowish, margin and fringe dark.

MALE GENITALIA. Aedeagus as in Figs. 323, 324, indentation at outer margin of distiphallus somewhat variable but normally pronounced; surstyli with a single bristle on inner corner; epiphallus and aedeagal hood as in *trifolii* (cf. Fig. 331).

EARLY STAGES. Puparium pale yellowish-orange, posterior spiracles on pronounced conical projections, each with 3 distinct bulbs, the 2 end ones elongate (Fig. 325).

HOST-PLANTS. Crop plants attacked are mainly in the three families Cucurbitaceae, Leguminosae and Solanaceae, as follows: Cucurbitaceae: *Cucumis melo* L. (cantaloupe), *C. sativus* L. (cucumber), *Cucurbita pepo* L. (squash).

Leguminosae: *Cajanus cajan* (L.) Millsp., *Lupinus* sp., *Medicago sativa* L. (alfalfa), *Melilotus alba* Desr., *Phaseolus lunatus* L. (lima bean), also 'pink bean' (California), 'black bean' (Venezuela), 'pole-bean' (Guam), *Vigna repens* Baker, *V. sinensis* Endl.

Solanaceae: *Capsicum annuum* L. (pepper), *Lycopersicum esculentum* L. (tomato), *Solanum melongena* L. (egg-plant), *S. tuberosum* L. (potato).

Less commonly *Brassica* spp. (Cruciferae) and *Gossypium* (Malvaceae) serve as hosts.

Other recorded hosts include Compositae: *Aster, Calendula, Dahlia, Galinsoga, Lipochaeta, Verbesina, Zinnia;* Euphorbiaceae: *Ricinus*; Hydrocotylaceae: *Hydrocotyle*; Labiatae: *Stachys*; Leguminosae: *Bauhinia, Cassia, Indigofera, Vicia*; Malvaceae: *Anoda, Sida*; Moringaceae: *Moringa*; Passifloraceae: *Passiflora*; Plantaginaceae: *Plantago*; Solanaceae: *Cestrum, Datura* (FRICK, 1952b, 1957a; SPENCER, 1963a; STEGMAIER, 1966b).

OATMAN (1959b) listed a number of additional hosts accepted experimentally by his 'melon leaf-miner'. This population came originally from Cranshaw melons from Patterson, California but was subsequently maintained on *Phaseolus lunatus*, Henderson bush variety. In view of the difficulties in ascertaining the exact status of this population, there seems little point in reproducing here OATMAN's list of host-plants.

BIOLOGY. The leaf-mine (Fig. 326) is irregularly linear, with frass in conspicuous black strips at alternate sides of the channel. Populations are frequently very large and STEGMAIER (1966b) reports up to 80 larvae in a single leaf of *Ricinus*. Leaf-mines on tomato in Venezuela are shown in Plate 7.

There are several generations in a year, breeding probably only being limited by the availability of fresh growth in suitable host-plants. In southern Florida plants can be attacked at any time but infestations are heaviest in the coolest parts of the year. In December and January the life cycle requires from 24–28 days (WOLFENBARGER, 1947).

DAMAGE CAUSED. Although the individual larva and leaf-mine are small, *L.*

Plate 7: *Liriomyza sativae:* mines on tomato, Venezuela (L. CERMELI, Servicio Shell, Venezuela).

sativae periodically occurs in very large numbers and the resulting destruction of leaf tissue can severely weaken or even destroy not only young but even mature plants. Yields of such diverse crops as 'beans, peas, cabbage, turnips, okra, squash, tomatoes, potatoes' can be affected. Further, the mines make infested leaves more susceptible to wind damage and possibly also to attacks by plant pathogens. Feeding punctures made by the females can cause the leaves to appear yellowish and give a stippled appearance (WOLFENBARGER, 1947).

DISTRIBUTION. Argentina, Peru, Venezuela; ?Central America; Barbados, Bahamas: Andros, Jamaica; United States: California, Alabama, Texas, Florida, S. Carolina, Ohio, Hawaii; Guam, Tahiti.

It is noteworthy that there are no records of *L. sativae* from Brazil. There are also no confirmed records from the eastern United States, though it is believed that *L. phaseolunata* (Frost) from New Jersey may represent this species (p. 74). Many specimens have been seen from Costa Rica probably referable to *sativae* but these have not been positively identified.

Records from Hawaii, Guam and Tahiti presumably represent relatively recent introductions. The absence of records from other Pacific islands suggests that the species is not generally distributed in the Pacific area.

PARASITES. In Florida STEGMAIER (1966b) records *Opius* sp. (Braconidae); *Chrysocharis* sp., *Closterocerus cinctipennis* Ashmead, *Derostenus* sp. (Eulophidae); *Halticoptera patellana* (Dalman) (Pteromalidae). Other records from Arizona (HILLS and TAYLOR, 1951), Texas (HARDING, 1965) and California (MICHELBACHER *et al.*, 1951; OATMAN, 1959a; JENSEN and KOEHLER, 1970) refer either only in part to *sativae* (as *munda*) or the identification of the host is open to doubt. The parasites recorded by these authors are therefore given in the introduction above, p. 203. One species recorded from Hawaii is *Opius dissitus* Muesebeck, 1963.

Discussion

TAXONOMY. *Liriomyza sativae* is readily distinguishable from the two other polyphagous species, with at least partially overlapping distributions, *L. huidobrensis* and *L. trifolii*; the important characters are the brilliantly shining black mesonotum, consistently small, entirely yellow third antennal segment and small discal cell. It can be confused with paler specimens of *L. brassicae* which do not have darkened orbits, and the male genitalia provide the only means of separating such specimens. I have also seen further undescribed species from Peru and Hawaii which can also only be distinguished by the genitalia.

The colour shows some variation. The extent of the darkening of the hind-margin of the eye and of the mesopleura is not entirely constant and the attachment of too much importance to these characters led FRICK (1952b) to describe the three species in Hawaii, *L. pullata*, *L. canomarginis* and *L. minutiseta*. However, the hind-margin of the eye is always darker than in *trifolii* and the mesopleura are always paler than in *huidobrensis*.

STEYSKAL (1964) synonymized *L. guytona* Freeman with *munda* Frick. It is of interest to compare his illustrations of the aedeagus with those of SPENCER (1963a; 1965a) and SASAKAWA (1964). I have examined a paratype of *Liriomyza minutiseta* Frick from Hawaii and the aedeagus appears substantially different from the illustration given by SASAKAWA (1964: Fig. 1). These examples clearly indicate the difficulty of viewing and drawing such minute structures objectively. The same thing appears slightly different to different workers and personal interpretations become very apparent in the drawings of individual workers. It is hoped that the further illustrations given here will provide additional help in the identification of this species.

In both larvae and puparia the three pores on each of the posterior spiracles immediately separate this species from *L. huidobrensis* but on larval characters *sativae* is not separable from *trifolii* and *trifoliearum*, all of which have some hosts in common.

FRICK (1957a) described *L. munda* in California from tomato and other hosts but made no attempt to compare or contrast this species with the three species he had earlier described from Hawaii, although the holotype of *minutiseta* was also from tomato and that of *pullata* was from another solanaceous genus, *Datura*. In the same paper he attempted to differentiate what has subsequently been referred to as the 'melon leaf miner' and which was unfortunately misidentified as *pictella* (Thomson). I have the holotype of *pictella* before me and, without having examined the male genitalia, FRICK quite reasonably accepted this as the darkest form he knew in California, from melon. However, in the true *pictella* the mesopleura are almost entirely deep black with only a small U-shaped area yellow in the centre below the upper margin. The male genitalia are also entirely distinct (cf. SPENCER, 1965a: Figs. 5A, B).

The differences given by FRICK between *munda* and his '*pictella*' can be tabulated as follows:

	munda	'*pictella*'
femora	yellow, each with a small dark spot distally	yellow, with streaks of black, appearing smoky
orbits	yellow	slightly darkened to lower ori
mesopleura	only small area dark (Fig. 327)	up to $\frac{2}{3}$ dark (Fig. 328).

However, these differences are not constant and FRICK himself was not able reliably to distinguish specimens of *munda* from tomato and '*pictella*' from melon (OATMAN, 1961).

ECONOMIC IMPORTANCE. *Liriomyza sativae*, under different names, has been recognized as a serious pest on a variety of crops in Argentina, Peru, Venezuela, Alabama, California, Florida, Texas and Hawaii. It is possibly the most injurious species known in both North and South America. Its significance as a pest can vary greatly from year to year and also from locality to locality but it seems to be agreed that the damage caused has appreciably increased since the first widespread use of DDT in the immediate post-war years and the subsequent general use of chlorinated hydrocarbons and organo-phosphorous insecticides.

In South America substantially greater damage appears to be caused in Venezuela than in either Argentina or Peru. In Argentina heavy attacks have been noted on *Medicago sativa*, and in one field an infestation rate of up to 80% of all plants has been observed; mines or adults have also been confirmed on *Dolichos lablab* and *Phaseolus vulgaris*. In Peru *L. sativae* is common on beans but no doubt occurs on other hosts. In Venezuela recorded hosts are tomatoes, potatoes, egg-plant and pepper; beans (*Phaseolus*) and peas; and cucurbits (cucumber, cantaloupe, watermelon, squash). Severe damage is caused to tomatoes and potatoes. On beans economic damage is particularly serious when plants are young. It is a serious pest on all four main cucurbit crops, damaging not only young but even mature plants (L. CERMELI, personal communication).

The main crop suffering damage in Florida is the tomato but in years of particularly heavy infestations, such as 1947, economic damage has been caused to a wide variety of truck crops, including beans, peas, okra, squash and potatoes. 'Large fields of nearly grown potatoes were injured so severely that they appeared brown and burnt. Individual plants occasionally died from the insect attacks' (WOLFENBARGER, 1947).

This species has caused considerable damage to peppers, cantaloupes, blackeyed peas and tomatoes in the Lower Rio Grande Valley in Texas. Since 1952 it has been considered the major problem in the production of peppers in Texas (WENE, 1953, identified as *Liriomyza subpusilla* Frost). WOLFENBARGER and GETZIN (1963) quantify the improved yields of the pepper crop following application of insecticides, as follows:

	Weekly applications	*One application*	*No. applications*
Yield (ton/acre)	6.29	5.80	4.75

WENE (1953, 1955) confirmed experimentally that certain insecticides exacerbated rather than improved the leaf-miner problem, by being more toxic to the hymenopterous parasites than the leaf-miners themselves. GETZIN (1960) followed up this work in attempting to find selective insecticides which were effective against *L. munda* but permitted survival of the parasites (see Chapter 1, p. 37).

In Alabama FREEMAN (1958) reported that in 1956 *L. sativae* (as *L. guytona*) caused serious damage to foliage of tomato, bean, squash, cucumber and other garden and greenhouse plants. In Hawaii severe damage is caused to tomatoes, beans, squash and eggplants (D. E. HARDY, personal communication).

Extensive damage to a number of cultivated plants is caused by this species in California but the exact degree of damage ascribable to *L. sativae* is obscured by uncertain identifications. Most references to *L. pictella* probably refer to this species and there may also have been confusion with *L. huidobrensis* (as *langei* Frick). The main crops affected appear to be tomatoes, beans, alfalfa and cucurbits.

In Barbados *Liriomyza sativae* has recently become an important pest on a number of vegetables. Between January and March, 1972, large populations were present on *Lactuca sativa*, *Cucumis sativus*, *Cucurbita moschata*, *Lycopersicum esculentum* and *Phaseolus vulgaris*.

Liriomyza trifolii (Burgess, 1880)

Oscinis trifolii Burgess, 1880: 201. Holotype lost, neotype designated by SPENCER, 1965a: 37.

Liriomyza trifolii de Meijere, 1925: 282.

Liriomyza alliovora Frick, 1955, **syn. nov.**

ADULT. Small species with conspicuously mat greyish mesonotum and both vertical bristles on yellow ground.

HEAD. Orbits entirely yellow, black of occiput approaching and frequently reaching eye margin beyond outer vertical bristle; all antennal segments bright yellow, third only finely pubescent.

THORAX. Blackish-grey, entirely mat, acrostichals in 3 or 4 rows in front, reduced to two rows behind, conspicuous yellow patch at each hind-corner of mesonotum adjoining scutellum; mesopleura with small blackish-grey patch on lower margin, sternopleura largely black, upper margin yellow.

LEGS. Coxae yellow, femora largely so but with slight, variable brownish striation; tibiae and tarsi darker, brown.

MALE GENITALIA. Aedeagus as in Figs. 329, 330, distiphallus distinctly constricted apically, neck behind adjoining and relatively long, little shorter than distiphallus proper; epiphallus conspicuously narrowing distally (Fig. 331); surstyli with single spine at end.

EARLY STAGES. Larval posterior spiracles each with three pores, the outer two elongate; puparium yellowish-brown.

HOST-PLANTS. Forty-seven genera in ten families have been recorded by STEGMAIER (1966a) as providing hosts for this species. Among crop plants, hosts are found in the families Chenopodiaceae, Compositae, Cucurbitaceae, Leguminosae, Liliaceae, Malvaceae, Solanaceae and Umbelliferae; the two other families recorded are Caryophyllaceae and Zygophyllaceae. Ornamentals attacked are discussed separately in Chapter 11.

The following summary can be given from STEGMAIER's list:

Chenopodiaceae: *Beta vulgaris* L., *Spinacia oleracea* L.

Compositae: *Lactuca sativa* L.

Cucurbitaceae: *Cucumis melo* L. (melon), *C. sativus* L. (cucumber), *Cucurbita pepo* L. (squash).

Leguminosae: *Phaseolus lunatus* L. (lima bean). *Phaseolus* sp. (string beans), *Pisum sativum* L. (pea), *Trifolium repens* L., *Vigna sinensis* Endl.

Liliaceae: *Allium cepa* L. (onion); also *A. schoenoprasum* L. (chives), Venezuela.

Malvaceae: *Hibiscus esculentus* L. (okra).

Solanaceae: *Capsicum* sp. (pepper), *Lycopersicum esculentum* Mill. (tomato), *Solanum melongena* L. (eggplant), *S. tuberosum* L. (potato).

Umbelliferae: *Apium graveolens* L. var. *dulce* DC. (celery), *Daucus carota* L. (carrot).

BIOLOGY. Studies on the life history of this species have been made by SMITH, BOSWELL and WAVE (1962), as *Liriomyza* sp.n. on Chrysanthemums, and by HARRIS and TATE (1933), as *Agromyza pusilla* Meigen on onions.

In Maryland, the egg hatches in 2 days and the larva completes its feeding in 7 or 8 days. The pupal stage in the laboratory covered about 10 or 11 days in February and 7 days in May. Most adults live for 2 weeks on caged chrysanthemums, and a few have been known to live for 3 weeks.

Adult flies and larval leaf mines have been found throughout the year in southern Florida but the first main generation appears to occur in April, followed by at

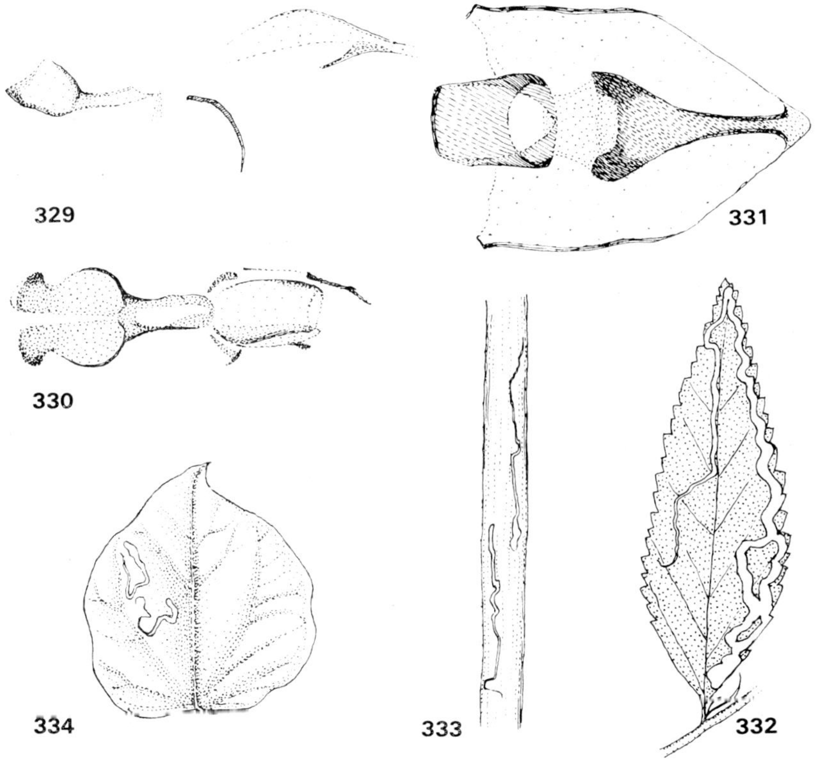

Figs. 329–334. *Liriomyza trifolii* (Burgess): 329, aedeagus, side view; 330, same, ventral view; 331, aedeagal hood and epiphallus; 332, leaf-mine on *Bidens*; 333, mines on *Allium*, Venezuela; 334, mines on cotton, South Carolina.

least two or three complete and several further partial generations. The leaf-mine varies in form with the host-plant but when adequate space is available it is normally long, linear, narrow and not greatly widening towards the end, as on *Bidens pilosa* (Fig. 332), with frass in well defined strips at alternate sides of the channel or towards the end somewhat more irregularly in grains or connected threads. In very small leaves the limited area for feeding results in the formation of a secondary blotch, as on *Tribulus terrestris*. Pupation takes place externally. Leaf-mines on

Allium are shown in Fig. 333. The leaf-mines on cotton in South Carolina discussed by McGregor (1914) are shown in Fig. 334.

Harris and Tate (1933) reported that the life cycle was completed in three weeks and made a count indicating 600,000 larvae per acre attacking onion in Iowa.

Damage caused. Actual leaf damage caused by a single larva is slight but very large populations are frequently present in Florida which are capable of destroying leaves and appreciably affecting the growth of young plants. Both reduction in yield of truck crops and loss of market value of ornamentals result from heavy infestations of this species.

Distribution. United States: Florida, particularly in Miami area, Delaware District of Columbia, Iowa, New Jersey, Ohio, Pennsylvania; Canada: Ontario; Bahamas: Eleuthera; Barbados; Guyana; Venezuela.

Parasites. Florida (Stegmaier, 1966a): Braconidae – *Opius* sp.; Eulophidae – *Chrysocharis* sp., *Closterocerus cinctipennis* Ashm., *Derostenus* sp., *D. agromyzae* Crawford, *D. variipes* Cwfd., *Diglyphus* sp., *Mirzagrammosoma lineaticeps* Girault; Pteromalidae – *Halticoptera patellana* (Dalman), formerly *aenea* Walker. Venezuela: *Chrysocharis* sp.

No information is available on the relative importance of the various parasites in controlling populations of *Liriomyza trifolii*. Harris and Tate (1933) mention that in Iowa the onion leaf-miner is 'subject to heavy parasitism, the percentage rate increasing progressively with the season so that the preponderance of insects emerging from infested leaves collected late in the fall are hymenopterous parasites'.

Discussion

Taxonomy. This species is instantly distinguishable from *L. sativae* Blanchard, with which it partially overlaps both in distribution and host-range, by the generally paler coloration, with both vertical bristles on yellow ground, and by the mat greyish mesonotum. It is distinguishable from *Liriomyza fricki* Spencer, 1965a, which also has these characters but is apparently limited to Washington State and in Canada: Alberta, Ontario and Quebec by having acrostichals in 3–4 rows (only in 2 rows in *fricki*). The larvae of *trifolii* and *sativae* are not at the moment distinguishable but the leaf-mines of *trifolii* are generally longer and narrower and not distinctly widening at the end, as frequently occurs in *sativae* (Fig. 326).

Frick (1955) described as *alliovora* the species which occurred in large numbers on onion (*Allium*) in Iowa in 1932 (Harris and Tate, 1933). Examination of the genitalia of a male paratype has confirmed that this species represents *trifolii*, with which it is now formally synonymized. Specimens mining onion have also been seen from Florida, Guyana and Venezuela.

Host-range. *Liriomyza trifolii* is one of the few truly polyphagous species in the Agromyzidae. Although its name is derived from a host in the Leguminosae, the evidence from Florida indicates that its most favoured host family is the

Compositae. In this family 23 genera are known as hosts and only 6 genera in the Leguminosae. It is perhaps surprising to find among its hosts such diverse genera as *Beta* and *Spinacia* (Chenopodiaceae) and *Allium cepa* (Liliaceae) but it is interesting to note that both in California and Venezuela *Liriomyza huidobrensis* (Blanchard) also occurs on these genera (as *langei* Frick or *pictella* Thomson). In Europe the polyphagous species, *Phytomyza horticola* Goureau, is also found, although infrequently, on *Allium*. This suggests that this monocotyledon genus contains some attractive substance and suitable nutrients, which facilitate colonization by species normally feeding on dicots.

DISTRIBUTION. The centre of distribution of *L. trifolii* at the present time appears to be Florida. Its range extends northwards through the eastern United States as far as Ontario in Canada and it seem sprobable that the population in the Bahamas has reached Eleuthera Is. from Florida. The only records in South America are from Guyana and Venezuela but further collecting will almost certainly reveal a wider distribution.

ECONOMIC IMPORTANCE. Although *L. trifolii* has a host range which is as extensive as that of *L. huidobrensis* and *L. sativae*, there are few records of serious economic damage definitely ascribable to this species on crop plants. It was only considered as a minor pest on onions in Iowa by HARRIS and TATE but in Venezuela serious outbreaks on onions occurred at Quibor, El Tocuyo and Siquisique in the State of Lara in 1960 and 1961 (L. CERMELI, personal communication). It is possible that some of the damage caused in Florida is due to *trifolii* rather than *sativae* and it is to be hoped that a positive identification will be made in future cases of leaf mining damage.

In Florida and farther north in Maryland and other adjoining States *Liriomyza trifolii* is considered a serious pest in Chrysanthemum nurseries (see Chapter 11).

GENUS PHYTOMYZA FALLÉN

Phytomyza horticola Goureau, 1851

ADULT. Mesonotum distinctively mat, ash-grey, without acrostichals.

HEAD (Fig. 335). Frons broad, twice width of eye, not projecting above eye in profile; 2 equal ors, 1 incurved ori, orbital setulae in single row, proclinate; eye round, jowls deep up to $\frac{1}{3}$ height of eye; third antennal segment small, round, arista appearing bare.

MESONOTUM. 3 + 1 strong dc, 2nd, 3rd and 4th approximately equal, acrostichals lacking (at most one or two isolated hairs present).

WING (Fig. 336). Length from 2.2–2.6 mm, second costal section short, $1\frac{2}{3}$–2 times length of fourth.

COLOUR. Frons yellow or slightly darker, more orange, orbits always somewhat paler; third antennal segment and palps black; face greyish-black; mesonotum ash-grey, sides of thorax similar; legs: coxae black, femora black with all knees

yellow, tibiae and tarsi black; veins pale, squamae grey, margin and fringe black.
MALE GENITALIA. Aedeagus (Figs. 337, 338), diverging arms above strongly developed, V-shaped.
EARLY STAGES. The early stages cannot be differentiated from those of *P. syngenesiae* (p. 233). Descriptions have been given by MELIS (1935) in Italy and KURODA (1960) in Japan (as *atricornis* Meigen).
HOST-PLANTS. Cultivated plants serving as hosts, apart from ornamentals, are found in eight families, as follows:
Compositae: *Lactuca sativa* L. (India).
Cruciferae: *Brassica campestris* L. (Europe, Eritrea, India, China), *B. napus* L. (India), *B. oleracea* L. (India, Eritrea), *B. rapa* L. (Egypt).
Cucurbitaceae: *Citrullus vulgaris* L. (Europe), *Cucumis sativus* L. (Europe, Japan), *Cucurbita pepo* L. (Europe, Japan).
Labiatae: *Mentha* sp. (Europe, India, Japan).
Leguminosae: *Cicer arietinum* L. (Europe), *Lens esculenta* Moench (India), *Pisum arvense* L. (India), *Pisum sativum* L. (Europe, Egypt, India, China), *Trifolium alexandrinum* L. (Egypt, India), *Vicia faba* L. (Europe, Libya, Japan), *Vigna sinensis* Endl. (U.S.S.R.: Rostov).
Liliaceae: *Allium cepa* L. (Europe, India, Japan).
Linaceae: *Linum usitatissimum* L. (Europe, China, India, Japan).
Solanaceae: *Lycopersicum esculentum* Mill. (Japan), *Solanum melongena I.* (India, Egypt), *S. tuberosum* L. (India).

GRIFFITHS (1967c) records in all 34 host families, the following being additional to those given above: Amaranthaceae, Anacardiaceae, Boraginaceae, Campanulaceae, Capparaceae, Caryophyllaceae, Chenopodiaceae, Convolvulaceae, Euphorbiaceae, Hydrophyllaceae, Malvaceae, Onagraceae, Papaveraceae, Plantaginaceae, Polemoniaceae, Polygonaceae, Ranunculaceae, Resedaceae, Rutaceae, Scrophulariaceae, Tropaeolaceae, Umbelliferae, Valerianaceae, Verbenaceae, Violaceae.

The main ornamentals which are regularly attacked are discussed in Chapter 11.
BIOLOGY. Detailed descriptions of the life history of this species (as *atricornis* Meigen) have been given by MELIS (1935) in Italy and AHMAD and GUPTA (1941) in India.

AHMAD and GUPTA, in experiments in March, 1938, found that three females laid 355, 358 and 329 eggs respectively. The maximum number of eggs laid in a single day was 50. MELIS found as many as 154 feeding punctures in a single leaf of *Pisum*, while the maximum found by AHMAD and GUPTA was 38. The larva hatches in from 2–4 days.

The larval leaf-mine (Fig. 339) is irregularly linear. In India the larval period was found to be as short as 5.2 days at 23–28°C and will be proportionally longer at lower temperatures. The pupal stage can last from 7–15 days but may last several months during aestivation or hibernation.

SASAKAWA, YOSHIDA and KANAYAMA (1970) investigated the effect of temperature and maternal age on the survival and fecundity of female flies and viability of their eggs. The hatchability of eggs of middle-aged parents was found to be higher than those of young and of old parents.

In N.W. Europe there are three or four generations between March and late autumn, while in the Mediterranean area the main period of activity is during the cooler, damper winter months. In India the flies are active from January to April.
DAMAGE CAUSED. Normal leaf-mining damage is caused by the larvae, leading to withering and premature dropping of leaves. Some damage also results from female feeding punctures.
DISTRIBUTION. Europe; N. Africa: Egypt, Eritrea, Libya, Morocco; S. Africa,

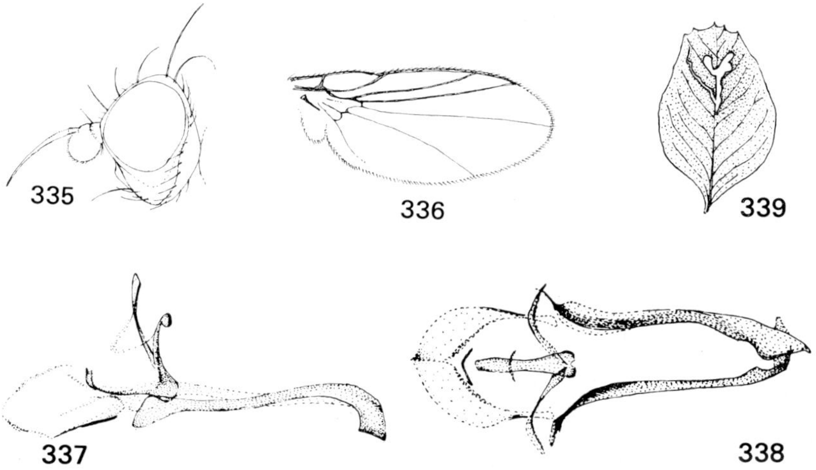

Figs. 335–339. *Phytomyza horticola* Gour.: 335, head; 336, wing; 337, aedeagus, side view; 338, same, ventral view; 339, leaf-mine on *Pisum*.

Madagascar, W. Africa: Cameroon; E. Africa: Kenya; India, China, Formosa, Japan.
PARASITES. GOUREAU (1851) described *Cirrospilus cyanops* and *Omphale stigma*. Other parasite records in Europe all refer to '*atricornis*' and there is thus no certainty in most cases as to whether they refer to *horticola* or *syngenesiae*. Nevertheless it seems certain that *Chorebus sativi* (Nixon) from Morocco on *Pisum sativum* and *C. canariensis* Griffiths from Gran Canaria, Canary Islands on an unidentified plant refer to *horticola*, as in both areas this species is so much more common than *syngenesiae*. Also in Italy parasites obtained from '*atricornis*' on *Pisum* can be accepted as referring to *horticola*, as follows: *Dacnusa areolaris* Nees (Braconidae); *Closterocerus formosus* Wesw., *Derostenus* sp., *Pleurotropis* sp., *Hemiptarsenus* sp. (Eulophidae), GOIDANICH (1928); also *Entedon* sp., *Achrysocharis* sp. (MELIS, 1935).

Four species referable to *horticola* in Egypt (discussed as *atricornis*) have been recorded by ASSEM (1966, unpublished thesis): *Aphidius ervi* Hal., *Opius* sp. near *parvulus* Wesm. (Braconidae) *Chrysocharis* sp., *Diglyphus* sp. (Eulophidae).

In Iraq AL-AZAWI (1967, 1971) gives a number of parasite records of *P. atricornis* which are clearly correctly referable to *horticola*, as follows: Pteromalidae: *Thinodytes cyzicus* (Wlk.); *Sphegigaster orobanchiae* Kurdjumov; *Halticoptera circulus* Wlk.
Eulophidae: *Achrysocharella formosa* (Westw.); *Pediobius acantha* (Wlk.); *Diglyphus isaea* Wlk., *D. crassinervis* Erdös; *Cirrospilus vittatus* Wlk.; *Tetrastichus*, group *strobliana* Ratz.

In India AHMAD and GUPTA (1941) record *Solenotus* sp. (Eulophidae); V. K. SEHGAL has reported rearing *Tetrastichus* sp. (Eulophidae) and *Opius* sp. in the Punjab (personal communication).

Discussion

TAXONOMY. *P. horticola* forms a distinctive group with *P. syngenesiae* and several other species of no economic importance, characterized by the type of pupation, with the puparium remaining in the leaf at the end of the mine, with the anterior spiracles directed somewhat ventrally and penetrating through the leaf epidermis. HARDY (1849) placed his *syngenesiae* in the separate genus *Chromatomyia* based on this character. *Chromatomyia* is at the present time sunk in *Phytomyza* but may in due course be resurrected if and when the large genus *Phytomyza* is split into a number of smaller genera on the basis of clearly monophyletic groups detectable from male genitalia and host plant associations.

DISTRIBUTION. *P. horticola* is more common in the Mediterranean area than in northern Europe and occurs widely throughout Asia, including Japan. It is absent from Australia and New Zealand, and also from North America but occurs locally in East and West Africa and is common in South Africa, at least in the Cape area. It has been recorded from Eritrea by DE LOTTO (1948). The absence from entire continents such as Australia, North and South America suggests that recent introductions by man must have played a quite subsidiary role in the present distribution of this species.

DOVNAR-ZAPOLSKI (1969) in his survey of leaf-mines in the Kirghiz Republic of the U.S.S.R. records *atricornis* on 17 genera of Compositae but also on *Glaucium, Hesperis, Lepidium, Medicago, Ononis, Phlox, Plantago, Ranunculus* and *Sisymbrium*. Even if some of these records may be open to doubt, it is quite clear that *horticola* must be represented and this is an important extension of the known range of *horticola* in Asia.

HOST-RANGE. The polyphagy of *P. horticola* can be seen from Table VII, p. 206 to overlap remarkably closely with that of the two *Liriomyza* species, *bryoniae* (Kalt.) and *strigata* (Mg.). This is all the more surprising since these species in widely differing genera are in no way related and the polyphagous habit must have arisen quite independently in the two cases.

BUHR (1953) discussed the choice of host of this species (as *atricornis* Mg.) and noted that while *Vicia faba* L. rarely serves as host, the closely related *Vicia narbonensis* L. at the same locality, Rostock in Germany, was frequently and heavily attacked.

ECONOMIC IMPORTANCE. The main host suffering damage from *P. horticola* is *Pisum*. Severe damage was reported on peas in 1928 in Britain at Cheddar in Somerset and Bangor in N. Wales (unpublished M.A.F.F. Report). However, on mature plants even a relatively heavy infestation, as recently observed on a field of several acres in Lincolnshire, England, appears to have little or no effect on the crop. In Egypt more damage has been caused, with 30% infestations on young plants, which must appreciably affect growth and crop yield. In India SINGH and SOHI (1957) report *atricornis* (=*horticola*) as a serious pest of peas in the Punjab. According to AHMAD and GUPTA (1941) the greatest damage is caused during February and April but the insect remains active until May. In Iraq AL-AZAWI (1967) records that peas can be sufficiently badly infested to require control measures.

SASAKAWA (1966) reports that *atricornis* (=*horticola*) 'does severe damage to various agricultural crops in Japan'. *Pisum sativum*, also in Japan, is one of the crops on which most damage is caused. SASAKAWA found experimentally that of three leguminous plants tested, *Pisum sativum*, *Vicia faba* and *Vicia sativa*, *Pisum* was significantly more susceptible to attack.

The heavy natural parasitism of *P. horticola* appears to control population levels under normal conditions and the potential significance of this species as a pest, with its wide host range of cultivated and wild plants, is thus successfully restricted.

Some damage is caused to a number of ornamentals and this is discussed in Chapter 11.

Phytomyza syngenesiae (Hardy, 1849)

ADULT. External characters as in *P. horticola* (p. 229).
MALE GENITALIA. Aedeagus (Figs. 340, 341), diverging arms above slender, weakly chitinized, Y-shaped.
EARLY STAGES. The oval, white egg is smooth and shining and the chorion is without reticulation; it measures 0.36 × 0.15 mm. The full-grown larva can be up to 3 mm long; the posterior spiracles each have from 8–10 pores. The delicate puparium is whitish-grey, somewhat flattened, measuring up to 2.5 mm in length. The early stages have been described by COHEN (1936) in England and WATT (1923) in New Zealand (as *atricornis* Meigen).
HOST-PLANTS. Among crops, Compositae: *Cynara scolymus* L., *Lactuca sativa* L.; Leguminosae: *Pisum sativum* L.; Umbelliferae: *Daucus carota* L.; also ten further genera of the Compositae which are recognized ornamentals (see Chapter 11). In the United States additional hosts recorded are *Mentha* and *Stachys* (Labiatae), *Malva* (Malvaceae) and *Melilotus* (Leguminosae).
BIOLOGY. Oviposition can take place on either surface of the leaf but the upper surface is preferred. GURNEY (in G.C.R.I. Ann. Rep., 1960) observed that 'oviposition tends to be localized on foliage in the lower shaded portion of the plant or on completely shaded plants'. The larva hatches in from 4–6 days, depending

on the temperature. A long, irregularly linear mine is formed, with frass normally deposited in distinct, well-separated grains but there appears to be some variation in the form of frass deposits, possibly depending on the nature of the host-plant. Pupation takes places in the mine, in a 'pupal blister' which frequently adjoins a vein, with the anterior spiracles projecting through the leaf-epidermis. Leaf-mines on *Chrysanthemum* are shown in Plate 11.

Numerous generations follow throughout the year, limited only by temperature and availability of suitable food-plants. KELSEY (1937) confirmed 15 complete generations in New Zealand in a 12-month period. In England larvae can be

340

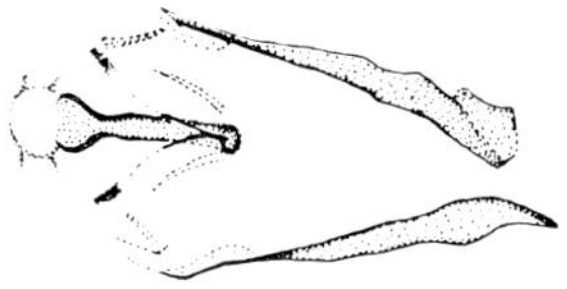

341

Figs. 340–341. *Phytomyza syngenesiae* (Hardy): 340, aedeagus, side view; 341, same, ventral view.

found feeding until December in a mild winter and the following spring generation can appear as early as March.

The biology has been discussed by SMULYAN (1914) in the United States (as *chrysanthemi* Kowarz), by COHEN (1936) in England (as *atricornis* Meigen) and by WATT (1923) in New Zealand.

DISTRIBUTION. Widespread throughout Europe, including Ireland, Scotland and Scandinavia, less common in south but isolated records from Spain and Canary Islands; North America: Canada, United States, mainly California, Oregon, Washington but also known from Massachusetts and Rhode Island.

PARASITES. None have been recorded specifically from *P. syngenesiae* but some known from '*atricornis*' certainly refer to this species (see Introduction above, p. 205). Four species reported by GURNEY (1960) in England can be taken to refer to *syngenesiae*, as follows: *Rhizarcha areolaris* (Nees); *Chrysocharis melaenis* Walk., *Cirrospilus vittatus* Walk., and *Diglyphus chabrias* (Walk.). *D. isaea* (Walk.) is also now known as an active parasite of this species.

WATT (1923) reports that in New Zealand 90% of larvae are parasitized but no species are recorded.

KLEINSCHMIDT (1970) records the following species in Australia for *atricornis* (=*P. syngenesiae*): Braconidae: *Apantales ruficrus* Hal.; *Opius oleracei* Fischer., Eulophidae: *Achrysocharis* sp., *Chrysocharis* sp., *Epilampus* sp., *Hemiptarsenus* sp., Pteromalidae: *Sphegigaster* sp., *Trigonogastra* sp.

FISCHER (1971) records two further *Opius* species in Australia: *O. atricornis* Fi. and *O. cinerariae* Fi. and also two species in N. America: *O. curticubitalis* Fi. and *O. mandibularis* Gahan.

Discussion

DISTRIBUTION. The main centre of distribution of this species is in western Europe. The localized records in the United States and Canada clearly indicate recent introduction by man rather than a natural holarctic distribution, as suggested by GRIFFITHS (1967c: 10 and SPENCER, 1969a: 278). Similarly, its occurrence in Australia and New Zealand must be without question the result of introduction by early settlers.

It will be of interest to establish the eastern limit of its range in the Eurasian land mass. In the U.S.S.R. there are no confirmed records of *syngenesiae* even from the Moscow district; DOVNAR-ZAPOLSKI (1969) records leaf-mines of *atricornis* on 17 genera of Compositae in the Kirghiz Republic but there is no certainty that any of these represent *syngenesiae* and it seems more probable that the species present in the area is *horticola* (p. 232). Among collections of Agromyzidae studied from Afghanistan and Mongolia neither *syngenesiae* nor *horticola* were represented.

HOST RANGE. *P. syngenesiae* is essentially a feeder on Compositae. It has been found naturally on *Pisum* in the United States, and in the laboratory in England has transferred readily to *Pisum* both from *Sonchus* and *Lactuca* in two separate experiments carried out by the author and Dr. N. W. HUSSEY in 1971.

American records of *Mentha*, *Stachys*, *Malva* and *Melilotus* (FRICK, 1959) are somewhat puzzling and appear to be outside the normal range of this species. It would be of interest to examine males from these hosts.

The only hosts recorded in Australia are in the Compositae (KLEINSCHMIDT. 1970), while in New Zealand two non-Composite hosts are attacked – *Plantago*, (HARRISON, 1959) and *Pisum* (WISE, 1953).

ECONOMIC IMPORTANCE. Serious damage has been caused to lettuce in England by *P. syngenesiae*, which is clearly one of its most favoured hosts. In 1924 at one unrecorded locality 15% of plants were attacked, while in 1966 1000 plants in a heated greenhouse in Norfolk were damaged. It is believed that in the latter case the pest had spread from nearby chrysanthemums (M.A.F.F., unpublished reports).

P. syngenesiae, well known as the chrysanthemum leaf-miner, is a serious pest on chrysanthemums in Europe and the United States and in most years causes serious commercial loss. It is also known as the cineraria leaf-miner and damage to this host is frequently considerable. The importance of *syngenesiae* as a pest of ornamentals is discussed in greater detail in Chapter 11.

IX. LEAF MINERS ON CEREALS AND OTHER GRAMINACEOUS CROPS

Agromyzidae feeding on cultivated cereals, including rice, maize (corn), wheat, rye, barley, oats and also sugar, are found virtually throughout the world and belong to five different genera – *Agromyza*, *Liriomyza*, *Pseudonapomyza*, *Phytomyza* and *Cerodontha*. Most Continents have their own species but there is a small degree of overlap between Africa and India, Europe and North America and between North and South America. Only a few species are monophagous, most showing an extensive degree of oligophagy, occurring on a number of genera of Gramineae.

No attempt has been made to review here all species known to feed on Gramineae, as the majority have either been reared only from wild hosts or at least have not been recorded on cereals. Only those species are discussed, therefore, which are known to occur regularly on cultivated cereal or pasture crops. An exception has been made in the case of the genus *Pseudonapomyza*, since 10 undescribed species have been discovered during the course of this study and for practical reasons it has seemed preferable to deal with these together.

Leaf-mining damage by this group is normally not severe but causes concern when very young plants are attacked or when populations become abnormally large. *Pseudonapomyza spicata* (Malloch) is considered a serious pest in Hawaii and other Pacific islands, where it can occur in large numbers and cause appreciable damage to young corn plants. *Ps. asiatica* Spencer, a widespread species in Asia, regularly attacks young rice seedlings. In southern Germany in 1964 and 1965 *Agromyza megalopsis* Hering suddenly and unexpectedly changed its status from an uncommon species known from a few isolated localities in eastern Germany to a major pest of barley throughout Württemberg and Bavaria. Populations have since receded to their normal level and the species is currently causing no problems. *Cerodontha* (*Poemyza*) *lateralis* (Macquart), which normally occurs on cereals in Europe in small numbers, was reported with large populations on barley and wheat in the vicinity of Moscow in 1886 (LINDEMANN, 1886) and also flared up in the Nishni-Novgorod area of the Ukraine in 1929 to cause some damage (ESTERBERG, 1931). It has not since been reported as a pest of any consequence.

These examples illustrate the fluctuating significance of many leaf-miners on this group of crops. However, all species discussed below, which are exclusively feeders on Gramineae, must all be considered as actual or potential pests.

GENUS AGROMYZA FALLÉN

Cereals attacked by *Agromyza* species include barley, oats, rye, wheat and corn (maize). Species in this genus have come to notice on cultivated crops in Europe, Africa, India and the United States.

In the past there has been considerable confusion regarding the identity of grass-feeding *Agromyza* species, both in Europe and in North America, owing to the many very similar species occurring on the same hosts. GRIFFITHS (1963) clarified a number of species in the *nigripes*-group and SPENCER (1966c) clarified five further species, three of which, *A. megalopsis* Hering, *A. nigrella* Rondani and *A. rondensis* Strobl, are discussed below.

In North America there have been many references in the literature to *Agromyza parvicornis* Loew. In a study of Canadian Agromyzidae SPENCER (1969a) described ten further species in this group. The host is unknown for these but several may be found to feed on cultivated cereals. However, in view of the absence of any concrete evidence either of the hosts or of damage to cereals by these species, it is not proposed to discuss them here.

A number of species have been described from South America but their hosts are unknown and their significance remains to be established. Two undescribed

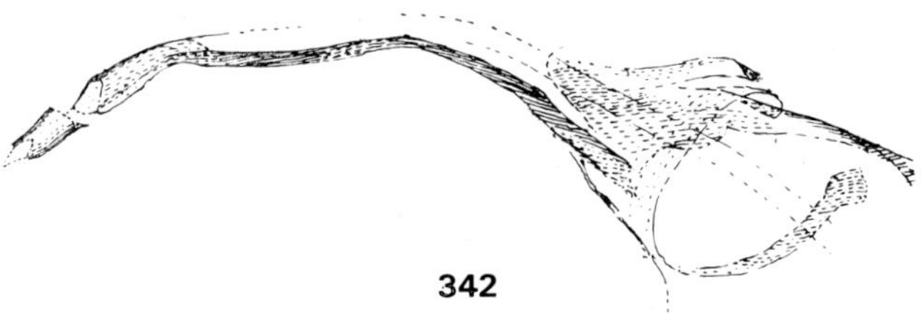

342

Fig. 342. *Agromyza cinerascens* Macq.: aedeagus.

species have been seen from Northern Nigeria and others certainly await discovery in Africa and India. None are likely to be found in Australia, where the genus *Agromyza* is poorly represented, with only a single species.

In addition to the 11 species discussed in detail below brief mention can be made of several other grass-feeding species which occur on cereals but are of lesser importance. *Agromyza cinerascens* Macquart, 1835 is occasionally found mining *Secale* in Europe and Japan but appears to prefer hosts among wild grasses, particularly *Dactylis glomerata*. It somewhat resembles *A. intermittens* Becker (p. 242) but the head is black; the aedeagus is shown in Fig. 342. VENTURI (1936) records *A. frontosa* Becker, 1908 mining *Triticum* in Italy but there is some doubt about this identification. *A. mobilis* Meigen, 1830, is a common species in Europe but its hosts are not properly known. VENTURI (1936) also records this species on *Triticum* but the identification requires confirmation.

In Japan *Agromyza yanonis* (MATSUMURA, 1916) feeds on *Hordeum hexastichon* L., *H. vulgare* L. and *Triticum aestivum* L., also on wild grasses such as *Alopecurus aequalis* Sobal. It is a large species, with wing length of up to 3.6 mm in the female and the costa ends at vein R_{4+5}. The posterior spiracles of the larva are distinctive, in having four small patches of hairs around the three spiracular pores. Although *A. yanonis* is apparently not uncommon in Japan, there are no reports of any significant damage having been caused to barley or wheat. SASAKAWA (1961) illustrates the head, wing, male genitalia, leaf mine and early stages.

Agromyza albipennis Meigen, 1830 – Europe, Japan, North America

ADULT. Medium-sized, black species, with white squamal fringe and costa extending to vein M_{1+2}.
HEAD. Frons broad, from 1.2–2 times width of eye, not significantly projecting above eye in profile; 2 ors, 2 or 3 ori, orbital setulae numerous, reclinate; jowls

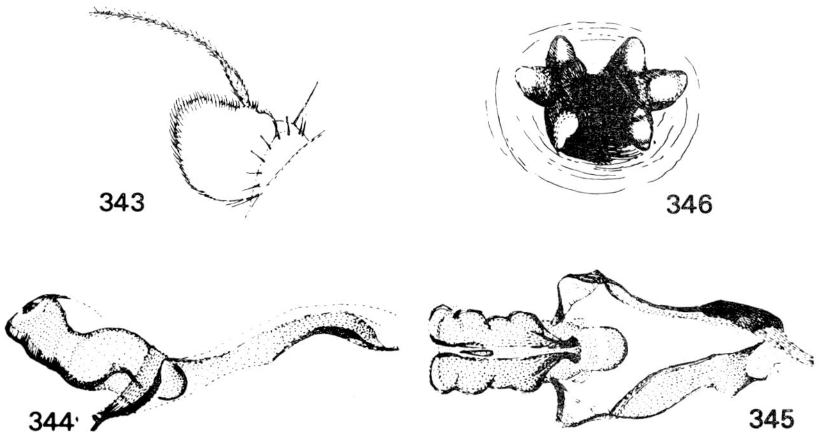

Figs. 343–346. *Agromyza albipennis* Mg.: 343, third antennal segment; 344, aedeagus, side view; 345, same, ventral view; 346, posterior spiracles of puparium.

narrow in centre below eye, deeper at rear; third antennal segment (Fig. 343) conspicuously cut away below.
MESONOTUM. 4 or 5 dc, greatly decreasing in size, acr in 6–8 rows.
WING. Length from 2–3 mm, last section of vein M_{3+4} about $\frac{2}{3}$ penultimate.
COLOUR. Frons generally dark, brownish-black, rarely slightly paler; orbits more distinctly black; third antennal segment black, first and second segments usually similar, sometimes paler, yellowish-brown; mesonotum shining black, appearing slightly grey-dusted viewed from front; legs most frequently entirely black, in palest specimens knees slightly yellowish, particularly on fore-legs; squamae white, margin pale yellowish, fringe silvery.
MALE GENITALIA. Aedeagus as in Figs. 344, 345.
EARLY STAGES. Larva up to 4.5 mm long (GRIFFITHS, 1963), 6 mm (SASAKAWA, 1961); anterior spiracles each with 12–16 minute pores, posterior spiracles adjoining, each with 3 pores. Puparium variable in colour, from dark reddish-brown to almost black, posterior spiracles as in Fig. 346.

HOST-PLANTS. In Europe *Hordeum vulgare* L., *Secale cereale* L.; in Japan *Hordeum hexastichon* L., *Oryza sativa* L., *Secale cereale*, *Triticum aestivum* L. In Europe the most favoured host appears to be *Phalaris arundinacea* L. Other genera of wild grasses regularly serving as hosts are *Poa* (Europe), *Agropyron*, *Phleum* (Japan). There are also isolated records on other genera.

BIOLOGY. The leaf-mine is formed on the upper leaf surface. The egg is laid near the edge of the leaf and the young larva at first feeds towards the apex, later turning and forms a broad blotch running downwards. Pupation can take place either inside the mine or outside, with the puparium loosely glued to the leaf some distance from the end of the mine.

The two main generations occur in June/July and September/October.

NISHIJIMA *et al.* (1963) investigated the mortality of the immature stages in Hokkaido, Japan. 'The total mortality attributable to various factors . . . was always high in summer, indicating that controlling factors have strongly operated on the first generation, the average mortalities in the egg, larva and pupa being 14.7, 52.0 and 91.5% in summer, 9.1, 18.2 and 49.1% in autumn respectively.' No egg parasites were observed. The main factor causing larval death was parasitism by Eulophids, with a rate of 55%. Pupal mortality was caused mainly by two species of Braconids and two Pteromalids. 'The average percentages of parasitism from 1958 to 1960 were 99.8, 79.7 and 95.1 in summer, 5.6, 69.0 and 72.8 in autumn respectively.'

DAMAGE CAUSED. A substantial area of leaf is consumed by each larva.

DISTRIBUTION. Widespread in Europe; E. Siberia, Japan; Canada, United States.

PARASITES. In Europe, *Exotela flavicoxa* (Thomson), *Chorebus lugubris* (Nixon), *C. nydia* (Nixon) (Dacnusinae); *Opius ambiguus* Wesm., *O. celsus* Hal., *O. parvungula* Th., *O. rex* Fi. (Opiinae); *Merismus* (*Merismus*) *splendens* Graham (Pteromalidae). In Japan: five species of Eulophidae, two Braconids and two Pteromalids are mentioned by NISHIJIMA, HONMA and OKUYAMA (1963) but specific names are not given.

Discussion

TAXONOMY. Important characters are the shape of the third antennal segment, the generally black legs and the silvery-white squamal fringe. The male genitalia are entirely distinctive. The species was studied in detail by GRIFFITHS (1963).

DISTRIBUTION. *A. albipennis* has a true Holarctic distribution and, although misidentified in North America until its confirmation in Canada by SPENCER (1969a), it appears to be as widespread there as in Europe.

ECONOMIC IMPORTANCE. In Europe *A. albipennis* occurs only rarely on cultivated cereals and has never been known to cause damage. However, in Japan the species is considered as a serious pest of wheat and barley and is known there as the 'Black Wheat Leaf-miner'. In Hokkaido the damage has been reported as 'chronically severe' (NISHIJIMA *et al.*, 1963).

Agromyza ambigua Fallén, 1823a – Europe, Canada, U.S.A.

Agromyza reptans var. d. *ambigua* Fallén, 1823a.
Agromyza niveipennis Zetterstedt, 1848.
ADULT. Medium-sized, shining black species, with conspicuously angulate third antennal segment (Fig. 347), silvery-white wings and costa ending at vein R_{4+5}.
MALE GENITALIA. Aedeagus as in Figs. 348, 349.

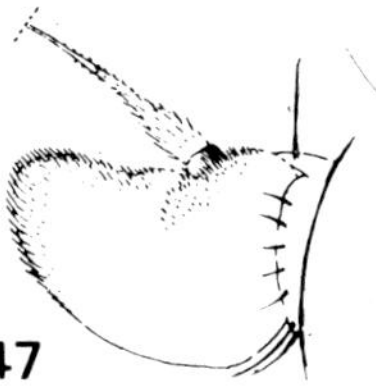
347

348

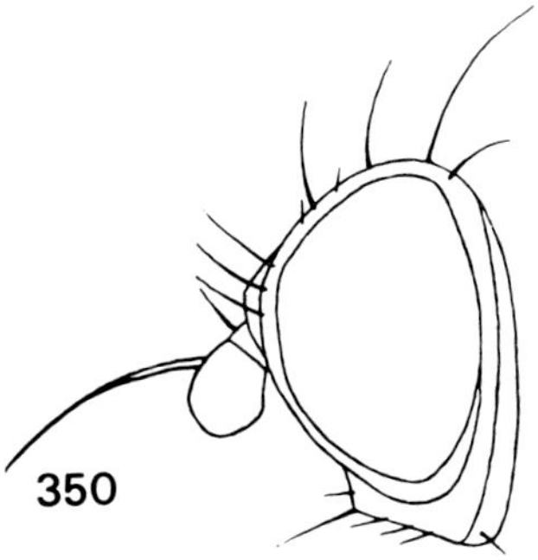
350

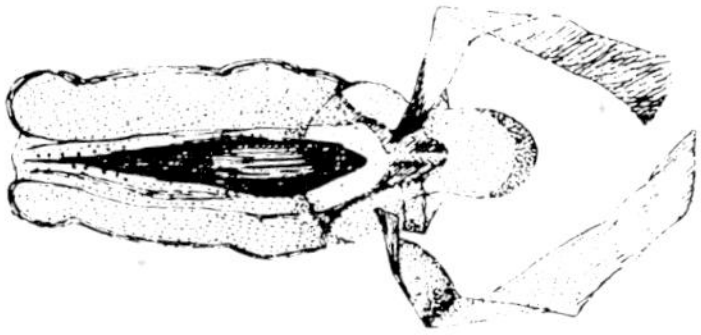
349

Figs. 347–349. *Agromyza ambigua* Fall.: 347, third antennal segment; 348, aedeagus, side view; 349, same, ventral view.

Fig. 350. *Agromyza hordei* Sp.: head.

EARLY STAGES. Larva with posterior spiracles widely separated, each with 3 pores, as in *nigrella* Rond. (cf. Fig. 361). Puparium pale yellowish-brown.
HOST-PLANTS. *Avena sativa* L., *Hordeum vulgare* L., *Secale cereale* L., *Triticum aestivum* L.
BIOLOGY. The leaf-mine, according to HERING (1957c), is characterized by being relatively short and broad, running down the leaf, with no evidence of any early mine running apically. The frass is in distinct black grains, never greenish. VENTURI (1939) states that there is only a single generation in Italy, with adults emerging in April and May. The entire summer and winter is said to be passed in the pupal stage.
DISTRIBUTION. Widespread in Europe from Sweden to Italy; United States, Canada (SPENCER, 1969a).

PARASITES. VENTURI (1939) records *Derostenus* sp. and *Solenotus* sp. in Italy; FISCHER (1971) records *Opius maculipes* Wesm. and *O. parvungula* Th. (under *niveipennis* Zett.).

Discussion

TAXONOMY. This species was known as *niveipennis* Zett., until the correct identity's of FALLÉN's *reptans* var. d. *ambigua* was established by SPENCER (1966c). The pale wings and angulate third antennal segment are distinctive characters of the species.

VENTURI (1939) discusses *niveipennis* in Italy but the illustrations he gives of the puparia clearly indicate rows of papilli and can only refer to *Pseudonapomyza atra* (Mg.) or a member of this group (cf. p. 266). The adults of the two species have entirely different wing venation and can scarcely have been confused and it is therefore accepted that VENTURI's discussion does in fact refer to *ambigua* (as *niveipennis*).

ECONOMIC IMPORTANCE. HERING (1957c) states that *Agromyza niveipennis* occurs as a pest on *Secale cereale* but no details are given. VENTURI only found the species on *Avena sativa* in Italy and does not suggest that any significant damage was caused. In fact *A. ambigua*, although widespread, is a fairly uncommon species and no records are known of its occurrence in large numbers.

Agromyza hordei Spencer, 1961b – India

ADULT. Large species, with brownish-black frons, mat greyish-black mesonotum, costa ending at vein R_{4+5} and white squamal fringe.

HEAD (Fig. 350). Frons broad, twice width of eye, projecting above eye in lower half; 2 strong ors, 3 largely incurved ori; orbital setulae fine, sparse, reclinate; jowls deeply extended at rear, in centre below eye $\frac{1}{8}$ vertical height of eye, cheeks forming broad ring below eye; third antennal segment large, arista appearing bare.

MESONOTUM. 4 strong dc, fourth immediately before suture, preceded by 2 or more minute further bristles, scarcely stronger than acrostichals, these in 6–8 irregular rows.

WING. Length in female 2.8 mm, costa ending immediately beyond vein R_{4+5}, last section of M_{3+4} short, little more than $\frac{1}{2}$ length of penultimate, first cross-vein at midpoint of discal cell.

COLOUR. Frons mat black with brownish undertone, orbits more distinctly black; antennae and remainder of head black; mesonotum largely mat, greyish-black; legs entirely black; squamae and fringe pale, white.

MALE GENITALIA. Not known.

EARLY STAGES. Described and illustrated by BERI (1971b). The anterior spiracles of the larva each have 10 pores, the posterior 3. The puparium is brown.

HOST-PLANT. *Hordeum vulgare* L.

BIOLOGY. Leaf-mines were found in February, mainly towards the apex of leaves. Several eggs are usually laid in one leaf. Pupation takes place on the ground (BERI, 1971b).
DISTRIBUTION. India: Delhi, Agra.
PARASITES. None recorded.

Discussion

TAXONOMY. *Agromyza hordei* superficially resembles the widespread European species, *A. rondensis* Strobl (cf. p. 257), but it will not be possible to assess its affiliations more accurately until the male genitalia can be examined.
ECONOMIC IMPORTANCE. This species is now known from several localities in India but is apparently not common and has hitherto caused no damage of significance.

Agromyza intermittens (Becker, 1907) – Europe, North Africa, Japan

ADULT. Medium-sized species with reddish, projecting frons and single ors.
HEAD (Fig. 351). Frons conspicuously projecting above eye; 1 ors directed upwards and outwards, 3 or 4 slender incurved ori; jowls deeply extended at rear; third antennal segment slender, elongate, rounded at end; mouth-margin with broad epistoma (Fig. 352).
MESONOTUM Up to 5 dc greatly decreasing in size, acr in 4 rows.
WING. Length from 2.3–2.75 mm, costa extending only to vein R_{4+5}, second cross-vein not infrequently lacking or vestigial.
COLOUR. Frons orange-red, orbits black; antennae partially pale but third segment largely black; mesonotum mat, greyish-black, legs black but all knees yellow; squamae white, margin darker, fringe white to ochrous.
MALE GENITALIA. Aedeagus as in Fig. 353.
EARLY STAGES. Puparium reddish-brown, posterior spiracles arising from small protuberance, separated by their own diameter, each with 3 well-defined pores.
HOST-PLANTS. *Hordeum vulgare* L., *Secale cereale* L., *Triticum aestivum* L. (HERING, 1962: 59 and 65); also *Bromus mollis* L. and probably other wild grasses.
BIOLOGY. Leaf-mine initially linear, later developing into a broad blotch, appearing greenish. Little is known about the detailed life history of this species. The first generation appears early, in April, and there is apparently only a partial second, autumn generation.
DISTRIBUTION. Tunis; England, Germany, Sweden, ?Italy; Japan.
PARASITES. None recorded.

Discussion

TAXONOMY. The species accepted by HENDEL and subsequent authors as *intermittens* (type locality Tunis) has a distinctive form of aedeagus as illustrated. The male genitalia indicate a close relationship with *A. cinerascens* Macquart (Fig. 342). These two species possibly deserve at least sub-generic status within the genus *Agromyza*.

I am here following HENDEL's interpretation of *intermittens*. The only significant difference between *intermittens* and *frontosa* Becker, 1908 from the Canary Islands is the presence of mid-tibial bristles in the latter which are lacking in the former species. Specimens recently seen from Egypt also lack the mid-tibial bristles but have the typical aedeagus of grass-feeders (cf. Figs. 359, 376). If *frontosa* is accepted as distinct (no males from the type locality are known), then the Egyptian

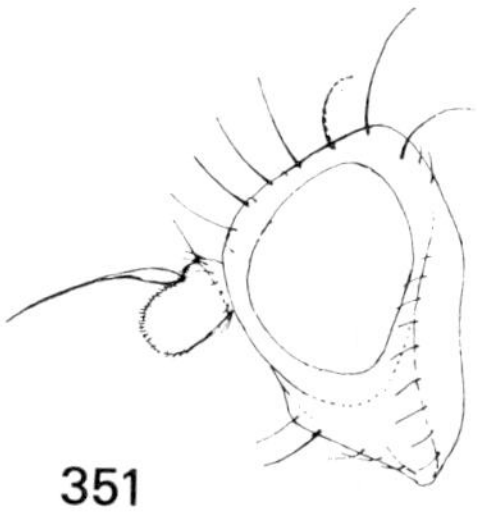

351

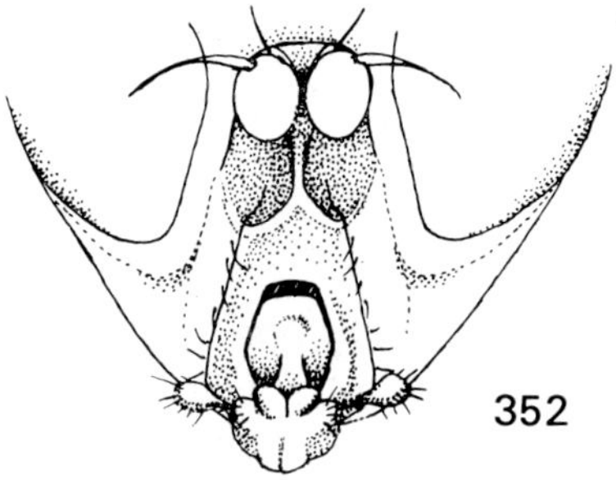

352

353

Figs. 351–353. *Agromyza intermittens* Beck.: 351, head; 352, head showing epistoma; 353, aedeagus.

specimens must represent a further undescribed species. A more detailed revision of this group is thus clearly necessary but more material is required before this can be satisfactorily undertaken.

VENTURI (1936) refers to *frontosa* mining *Triticum vulgare* in the Bologna area of N. Italy. It has not been possible to confirm the identity of this species.

ECONOMIC IMPORTANCE. *A. intermittens*, as here defined, has not yet been recorded as causing any significant damage to cereals but with wheat, barley and rye as frequent hosts, the species must be treated as a potential pest.

Agromyza megalopsis Hering, 1933 – Europe, Iraq

ADULT. Medium-sized black species, costa ending at vein R_{4+5}, squamal fringe white.

HEAD. Frons distinctly projecting above eye towards base of antennae, jowls extended at rear, $\frac{1}{4}$ height of eye; third antennal segment rounded, slightly longer than broad.
MESONOTUM. Normally 4 fully developed dc, greatly decreasing in size, presutural one thus very small; acrostichals irregularly in 6 rows; pre-scutellars strong.
WING. Length from 2.5 mm in male to 2.7 mm in female, costa ending at vein R_{4+5}, first cross-vein at or beyond midpoint of discal cell, last section of M_{3+4} $\frac{2}{3}$ penultimate.

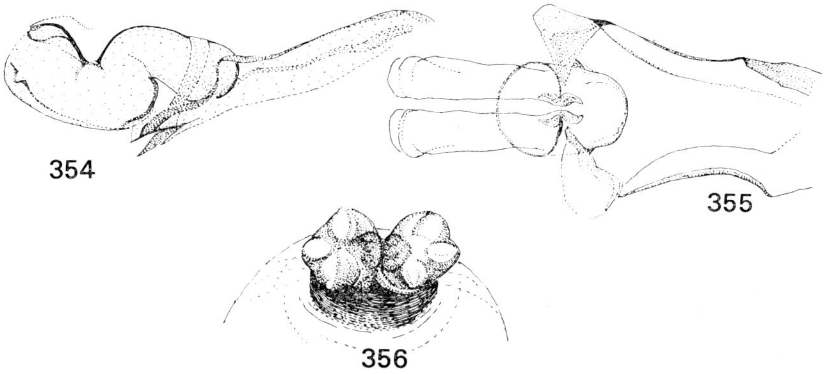

Figs. 354–356. *Agromyza megalopsis* Her.: 354, aedeagus, side view; 355, same, ventral view; 356, posterior spiracles of puparium.

COLOUR. Frons brownish-black, darker in front, antennae entirely black; mesonotum black, only moderately shining, legs entirely black; squamae and fringe yellowish-white.
MALE GENITALIA. Aedeagus as in Figs. 354, 355.
EARLY STAGES. Puparium reddish- or blackish-brown, posterior spiracles (Fig. 356) adjoining, each with 3 pores.
HOST-PLANTS. *Hordeum distichon* L., *H. vulgare* L.; possibly also *Secale cereale* L.
BIOLOGY. According to HERING (1962) oviposition normally takes place towards the base of a leaf. The young larva forms a narrow mine running towards the apex of the leaf and then turns, feeding downwards, forming a broad blotch mine. Several larvae can feed together in a single leaf. Larvae feeding are shown in Plate 8.

The biology has also been studied by GEIGENMÜLLER (1966), SCHERNEY (1966), WEIGAND (1966), and SCHERNEY and WEIGAND (1967). In Württemberg oviposition lasts from the end of May until the end of June. As many as 30–40 eggs were frequently observed in a single leaf. One female can lay from 20–30 eggs. Fresh mines are pale green, later turning brownish-yellow, and appearing darker at some points from the frass deposits. Pupation can take place in the mine but more frequently the larva leaves the mine to pupate on the ground.

Only a partial emergence takes place in the late summer or autumn, the majority of puparia overwintering and producing adults the following May.

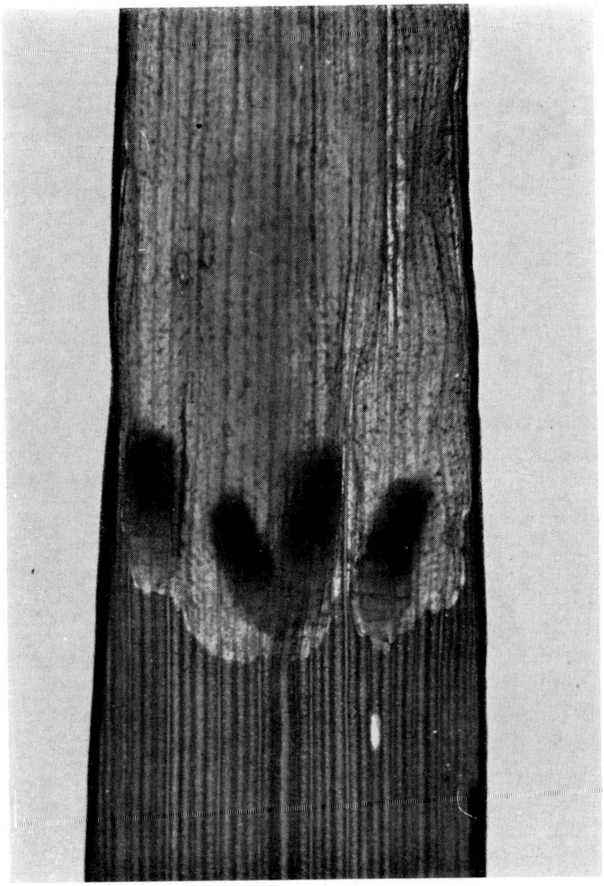

Plate 8: *Agromyza megalopsis:* larvae feeding in *Hordeum distichon*, Württemberg (MARIA GEIGENMÜLLER, Landesanstalt für Pflanzenschutz, Stuttgart).

DAMAGE CAUSED. Larval feeding leads to leaves turning partially or completely yellow and in cases of heavy infestation they die.

DISTRIBUTION. Germany: Saxony, Württemberg, Bavaria; Czechoslovakia; Iraq.

PARASITES. SCHERNEY and WEIGAND (1967) report a rate of parasitism in Franken in 1964 of 30% which in 1965 had risen to 75%. In the Regensburg area the parasitism was 49% in 1964, increasing to 81% in 1965. Several species of Chalcids and Braconids were represented but the only identification given was *Gyrocampa* sp. (Braconidae), which was the species most frequently encountered.

Discussion

TAXONOMY. *A. megalopsis* cannot be distinguished on external characters from *A. nigrella* Rond. but the two species are immediately separable by the larval (or puparial) characters, cf. Figs. 356 and 361. SCHERNEY (1966), WEIGAND (1966) and SCHERNEY and WEIGAND (1967), in their useful papers discussing this species, refer to it as *nigrella* but I am satisfied that this was based on a misidentification.

ECONOMIC IMPORTANCE. For 30 years after its description in 1933 *Agromyza megalopsis* was only known in small numbers in eastern Germany (HERING, 1962: 32). However, in 1964 and 1965 the barley crop in large areas of southern Germany, particularly in Württemberg, Mittelfranken and Oberpfalz, and in parts of Czechoslovakia was heavily infested. Investigations showed that 32% of the summer barley in Bavaria, which makes up almost 28% of the total area under grain cultivation, was affected. During the first heavy outbreaks in 1964 farmers feared a crop loss of up to 20%.

Experiments carried out in Bavaria in 1965 (WEIGAND, 1966) showed that with suitable applications of insecticides both eggs and larvae could be controlled but any efficient control of the adults was not practical owing to their long period of flight. The build up in the rate of parasitism which was noted in all affected localities from the year 1964 to 1965 in fact gave the most effective control and this leaf-miner has not subsequently been found in any significant numbers.

Tests of the yield of treated and untreated crops showed only a marginally better yield in the treated areas. It is deduced from this that despite the initial destruction of leaf tissue, at least 50% of the assimilation tissue remains intact in the leaf sheath, and this suffices for the plants to mature and produce a normal yield.

Agromyza nigrella (Rondani, 1875) – Europe

ADULT. Large black species with costa ending at vein R_{4+5}.

HEAD. Frons distinctly projecting above eye, jowls deeply extended at rear; third antennal segment quadrate above, more rounded below (Fig. 357); mesonotum with 5 dc, greatly decreasing in size, at least 2 differentiated in pre-sutural area; wing length from 2.5–3.1 mm, costa ending at vein R_{4+5}, last section of M_{3+4} about $\frac{2}{3}$ penultimate, first cross-vein generally at midpoint of discal cell (Fig. 358).

COLOUR. Frons dark, brownish to black; mesonotum shining black but appearing somewhat more mat viewed from front; legs entirely black; squamae white, margin only slightly differentiated, yellowish, fringe white to ochrous.

MALE GENITALIA. Aedeagus with elongate distiphallus, as in Figs. 359, 360.
EARLY STAGES. Larval posterior spiracles on pronounced protuberance but widely separated (Fig. 361), each with 3 pores. Puparium reddish-brown.
HOST-PLANTS. *Avena sativa* L. (oats), *Hordeum vulgare* L. (barley), *Secale cereale* L. (rye), *Triticum aestivum* L. (wheat); also *Dactylis, Festuca, Glyceria, Holcus, Lolium, Phleum, Poa, Setaria* and *Trisetum* (HERING, 1957c).
BIOLOGY. The larval leaf-mine starts as a narrow channel running towards the apex of the leaf, which then turns and develops into a broad blotch; this can

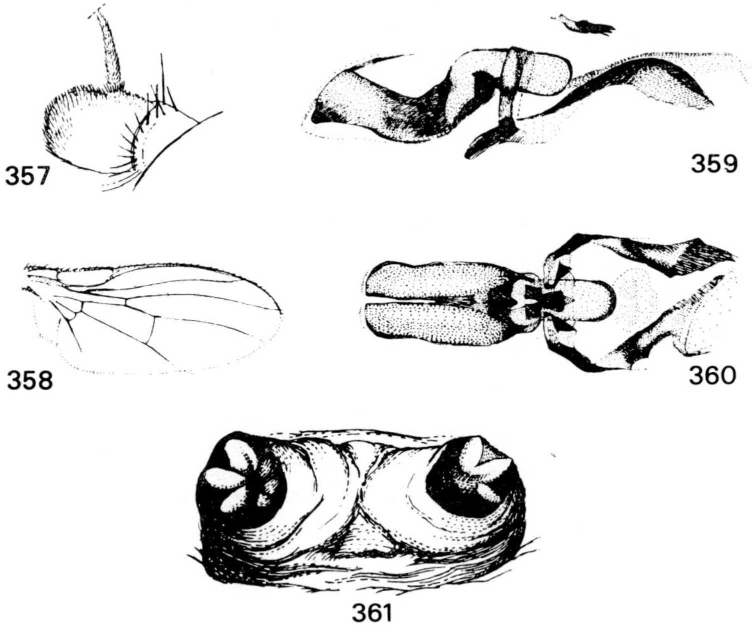

Figs. 357–361. *Agromyza nigrella* Rond.: 357, third antennal segment; 358, wing; 359, aedeagus, side view; 360, same, ventral view, 361, posterior spiracles of puparium.

almost fill the leaf. Mines can be up to 9 cm long (DUTHOIT, 1968) or 6–8 inches (CAMERON, 1945). The frass is largely diffused, giving the mine a characteristic greenish appearance. Pupation takes place externally, on the ground.

In England flies have been observed early in June and larval feeding takes place later in June and in July. There is almost certainly a second, autumn generation.
DAMAGE CAUSED. A substantial mine is formed by the larva of this species, which can lead to the blanching and wilting of the leaves. The flag leaf is frequently affected and as this is important in the translocation of food to the developing grain, damage to the flag leaves could affect the ultimate yield (DUTHOIT, 1968).
DISTRIBUTION. Widespread in Europe; in Britain recorded also in Scotland.

PARASITES. In Britain two confirmed species of Dacnusinae are *Dapsilarthra balteata* (Thomson) and *Protodacnusa tristis* (Nees). FISCHER (1971) records four species in Europe: *Opius cingulatus* Wesm., *O. maculipes* Wesm., *O. parvungula* Th. and *O. propodealis* Fi.

Discussion

TAXONOMY. *Agromyza nigrella* has in the past been confused with *A. ambigua* Fallén, until the types of the latter were examined and clarified by SPENCER (1965d). This species most closely resembles *A. megalopsis* Hering; the two can be separated by the male genitalia but most easily in the larval or puparial stage by the differing arrangement of the posterior spiracles (cf. Figs. 356, 361).

ECONOMIC IMPORTANCE. *A. nigrella* periodically occurs in large numbers and the young leaves of cereals are appreciably damaged. A heavy attack occurred in Scotland in 1944 and CAMERON (1945) considered that 'although such plants may not be irreparably injured, the damage may be greater than that due to mildew and rust together'.

In July, 1965, widespread damage was caused to winter wheat in the South-east counties of England. 'One site examined, in Buckinghamshire, had a 100% attack and a 30% leaf length loss.' (DUTHOIT, 1968.) Heavy attacks continued in 1966 but by 1967 the infestations had become insignificant.

It has not been proven experimentally that early leaf-mining damage by this species has any direct effect on the grain yield. Nevertheless, the periodical mass outbreaks which can be expected will always be a cause of concern to growers and more detailed investigations into the effective damage caused by this species seem called for.

Agromyza nigrociliata Hendel, 1931 – Europe

ADULT. Large species, wing length 2.9–3.5 mm, costa extending only to vein R_{4+5}; frons projecting above eye, jowls deeply extended at rear; frons black, mesonotum shining black, with 5 or more dc greatly decreasing in size; squamal fringe brown to black.

MALE GENITALIA. Aedeagus as in Fig. 362.

EARLY STAGES. Larval mouth-parts as in Fig. 363, posterior spiracles each with 3 somewhat elongate pores; puparium reddish-brown.

HOST-PLANTS. *Secale cereale* L., *Triticum aestivum* L.; also other wild grasses.

BIOLOGY. The first larvae appear in the beginning of June. The mine at first runs towards the apex of the leaf and then turns, running down towards the base. Several eggs are often laid in one leaf and a large communal mine is then formed. The frass is somewhat diffused, greenish. The puparium is dark reddish-brown and pupation takes place externally. There is only a single generation, with the adult flies emerging in May or June (HERING, 1953).

DAMAGE CAUSED. With several larvae feeding together, the large leaf-mine causes considerable reduction in the assimilation of the plant, particularly when the upper leaf is attacked. HERING (1953 : 189) considers this will lead to a reduction in yield.

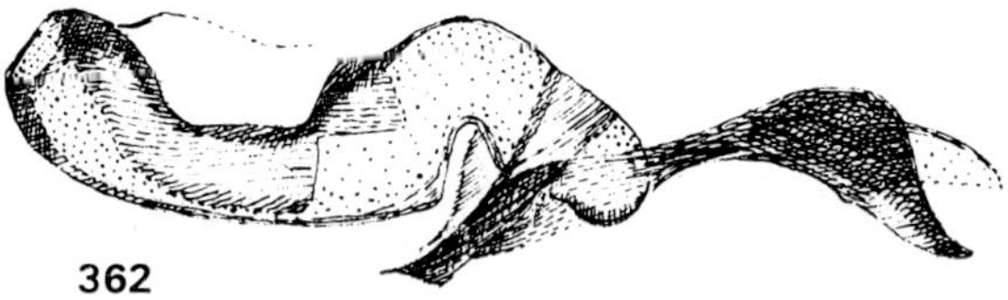
362

363

Figs. 362–363. *Agromyza nigrociliata* Hd.: 362, aedeagus; 363, larval mouth-parts.

DISTRIBUTION. Europe, confirmed in Austria, Britain, Germany, Hungary Sweden.
PARASITES. None recorded.

Discussion

TAXONOMY. This species resembles both *nigrella* Rondani and *mobilis* Meigen, in having the costa terminating at vein R_{4+5}. However, the squamal fringe is darker than in *nigrella* and can be differentiated from *mobilis* by the greatly projecting frons. The male genitalia closely resemble those of both *nigrella* and *mobilis* but the aedeagus is different in detail.
ECONOMIC IMPORTANCE. HERING (1953) records a mass outbreak of this species in the Görlitz area of Saxony in eastern Germany on both wheat and rye. Other wild grasses in the vicinity which were also attacked included *Agropyron caninum* (L.) P.B., *Apera spica-venti* (L.) P.B., *Arrhenaterum elatius* (L.) Presl. and *Hordeum murinum* L. Previously only isolated specimens had been known. Thus, although normally occurring in small numbers, presumably controlled by parasites, this species like others in this group, periodically undergoes a population build-up which can lead to severe leaf-mining damage. It deserves recognition as a minor pest.

Agromyza oryzae (Munakata) – Japan, Siberia

Oscinis oryzae Munakata, 1910.
Oscinis oryzella Matsumura, 1915.
Agromyza oryzae Hendel, 1931.
Agromyza oryzella Yuasa and Koyama, 1938.

ADULT. Medium-sized black species, with brown to black squamal fringe.
HEAD (Fig. 364). Frons narrowly projecting above eye, up to $1\frac{1}{2}$ times width of eye; 2 equal ors, 3–5 ori, normally 4, the lowest always small; orbital setulae

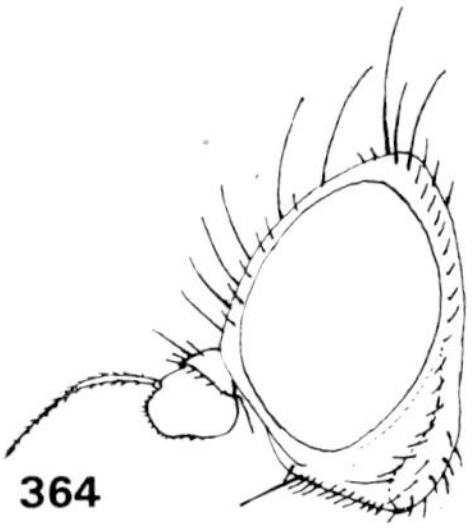
364

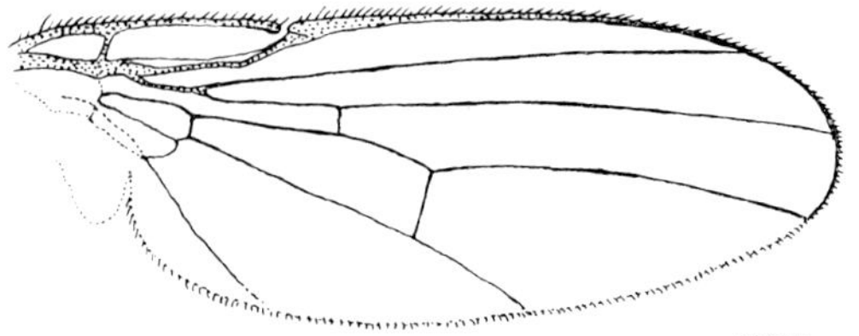
365

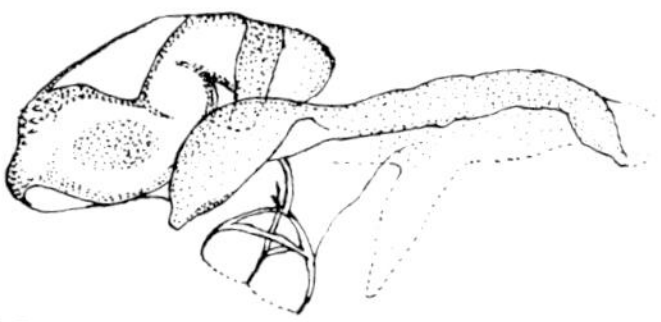
366

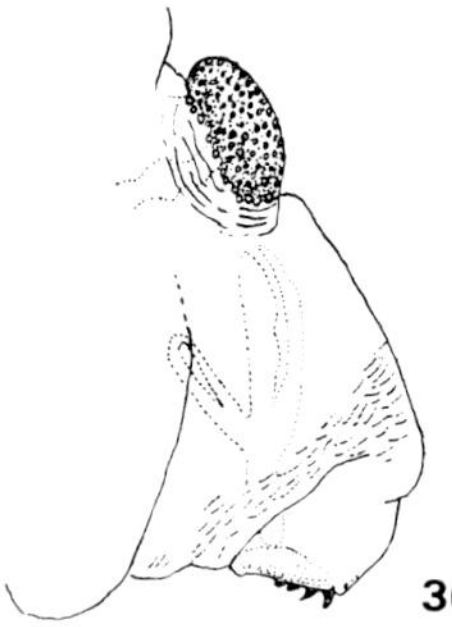
367

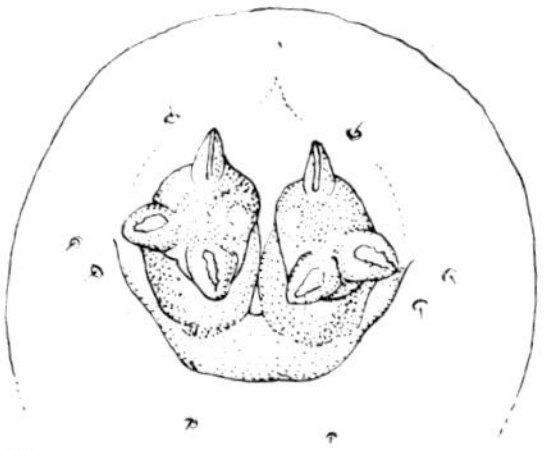
368

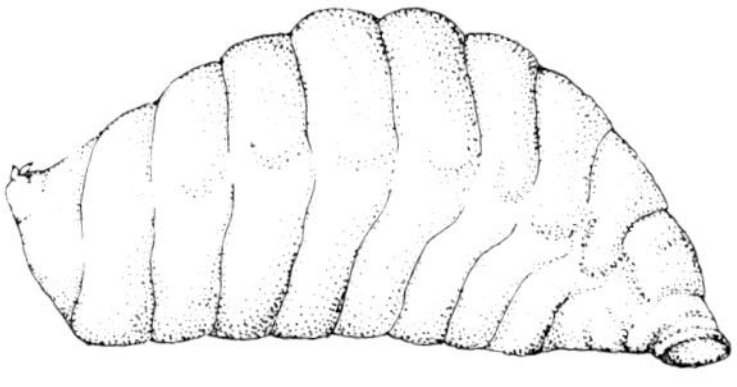
369

Figs. 364–369. *Agromyza oryzae* (Mun.): 364, head; 365, wing; 366, aedeagus; 367, anterior spiracles and head of larva; 368, posterior spiracles; 369, puparium (all after KATO).

sparse, reclinate; jowls broad, $\frac{1}{6}$ height of eye, deepest at rear; third antennal segment axe-shaped, arista slightly pubescent.
MESONOTUM. 4 + 2 dc, the two pre-sutural ones conspicuously shorter, acr in 6 rows, ending before pre-scutellars.
WING (Fig. 365). Length 2.5–3 mm, costa extending strongly to vein M_{1+2}, last section of M_{3+4} about $\frac{2}{3}$ length of penultimate, first cross-vein slightly beyond midpoint of discal cell.

LEGS. Mid-tibia with 2 strong lateral bristles.
COLOUR. Frons mat black, ocellar triangle and orbits moderately shining; mesonotum black, slightly greyish, particularly when viewed from front; legs entirely black, abdomen shining; squamae yellowish-white, margin pale brown, fringe darker, brown to black.
MALE GENITALIA. Aedeagus as in Fig. 366.
EARLY STAGES. 'Full grown larva pale yellowish white, about 5–5.5 mm in length, mandibles unequal, teeth distinctly alternate, each mandible with five teeth; anterior spiracles very large, somewhat hemispherical, each with 88–133 spiracular bulbs (Fig. 367), posterior spiracles each having three bulbs (Fig. 368); four pairs of minute sensory papillae surrounding posterior spiracles.' (KATO, 1956).

A notable feature of this species is that the puparium shows a marked difference between the hibernating and non-hibernating generations. In the hibernating form it is smaller, darker, not shining, more convex; the non-hibernating form is shown in Fig. 369.
HOST-PLANTS. *Oryza sativa* L., *Zizania latifolia* Turez.
BIOLOGY. The following account of the life history is based on investigations undertaken in Hokkaido by KUWAYAMA (1950). Adults begin to emerge at the end of May or early June and emergences continue until the end of June, although these times can show some variation depending on the weather from year to year. Oviposition takes place towards the tips of young leaves. The larva mines downwards, as it matures forming a large blotch. The mine is white or pale green at first but later becomes brown or ochrous. The maximum length of mine found was 115 mm, the maximum width 9 mm.

The larval stage lasts from 10–44 days. Pupation normally takes place outside the mine but on the leaf, either on the upper or lower surface or near the leaf sheath. The pupal period of the first generation is short, from 7–18 days, with emergences taking place from the end of June to mid-July. Larvae of this second generation are abundant from mid-July and the resulting puparia soon fall from the plant and remain on the ground throughout the winter. A small number of these puparia produce a partial third generation and mines can be found from mid-August to mid-September.
DAMAGE CAUSED. Although most leaves contain only a single larva, as many as three can occur together. Mining on this scale destroys the leaves and young plants are seriously weakened.
DISTRIBUTION. Northern Japan, from Kamikawa district of Hokkaido in the north to Toyama Prefecture of Honshu (36.5°N to 44°N); eastern Siberia.
PARASITES. No information available.

Discussion

TAXONOMY. In view of the importance of this species, its full synonymy is given above. *A. oryzae* belongs to the large group of closely related Holarctic grass-feeders, which can frequently only be reliably separated by their male genitalia.

The greatly enlarged anterior spiracles appear to be a modification facilitating survival of the puparia in water; a similar form is only known in the European species *alunulata* Hendel (=*distorta* Griffiths) feeding on *Glyceria maxima* (Hartm.) Holmb. (cf. SPENCER, 1972b: 143).

The species has been re-described in detail by KATO (1956) and SASAKAWA (1961).

ECONOMIC IMPORTANCE. According to KUWAYAMA (1950) *Agromyza oryzae* is among the more serious of the 190 insect species known to feed on rice in Japan. Damage is caused particularly to young plants.

The secondary host, *Zizania latifolia*, is a wild plant in Hokkaido where *A. oryzae* causes most damage, but is cultivated farther south and also in Manchuria, China and Korea, where the young shoots are eaten as a vegetable.

Agromyza parvicornis Loew, 1869 – United States, Canada, South America

ADULT. Large species with brown frons and costa extending to vein M_{1+2}.

HEAD. Frons brown, broad, almost twice width of eye, distinctly projecting above eye anteriorly; 2 strong reclinate ors, 2 weaker, incurved, ori; jowls deeply extended at rear, $\frac{1}{3}$ vertical height of eye, in centre $\frac{1}{6}$ eye height; third antennal segment round, virtually bare, pale brown, arista distinctly pubescent.

MESONOTUM. Shining black, with 4 or 5 dc, greatly decreasing in size.

WING. Length from 2.5 mm in male to 3.1 mm in female, costa extending strongly to vein M_{1+2}, last section of M_{3+4} $\frac{2}{3}$ penultimate; squamae and margin yellowish-white, fringe either entirely white or slightly darker, ochrous-brown. Legs black, fore-knees paler, all tibiae distinctly brownish.

MALE GENITALIA. Aedeagus (Figs. 370, 371) with conspicuously short and broad distiphallus; surstyli (Fig. 372) with a group of 12–15 stout bristles on inner margin.

EARLY STAGES. Puparium reddish-brown, posterior spiracles on a small subsidiary projection arising from a large protuberance, the two processes slightly separated, each with 3 pores (Fig. 373).

HOST-PLANTS. *Zea mays* L., *Panicum miliaceum* L.; *Echinochloa crus-galli* L. Probably other cultivated cereals and wild grasses.

BIOLOGY. Oviposition can take place either in the upper or lower surface of the leaf, near the apex of the leaf in young plants but in more mature plants at any point in the leaf. During summer larvae can complete their development in as little as four days, 'while in the cooler weather of spring or fall they will require ten days or more'. The larva leaves the mine to pupate just below the surface of the ground. In summer the pupal stage lasts only 14 days but will last for several months in the overwintering generation.

Adults appear in numbers by May. There are at least four complete generations. In Indiana adults and larvae are killed by the first frosts in November but during mild winters in Florida some larval activity can continue throughout the winter. Adults emerge earlier than farther north and were found to be abundant by early February (PHILLIPS, 1914).

Damage caused. The larval mine develops into a broad blotch. 'One larva is sufficient to destroy a young corn leaf, although as many as four have been found in a single leaf in the field. After a few days they turn brown and curl up.' (Phillips, 1914.) In young leaves the entire parenchyma can be consumed, leaving only the upper and lower epidermis intact. However, in older leaves mines on the lower surface leave the upper surface parenchyma intact. Such mines are scarcely visible from the upper leaf surface and the damage caused is proportionately less.

Distribution. United States: Florida, Texas, Washington, DC, Utah, Wisconsin, probably many other states; Canada: British Columbia, Ontario.

Parasites. Phillips states that there are 18 parasites known to attack this species in the United States – 3 Braconids and 15 Chalcids. The most important

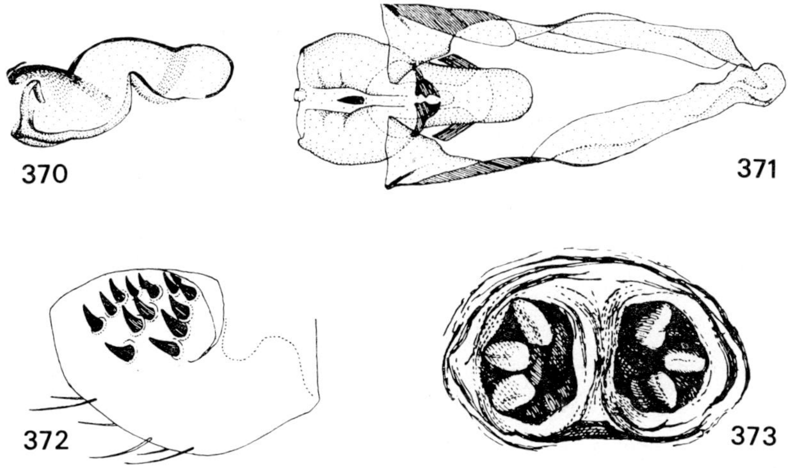

Figs. 370–373. *Agromyza parvicornis* Loew: 370, aedeagus, side view; 371, same, ventral view; 372, surstylus; 373, posterior spiracles of puparium.

is *Derostenus diastatae* (Howard) and as many as 8 larvae have been found in a single larva of *parvicornis*. Other Chalcids recorded are *Derostenus punctiventris* Cwfd., *Diaulinus begini* Ashm., *D. pulchripes* Cwfd., *D. Websteri* Cwfd., *Zagrammosoma multilineata* (Ashmead), *Sympiesis* sp., *Closterocerus tricinctus* (Ashm.), *C. utahensis* Cwfd., *Notanisomorpha ainsliei* Cwfd., *Cirrospilus flavoviridis* Cwfd., *Chrysocharis ainsliei* Cwfd., *C. parksi* Cwfd., *Pleurotropis utahensis* Cwfd. (Chalcididae); *Opius diastatae* Ashm., *O. succineus* Gahan, *O. utahensis* Gahan (Braconidae) and 2 Pteromalidae.

Discussion

Taxonomy. *A. parvicornis* belongs to a large complex of grass-mining species and has certainly been frequently misidentified in the past. Ten new species in this group were described from Canada (SPENCER, 1969a) and there is little doubt that many others await description in America.

Diagnostic characters of this species are the pale antennae, shining black mesonotum, costa extending to vein M_{1+2} and the pale squamal fringe. The male genitalia are very distinctive. FRICK (1959: 357) stated that there are only 2 dc; this is not entirely accurate, as there are at least 2 or 3 smaller ones.

Economic importance. *Agromyza parvicornis* occurs periodically in large numbers in restricted localities and causes appreciable local damage. RUHMANN (1935) reports that 80% of the maize grown for fodder in one area of British Columbia was infested and in the spring of 1935 in Kansas there was an extreme outbreak attacking maize in the 2–4 leaf stage, retarding its growth by the destruction of one or more leaves (ANON. b, 1937).

Whenever occurring in large numbers this species will always be a potentially serious pest. Almost certainly effective control is exercised by its numerous parasites. PHILLIPS (1914: 26) correctly appreciated the significance of this parasite complex, writing: 'With this army of parasites in the field it will be readily seen that the chances that the miner will do serious mischief are reduced to a minimum. Only under exceptional circumstances would *Agromyza parvicornis* be able to elude this array of enemies. This is a very good illustration of the holding in check by its natural enemies of what would otherwise in all probability be a pernicious insect, thus showing that it is entirely possible for a group of parasites under favourable conditions to control their host insect.'

Agromyza penniseti Spencer – West, East and South Africa

Agromyza penniseti Spencer, 1959. Holotype from Cameroon in Zoologisches Museum, Berlin.

Agromyza graminivora Spencer, 1960, **syn. nov.** Holotype ♂ from Senegal in British Museum (Natural History).

ADULT. Medium-sized dark species with the costa extending to vein M_{1+2}.

HEAD. Frons little more than width of eye, not projecting above eye in profile; 2 reclinate ors, the upper somewhat stronger, 2 or 3 incurved ori; jowls narrow below centre of eye but deeper at rear, there almost $\frac{1}{5}$ height of eye; third antennal segment slightly lower than broad, rounded at end, arista slightly pubescent.

MESONOTUM. 3 distinct dc, those beyond little differentiated from acr, acr in 8 rows.

WING. Length from 2.4 mm in male to 2.8 mm in female; costa extending to vein M_{1+2}, last section of M_{3+4} short, $\frac{1}{2}$ length of penultimate, first cross-vein at midpoint of discal cell.

COLOUR. Frons brownish-black, ocellar triangle and base of orbital bristles moderately shining; antennae black, third segment slightly paler, brownish, on inside; mesonotum shining black, pleura black, legs largely black but tibiae and tarsi slightly paler, brownish-black; wings hyaline, veins brown; squamae white, fringe variable, from white to distinctly black.

MALE GENITALIA. Aedeagus as in Figs. 374, 375; surstyli with an oval chitinized area on lower, inner corner bearing an irregular group of 10–12 strong bristles.
EARLY STAGES. Puparium shining, reddish-brown, deeply segmented, hind-end conspicuously bending ventrally; posterior spiracles closely adjoining, each with 3 pores. DE MEIJERE (1940b: 163) discusses and illustrates puparia from Mt. Cameroon.
HOST-PLANTS. *Pennisetum purpureum* Schum. (Elephant grass), *P. pedicillatum* Trin.
BIOLOGY. Details of the leaf-mine and life cycle are not known but the larva pupates in the mine, with the anterior spiracles of the puparium projecting through the upper epidermis and the posterior spiracles through the lower.

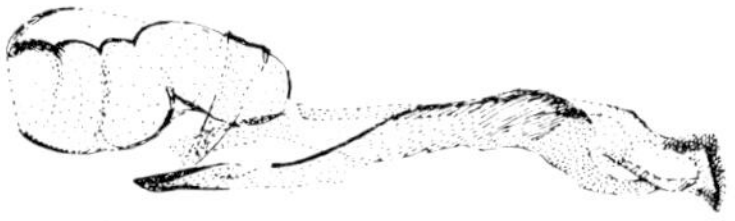
374

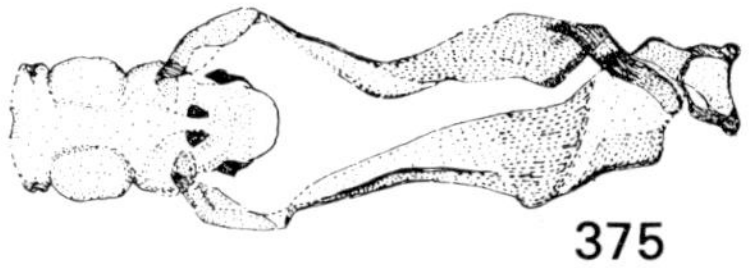
375

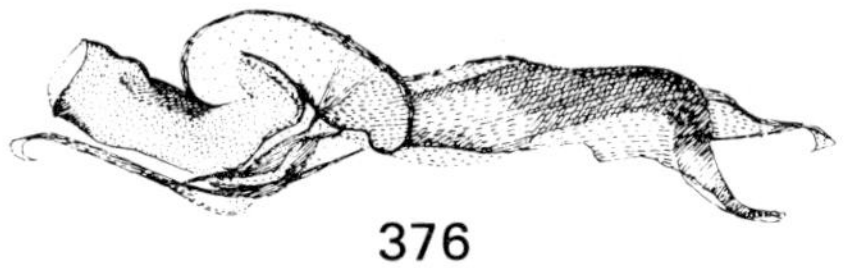
376

Figs. 374–375. *Agromyza penniseti* Sp.: 374, aedeagus, side view; 375, same, ventral view.

Fig. 376. *Agromyza pennisetivora* Sp.: aedeagus, side view.

DISTRIBUTION. West Africa: Cameroon, N. Nigeria; East Africa: Uganda; South Africa: Natal.
PARASITES. In N. Nigeria an *Opius* sp. (Braconidae).

Discussion

TAXONOMY. This species belongs to a large group of shining black grass-feeders. It is unusual in the variability of the colour of the squamal fringe and this led to the description, as distinct, of *A. graminivora*. Re-examination of the types of *A. penniseti* and *A. graminivora* and also of males from South Africa with both a white and a black squamal fringe shows the aedeagus to be similar in all specimens. *A. graminivora* is therefore synonymized with *penniseti* herewith.

A male from India: Namkum, bred from an unidentified grass was previously identified as *graminivora* (Spencer 1966f: 515), and *papillata* Sasakawa, 1963a from Vietnam was synonymized with it. I have now seen a puparium of the Indian specimen and this is of normal shape, without the conspicuous ventral curvature of *penniseti*. It must therefore be treated as distinct from *penniseti* and may well represent *papillata*. However, the identity of *papillata* can only be properly established when the male genitalia of specimens from the type locality can be examined.

ECONOMIC IMPORTANCE. *Pennisetum purpureum* is an important fodder crop in Africa and is also used in the manufacture of paper. Larval leaf-mines can largely destroy individual young leaves but this is such a robust grass that no significant damage will be caused, unless populations of *Agromyza penniseti* increase far beyond existing levels.

Agromyza pennisetivora Spencer, 1961b – India

ADULT. Large species with reddish-brown frons and costa extending only to vein R_{4+5}.

HEAD. Frons $1\frac{1}{2}$ times width of eye, conspicuously projecting above eye in lower half; orbital bristles short, slender, 1 reclinate ors, 2 incurved ori; jowls angular, deeply extended at rear, in ratio 8:20 with height of eye, cheeks forming conspicuous ring below eye, broad epistoma present.

MESONOTUM. With 5 dc, greatly decreasing in size, acr irregularly in 6 rows.

WING. Length 2.7 mm, costa extending only to vein R_{4+5}, last section of M_{3+4} $\frac{2}{3}$ length of penultimate, first cross-vein at or beyond midpoint of discal cell.

COLOUR. Frons, jowls and face reddish-brown; hind-margin of eye and upper orbits black, facial keel black; first and second antennal segments reddish (third missing in all available specimens but probably at least partially pale); all bristles on head slightly but distinctly reddish; mesonotum mat black, legs entirely black; wings and all veins pale, squamae and fringe whitish-yellow.

MALE GENITALIA. Aedeagus as in Fig. 376, surstyli with an angular area of chitinization but without any developed bristles.

EARLY STAGES. Not known.

HOST-PLANT. *Pennisetum typhoideum* Rich. ('bulrush, spike or pearl millet').

BIOLOGY. No details recorded.

DISTRIBUTION. India: New Delhi.

PARASITES. None recorded.

Discussion

TAXONOMY. The distinctive characters of this species are the reddish frons and the costa ending at vein R_{4+5}. The male genitalia are typical of feeders on Gramineae.

ECONOMIC IMPORTANCE. The host-plant of *A. pennisetivora* is widely cultivated in India but no information is available regarding the degree of damage caused. The larvae will almost certainly occur on other cultivated species of *Pennisetum* and possibly also on other related genera.

Agromyza rondensis Strobl, 1900a – Europe

Agromyza nigripes var. *rondensis* Strobl, 1900a: 61.
Domomyza occellaris Hendel, 1920: 124.
Agromyza nigrifemur Hendel, 1931.
Agromyza veris Hering, 1951b.

ADULT. Medium-sized species, with brown frons, elongate third antennal segment and pale squamal fringe; costa ending at vein R_{4+5}.

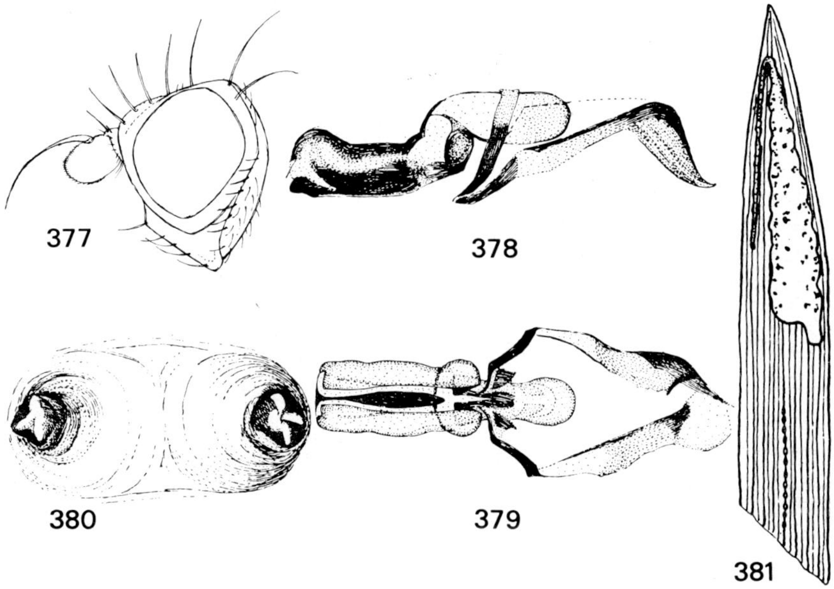

Figs. 377–381. *Agromyza rondensis* Strobl: 377, head; 378, aedeagus, side view; 379, same, ventral view; 380, posterior spiracles of puparium; 381, leaf-mine on *Triticum* (Hering).

HEAD (Fig. 377). Frons broadly projecting above eye, particularly in front, jowls deeply extended at rear, third antennal segment conspicuously elongate.

MESONOTUM. 3 + 2 dc, decreasing in size substantially but the two pre-sutural bristles still relatively long; acrostichals irregularly in three rows in front, two only behind.

WING. Length from 1.75 mm in male to 3 mm in female, costa ending at vein R_{4+5}, last and penultimate sections of M_{3+4} approximately equal but last may be somewhat longer or shorter.

COLOUR. Frons distinctly brownish, hind-margin of head and orbits black; mesonotum conspicuously grey, mat; squamal fringe white.

MALE GENITALIA. Aedeagus as in Figs. 378, 379; surstyli elongate.

EARLY STAGES. Larval mouth-hooks with upper tooth substantially longer than the lower; anterior spiracles on a short projection, each with an ellipse of 7 pores; posterior spiracles (Fig. 380) each with 3 pores. The puparium is reddish-brown.

HOST-PLANTS. *Hordeum vulgare* L. (barley), *Secale cereale* L. (rye), *Triticum aestivum* L. (winter wheat); also on *Arrhenaterum*, *Bromus*, *Calamagrostis*, *Dactylis* and *Poa*.

BIOLOGY. According to HERING (1951b) this species has only a single generation, with larval feeding taking place from the middle of April till the middle of May. However, there are apparently two further generations in the summer and autumn which he overlooked. In England leaf-mines collected in July produced flies in early September, which will certainly have given rise to at least a partial further generation of larvae to provide the overwintering puparia.

Oviposition takes place towards the apex of the leaf and the young larva initially forms a narrow mine running up the leaf which then turns to form a broad blotch running downwards (Fig. 381). Normally only a single larva feeds in one leaf.

DAMAGE CAUSED. Normal leaf-mining damage.

DISTRIBUTION. Europe: Germany, Austria, England, Spain.

PARASITES. None recorded.

Discussion

TAXONOMY. There has been considerable confusion regarding the identity of this species in the past and it has been described under four separate names by three different workers. *Domomyza occellaris* Hendel, *Agromyza nigrifemur* Hendel and *A. veris* Hering were synonymized with *rondensis* Strobl by SPENCER, 1966c.

The distinctive characters of *rondensis* are the elongate third antennal segment, the brownish frons and the mat greyish mesonotum. It closely resembles *A. conjuncta* Spencer, 1966c known from Italy, Spain and Yugoslavia; this species has a similarly elongate third antennal segment but the frons is black and the mesonotum shining black.

ECONOMIC IMPORTANCE. If, as HERING (1951b) maintains, only a single larva feeds in a leaf, the damage to cereal plants must be negligible. However, HERING emphasizes that larval feeding in the first generation occurs early in the year on young plants, when any damage to the leaves will have the maximum effect on the growth of the plants. At this time of the year the number of parasites is always small and would not be able to exercise any effective control. Any unusually large population might therefore cause significant damage to spring cereals. No such outbreak, however, has yet been reported.

GENUS LIRIOMYZA MIK

Liriomyza species are not a serious problem on cereals or pasture crops, although many grass-feeders are known in the northern hemisphere. The genus is poorly represented in Africa, Asia and Australia. A number of species are known in South America but only one of the two species discussed below occurs there as a pest of cereals, attacking oats in Peru.

Identification of species in this genus is frequently difficult and in many cases only possible after examination of the male genitalia. Any species coming to notice which are not readily identifiable as one of the two dealt with here will therefore require examination by a specialist.

Liriomyza flaveola (Fallén, 1823a) – Europe

ADULT. Medium-sized species, with entire mesonotum black and conspicuously yellow knees.

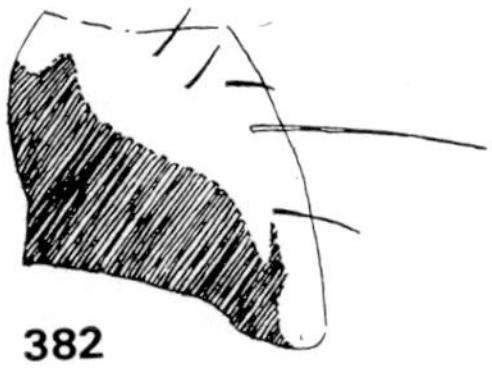

382

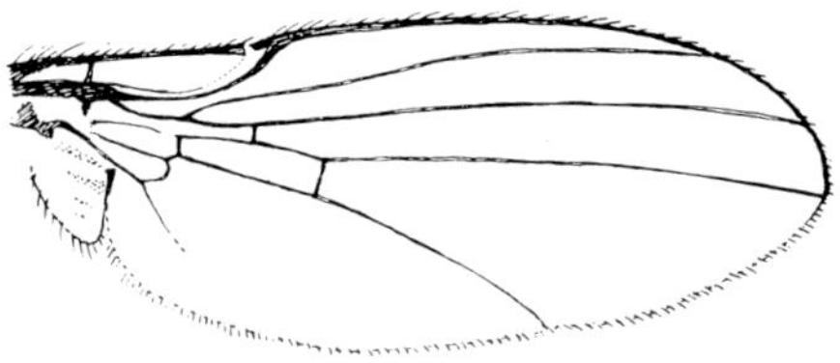

383

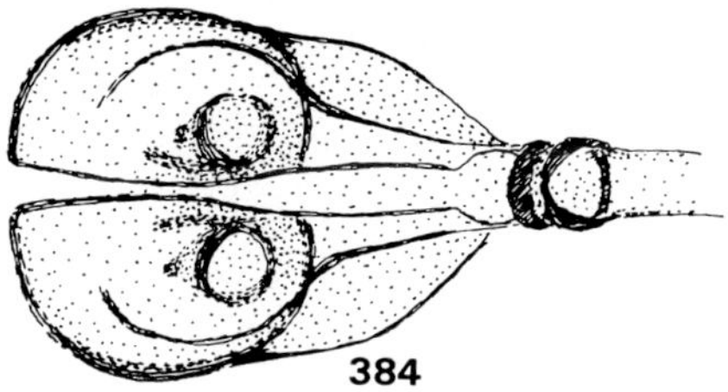

384

Figs. 382–384. *Liriomyza flaveola* (Fall.): 382, mesopleuron; 383, wing; 384, distiphallus.

HEAD. Frons not significantly projecting above eye, with 2 strong equal ors and 2 or 3 ori; orbits and frons entirely yellow, hind-margin of eye dark at least to base of inner vertical; all antennal segments yellow.
THORAX. Mesonotum deep black, though distinctly mat, with 3 + 1 strong dc, acr irregularly in 4 rows; mesopleura black along front and lower margins, mesopleural bristle on yellow ground (Fig. 382).
LEGS. Femora largely black but conspicuously yellow distally.
WING (Fig. 383). Length from 2.1–2.7 mm, last section of vein M_{3+4} approximately twice length of penultimate.
MALE GENITALIA. Aedeagus with distinctive distiphallus, as in Fig. 384.

EARLY STAGES. Puparium reddish-brown, posterior spiracles each with 3 pores, one longer, almost hook-like.
HOST-PLANTS. *Avena sativa* L., *Hordeum vulgare* L.; many pasture and wild grasses, frequently *Bromus*, *Dactylis*, *Holcus* and *Poa*.
BIOLOGY. The larval leaf-mine is whitish, narrow, not greatly widening, running down the leaf from the apex. Pupation invariably takes place outside the mine; puparia found in the mine will normally prove to be parasitized.

Adults are most common in early summer but there are certainly one and probably two later generations.
DISTRIBUTION. Europe, reported as far north as Swedish Lappland, less common in Mediterranean area, eastern limit not known, not present in Japan.
PARASITES. *Dacnusa adducta* (Haliday), *Chorebus aphantus* (Marshall); *Opius levis* Wesm.

Discussion

TAXONOMY. *Liriomyza flaveola* closely resembles two other grass-feeding species, *L. pedestris* Hendel and *L. phryne* Hendel. The former is darker, with the mesonotum normally more shining and the latter has the femora yellow distally for between $\frac{1}{2}$ and $\frac{1}{3}$ of their length (cf. SPENCER, 1972b).
ECONOMIC IMPORTANCE. Cultivated cereals do not appear to be favoured hosts of this species. Individual mines are not uncommon on barley and oats but wheat and rye have not been recorded as hosts. Adults are caught in considerable numbers in meadows but damage caused to the hay crop has never been reported as serious.

Liriomyza marginalis (Malloch, 1913a) – United States, Caribbean, South America

ADULT. Small yellow species, with mesonotum variably brownish or black but always yellow centrally adjoining scutellum (Fig. 385).
HEAD. Frons not projecting above eye; third antennal segment round, arista long, distinctly pubescent; frons and orbits uniformly yellow, with both vertical bristles on yellow ground; all antennal segments yellow.
MESONOTUM. 3 + 1 or 2 + 1 dc, acr irregularly in 4 rows; central area and lateral bands varying from deep shining black to pale reddish-yellow or intermediate (Fig. 385), large area adjoining scutellum always yellow.
LEGS. Coxae and femora bright yellow, tibiae and tarsi only somewhat darker, brownish.
WING. Length from 1.8–1.9 mm, last section of M_{3+4} variable but normally $1\frac{1}{2}$ times length of penultimate.
MALE GENITALIA. Aedeagus as in Figs. 386, 387; epandrium with a conspicuous chitinized bar on inner margin (Fig. 388).

EARLY STAGES. Puparium pale yellowish-brown, anterior spiracles long, black, divided distally (Fig. 389); posterior spiracles arising from rounded projection, each with 3 pores.

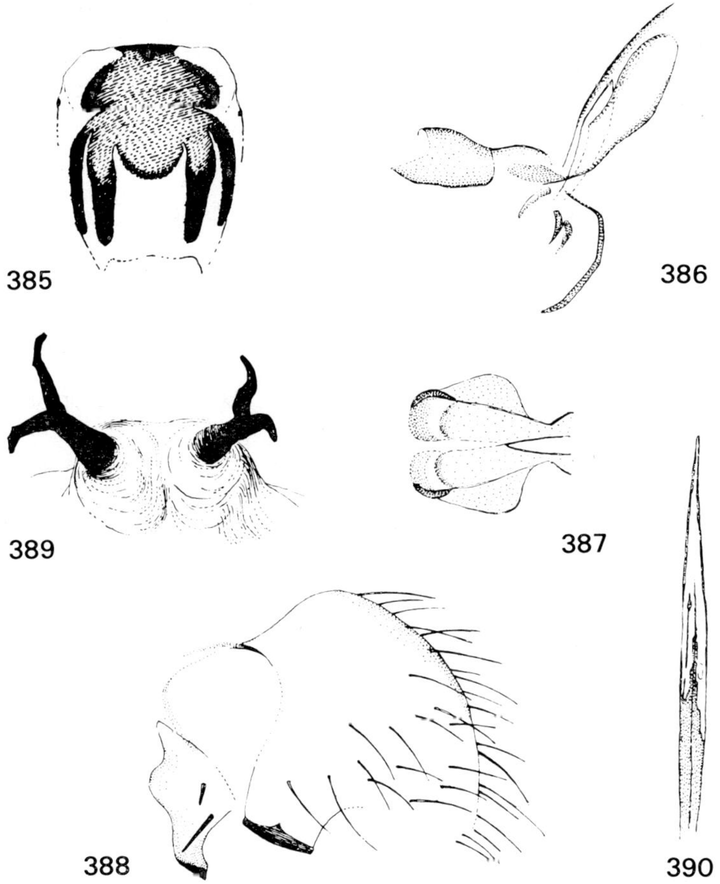

Figs. 385–390. *Liriomyza marginalis* (Mall.): 385, mesonotum (Dominica); 386, aedeagus, side, view; 387, distiphallus, ventral view; 388, epandrium; 389, anterior spiracles of puparium; 390, leaf-mine on unidentified grass, Jamaica.

HOST-PLANTS. *Avena sativa* L. and probably other cereals; *Panicum miliaceum* L., particularly common on *Paspalum* and probably occurring on many other wild grasses.

BIOLOGY. The larva forms a long whitish mine in the leaf-blade (Fig. 390), with the anterior spiracles projecting through the epidermis. There are probably several generations in a year.

DISTRIBUTION. United States: Florida, South Carolina and certainly present in other southern States; Caribbean: Bahamas, Dominica (W.I.), Jamaica; South America: Brazil, Peru.
PARASITES. STEGMAIER (1967c) reared a number of parasites from '*Liriomyza sorosis* Williston', a name used at that time for several species which have now been split up, including *L. marginalis* (SPENCER and STEGMAIER, 1973). These species are recorded below but it will be necessary to confirm which parasitize the true *marginalis*.
EULOPHIDAE. *Chrysocharis majoriani* (Girault), *Chryocharis* sp., *Derostenus agromyzae* Cwfd., *Diaulinopsis callichroma* Cwfd.; Pteromalidae: *Heterochema* sp.

Discussion

TAXONOMY. There are a number of species in the southern States of America and in South America with the similar distinctive coloration of *L. marginalis* but this name is now restricted to the species occurring commonly on *Paspalum*, with the aedeagus and pupal spiracles as illustrated.
ECONOMIC IMPORTANCE. Mines were found by the author in Florida on very young plants of *Panicum miliaceum* only 5 cm high. The leaves in such seedlings are largely destroyed and a large population attacking cereals in this way would cause significant damage.

L. marginalis has been confirmed in Peru attacking oats but the species is only considered there as a secondary pest (C. A. KORYTKOWSKI, personal communication).

GENUS PSEUDONAPOMYZA HENDEL

This genus is divided into two distinct species groups, one feeding on Gramineae, the other on Acanthaceae. The close relationship between these two groups is clearly indicated by the generally similar external morphology, by similarities in male genitalia and by the characteristic papilli on larva and puparium in a number of species in both groups. A distinctive difference between the groups is in the angulate third antennal segment of the Gramineae feeders. The species on Acanthaceae probably represent the ancestral group but the world-wide distribution and differentiation of the Gramineae feeders suggests that the transference to and subsequent evolution on this family is certainly not recent. The two species groups could therefore logically be treated as distinct genera but this step is not proposed at the present time.

Among the Gramineae feeders only three species have hitherto been recorded as pests – *Ps. atra* (Mg.) feeding on a number of cereals in Europe, *Ps. asiatica* Spencer feeding on rice in Formosa, and *spicata* (Malloch) feeding on corn (maize) and sugar in Hawaii and Guam, and on corn and barley in Egypt and the Sudan. *Ps. spicata* has in addition been confirmed in the Pacific from Bonin Islands, Iwo Jima, Yap and the Marshall Islands, and in the Middle East from Iraq and Israel. Among specimens previously identified as *spicata*, it has now been found that there is a second, undescribed species of wide distribution which is described below as *spinosa* sp.n. This species has been confirmed on *Hordeum vulgare* L. in Egypt and on *Triticum aestivum* L. in Nigeria. It is also known from Samoa, Gilbert Islands, Australia, India and South Africa.

ALY (1971) has made a detailed study of *Ps. spicata* in Egypt. However, it seems very probable that some or even much of the material he studied, particularly from *Hordeum*, will have referred to *Ps. spinosa*. It is not even certain that *Ps. spinosa* may not also feed on maize, the host on which most of his observations are based. Nevertheless, although his information cannot be ascribed with certainty to either of these two species, it is of considerable interest. His '*spicata*' was found to attack not only maize and barley but also four wild grasses – *Cynodon dactylon* Pers., *Echinochloa colonum* Link., *E. crus-galli* var. *longisetum* Tackh. and *Panicum repens* L.

ALY noted a close correlation between the age of the plant and the degree of infestation. The progress in the age of the plant from 15 to 45 days is accompanied by an equivalent rise in the infestation rate. In 1968 the mean counts of mines per infested plant were 4–8.2, 20.4–41.8 and 27.6–79.8 as the plants were 15, 30 and 45 days old respectively. The density of infestation on successive maize plant leaves is also related to the age of the plant. In seedlings less than 15 days old 1.2–2.8 mines per leaf were noted on the three lower leaves. As the plants mature 20.7 and 24.6 mines per leaf were found on the fifth and sixth leaves, when the plants are between 30 and 45 days old.

A typical leaf-mine of an unidentified species in this group on corn in the Philippines is shown in Fig. 391.

Examination of all available material has now shown that in all no less than 21 species can be differentiated in this group, of which 10 new species are described below. Only a single species, *atra* (Meigen), has hitherto been known in Europe. It has now been found that this is the sole species present in northern and western Europe but in addition seven further species have been confirmed in Europe. Six of these, from southern and eastern Europe, are described as new and one, *lacteipennis* (Malloch), has been identified in Britain, representing the first palaearctic record of this well-known North American species. A key to the eight European species is given below.

A species from Mongolia previously provisionally identified as *atra* has now been found to be distinct. A new species feeding on *Zea mays* is described from Ghana and also a new species from Malaya. The species recorded by BEZZI (1928) from Fiji as *atra* is accepted as *dilatata* Sasakawa, 1963b, described from Samoa. Other species known are *philippinensis* Spencer, 1961b and *trilobata* Sasakawa, 1963a from the Philippines, *cingulata* Sasakawa, 1963c from New Guinea, *quatei* Sasakawa, 1963a from Vietnam; *nigralis* Spencer, 1961c from Ethiopia, *hohmanni* Spencer, 1965b from South and S.W. Africa; and finally *nitidula* Malloch, 1913b from the United States, which should be treated as distinct from the European *atra*.

It is felt desirable to record here known species in this group and also briefly to describe the new species which have been discovered, as apart from wild grasses, many or most of them may also feed on cereal or fodder crops. All are almost entirely black, with characteristic wing venation (Figs. 398, 416), slightly variably angulate third antennal segment (Figs. 392, 397) and wing length from 1.2–1.75 mm. Only differentiating characters are therefore mentioned for most of the species discussed below. The following key will permit identification of the eight species known in Europe:

Key to European *Pseudonapomyza* species

1	Squamal fringe black	*balkanensis* sp.n.
—	Squamal fringe silvery-white	2
2	Wings conspicuously white; veins R_{2+3} and R_{4+5} pale, yellowish-brown	3
—	Wings hyaline, normal; veins R_{2+3} and R_{4+5} darker brown	4
3	Mesonotum mat grey, tibiae and tarsi yellowish	*lacteipennis* (Malloch)
—	Mesonotum black, legs entirely black	*hungarica* sp.n.
4	Mesonotum mat, greyish-black	*hispanica* sp.n.
—	Mesonotum shining black	5
5	Veins pale brown	*strobliana* sp.n.
—	Veins dark brown to black	6
6	Second costal section distinctly longer than fourth	*atra* (Meigen)
—	Second costal equal to or only marginally longer than fourth	7
7	Aedeagus as in Figs. 408, 409	*europaea* sp.n.
—	Aedeagus as in Figs. 433, 434	*vota* sp.n.

Pseudonapomyza asiatica Spencer, 1961b – Asia, Africa

ADULT. Very small species, with bluntly angulate third antennal segment and mat mesonotum.

HEAD. Frons not projecting above eye, orbits with 1 reclinate ors and 3 incurved ori (or 2 + 2); orbital setulae minute, upright or reclinate; jowls extended at rear, $\frac{1}{5}$ height of eye; third antennal segment (Fig. 392) with blunt angle at upper corner (never finely pointed).

MESONOTUM. 3 dc, third at most half length of second, acrostichals irregularly in 4 rows, extending only to level of second dc.

WING. Length from 1.6 mm in male to 1.75 mm in female; costa ending at vein R_{4+5}, second costal section just more than $1\frac{1}{2}$ times length of fourth.

COLOUR. Frons mat black, orbits slightly paler, more greyish; mesonotum distinctly mat, greyish-black, but with some subshine; sides of thorax black; wing base pale, whitish, veins R_{2+3} and R_{4+5} brownish, M_{1+2} and M_{3+4} largely colourless; squamae and fringe conspicuously silvery-white, halteres white.

MALE GENITALIA. Aedeagus as in Figs. 393, 394.

EARLY STAGES. Puparium 1.6 × 0.75 mm, mat, dark-brown, each segment boundary marked by a band of 4 or 5 rows of irregularly placed black tubercles, surface otherwise smooth, without papilli (Fig. 395); posterior spiracles each on a short conical projection, with 4–6 pores.

HOST-PLANTS. *Zea mays* L. (maize, corn); also *Cynodon dactylon* (L.) Pers., *Eragrostis* sp. and probably other grasses.

BIOLOGY. The larva forms a conspicuous whitish leaf-mine at the apex of young leaves (Fig. 396). Each mine contains only a single larva. Pupation takes place

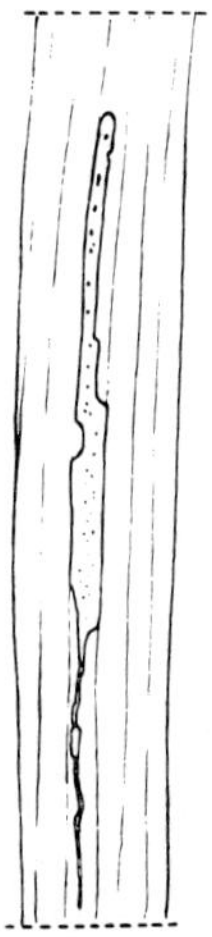

391

392

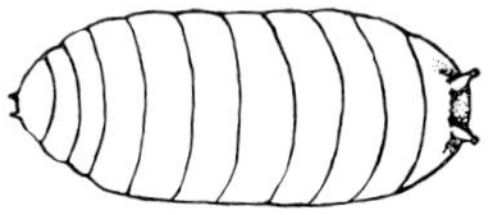

395

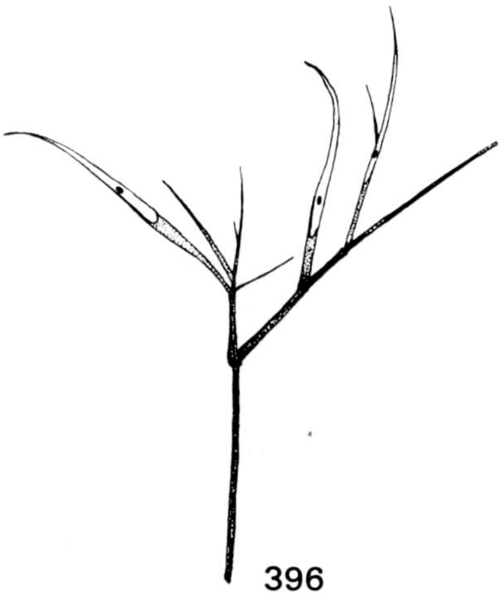

396

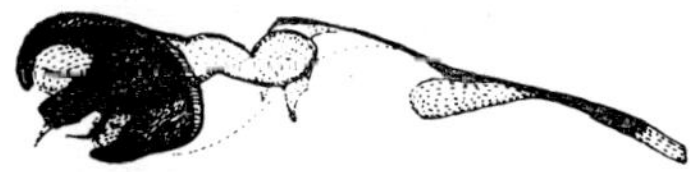

393

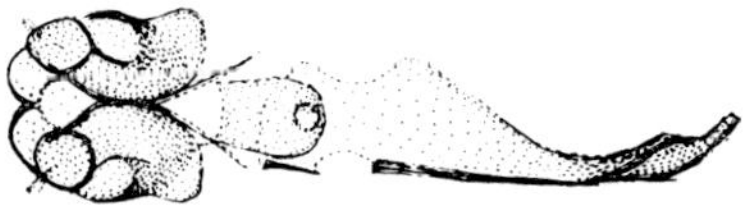

394

Fig. 391. *Pseudonapomyza* sp.: leaf-mine on *Zea mays* (Philippines).

Figs. 392–396. *Pseudonapomyza asiatica* Sp.: 392, third antennal segment. 393, aedeagus, side view; 394, same, ventral view; 395, puparium; 396, leaf-mine on *Cynodon dactylon*.

within the mine and the puparium is firmly glued to the lower leaf surface by a mass of black frass.

DAMAGE CAUSED. Young growing shoots are largely eaten out but the damage is normally not sufficient to impair subsequent growth of the plants.

DISTRIBUTION. Singapore, Formosa, Philippine Is.: Mindanao; India: Chandigarh and Hyderabad; Ethiopia: Addis Ababa; Zululand; Cape Verde Is.

PARASITES. None recorded.

Discussion

TAXONOMY. *Ps. asiatica* partially overlaps in distribution with the better known species, *Ps. spicata* (Malloch) and also with *Ps. spinosa* sp.n. but is readily distinguishable by the more obtuse angle of the third antennal segment and the more mat mesonotum. In the Philippines, *Ps. philippinensis* Spencer closely resembles *asiatica* but the biology is different, with many larvae feeding together in a large mine on *Rottboelia exaltata* L.f., there are distinct papilli on the larva and puparium (SPENCER, 1961b: Fig. 33) and the male genitalia are entirely distinct (SPENCER, 1961b: Fig. 47). In Africa, *Ps. asiatica* might be confused with *Ps. zeae* (see p. 278 below) but this new species has a more distinct projection on the third antennal segment rather than a mere angle and there are also papilli on larva and puparium.
ECONOMIC IMPORTANCE. On Formosa *Ps. asiatica* occurs very commonly on young rice plants but according to Dr. SUNG-YANG LEE (personal communication) has not yet caused any significant damage. Any substantial increase in population levels might nevertheless threaten young plants and this species deserves to be noted as a potential pest of rice.

Pseudonapomyza atra (Meigen, 1830) – Europe

ADULT. Frons narrow, little wider than eye width; third antennal segment slightly concave (Fig. 397); distance between first and second dc only slightly more than between second and third, third only slightly shorter than second; wing as in Fig. 398, length from 1.3 mm in male to 1.7 mm in female; frons mat black, orbits only weakly shining; mesonotum deep black, brilliantly shining viewed from rear, more mat viewed from front; squamae and fringe silvery-white.
MALE GENITALIA. Aedeagus as in Figs. 399, 400.
EARLY STAGES. Posterior spiracles of larva with 7 pores (Fig. 401), puparium shining reddish-brown, segments bearing characteristic papilli (Fig. 402).
HOST-PLANTS. *Avena sativa* L., *Hordeum vulgare* L., *Triticum aestivum* L.; also wild grasses.
BIOLOGY. Leaf-mine normally at apex of leaf, short, narrow but deep.
DISTRIBUTION. Europe, confirmed in England, Germany, Hungary, Sweden, Yugoslavia. Apparently uncommon in Mediterranean area and eastern limit of range not known.

Discussion

TAXONOMY. This species agrees closely with *Ps. europaea* sp.n. but the frons is marginally wider. With available records there is only slight overlapping in distribution between the two species but in doubtful cases a positive identification will only be possible from examination of the male genitalia.

Economic importance. *Ps. atra* is known as a leaf-miner on barley in England (Goodliffe, 1941) and Germany, and is also recorded on oats and wheat

397

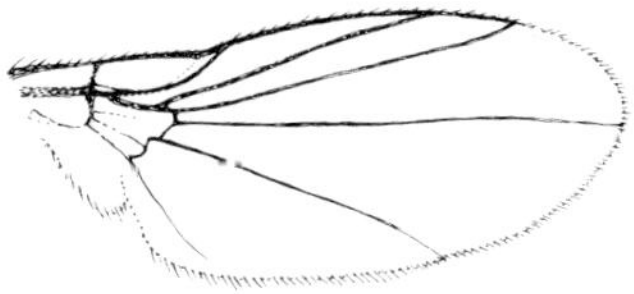
398

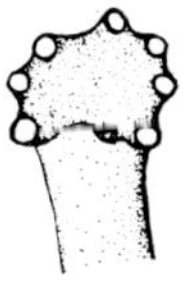
401

399

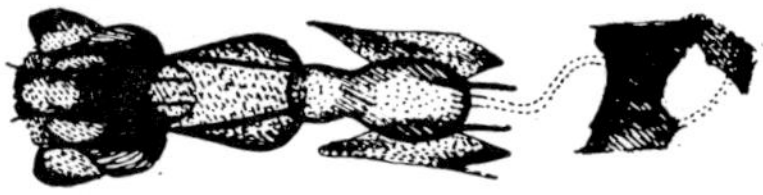
400

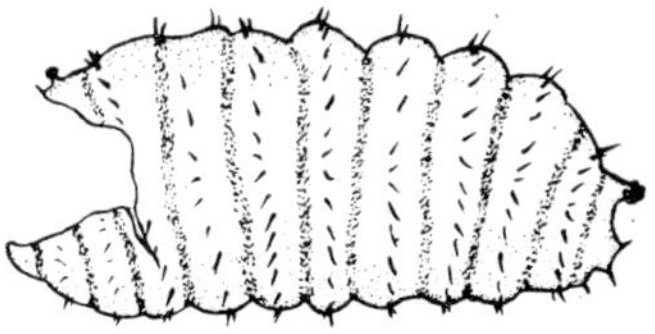
402

Figs. 397–402. *Pseudonapomyza atra* (Mg.): 397, third antennal segment; 398, wing; 399, aedeagus, side view; 400, same, ventral view; 401, anterior spiracular process of larva; 402, puparium (Figs. 401, 402, Hering).

(Hendel, 1931–1936: 304). It has never been reported in large numbers and the damage caused by isolated specimens is negligible.

Pseudonapomyza balkanensis sp.n. – Yugoslavia

Adult. Third antennal segment conspicuously angulate; 1 strong ors, 3 or 4 weaker incurved ori; third dc strong, $\frac{2}{3}$ length of second, acr irregularly in 5 or 6 rows; wing length from 1.5–1.7 mm, second costal section slightly longer than fourth.
Colour. Frons sooty black or somewhat more brownish; ocellar triangle brilliantly, orbits weakly shining; mesonotum shining black; wing generally dark, veins R_{2+3} and R_{4+5} black, M_{1+2} and M_{3+4} fully developed; squamae grey, margin and fringe black.
Male genitalia. Aedeagus distinctive, as in Figs. 403, 404.

HOLOTYPE ♂, Yugoslavia, Dalmatia: Novi Grad, 'polje in valley', 27–31.v.1958; paratypes: 1♂, 1♀, Macedonia, Prespa Geul, Otesevo, swept from vegetation by lake, 9–21.vi.1958 (all R. L. COE). Holotype and female paratype in British Museum (Nat. Hist.), male paratype in author's collection.

Discussion

This species is immediately distinguishable from the others known in Europe by the dark squamal fringe, the darker wings and slightly larger size. The male genitalia are highly distinctive.

Pseudonapomyza dilatata Sasakawa, 1963b – Fiji, Samoa

ADULT. Third antennal segment bluntly angulate (Fig. 405); third dc small, $\frac{1}{2}$ length of second; wing length 1.3–1.6 mm, second costal section $1\frac{1}{2}$ times length of fourth or slightly longer.
COLOUR. Frons blackish-brown, mesonotum greyish-black, weakly shining viewed from behind; squamae and fringe silvery-white.
MALE GENITALIA. Aedeagus as in Figs. 406, 407.
EARLY STAGES. Puparium reddish-brown, segments with indications of but not fully developed papilli (in only example seen).
HOST-PLANTS. *Zea mays* L., *Saccharum officinarum* L., *Coix Lachryma-Jobi* L. (Job's tears); also *Cynodon dactylon* Pers. (Bermuda grass), *Eleusine indica* L., *Miscanthus* sp. and probably other wild grasses (SWEZEY, 1941).
DISTRIBUTION. Fiji: Lautoka (sugar-cane), Labasa (corn); Samoa.
PARASITES. Chalcids have been reared both on Fiji and Samoa. LEVER (1944) identified the species on Fiji as *Hemiptarsenus semialbiclavus* (Gir.) but SWEZEY (1941) merely writes: 'Chalcid parasites were reared on several occasions'.

Discussion

TAXONOMY. This species was misidentified on Fiji as *atra* Mg. by BEZZI (1928) and as *spicata* (Malloch) by LEVER (1942, 1944). It was also identified as *spicata* on Samoa by SWEZEY (1941).

SASAKAWA (1963b) described *Ps. dilatata* from a single female reared from *Coix Lachryma-Jobi* on Samoa. I have seen a series of specimens reared from the same host and also others reared from maize on Fiji which clearly represent the same species. The blunt angle of the third antennal segment and the mat greyish mesonotum immediately distinguish this species from *spicata* and *spinosa*. In both these characters it closely resembles *Ps. asiatica* Spencer and the two species can only be reliably separated by their distinctive male genitalia (cf. Figs. 393, 394 and 406, 407).

DISTRIBUTION. Although the only known records are from Fiji and Samoa, the species may be present on other islands in the Pacific but additional collecting will be necessary before this can be established.

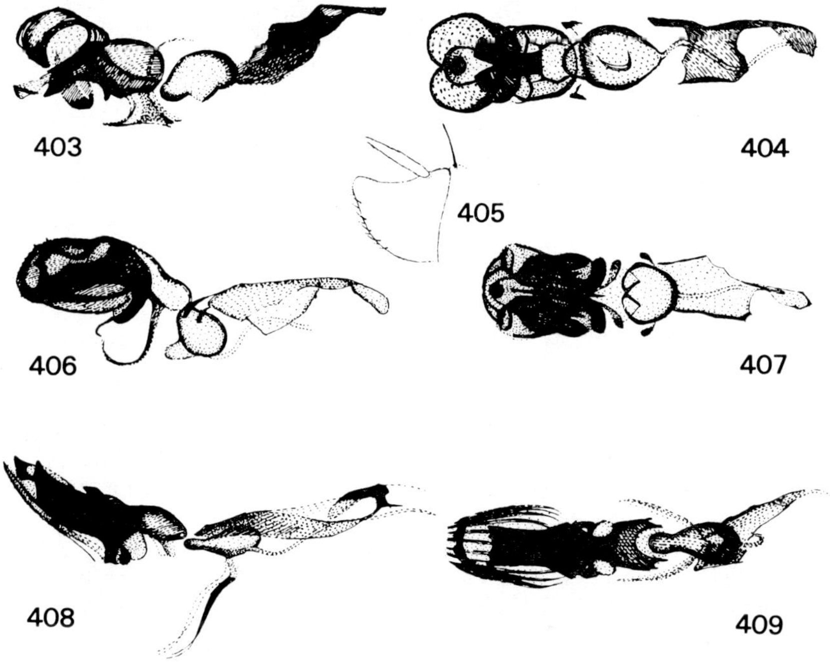

Figs. 403–404. *Pseudonapomyza balkanensis* sp.n.: 403, aedeagus, side view; 404, same, ventral view.

Figs. 405–407. *Pseudonapomyza dilatata* Sas.: 405, third antennal segment; 406, aedeagus, side view; 407, same, ventral view.

Figs. 408–409. *Pseudonapomyza europaea* sp.n.: 408, aedeagus, side view; 409, same ventral view.

ECONOMIC IMPORTANCE. LEVER (1942) reports that on Fiji leaf-mining damage on maize can be severe enough to require spraying.

Pseudonapomyza europaea sp.n. – Central Europe

ADULT. Frons narrow, equal to width of eye, orbits pronounced, wing length from 1.3 mm in male to 1.7 mm in female.
COLOUR. Frons brownish, orbits shining black, particularly above and extending to hind-margin of eye; mesonotum shining black; squamae and fringe white.
MALE GENITALIA. Aedeagus as in Figs. 408, 409.

HOLOTYPE ♂, Italy: Sestriere, 1,800 m, 6.vii.1960 (K.A.S.); paratypes: Switzerland: Stalden, Valais, 1♂, 3.viii.1956 (K.A.S.); Austria: Steiermark, Admont area, 1♂, 1♀, no date (G. STROBL, 1910); Yugoslavia, Slovenia: Postojna, highland meadow, 1♂, 3♀, 13.vii–1.viii.1958; Macedonia: Lake Ochrid, Hill to N. of town, 2♀, 14–19.vi.1958; Prespa Geul, Otesevo, edge of oak wood above lake, 1♂, 20–27.vi.1958; Serbia, Morovic, 1♀, 3–7.vii.1958 (all R. L. COE). Holotype in author's collection, paratypes in British Museum (Natural History) and coll. STROBL, Admont, Austria.
REMARKS. This species closely resembles *Ps. atra* which is widespread in Europe north of the Alps but has also been recorded from Hungary and Macedonia; the frons is slightly narrower but at adjacent limits of distribution of the two species it will be necessary to examine male genitalia to make a reliable identification.

Pseudonapomyza hispanica sp.n. – Spain, Israel

ADULT. Frons broad, 1½ times width of eye; wing length 1.6–1.7 mm, second costal section only slightly longer than fourth; third dc somewhat more than half length of second, acr regularly in 4 rows.
COLOUR. Frons black, mesonotum entirely mat black, appearing more greyish viewed from front; squamae and fringe white.
MALE GENITALIA. Aedeagus as in Figs. 410, 411.
HOLOTYPE ♂, Spain, Prat de Llobregat, nr. Barcelona, 4.xi.1963 (K.A.S.); paratypes: Albarracin, 1♀, 2.iii.1933 (E. M. HERING); Sierra Nevada, Highway, 1200 m, 27.vii.1960, 1♂ (J. R. VOCKEROTH); Israel: Givat-Brenner, 1♂, 14.vi.1971, on *Sorghum halepense* (L.) Pers. (J. KUGLER). Holotype and female paratype in author's collection, male paratypes in Canadian National Collection, Ottawa and Dept. of Zoology, Tel Aviv University.
REMARKS. This species is immediately distinguishable from the three others known in Europe by the broader frons and mat mesonotum. The genitalia indicate a close relationship with *Ps. europaea*. The specimen from the Sierra Nevada was originally identified as *Ps. atra* (Mg.) by SPENCER (1972a).

It is reasonably certain that *Sorghum halepense*, an important fodder crop, is the host of the specimen from Israel. A female from Tel Aviv, 7.i.1970 (J. KUGLER) possibly represents this species but the mesonotum appears to be somewhat less mat and less greyish.

Pseudonapomyza hungarica sp.n. – Hungary

ADULT. Agreeing in external characters exactly with *Ps. strobliana* (p. 277), apart from the unusually pale, milky wings and yellowish-brown veins and wing base.
MALE GENITALIA. Aedeagus as in Fig. 412, 413.

HOLOTYPE ♂, Hungary: Gyón, 15.vii.1905 (KERTÉSZ); paratypes: Gyón, 1♂, 18.vii.1903 and 1♀, 30.vi.1917 (KERTÉSZ); 1♀, 11.vi.1921 (UJHELYI). Holotype and paratypes in Naturhistorisches Museum, Vienna.
REMARKS. The distinctly pale wings are the characteristic feature of this species. In this character it agrees with *P. lacteipennis* (Malloch) but the two species cannot be confused, as the colour of the mesonotum is entirely different. The aedeagus of

410 411

412 413

Figs. 410–411. *Pseudonapomyza hispanica* sp.n.: 410, aedeagus, side view; 411, same, ventral view.

Figs. 412–413. *Pseudonapomyza hungarica* sp.n.: 412, aedeagus, side view; 413, same, ventral view.

Ps. hungarica closely resembles that of *Ps. strobliana* (cf. Figs. 430, 431) but the detail is significantly different in the two species.

Pseudonapomyza lacteipennis (Malloch, 1913b) Britain, N. America

ADULT. A distinctive species with shining white wings, pale veins, mat grey mesonotum and slightly yellowish tarsi. The wing length is 1.6–1.7 mm.
MALE GENITALIA. Aedeagus as in Figs. 414, 415.
HOST-PLANT AND EARLY STAGES. Unknown.

Discussion

P. lacteipennis has similarly pale wings to *P. hungarica* but is readily distinguishable by the mat grey mesonotum, paler tarsi and larger size.

It has hitherto only been known from the United States and Canada but recently I have seen a male and female, which can only be identified as this species, from England: S. Devon, Dawlish, 25.viii.1960 (J. R. VOCKEROTH). This is the first record for the Palaearctic Region and the male is the only one known.

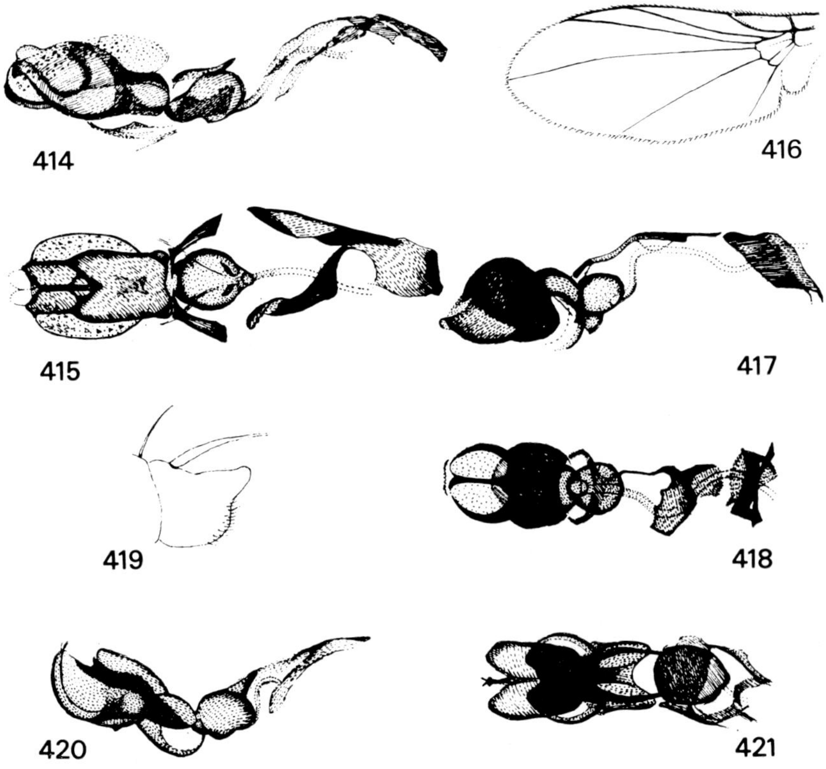

Figs. 414–415. *Pseudonapomyza lacteipennis* (Mall.): 414, aedeagus, side view; 415, same, ventral view.

Figs. 416–418. *Pseudonapomyza malayensis* sp.n.: 416, wing; 417, aedeagus, side view; 418, same, ventral view.

Figs. 419–421. *Pseudonapomyza mongoliensis* sp.n.: 419, third antennal segment; 420 aedeagus, side view; 421, same, ventral view.

Pseudonapomyza malayensis sp.n. – Malaya

ADULT. Third antennal segment with fine point at upper corner; third dc small; wing length in male 1.2 mm, second costal section unusually short, equal to fourth (Fig. 416).
COLOUR. Frons brownish-black, mesonotum black but only moderately shining; squamae and fringe white.
MALE GENITALIA. Aedeagus as in Figs. 417, 418.

HOST. *Cynodon dactylon* Pers. (Bermuda grass).
HOLOTYPE ♂, Malaya: Kuala Lumpur, 30.x.1936 ex Bermuda grass (Entom. Div., Agric. Dept.), in British Museum (Nat. Hist.).

Discussion

This is the smallest species known in this group. It has the finely pointed third antennal segment of *Ps. spicata* and *Ps. spinosa* but the mesonotum, although black (not grey as in *asiatica* and *dilatata*), is only moderately shining. It is also distinguished by the very short second costal section.

Although the only recorded host is *Cynodon dactylon*, this species may well also feed on cultivated cereals.

Pseudonapomyza mongoliensis sp.n. – Mongolia

ADULT. Frons 1½ times width of eye; third antennal segment (Fig. 419) conspicuously angulate; wing length 1.3–1.5 mm.
COLOUR. Frons brownish-black, orbits and cheeks distinctly shining; mesonotum mat black; wing with veins pale and base yellowish, squamae and fringe white.
MALE GENITALIA. Aedeagus as in Figs. 420, 421.
HOLOTYPE ♂, Mongolia: Chentej aimek, Candagan tal, 40 km E. of Somon Zargaltchean, 1300 m, 28.vii.1965 (No. 316); paratype ♀, Suchebaator aimek, 45 km N. of Somon Erdenezagan, 900 m, 9.viii.1965 (No. 381), both leg. Z. Kaszab. Holotype and paratype in Hungarian National Museum, Budapest.
REMARKS. This species is immediately distinguishable from the western palaearctic species, *atra* (Mg.), by the mat mesonotum. In this character it resembles *Ps. hispanica* but the male genitalia of the two species are entirely distinct. The aedeagus of *mongoliensis* somewhat resembles that of *Ps. vota* sp.n. from Spain but the shape of the third antennal segment (Figs. 419, 432) is very different in the two species.

A further female from Mongolia: Central aimek, Zuunchara, 850 m, 8.vii.1964 (No. 281) has a more shining mesonotum and appears to represent a different species.

Pseudonapomyza spicata (Malloch, 1914) – Formosa, Pacific, Middle East

ADULT. Third antennal segment with fine point at upper corner (Fig. 422); 3 dc, distance between first and second twice that between second and third, third small, half length of second; acrostichals in 4 rows; wing (Fig. 423) with second costal section from $1\frac{1}{3}$–$1\frac{1}{2}$ times length of fourth, length from 1.3 mm in male to 1.6 mm in female.

COLOUR. Frons brownish-black, ocellar triangle conspicuously shining; mesonotum black, shining viewed from rear, somewhat more mat viewed from front; squamae and fringe silvery-white.
MALE GENITALIA. Aedeagus as in Figs. 424, 425, distiphallus small, pale, clearly divided into separate sections.
EARLY STAGES. Puparium (Fig. 426), pale brownish-orange, segments with well-defined bands of papilli; posterior spiracles each with 3–5 pores.
HOST-PLANTS. *Zea mays* L. (Guam, Hawaii, Egypt), *Triticum aestivum* L. (Sudan), *Panicum miliaceum* L. (Iraq), *Saccharum officinarum* L. (Hawaii); also 14 wild grasses in Hawaii.
BIOLOGY. The larva forms narrow leaf-mines, particularly at the tip of the leaves in young plants. SWEZEY (1946) reports as many as 100 larvae per leaf on young corn (*Zea*) plants in Guam.
DISTRIBUTION. Pacific: Guam, Hawaii (Oahu), Bonin, Palau, Iwo Jima, Yap, Marshall Is.; Iraq: Baghdad; Israel: Jerusalem; Egypt: Assuit Distr., Giza; Sudan: Wad en Naim.
PARASITES. Guam, Hawaii: *Hemiptarsenus semialbiclavus* (Girault), Eulophidae, also 2 further chalcids (SWEZEY, 1946, 1948).

Discussion

TAXONOMY. The only apparent external difference between this species and *Ps. spinosa* with which it has hitherto been confused is the distinctly longer costal section. Measurements in six specimens are as follows:

Guam, Hawaii	: 1.6:1.1 (ratio of second costal section to fourth)
Bonin Is.	: 1.7:1.3
Baghdad	: 1.5:1.1
Egypt	: 1.8:1.4
Sudan	: 1.7:1.3

There appears to be no overlapping in this character with *Ps. spinosa*, in which the second costal section is little longer than the fourth. The male genitalia of the two species are entirely distinct.
DISTRIBUTION. The apparently disjunct areas of distribution of *Ps. spicata* in the Pacific on the one hand and the Middle East on the other are probably of no significance and merely represent inadequate collecting in the intervening areas.
ECONOMIC IMPORTANCE. This species is considered a serious pest on *Zea mays* in Guam and Hawaii. SWEZEY (1946) states that the larvae are very injurious to the growth of young corn plants. In Hawaii it is particularly widespread on Oahu and present on all the islands. It has been recorded mining young sugar-cane seedlings (SWEZEY, 1948) but it is not known whether this host is as susceptible as corn.

In Egypt considerable damage is caused to young plants of maize and other cereals (HAMMAD, personal communication) but it remains to be clarified whether the species mainly responsible is *spinosa* or *spicata*, which overlap in distribution in the Nile Valley (see also p. 276). It has been recorded in Egypt, as *atra* Meigen, by HAMMAD (1956).

It seems probable that comparable damage is caused in Iraq, where the only known host is *Panicum miliaceum* (millet).

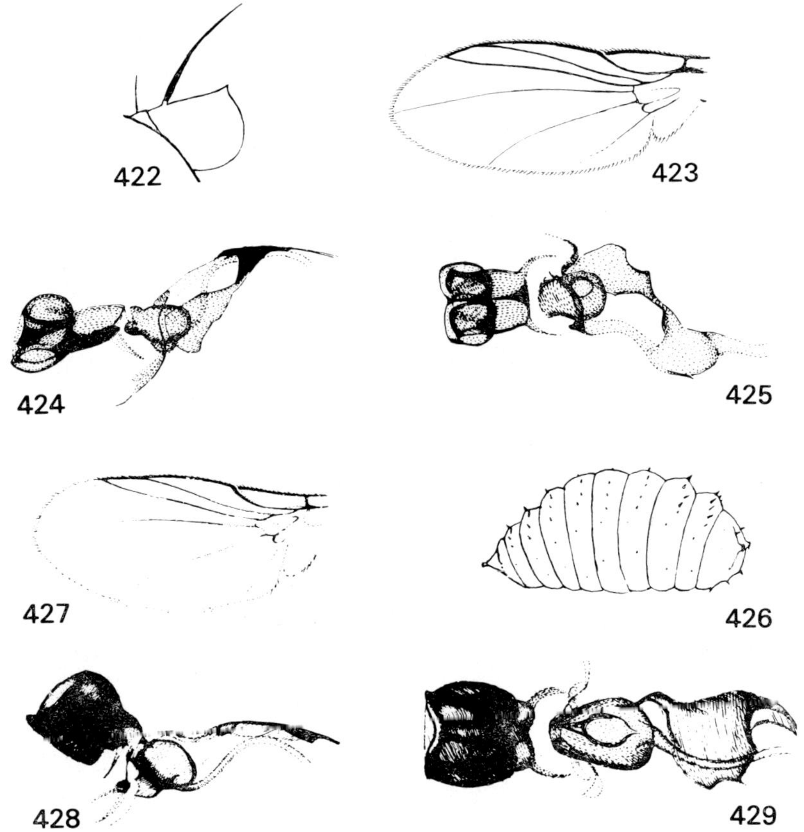

Figs. 422–426. *Pseudonapomyza spicata* (Mall.): 422, third antennal segment; 423, wing; 424, aedeagus, side view; 425, same, ventral view; 426, puparium.

Figs. 427–429. *Pseudonapomyza spinosa* sp.n.: 427, wing; 428, aedeagus, side view; 429, same, ventral view.

Pseudonapomyza spinosa sp.n. – Africa, India, Australia, Pacific

ADULT. Third antennal segment with fine point at upper corner; distance between first and second dc substantially greater than between second and third, third small, less than half length of second.

WING (Fig. 427). Length from 1.3–1.6 mm, second costal section little longer than fourth.
COLOUR. Frons distinctly brown, ocellar triangle brilliantly shining black, orbits only weakly shining; mesonotum shining black but appearing slightly grey-dusted viewed from front; squamae and fringe silvery-white.
MALE GENITALIA. Aedeagus as in Figs. 428, 429, distiphallus large, conspicuously black, appearing largely entire.
EARLY STAGES. Larval mouth-hooks each with 2 teeth, regularly alternating; anterior spiracles with 5 pores, posterior spiracles each with 3–5, on short projections. The brownish puparium is covered with conspicuous papilli (cf. Fig. 426).
HOST-PLANTS. *Hordeum vulgare* L. (Egypt), *Triticum aestivum* L. (Nigeria); also *Eleusine indica* Gaertn. (Australia).
BIOLOGY. Leaf-mines are usually at the tips of leaves and several larvae can feed together, thus forming a substantial secondary blotch. Pupation takes place externally. KLEINSCHMIDT (1970) gives two photographs of mines on *Eleusine indica* in Australia (as *spicata* Mall.).
PARASITES. None recorded.
HOLOTYPE ♂, Egypt: Alexandria, 21.xi.54, ex *Hordeum vulgare* L. (S. M. HAMMAD); paratypes: 1♂, same data; N. Nigeria: Samaru, Zaria, 2♂, 2♀, 4.iv.67, ex leaf-mines on Wheat found 27.iii.67 (J. C. DEEMING); S. Africa: Pretoria, 1♂, 1♀, Jan. 1967, ex suction trap (Parasite Lab.); Basutoland, Maseru Dist., Maloti Mtns., Bushman's Park, 2125–2250 m, 1♂, 8–14.i.1963 (B. and P. STUCKENBERG); India: Namkum, 1♂, 7.vi.1963, ex unidentified grass (V. K. SEHGAL); Australia: Sydney, 2♂, 4♀, 4.ii.61, ex *Eleusine indica* (L.) Gaertn. (K.A.S.); Micronesia: Gilbert Is., Tarawa Atoll, 1♂, 1.xi.1957 (N. L. H. KRAUSS); Samoa: Swain's Is., 1♂, 6.iv.1933 (E. H. BRYAN, JR.). Holotype and paratypes in author's collection, further paratypes in British Museum (Natural History) and Bishop Museum, Honolulu.

Discussion

TAXONOMY. This species has the finely pointed third antennal segment and the shining black mesonotum in common with *spicata* (Mall.) but the costal ratio is consistently less in *spinosa*. Measurements of five specimens are as follows:

Australia – 1.4:1.3 (ratio of second to fourth costal segments)
India – 1.5:1.3
Egypt – 1.3:1.3
S. Africa – 1.4:1.5
Nigeria – 1.5:1.3

DISTRIBUTION. *P. spinosa* is less widely distributed in the Pacific than *P. spicata* but its complete range is certainly not yet known. It appears to be absent from Hawaii. It is less common in the Nile Valley than *P. spicata* but is represented elsewhere in Africa where *spicata* is lacking.

Its present known range probably largely reflects natural distribution but recent introduction by man may be responsible for its presence on some Pacific islands.

HOST RANGE. Both *Ps. spinosa* and *spicata* feed on *Triticum* and there are probably other hosts in common but the exact host range of these two species remains to be established.

ECONOMIC IMPORTANCE. In Egypt considerable damage is caused to young cereal crops but since it is now established that both *spicata* and *spinosa* occur there with an overlapping distribution, further breeding will be necessary to clarify which of the two species causes the greater damage (see p. 274). In Nigeria young wheat crops are regularly heavily infested by *spinosa* (J. C. DEEMING, personal communication).

Pseudonapomyza strobliana sp.n. – Yugoslavia

ADULT. Very small species, wing length in male 1.5 mm, in female 1.6 mm; second costal section short, only marginally longer than fourth; third antennal segment conspicuously angulate.

COLOUR. Frons black, mesonotum shining black; wings hyaline, normal, veins brown, squamae and fringe silvery-white.

MALE GENITALIA. Adeagus as in Figs. 430, 431.

HOLOTYPE ♂, Yugoslavia: Dalmatia, Split, no date; paratypes: 1♂, same data; 2♀♀, 'Öster. Litorale' (all G. STROBL), in collection Strobl, Admont, Austria.

REMARKS. This species is not distinguishable from *Ps. europaea* on external characters but the male genitalia are entirely distinct.

The aedeagus closely resembles that of *Ps. hungarica* (cf. Figs. 412, 413) but the darker wings in *strobliana* and the differences in detail in the genitalia of the two species confirm that they are distinct.

Pseudonapomyza vota sp.n. – Spain

ADULT. Small species, wing length in male 1.4 mm, second costal section short, equal to length of fourth; frons narrow, equal to width of eye; third antennal segment conspicuously angulate (Fig. 432).

COLOUR. Frons brownish-black, orbits distinctly shining; mesonotum shining black; wings hyaline, veins dark, almost black, squamae and fringe silvery-white.

MALE GENITALIA. Aedeagus as in Figs. 433, 434.

HOLOTYPE ♂, Spain: Granada, 700 m, 19.vii.1960 (J. R. VOCKEROTH), in Canadian National Collection, Ottawa.

REMARKS. The aedeagus of this species most closely resembles that of *Ps. mongoliensis* but is distinctively different in detail. *Ps. vota* is not distinguishable from *Ps. europaea* on external characters but can immediately be separated by differences in the male genitalia.

Pseudonapomyza zeae sp.n. – Ghana

ADULT. Minute black species, with blunt projection at upper corner of third antennal segment.

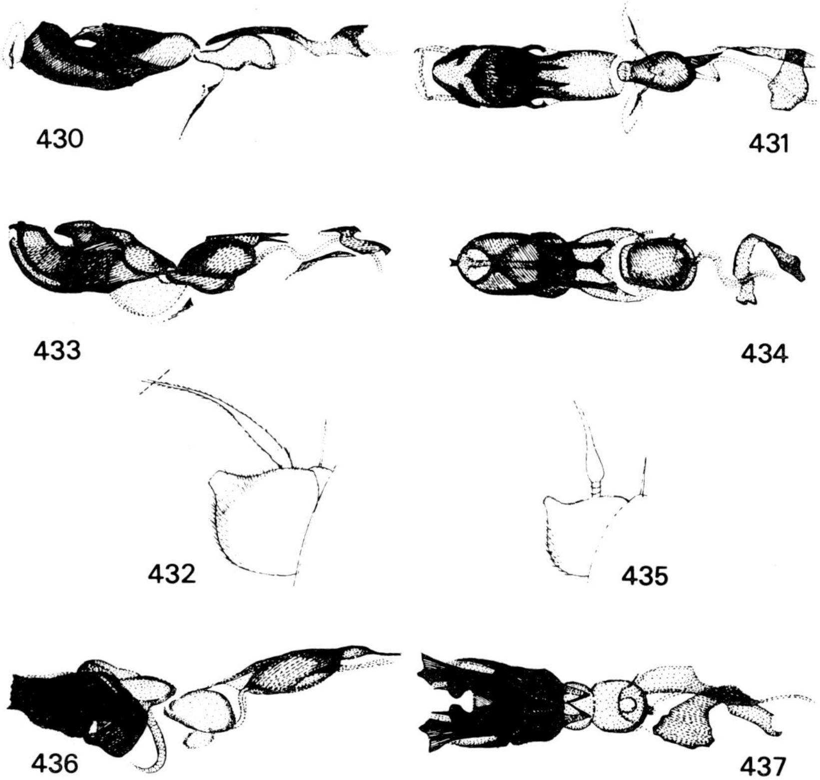

Figs. 430–431. *Pseudonapomyza strobliana* sp.n.: 430, aedeagus, side view; 431, same, ventral view.

Figs. 432–434. *Pseudonapomyza vota* sp.n.: 432, third antennal segment; 433, aedeagus, side view; 434, same, ventral view.

Figs. 435–437. *Pseudonapomyza zeae* sp.n.: 435, third antennal segment; 436, aedeagus, side view; 437, same, ventral view.

HEAD. Frons not projecting above eye; 4 orbital bristles, only the upper reclinate, the lower 3 incurved; orbital setulae sparse, basically reclinate but those below more upright; eye large, oval, jowls extended at rear, at deepest point $\frac{1}{5}$ height of eye; third antennal segment conspicuously angulate, with blunt projection at upper corner (Fig. 435).

MESONOTUM. 2 strong dorso-centrals, with third weak, less than ½ length of second, distance between first and second twice that between second and third; acrostichals irregularly in 4 or 5 rows, only isolated hairs at level of first dc.
WING. Length in male 1.3 mm, second costal section 1½ times length of fourth.
COLOUR. Frons mat black, ocellar triangle and orbits scarcely shining; antennae entirely black; mesonotum moderately shining but distinctly grey-dusted; legs black; wing with veins R_{2+3} and R_{4+5} dark but M_{1+2} conspicuously pale; squamae and fringe silvery-white; halteres white.
MALE GENITALIA. Aedeagus as in Figs. 436, 437; ejaculatory apodeme large, with distinct stalk.
EARLY STAGES. Puparium reddish-brown, all segments with a row of conspicuous papilli.
HOST-PLANT/BIOLOGY. *Zea mays* L. (Maize, corn), larva feeding as leaf-miner.
HOLOTYPE ♂, Ghana, Kumasi, 3.v.71, ex leaf-mine on maize leg. 22.iv.71 (G. SCHEIBELREITER, C.I.B.C.); paratype: 1, sex indet. (abdomen missing), S.E. of Kumasi, 2.v.71, ex leaf-mine leg. 25.iv.71.Holotype and paratype presented by the Commonwealth Institute of Biological Control to the British Museum (Natural History).

Discussion

TAXONOMY. Both the distinctive form of the third antennal segment and also the male genitalia confirm that this species is distinct from the other three grass-mining species of *Pseudonapomyza* known in Africa. The author's (1959) key to Ethiopian species can be revised as follows to include this new species and also *spinosa* sp.n.:

1	Squamal fringe white	2
—	Squamal fringe black	3
2	Third antennal segment with fine point at upper corner, mesonotum shining black	2a
—	Third antennal segment with elongated angle or blunt projection at upper corner; mesonotum at least slightly mat	2b
2a	Second costal section little longer than fourth	*spinosa* sp.n.
—	Second costal section 1⅓–1½ times length of fourth	*spicata* (Mall.)
2b	Orbits and cheeks brilliantly shining black	*hohmanni* Spencer
—	Orbits and cheeks not conspicuously shining	2c
2c	Third antennal segment with obtuse angle at upper corner (Fig. 392)	*asiatica* Spencer
—	Third antennal segment with blunt projection at upper corner (Fig. 435)	*zeae* Spencer

Ethiopian grass-feeding species in this genus have been discussed by SPENCER (1961c; 1961d; 1965b).

ECONOMIC IMPORTANCE. No information is available on the significance of the leaf-mining damage caused by this species. However, it clearly deserves further study as a potential pest of maize.

GENUS PHYTOMYZA FALLÉN

Although this is the largest genus in the family, very few *Phytomyza* species have successfully colonized the Gramineae. Apart from the two species considered below, *Phytomyza milii* Kaltenbach occurs commonly on *Milium effusum* L. in Europe and is also known in North America but has never been recorded on cereals.

The genus *Phytomyza* is poorly represented in the tropics and the southern hemisphere and no species are known as grass-feeders outside Europe, temperate Asia, Japan and North America.

Phytomyza fuscula Zetterstedt – Europe, Canada

Phytomyza fuscula Zetterstedt, 1838: 794; 1848: 2831. Holotype ♀ in Zoological Museum, Lund.

Phytomyza avenae de Meijere, 1926: 251. Types in Zoological Museum, Amsterdam.

ADULT. Medium-sized species, with yellow frons, mat grey mesonotum and sparsely haired eyes.

HEAD. 2 equal ors, 1 incurved ori, with sometimes an additional minute hair beyond, orbital setulae long, proclinate; eye round, distinctly pilose in male, less so in female; jowls extended at rear, cheeks broad, $\frac{1}{2}$ depth of jowl; third antennal segment round, arista short.

MESONOTUM. 3 + 1 strong dc, acrostichals sparse, irregularly in 2 rows.

WING. Length from 2.3 mm in male to 2.6 mm in female, second costal section twice length of fourth.

COLOUR. Frons orange-yellow, orbits differentiated, either paler yellow or slightly greyish; face greyish-yellow, antennae black; mesonotum greyish-black with only weak subshine, pleura almost entirely black, mesopleura with only very narrow yellowish-orange upper margin; legs black but all knees distinctly yellow; squamae yellowish-grey, margin and fringe black.

MALE GENITALIA. Aedeagus as in Fig. 438; ejaculatory apodeme with asymmetrical blade (Fig. 439).

EARLY STAGES. Larval anterior spiracles each with 8 pores, on two short stalks, arranged in a V-shape (Fig. 440); posterior spiracles each with 4–6, rarely with 10 pores; puparium pale yellowish, with darker markings extending laterally along ventral surface of each segment.

Host-plants. *Avena sativa* L., *Hordeum vulgare* L., *Secale cereale* L.; also *Arrhenaterum elatius* (L.) J. and C. Presl, *Phleum pratense* L. (Timothy) and certainly other wild grasses.

Biology. The larva forms a whitish linear mine (Fig. 441), with frass in distinct black lumps; pupation takes place in the mine, with the anterior spiracles normally

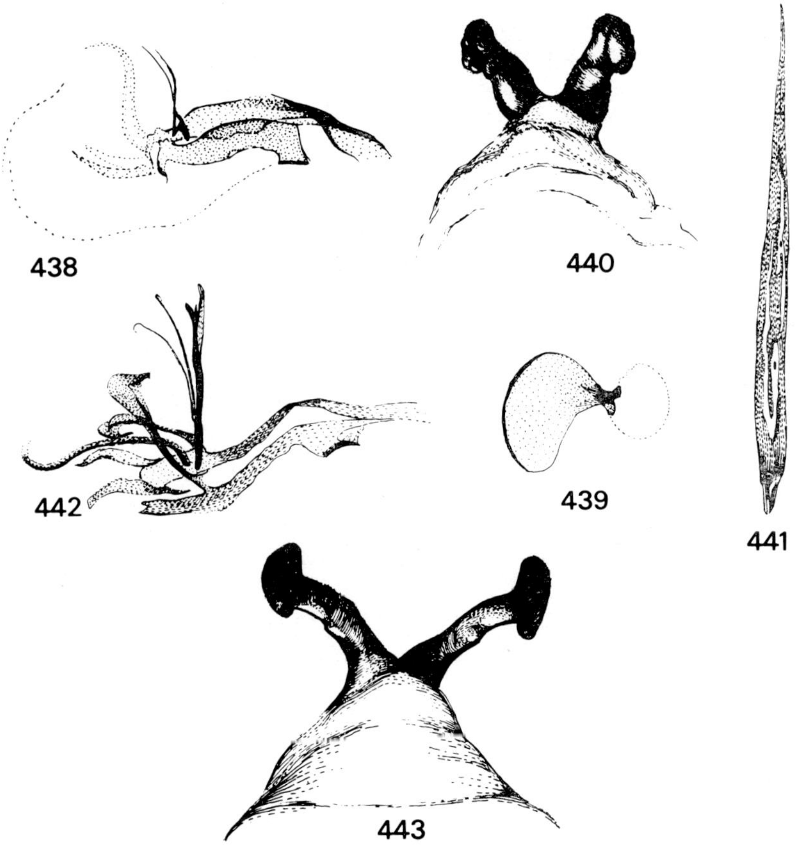

Figs. 438–441. *Phytomyza fuscula* Zett.: 438, aedeagus; 439, ejaculatory apodeme; 440, anterior spiracles of puparium; 441, leaf-mine on *Arrhenatarum* (after Mühle).

Figs. 442–443. *Phytomyza nigra* Mg.: 442, aedeagus; 443, anterior spiracles of puparium.

just penetrating through the epidermis. There are at least two generations in the year but the life cycle has not been studied in detail.

Distribution. Europe, more common in north; widespread in Canada, particularly common in Ottawa area; Greenland.

PARASITES. *Dacnusa maculipes* Thoms.; *Chrysocharis chilo* (Walk.), *C. pubicornis* (Zett.), *C. polyzo* (Walk.), *Cirrospilus vittatus* Walk., *Pediobius acantha* (Walk.), *Hemiptarsenus dropion* (Walk.).

Discussion

TAXONOMY. This species was described in 1838 from a single female from the area of Tornea in Swedish Lappland at the head of the Gulf of Bothnia. Later, ZETTERSTEDT (1848: 2831) treated *P. nigritula*, 1838: 793 as a variety of *fuscula*. HENDEL (1931–1936: 407) accepted this and showed *nigritula* as a synonym of *fuscula*. I have examined the female holotype of *nigritula* and can now confirm that it is certainly distinct from *fuscula*, representing a species in the *robustella* group. With further study a positive identification may be possible.

HENDEL correctly synonymized *P. avenae* de Meijere with *fuscula*. DE MEIJERE's illustration of the aedeagus of *avenae*, although lacking detail, clearly shows the S-shaped tubule which is characteristic of this species.

P. fuscula has frequently been confused with *P. nigra*. The most reliable character for separating adults is the greater pilosity of the eye in *nigra* but this is not a satisfactory character for identifying females. The male genitalia and larval spiracles of the two species are entirely distinct (cf. Figs. 438, 442, and 440, 443).

ECONOMIC IMPORTANCE. *P. fuscula* normally occurs in small numbers and larval damage can be considered of no significance. However, in northern Europe substantially increased populations sometimes occur and the amount of leaf damage to young cereals can cause concern to growers. During the years 1959–1960 mines were unusually numerous on oats, barley and rye in central and northern Finland (VAPPULA, 1965). MÜHLE (1971) records *P. fuscula* as damaging *Arrhenaterum elatius*, an important meadow grass, in eastern Germany.

Phytomyza nigra Meigen, 1830 – Europe, N. America

ADULT. Generally similar to *Phytomyza fuscula*, with following points of difference: frons sometimes darker, with more greyish undertone, eye more conspicuously pilose, acrostichals sparser.

MALE GENITALIA. Aedeagus as in Fig. 442.

EARLY STAGES. Larval anterior spiracles on relatively long stalks (Fig. 443), each with up to 18 pores; posterior spiracles on short stalks, each with 10–13 pores; puparium yellowish-brown.

HOST-PLANTS. *Avena sativa* L., *Hordeum vulgare* L., *Secale cereale* L., *Triticum aestivum* L.; also on many wild grasses.

BIOLOGY. The larva forms a narrow, whitish linear mine, pupating in the mine, with the long anterior spiracles projecting through the epidermis. Mines can be found from early spring until late in the autumn.

DISTRIBUTION. Widespread throughout Europe but apparently absent from

northern Scandinavia and Scotland; only known in Canada from Vancouver (SPENCER, 1969a) and in the United States from Oregon and Washington (FRICK, 1952b).
PARASITES. *Dacnusa areolaris* (Nees), *D. maculipes* Thomson.

Discussion

This species, although common and widespread in Europe, occurs less frequently on cereals than *P. fuscula*. In Madeira there appears to be a local form which is substantially larger than the main European population. I have seen a female with wing length of 3.4 mm (P. BRINCK).

In Japan *P. nigra* is considered a serious pest of cereals. SASAKAWA (1953) writes: '*Phytomyza nigra* Meigen is one of the most important economic species in the corn field in Japan and infests the leaves of cultivated barley, wheat, rye and many other weeds of the family Gramineae. This species is widely distributed in Japan.' SASAKAWA (1954a) investigated the susceptibility of attack from *P. nigra* of 14 varieties of barley and two of wheat. He found that more eggs were laid on the early maturing varieties. The leaf colour, hardness and softness of the leaf tissue have no influence on oviposition. The mortality of the larvae was found to be lower in the varieties containing smaller amounts of crude protein.

Damage to wheat, barley and oats in Japan is also recorded by ISITANI (1938).

GENUS CERODONTHA RONDANI

This genus feeds exclusively on monocotyledons, in addition to the Gramineae also on the families Cyperaceae, Iridaceae and Liliaceae. Species in two of the sub-genera of *Cerodontha* – *Poemyza* and *Cerodontha* – have hosts among cereals and pasture crops. Five *Cerodontha* species, from Australia, New Zealand, Europe, North and South America, and four *Poemyza* species, from Europe, Japan, Africa and Malaya are discussed below.

Re-examination of material from Australia and New Zealand has confirmed that *Cerodontha australis* Malloch occurs not only in Australia but also in New Zealand. It has further been found that the species previously considered to represent *australis* in Australia (SPENCER, 1963b: Fig. 42) is an undescribed species, described below as *milleri* sp.n. It is possible that yet another species in this complex is present in New Zealand (cf. p. 291). The European species *C. denticornis* (Panzer) has been mistakenly referred to in New Zealand literature; however, it does not occur outside Europe, North Africa and possibly temperate Asia.

Poemyza incisa (Meigen) occurs rarely on barley, oats, rye and wheat in Europe and has been recorded, as *okazakii* (Matsumura), on barley in Japan (SASAKAWA, 1961: 372). Examination of a male from Kyoto has satisfied me that *okazakii* is synonymous with *incisa*, as suggested by NOWAKOWSKI (1967). A closely related species, *P. pygmaea* (Meigen), is very common in Europe and Canada on wild grasses and occurs only rarely on cereals. It closely resembles *incisa* but is immediately distinguishable by its black squamal fringe (yellow in *incisa*).

Corn (maize) is attacked by *Poemyza orbitona* Spencer in Ghana and rice by *P. oryzivora* Spencer in Malaya.

Cerodontha (Poemyza) incisa (Meigen, 1830) – Europe, Japan, N. America

Cerodontha (*Poemyza*) *okazakii* (Matsumura, 1916).

ADULT. Black species, with yellow squamal fringe.

HEAD. Frons broad, twice width of eye, black, orbits moderately shining black, not paler on inner margins; lunule high, narrow, upper margin at level of lower ors; third antennal segment small, round, black.

THORAX. Mesonotum greyish-black, only slightly shining, with 3 + 1 dc, acr coarse in 6 rows; pleura black, upper margin of mesopleura only narrowly yellow; legs black, only fore-knees yellow; wing length from 2.3–2.75 mm, last section of vein M_{3+4} twice length of penultimate; wing base, including squamae and fringe, conspicuously yellow.

MALE GENITALIA. Aedeagus as in Fig. 444.

EARLY STAGES. Posterior spiracles of larva arising from conical projection, each with 3 pores (Fig. 445); puparium deeply segmented, shining black, with bluish or purplish reflections.

HOST-PLANTS. *Avena sativa* L., *Hordeum vulgare* L., *Secale cereale* L., *Triticum aestivum* L.; also many wild grasses, including *Bromus*, *Calamagrostis*, *Lolium* and *Phalaris*.

BIOLOGY. The larvae form a long whitish mine in the leaf-blade, frequently several feeding together; pupation takes place in the mine, the puparia being lightly attached to the leaf with silk. There are at least two generations in a year.

DISTRIBUTION. Widespread in Europe from northern Scandinavia to the Mediterranean; known from Moscow (K.A.S.); Japan; Canada.

PARASITES. *Exotela flavicoxa* (Thomson), *Chorebus difficilis* Griffiths, *C. poemyzae* Griffiths, *C. talaris* (Haliday); *Chrysocharis polyzo* (Walk.), *Pediobius acantha* (Walk.); *Opius ambiguus* Wesm., *O. rex* Fi., *O. variegatus* Szépl.; *Eurytenes abnormis* Wesm.

Discussion

TAXONOMY. *P. incisa* is clearly closely related to *C. pygmaea* and is not readily distinguishable by male genitalia or larval characters. However, the two are immediately separable by the squamal fringe which is yellow in *incisa* and black in *pygmaea*. The hosts of the two species entirely overlap.

SASAKAWA (1961) retained *P. okazakii* (Matsumura), which feeds on barley in Japan, as distinct from *incisa*. NOWAKOWSKI (1967) tentatively synonymized the two. I have now examined a male from Kyoto bred from *Phalaris* and am satisfied that NOWAKOWSKI's proposed synonymy is correct.

ECONOMIC IMPORTANCE. With the wide range of hosts available among wild grasses, *P. incisa* has not so far occurred in significant numbers on cultivated cereals. However, it does regularly occur on barley, oats, rye and wheat and there is every probability that somewhere, sometime, it may occur in pest proportions

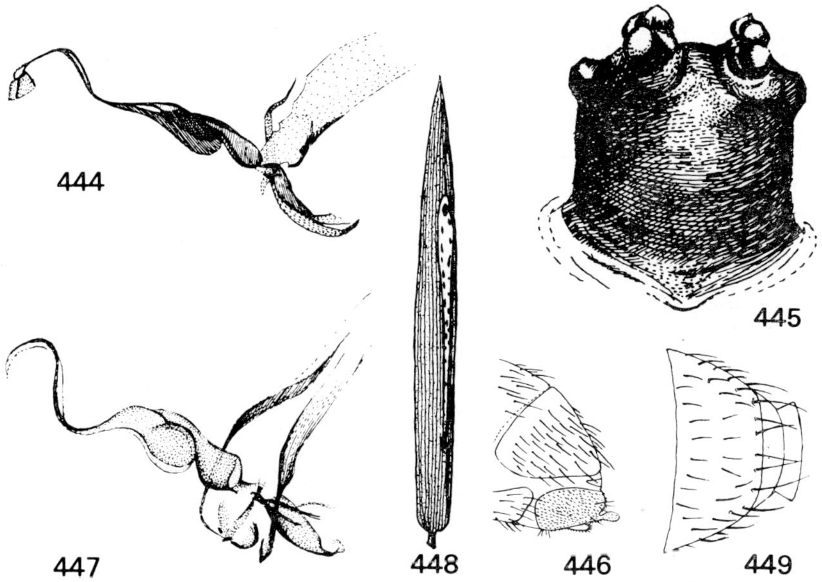

Figs. 444–445. *Cerodontha* (*Po.*) *incisa* (Mg.): 444, aedeagus; 445, posterior spiracles of puparium.

Figs. 446–448. *Cerodontha* (*Po.*) *lateralis* (Macq.): 446, epandrium; 447, aedeagus; 448, leaf mine on *Agropyron* (Hering).

Fig. 449. *Cerodontha* (*Po.*) *superciliosa* (Zett.): epandrium.

on one or other of these crops, as has happened with *P. lateralis* in the Ukraine (cf. p. 286).

Cerodontha (Poemyza) lateralis (Macquart, 1835) – Europe

ADULT. Small species, with high lunule; frons and notopleura conspicuously yellow.

HEAD. 2 equal ors, 2 ori directed inwards and upwards; frons mainly yellow but somewhat darkened centrally towards lunule; orbits yellow above, shining black below; all antennal segments black.

THORAX. Mesonotum mat greyish-black; notopleura bright yellow, mesopleura black, only upper margin narrowly yellow.
LEGS. Black but all femora with yellow knees.
WING. Length from 2–2.5 mm, costa extending to vein M_{1+2}, last section of M_{3+4} twice length of penultimate. Epandrium in male with conspicuous knob above cerci (Fig. 446).
MALE GENITALIA. Aedeagus as in Fig. 447.
EARLY STAGES. Larva yellowish-white, up to 3 mm long; anterior spiracles short, each with 7–8 pores, posterior spiracles arising from elongate protuberance, each with 3 elongate pores; puparium brownish-black.
HOST-PLANTS. *Hordeum vulgare* L., *Secale cereale* L., *Triticum aestivum* L.; probably other wild grasses.
BIOLOGY. According to VENTURI (1935) the larva normally mines towards the apex of the leaf, the early linear mine widening to a distinct blotch (Fig. 448); pupation always takes place in the mine.

The pupal period varies from 13–15 days in the summer to 20 days in colder weather; pupal diapause lasts for several months from the last autumn generation to emergence of adults in early spring. VENTURI has noted as many as six generations in Italy.
DISTRIBUTION. Europe, from Scandinavia to the Mediterranean and eastwards to Russia and Japan.
PARASITES. In Italy six species have been recorded (VENTURI, 1935): *Gyrocampa affinis* Nees (Braconidae); *Cirrospilus* sp., *Eulophus* sp., two species of *Pleurotropis* and one further sp. and gen. indet.; in France *Diglyphus isaea* (Walk.). GRIFFITHS (1968) records *Chorebus fordi* (Nixon) from Poland.

Discussion

TAXONOMY. *Poemyza lateralis* has in the past been confused with *P. superciliosa* (Zett.), until NOWAKOWSKI (1967) separated the two species. In *lateralis* the ori are directed inwards and upwards and in the male the epandrium bears a knob-like process (Fig. 446), while in *superciliosa* the ori are distinctly proclinate and the epandrium bears a conspicuous plate (Fig. 449). *P. superciliosa* has not been recorded on cultivated cereals.
ECONOMIC IMPORTANCE. Larval damage by this species is normally negligible. VENTURI ascribes this to effective control by parasites and the cutting and destruction of wild hosts adjoining fields of cereals, which will destroy the puparia remaining in the leaves. However, *P. lateralis* is capable of a rapid population build-up, which can result in a mass outbreak of mining activity. This situation developed in the Ukraine in 1929 (ESTERBERG, 1931).

Cerodontha (Poemyza) orbitona (Spencer), **comb. nov.** – South and West Africa

Phytobia (*Poemyza*) *orbitona* Spencer, 1960a. Holotype ♂ from South Africa in Division of Entomology, Pretoria.

ADULT. Small species with conspicuously broad, yellow orbits.
HEAD (Fig. 450). Orbits narrowly projecting above eye, exceptionally broad, widening anteriorly, at widest point occupying $\frac{3}{4}$ area between eye margins; 2 slender, equal ors, 2 ori, the lower small, incurved; lunule high and narrow; jowls deepest at rear, $\frac{1}{5}$ height of eye; third antennal segment small, round.
MESONOTUM. 3 + 1 dc, third and fourth small, $\frac{1}{2}$ length of second, equidistant each side of suture; acr in 4 rows, extending to first dc.
WING. Length from 1.9 mm in male to 2.2 mm in female, costa extending to vein M_{1+2}, last and penultimate section of M_{3+4} approximately equal.
COLOUR. Orbits lemon yellow, frons brownish-black, with yellowish undertone; third antennal segment variable, pale, yellow or brown on inside, darker, brownish-black on outside; mesonotum and scutellum shining-black; humeral callus, notopleura and upper margin of mesopleura bright yellow, pleura otherwise black; legs: coxae blackish at base, yellow anteriorly; femora black in basal third, yellow beyond, tibiae and tarsi brown; abdomen shining black, all tergites with narrow yellow margins; veins pale brown; squamae, including margin and fringe, uniformly yellowish.
MALE GENITALIA. Aedeagus exceptionally long, as in Fig. 451.
EARLY STAGES. Puparium reddish-brown, posterior spiracles at right-angles to anal segment, fused at base but then dividing into two rounded protuberances, each with 3 highly modified pores, the upper long, black, finger-like (Fig. 452).
HOST-PLANTS. *Oryza sativa* L.
BIOLOGY. A narrow linear mine is formed in the upper part of young leaves; pupation takes place externally.
DAMAGE CAUSED. Individual larvae cause insignificant damage.
DISTRIBUTION. Ghana; South Africa.
PARASITES. In Ghana six species have been recorded, as follows:
Ceraphronidae: *Ceraphron* cf. *fijiensis* Ferr.
Pteromalidae: *Callitula* sp.
Cynipoidea, Eucolidae: *Eucoilidea* sp.
Eulophidae: *Sympiesis* sp.; *Hemiptarsenus semialbiclavus* Gir.; *Enaysma* sp.

The *Sympiesis* and *Hemiptarsenus* species kill the larvae and pupate in the leaf-mine. The four other species permit the Agromyzid larva to pupate normally.

Discussion

TAXONOMY. This species is entirely typical of the sub-genus *Poemyza*. Species with similarly modified posterior spiracles are known from Europe.
DISTRIBUTION. The only previous record of *Po. orbitona* is from South Africa; it can now also be expected to occur in East Africa.
ECONOMIC IMPORTANCE. Large populations might affect the growth of young rice plants but it appears that the species is well controlled by its numerous parasites. As a new potential pest of rice, *Po. orbitona* deserves further study.

Cerodontha (Poemyza) oryzivora (Spencer, 1961b) – Malaya

ADULT. Very small species, with head entirely dark and unusually wide orbits. HEAD. Frons at foremost ocellus $1\frac{1}{2}$ times width of eye, orbits greatly widening

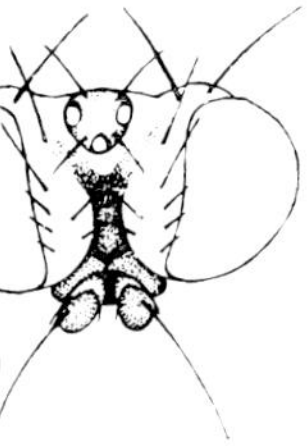
450

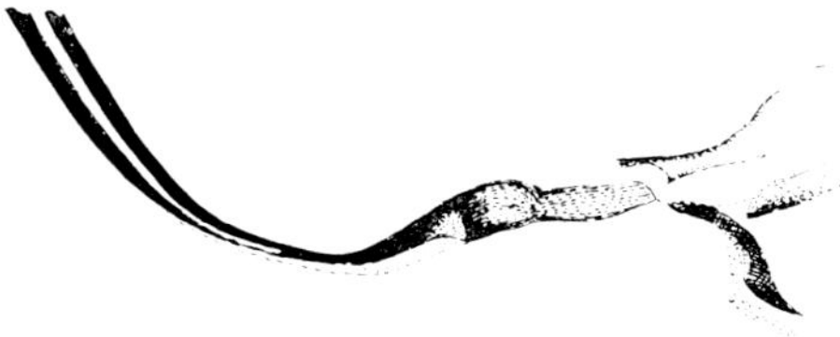
451

452

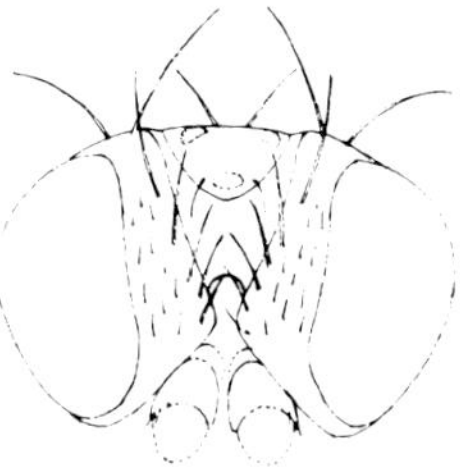
453

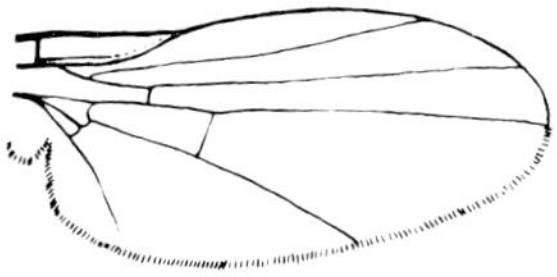
454

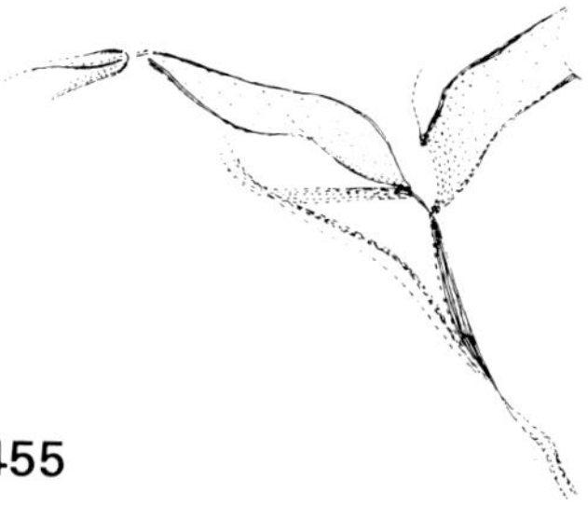
455

456

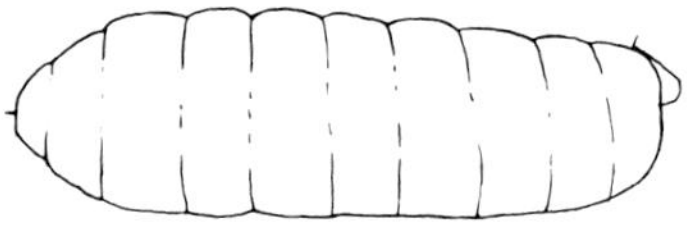
457

Figs. 450–452. *Cerodontha* (*Po.*) *orbitona* (Sp.): 450, head; 451, aedeagus; 452, posterior spiracles of puparium.

Figs. 453–457. *Cerodontha* (*Po.*) *oryzivora* (Sp.): 453, head showing orbits and lunule; 454, wing; 455, aedeagus, side view; 456, distiphallus, ventral view; 457, puparium.

anteriorly, at widest point together occupying $\frac{7}{8}$ distance between eyes (Fig. 453); lunule very high and narrow, height equal to half distance between base of antennae and foremost ocellus; 2 (on one side 3) ors, the upper slightly longer, 2 weak ori, both directed largely upwards; orbital setulae long, in 2 rows, reclinate; jowls

deepest at rear, cheeks forming only narrow ring below eye; third antennal segment small, rounded, arista long, conspicuously pubescent.
MESONOTUM. 3 well-developed dc, small pre-sutural 4th possibly present (mesonotum damaged in only available specimen); acrostichals coarse, in some 6 rows, not extending behind 1st dc.
WING (Fig. 454). Length in male 1.6 mm, last section of M_{3+4} $1\frac{1}{2}$ times length of penultimate, first cross-vein at midpoint of discal cell.
COLOUR. Head largely mat black, frons at apex of ocellar triangle and second antennal segment more brownish; mesonotum mat black, slightly more shining from behind; pleura uniformly black, though notopleural area more brownish; legs black, all knees bright yellow; squamae and fringe pale, yellowish.
MALE GENITALIA. Aedeagus as in Figs. 455, 456, distal tubules greatly reduced.
EARLY STAGES. Puparium (Fig. 457) bright orange-yellow, 1.9 mm long, segments deeply indented, with no apparent bands of tubercles; anterior spiracles in form of 2 minute, finger-like projections, posterior spiracles minute, visible only as a fine, needle-like projection at right-angles to anal segment; anus forming conspicuous rounded protuberance.
HOST-PLANT. *Oryza sativa* L. (rice).
BIOLOGY. Larva forming leaf-mine in November but no details available.
DAMAGE CAUSED. In view of the small size of the larva, the damage caused even to young leaves cannot be significant, as populations are apparently small.
DISTRIBUTION. Malaya: Kuala Lumpur.
PARASITES. None recorded.

Discussion

TAXONOMY. This is one of the smallest known *Poemyza* species. It has the enormously widened orbits found also in *P. javana* de Meijere (Java), *P. orbitona* Spencer (South and West Africa) and *P. setariae* Spencer (South Africa) but in these three species the orbits are bright yellow. Both the male genitalia and posterior larval spiracles diverge considerably from the form found in typical *Poemyza* species.
ECONOMIC IMPORTANCE. *Poemyza oryzivora* is apparently an uncommon species and at existing population levels its importance as a feeder on rice must be considered negligible.

Cerodontha (Cerodontha) angustipennis Harrison, 1959 – New Zealand

ADULT. Mesonotum brilliantly shining black, without acrostichals.
HEAD (Fig. 458). Frons only slightly projecting above eye anteriorly, 1 strong ors below centre of frons, 1 slightly incurved ori and rarely a smaller one below; orbital setulae sparse or lacking; inner vertical bristle displaced forwards, arising from upper orbits; third antennal segment small, slightly angulate at upper corner and with a fringe of long hairs but no spine.

THORAX. Mesonotum with 3 dc, the first (behind) weaker, the second strongest, the presutural stronger than first; acrostichals absent; 2 notopleural bristles (not 1 as in description).
WING. Length from 2–2.5 mm, unusually narrow.
COLOUR. Frons mainly bright yellow but distinctly darkened centrally, face entirely yellow; first and second antennal segments yellow, third black; pleura entirely shining black; coxae and femora bright yellow, tibiae and tarsi slightly darker, yellowish-brown; abdomen black.
MALE GENITALIA. Not known.
EARLY STAGES. Not known.
HOST-PLANTS. Certainly Gramineae but species not established.
DISTRIBUTION. New Zealand: North and South Island.

Discussion

TAXONOMY. This is a distinctive species, immediately distinguishable from the others in the genus known in New Zealand by the brilliantly shining black mesonotum.
ECONOMIC IMPORTANCE. *C. angustipennis* has been recorded from many localities in New Zealand and appears to be widespread in both North and South Islands. It has been found in January, February, May, September and December. CUMBER and HARRISON (1959) in their review of insects on sown pastures write: 'It is apparently an early species, and although numbers taken are small, it cannot be dismissed as unimportant.'

Cerodontha (Cerodontha) australis (Malloch, 1925) – Australia, New Zealand

ADULT. Medium-sized species, wing length up to 2.6 mm; essentially as in *C. milleri* described below, apart from larger size and in some specimens the more yellowish colouration of the pleura; third antennal segment as in Fig. 459.
MALE GENITALIA. Aedeagus as in Fig. 460, distal tubules short, mesophallus also short, broad.
EARLY STAGES. MORRISON (1938) reports that the puparium is 'barrel-shaped' and reddish-brown in colour.
HOST-PLANTS. *Hordeum vulgare* L., *Triticum aestivum* L.; also *Lolium perenne* L. (rye-grass), and probably other pasture grasses.
BIOLOGY. The larva feeds primarily in the leaf-sheath and puparia are always found singly. 'They are located between the centre stem and its ensheathing leaves and usually attached to the inner face of the latter. They normally occurred at a higher level on the plant than do the Hessian flies and they were not placed as a rule close to a node' (MORRISON, 1938). Puparia have been found in December in North Canterbury, which produced flies on 25 December. Adults were observed

emerging on 18 January at Hastings. Information given by HAMILTON (1947), as *denticornis*, is taken from LUGINBILL and URBAHNS (1916) who are discussing *C. dorsalis* in the United States; the figures reproduced by HAMILTON are also taken from the same paper, without acknowledgment!
DISTRIBUTION. Australia: New South Wales; New Zealand: widespread on North and South Islands, also Moko Hinan Is., Little Barrier Is., Mayor Is.,

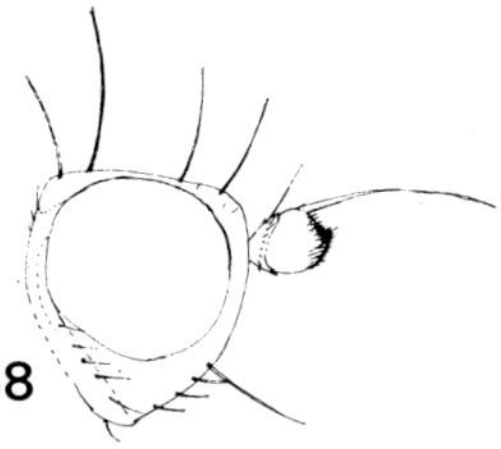
458

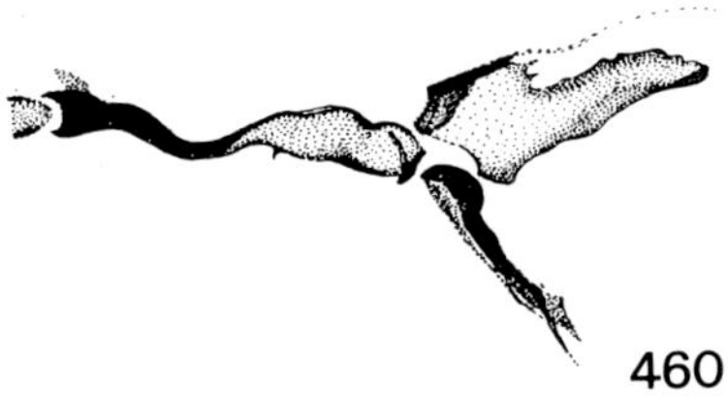
460

459

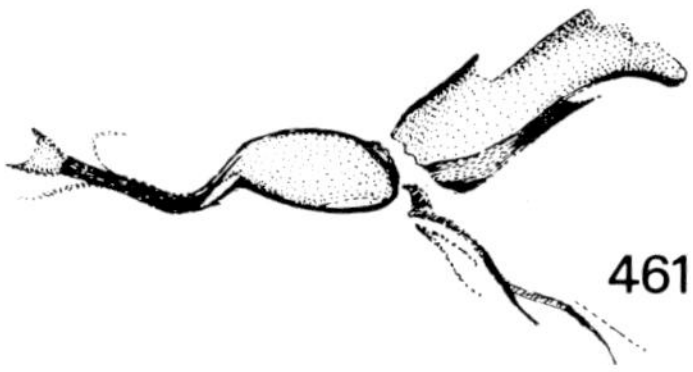
461

Fig. 458. *Cerodontha* (*Cer.*) *angustipennis* Harrison: head.

Figs. 459–461. *Cerodontha* (*Cer.*) *australis* (Mall.): 459, third antennal segment; 460, aedeagus; 461, aedeagus of ♂ from New Zealand.

Chatham Is. However, these records may refer to two distinct species in New Zealand, as discussed below.

Discussion

TAXONOMY. MALLOCH described this species from three females from Sydney; the wing length was given as 2.5 mm.

I am now satisfied that the illustration of the aedeagus given by SPENCER (1963b: Fig. 42) was not of *australis* but of the species described below as *milleri* sp.n. Although there appear to be no reliable external characters for separating these two species, except possibly the greater size in *australis*, the male genitalia are entirely distinct.

True *australis* has been confirmed by genitalia studies from Nihorupu, Auckland, 2.i.1950 (R. A. HARRISON); the aedeagus of this specimen and of one from the National Park, south of Sydney, agree exactly.

A single specimen examined from Dreamlands, Waitakore, 27.xii.1950 (K. P. LAMB) has a significantly different aedeagus (Fig. 461). This appears to represent a further species in this complex. However, it is not proposed to treat this as distinct until additional material can be examined.

HARRISON (loc. cit.) treats the New Zealand populations as the palaearctic species *C. denticornis*. This is without question distinct, as can be seen from its aedeagus (Fig. 462).

DISTRIBUTION. In Australia *C. australis* has only been recorded from New South Wales but very probably extends into both Victoria and Queensland. The numerous records in New Zealand may refer either to the true *australis* or to the possibly undescribed species discussed above.

ECONOMIC IMPORTANCE. HAMILTON (1947) reports some damage to rye-grass grown for seed. The larval feeding results in a premature whitening of the seed-head. Attacked plants are readily visible by their white straws among the otherwise green crop. A badly attacked paddock in North Canterbury in December, 1944, had up to eight white heads per square foot.

CUMBER and HARRISON (1959) refer to the wide distribution of *australis* in New Zealand (as *denticornis*) and in their survey of insects on pastures it proved to be the commonest insect they found. In 109 paddock and 82 roadside samples they obtained 5614 at 1266 specimens respectively. 'The greatest catch was in a paddock, sample from Waverley on 24 January, 1957, which yielded 720 specimens.' They write further: 'No great importance has been attached to its attacks on barley and wheat in New Zealand, and the damage to rye-grass has not been reported as severe. However, a closer study of this species may reveal that it is far more damaging than has hitherto been suspected, as the numbers taken from pastures are very high indeed.'

Cerodontha (Cerodontha) denticornis (Panzer, 1806) – Europe, North Africa, Japan

ADULT. Variable species in size and colour, with strong spine on third antennal segment.

HEAD. Frons strongly projecting above eye in profile, with 2 reclinate ors and 1 strong incurved ori; third antennal segment elongate, with short, sometimes blunt, spine at end.

MESONOTUM. 3 + 1 strong dc, no acrostichals.

WING. Length from 1.8 mm in male to 2.5 mm in female; costa extending strongly to vein M_{1+2}, last and penultimate sections of M_{3+4} about equal.

COLOUR. Frons, jowls and face orange-yellow, orbits variably darkened; first and second antennal segments yellow, third black; mesonotum mat, greyish-black, notopleura and humerus largely yellowish, mesopleura variable, from black to yellow; scutellum either entirely black or broadly yellow centrally; legs: coxae and femora yellow, tibiae and tarsi darker, from brownish to almost black.

MALE GENITALIA. Aedeagus as in Fig. 462, with distinctive curvature.

EARLY STAGES. Larva long and thin, whitish; anterior spiracles each with about 12 pores, posterior spiracles each with from 15–25, arranged irregularly in 2 or 3 groups; puparium yellow, narrow.

HOST-PLANTS. *Avena sativa* L., *Hordeum vulgare* L., *Triticum aestivum* L.; also many pasture and wild grasses, including *Agropyron, Dactylis, Festuca, Holcus, Phleum, Poa.*
BIOLOGY. The larva feeds mainly in the leaf-sheath; if the mine commences in the blade, it runs downwards straight to the leaf-sheath, where pupation takes place.

There are at least two generations. Adults can be found in January in Morocco, in April in southern Europe (Sicily) or in May and June further north. The species occurs in large numbers but the leaf-mines are inconspicuous, being largely out of view in the leaf-sheath.

The biology has been discussed in some detail by VENTURI (1946a) in Italy.
PARASITES. *Chorebus crenulatus* (Thom.); *Opius depeculator* Först., *O. parvungula* Thom.
DISTRIBUTION. Europe, from the Outer Hebrides off N.W. Scotland, southern Sweden and Finland to the Mediterranean, including Sicily (SPENCER, 1966g) and Morocco: Atlas Mountains (SPENCER, 1967).

Discussion

TAXONOMY. This species is notable for its colour variation. The darkest flies are normally found in the spring and early summer from April to June, the palest from June to September. In the majority of specimens the colouration is intermediate.

The name *denticornis* has been mistakenly used for the species occurring in Australia and New Zealand (see pages 291, 292). A similar but certainly distinct species, *C. heringiella* Spencer, 1961c, has been found in the Abyssinian Highlands.
ECONOMIC IMPORTANCE. Although adults of this species can occur in very large numbers, there are few reports of leaf-mining damage. GOODLIFFE (1941) reared specimens from barley in southern England and ISITANI (1938) records slight damage to the same host in Japan, in the Tokyo area. Damage to cereals and pasture grasses may be more widespread than appears, owing to the concealed feeding of the larvae in the leaf-sheath. A more detailed survey of the occurrence of the species on oats, wheat and barley could prove of interest.

Cerodontha (Cerodontha) dorsalis (Loew, 1863) – North and South America, Mongolia

ADULT. Yellow and black species with spine on third antennal segment.
HEAD (Fig. 463). Frons conspicuously projecting above eye towards base of antennae, 2 equal ors, 1 smaller incurved ori, orbital setulae sparse; jowls deeply extended at rear, $\frac{1}{3}$ height of eye; eye conspicuously slanting; third antennal segment elongate, with a fringe of white hairs and a short spine at upper corner, arista usually distinctly flattened. Mesonotum with 3 + 1 strong dc, acr lacking.
WING. Length from 2.4–2.7 mm, costa extending strongly to vein M_{1+2}, last section of M_{3+4} long, only slightly shorter than penultimate.

COLOUR. Frons orange-yellow, hind-margin of eye black, vte on black ground, vti normally just on yellow; jowls and face yellow; third antennal segment black, first and second yellowish; mesonotum mat greyish-black, with somewhat variable yellowish-grey patch centrally adjoining scutellum and two narrow yellowish bands laterally; scutellum yellowish centrally, darker at sides; pleura predominantly pale but mesopleura sometimes more greyish; legs: coxae and femora yellow, tibiae and tarsi darker, brownish.
MALE GENITALIA. Aedeagus as in Fig. 464, 465.
EARLY STAGES. No detailed descriptions are available. The puparium is shining yellowish-brown, somewhat flattened, with the posterior spiracles irregularly branching, on short stalks, arising from a low, rounded protuberance, each bearing some 15 minute bulbs (Fig. 466).
HOST-PLANTS. *Avena sativa* L., *Hordeum vulgare* L., *Panicum* sp. (millet), *Secale cereale* L., *Triticum aestivum* L., *Zea mays* L.; also fodder-grasses, such as *Phleum pratense* L. (Timothy). Wild grasses which have been confirmed as hosts have been recorded in the genera *Bromus*, *Dactylis*, *Eleusine*, *Elymus*, *Hordeum*, *Panicum*, *Phalaris* and *Poa* (LUGINBILL and URBAHNS, 1916).
BIOLOGY. The life history of this species has been studied in detail by LUGINBILL and URBAHNS (1916). Oviposition takes place towards the apex, usually on the upper side of the leaves or in very young plants also in the stem. The young larva feeds down the leaf and in the first day can form a mine up to 17 mm in length. The mine can follow a zig-zag course along the leaf, until it ultimately reaches the leaf sheath, in tender plants even penetrating the stem. The total length of mine can be as much as 20 inches.

A single female has been observed to lay 20 eggs within a 24 hour period but the average was 9. The maximum number laid by one individual was 188 and longest period of oviposition of one female was 33 days. Observations at various localities suggest that about 60% of the eggs deposited are fertile.

The length of larval life varies greatly with the season and locality and is from 9 to 24 days. The pupal stage varies similarly, dependent on temperature, and times also of from 9 to 24 days have been observed.

In California adults are active throughout the year but appear in larger numbers in grain fields from February to April. The long period of oviposition of the females leads to overlapping of generations, which continue until frosts stop all further growth of the host-plants. As many as six generations were observed in South Carolina, while in Pasadena, California, eight generations were noted. In southern California all stages may be found throughout the entire winter.
DAMAGE CAUSED. The numerous feeding punctures made by the females leads to a substantial loss of sap and young leaves 'assume a pale yellow, sickly appearance, first at the tips, but gradually extending down toward the base of the stem, and finally they curl up and wither away. When more than one leaf or if a majority of the leaves of the same plant are injured to this degree by the punctures, the whole plant may die from the injury or produce an inferior, worthless plant'.

Most damage, however, is caused by larval feeding. In young plants the larvae can feed down from the base of the leaf sheath into the heart of the plant near or even below the surface of the ground. When the leaves arising from the bud are severed, they dry up and the 'whole plant presents an appearance as though infested with the wheat bulb-worm (*Meromyza americana* Fitch) or some species of the genus *Oscinis*'. Damage of this kind has been observed on young oat plants in South Carolina and also on small barley plants at several localities in southern California (LUGINBILL and URBAHNS, 1916).

462

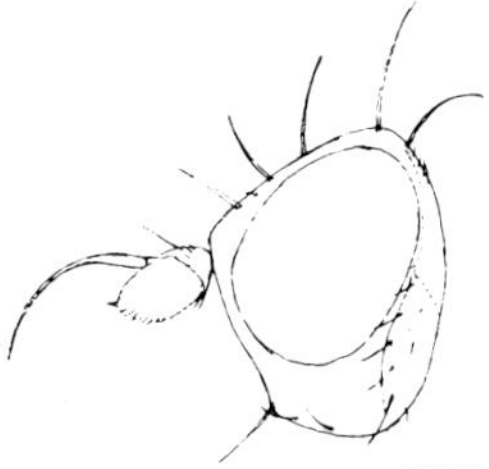

463

464

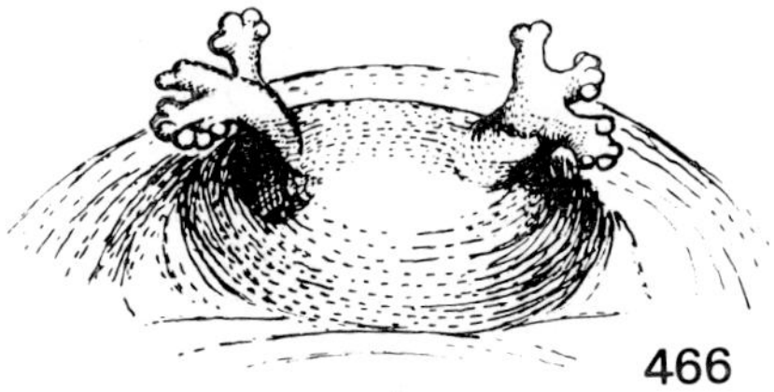

466

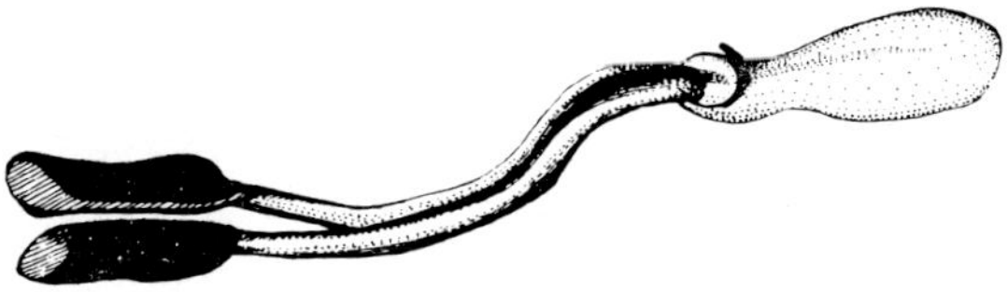

465

Fig. 462. *Cerodontha* (*Cer.*) *denticornis* (Pz.): aedeagus.

Figs. 463–466. *Cerodontha* (*Cer.*) *dorsalis* (Loew): 463, head; 464, aedeagus, side view; 465, same, ventral view; 466, posterior spiracles of puparium.

DISTRIBUTION. United States, widespread, particularly in southern States; Canada: Alberta, British Columbia; Argentina, Brazil, Peru, Costa Rica, Guatemala, Dominica (W.I.); Mongolia.

PARASITES. Nine species have been recorded in the United States by LUGINBILL and URBAHNS as follows: *Cirrospilus flavoviridis* Cwfd. (Nevada); *Cyrtogaster occidentalis* Ashm. (California to Florida); *Diaulinus websteri* Cwfd. (California); *Diaulinopsis callichroma* Cwfd. (Florida, Mississippi); *Polycystus foersteri* Cwfd. (California, Florida); *Chrysocharis parksi* Cwfd. (California); *Opius dimidiatus*

Ashm. (Indiana); *Opius aridis* Gahan (Florida); *Dacnusa* sp. (California); also *Opius coriaceus* Gahan (U.S.A.).

Discussion

TAXONOMY. The colour of this species is distinctly variable, with the paler, more yellowish form occurring mainly in the eastern United States and also in South America, while the darker form has been noted mainly in the western States of America (ALDRICH, 1918; FRICK, 1959). A substantially darker and larger species, *C. occidentalis* Sehgal, 1968, was recently described from western Canada. The host-plants of this species are not known.
DISTRIBUTION. *C. dorsalis* has a wide distribution in North and South America and has now also been confirmed in the eastern Palaearctic Region, in Mongolia (SPENCER 1969a: 143).
ECONOMIC IMPORTANCE. Although occurring frequently on cultivated cereals, *C. dorsalis* cannot be considered as a serious pest. In Arizona in 1915 a field of barley was recorded with 5% of the plants having one or more leaves mined (LUGINBILL and URBAHNS, 1916: 9). A more serious outbreak was reported in Kansas in the spring of 1935, when young maize plants were extensively attacked, with the growth of the plants retarded by the destruction of leaves. It has also been reared from maize in Argentina (P. WYGODZINSKI) and from oats in Peru (C. A. KORYTKOWSKI).

Apparently effective control is normally provided by the numerous parasites attacking this species.

Cerodontha (Cerodontha) milleri sp.n. – Australia

ADULT. Small species, with mat-grey mesonotum and at least some acrostichals in 2 rows.
HEAD. Frons distinctly projecting above eye in profile; 1 strong ors, 2 ori, orbital setulae short, sparse, not above upper ori; jowls deeply extended at rear; third antennal segment (Fig. 467) elongate, with short spine on upper corner.
MESONOTUM. 3 + 1 strong dc, at least one or two pairs of acr between third and fourth dc.
WING. Length in male 1.9–2.1 mm, female not known.
COLOUR. Head predominantly yellow, orbits paler than frons, which is frequently more brownish; first and second antennal segments yellow, third normally black but often paler, yellowish, particularly on inside; mesonotum mat, greyish-black, scutellum similar, without any trace of yellow; mesopleura blackish-grey with very narrow yellowish area along upper margin; notopleura grey, humerus at most slightly yellowish; abdomen blackish; legs: coxae and femora bright yellow but on hind- and mid-legs sometimes distinctly darker, more brownish; tibiae and tarsi brown.

MALE GENITALIA. Distal section of aedeagus (Fig. 468) curved, long, rather pale, the two tubules parallel, mesophallus long and slender.
HOST-PLANTS. Unknown but certainly Gramineae.

Figs. 467–468. *Cerodontha* (*Cer.*) *milleri* sp.n.: 467, third antennal segment; 468, aedeagus.

HOLOTYPE ♂, Australia, New South Wales: Bowral, 30.i.1961 (K.A.S.); paratypes: 1♂, same data; 1♂, Blue Mountains, Mt. Wilson, 2.v.1958 (D. K. McAlpine). Holotype and paratype in author's collection, 1 paratype in Australian Museum, Sydney.

Discussion

TAXONOMY. This species agrees completely in external characters with *australis* Malloch, apart from its smaller size. The aedeagus, however, is entirely distinct and was previously illustrated (SPENCER, 1963b: Fig. 42a) as *australis*. The two type specimens from Bowral were caught while I was enjoying the hospitality of Mr. J. MILLER and I have pleasure in naming the species after him.
ECONOMIC IMPORTANCE. No bred specimens are known but the larvae certainly feed in meadow grasses and possibly also cereals.

X. STEM-BORERS, GALL-CAUSERS AND LEAF-MINERS ON TREES

Relatively few Agromyzidae feed on trees. Those causing damage of any consequence feed internally as cambium-borers and belong to the genus *Phytobia* Lioy. A few species in the genera *Agromyza*, *Japanagromyza*, *Paraphytomyza* and *Phytomyza* feed on trees as leaf-miners but damage to the leaves is slight and the effect on the growth of the trees must be insignificant. Five species in the genus *Hexomyza* Enderlein feeding as gall-causers are also briefly discussed.

Some 50 species throughout the world are known in the genus *Phytobia* but the host has been discovered for only 9 of these, 5 in Europe and 4 in North America. Attempts to find the larvae of these species are inhibited by the inevitable damage caused to young trees in the process, and the larval feeding tracks in the wood are frequently not recognized by growers to be the result of feeding by dipterous larvae.

The first reference to a species in this genus was by RATZEBURG (1868) who found larvae feeding in twigs of *Acer platanoides* L. sent to him from St. Petersburg. No adults were reared but the larva was described and the name *Tipula suspecta* was used in the Index, although not in the text.

Prior to this, early nineteenth-century botanists who had recorded the so-called 'pith flecks' in the wood of various trees, including *Betula*, *Corylus avellana* L. (hazel) and *Alnus*, considered that the lines of modified cells making up the pith-flecks were a botanical character of each species of tree concerned. In fact the hollow larval feeding channel after one or two years becomes occluded with new callus-type cells, as can be clearly seen in the micro-photograph of *Salix alba*, attacked by *Phytobia cambii* in Norfolk, England (Plate 10, Photo, C. F. YEATES). NIELSEN (1905, 1906) was the first to breed specimens in this genus, from puparia found beneath *Alnus* trees in Denmark. He misidentified these as *Agromyza carbonaria* Zetterstedt but unfortunately the specimens are now lost and the species attacking this host is still not known. KANGAS (1935, 1937, 1949) studied these species in Finland and described species from *Betula*, *Populus tremula* L. and *Sorbus*. Additional species are known in Europe from *Crataegus*, *Malus*, *Prunus*, *Salix* and *Populus* (see below).

In North America GREENE (1914, 1917) discovered three species feeding in *Betula nigra* L., *Acer rubrum* L. and *Amelanchier canadensis* (L.) Medic. The typical feeding tracks of the larvae were also recorded in *Quercus rubra* L. and *Ilex monticola* A. Gray. GROSSENBACHER (1915) described *Agromyza pruni* feeding in the cambium of *Prunus avium* and *P. domestica* in New York but no specimens can now be traced and it has not been recorded since.

The largest species in the family belongs to this genus and was described from

Burma as *P. gigas* Spencer, 1966a. The wing length is 6.5 mm. Several large species are also known from Brazil (SPENCER, 1966e) and from Central America (SPENCER and STEGMAIER, 1973).

Among the leaf-miners, *Agromyza albitarsis* Meigen, 1830, forms blotch-mines on *Populus* in Europe and N. America and a further species on poplar, *A. populoides* Spencer, 1969a was recently described from Canada. In Japan SASAKAWA and FUKUHARA (1965a, 1965b) described *Agromyza morivora* forming linear blotch mines on cultivated *Morus* sp. (mulberry). *Agromyza trebinjensis* Strobl (cf. SPENCER, 1966c: 293) forms linear mines on *Celtis australis* L. in the Mediterranean area, and a closely related species, *A. varifrons* Coquillett (cf. SPENCER, 1969a: 59), feeds on *Celtis* spp. in North America, from Florida to Quebec. *Ulmus* has no Agromyzid leaf-miner in Europe but is attacked by *Agromyza aristata* Malloch in America and Canada (cf. SPENCER, 1969a). Both alder (*Agromyza alnivora* Spencer, 1966c) and birch (*Agromyza alnibetulae* Hendel, 1931) have their own species of leaf-miner in Europe and the distinctive leaf-mines are shown in Figs. 469, 470.

Four *Japanagromyza* species are known as leaf-miners on trees. *J. quercus* (Sasakawa), cf. SASAKAWA, 1961: 336, forms blotch-mines on *Quercus glauca* Thunb. in Japan; *J. viridula* Coquillett, 1902, attacks oaks in North America and is discussed below. In Australia the only known Agromyzid feeding on Eucalypts, *J. eucalypti* Spencer, 1963b, has been recorded as a leaf-miner on *Eucalyptus camaldulensis* Dehnhard. *J. salicifolii* (COLLIN, 1911), originally described from Egypt as a blotch-miner on willow, has now been found to occur from Alma-Ata in the Kazakh S.S.R. through Turkey and Israel to Spain and Portugal. It feeds most commonly on *Populus* spp. The aedeagus of *J. eucalypti* and of *J. salicifolii* is shown in Figs. 471, 472.

Species in the genus *Paraphytomyza* are small, usually occur singly and are of no real significance. However, *Paraphytomyza populi* (KALTENBACH, 1864) frequently comes to notice on poplar in Europe, where many mines can occur together in one leaf. ARRU (1966–1967) discusses this species, and also *P. populicola* (WALKER, 1853), in his review of insects attacking cultivated poplars in northern Italy. *P. heringi* (HENDEL, 1920) forms mines on *Fraxinus* in the late autumn in Europe.

The gall-causers were until recently included in the genus *Melanagromyza* but were segregated in the available genus *Hexomyza* Enderlein, 1936, by SPENCER, 1966b. *H. albicula* Spencer, 1969a, forms twig-galls on *Salix* sp. in Canada; similar galls on *Salix* are formed in Europe by *H. cecidogena* Hering, 1927b. *H. schineri* (GIRAUD, 1861) was described in Europe from *Populus alba* L. and also occurs on *P. tremula* L. This species is not uncommon in Canada on *P. tremuloides* Michx. and has also been recorded in California on *P. fremonti* Wats. (Cottonwood) by BROWN and EADS (1969). *H. simplicoides* (HENDEL, 1920) is also holarctic in distribution. It is known in the Palaearctic region from western Europe to the Kirghiz Republic of the U.S.S.R.; there are records in the United States from New York and Ohio (SPENCER, 1966b). COUDEN (1908) described *H. tiliae* in America from *Tilia americana* Marsh. which was 'showing evidence of a "peculiar disease" for which a remedy was desired' (Fig. 473). It is only known from Illinois and

Missouri (FRICK, 1959). A further species in this group, *H. websteri* (Mall.) forming galls on *Wistaria* in Japan and California is discussed in Chapter 11.

Five cambium-borers and three leaf-miners are discussed in greater detail below.

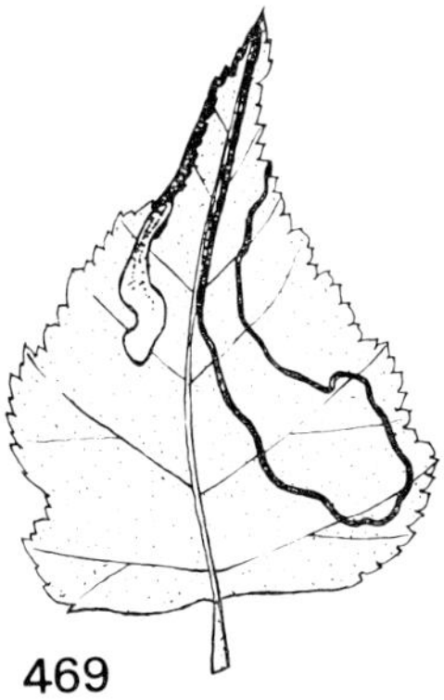

469

470

473

471

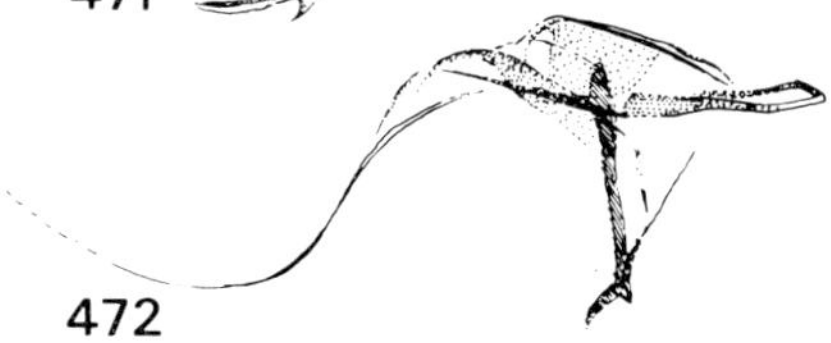

472

Fig. 469. *Agromyza alnibetulae* Hd.: leaf-mine on *Betula* (Hering).

Fig. 470. *Agromyza alnivora* Sp.: leaf-mine on *Alnus*.

Fig. 471. *Japanagromyza eucalypti* Sp.: aedeagus.

Fig. 472. *Japanagromyza salicifolii* (Collin): aedeagus.

Fig. 473. *Hexomyza tiliae* (Couden): galls on *Tilia americana* (Couden).

CAMBIUM-BORERS ON BETULACEAE, ROSACEAE AND SALICACEAE IN EUROPE AND N. AMERICA

Phytobia betulae (Kangas, 1935) – Scandinavia, Germany, ? England

ADULT. Very large species, with wing length up to 4 mm in female; frons black, lunule silvery; mesonotum mat-greyish (essentially as in ***P. cambii***, see below).

MALE GENITALIA. Aedeagus as in Fig. 474.
EARLY STAGES. Larval mandibles each with two teeth, the two on the left only about half as long as the two on the right; anterior spiracles each with a semicircle of about 10 minute pores; posterior spiracles each with 3 pores (Fig. 475). The adult larva measures from 1.5–2 cm. The puparium is pale, straw-coloured or whitish.
HOST-PLANTS. *Betula pubescens* Ehrh. and probably other *Betula* species.
BIOLOGY. Adults appear in the first two weeks of June. Eggs are laid in young shoots and the larva moves into the stem, normally first feeding downwards. KANGAS (1935) reports that a 2-week-old larva had tunnelled a distance of almost 2 metres. Feeding continues to the base of the stem and even into the root. The larva may then feed upwards and downwards in the stem, the length of channel necessary to complete its development depending largely on the quality of the wood being consumed. Larval feeding does not normally occur more than 6 m above the ground. On completion of feeding, a hole is cut through the bark and the larva falls to the ground to transform. The puparium remains just below the surface until the following spring (KANGAS, 1935).

VON TUBEUF (1897) studied the habits of larvae in birch north of Berlin and observed larval feeding between May and September. Most could be found near the base of the stem but some up to 10 m above ground. Trees growing on damp ground appear to be more favoured than those in dry, sandy localities.
DAMAGE CAUSED. Larval feeding channels can occur in large numbers in badly infested trees. KANGAS (1935: Fig. 13) illustrates a cross-section of a 6-year-old birch showing each annual growth ring almost entirely encircled with 'pith flecks'.
DISTRIBUTION. Europe: Finland, S. Sweden, Denmark, Germany; ?England (no adults but larval feeding seen in birch, Plate 9).
PARASITES. *Symphya ringens* Hal. Two-thirds of all puparia found in 1934–1935 in Finland proved to be parasitized and in other breeding trials more parasites were obtained than flies (KANGAS, 1935).

Discussion

TAXONOMY. Adults of *P. betulae* are not distinguishable on external characters from those of *P. cambii*. However, the male genitalia are obviously distinct, although of the same general form (Figs. 474, 481). The puparia of the two species are also very different – pale whitish in *betulae*, dark reddish-brown in *cambii*.

The male genitalia of *P. betulae* in Europe and of *P. betulivora* in N. America are substantially different and show that the two species are not closely related (Figs. 474, 477).
ECONOMIC IMPORTANCE. Birch is one of the main timbers used in furniture manufacture and the quality of the wood can be seriously impaired by the larval feeding of this species (KANGAS, 1937).

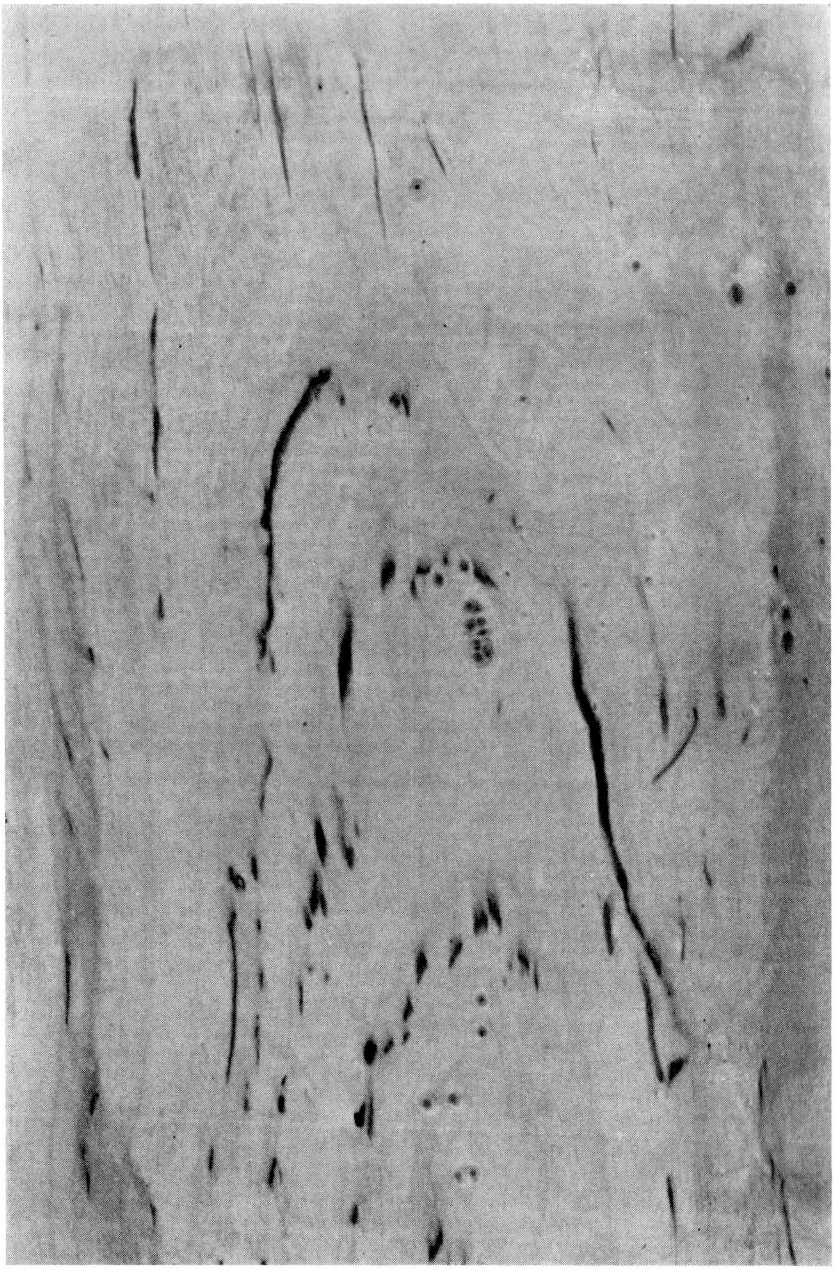

Plate 9: *Phytobia betulae:* larval feeding tunnels in timber of birch, near Norwich, England (C. F. YEATES, Norwich).

Phytobia betulivora Spencer, 1969a – N. America

ADULT. Large species, with reddish frons and mat black mesonotum.
HEAD (Fig. 476). Frons broad, twice width of eye, conspicuously projecting above eye in profile; jowls broad, $\frac{1}{4}$ vertical height of eye, mouth-margin with conspicuous epistoma.

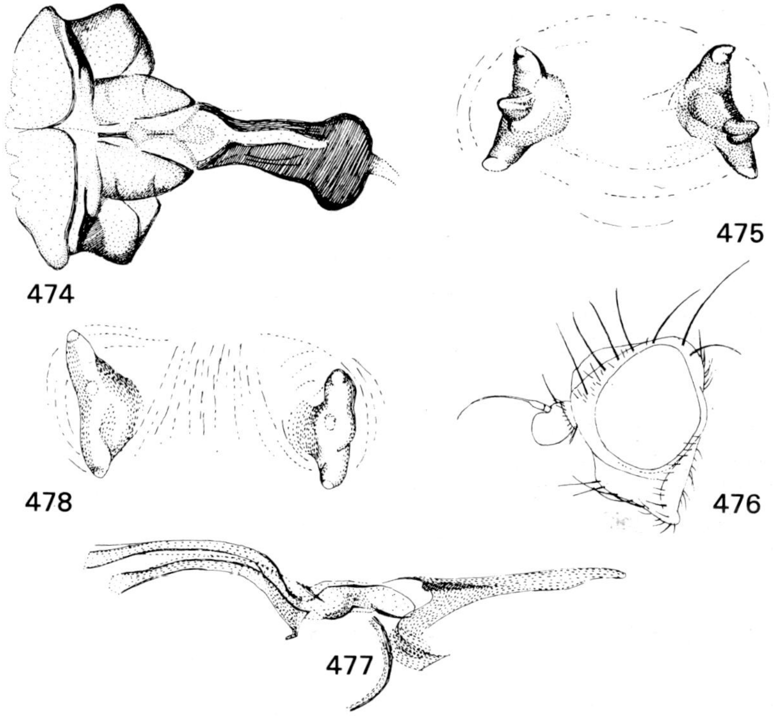

Figs. 474–475. *Phytobia betulae* (Kangas): 474, aedeagus; 475, posterior spiracles of puparium.

Figs. 476–478. *Phytobia betulivora* Sp.: 476, head; 477, aedeagus; 478, posterior spiracles of puparium.

MESONOTUM. 3 + 1 strong dc, acr in some 6 rows.
WING. Length in male 4.3 mm.
COLOUR. Frons, including orbits, also lunule, cheeks, jowls, all antennal segments and palpi reddish-orange; mesonotum blackish-grey; wing base brownish-orange, squamae pale grey, margin and fringe black.
MALE GENITALIA. Aedeagus as in Fig. 477, distal tubules long, essentially parallel.

EARLY STAGES. Larva white, long, up to 3 cm in length; puparium dark-brown, 4–5 mm long, posterior spiracles not greatly raised, each with 3 pores (Fig. 478).
HOST-PLANT. *Betula nigra* L. (river birch).
BIOLOGY. The exact point of oviposition has not been established but the larva tunnels both downwards and upwards in the stem, finally leaving the root or stem near ground level to pupate (GREENE, 1914).
DISTRIBUTION. United States: District of Columbia; Canada: Ontario.
PARASITES. *Symphya agromyzae* Rohwehr.

Discussion

TAXONOMY. The specimens reared from birch were identified by GREENE (1914) as *Agromyza pruinosa* Coquillett. However, it has now been established that this was not the species concerned and, as GREENE's original series clearly represented a new species, this was described as *betulivora* by SPENCER (1969a).

P. betulivora is closely related to *P. amelanchieris* (GREENE, 1917) on *Amelanchier* and to *P. setosa* (LOEW, 1869) on *Acer*. The male genitalia clearly indicate a less close relationship with *P. betulae* Kangas on *Betula* in Europe (cf. Figs. 474, 477).
ECONOMIC IMPORTANCE. This species can be very common on birch and much timber must be damaged by the boring of the larvae. GREENE (1914) reports that in August, 1912, every tree that was examined at Chain Bridge, DC 'contained new work'. The wood of *Betula nigra* is hard, strong and light and finds many industrial applications. *P. betulivora* must therefore be considered as a potential pest of some significance.

Phytobia cambii (Hendel, 1931) – W. Europe

Dizygomyza barnesi Hendel, in BARNES, 1933.
ADULT. Very large species, with black frons and contrasting silvery lunule.
HEAD (Fig. 479). Frons broad, twice width of eye, strongly projecting above eye in profile; orbital bristles variable but normally 2 reclinate ors and 4 more incurved ori; orbital setulae somewhat variable, upright or reclinate; lunule large, semi-circular; jowls about $\frac{1}{4}$ height of eye, cheeks forming broad ring below eye; third antennal segment rounded, arista long, only finely pubescent.
MESONOTUM. 3 + 1 strong dc, third and fourth equal; one pair of pre-scutellars normally present but these variable, sometimes only 1 bristle developed or 2 on same side; acrostichals numerous, in some 8 rows.
WING (Fig. 480). Length from 3.7 mm in male to 4.1 mm in female; costa extending strongly to vein M_{1+2}, last and penultimate sections of M_{3+4} approximately equal.
LEGS. Mid-tibia normally with one strong lateral bristle, this occasionally lacking.

COLOUR. Frons mat black, somewhat silvery dusted, lunule silvery, tending to brownish; ocellar triangle and orbits only weakly shining; all antennal segments black; mesonotum mat greyish, all pleura black; squamae grey, margin and fringe black; halteres yellowish-orange.
MALE GENITALIA. Aedeagus as in Fig. 481.
EARLY STAGES. The egg is cylindrical, opaque white, about 0.85 mm long and 0.2 mm wide and the chorion is irregularly reticulated. The larva (Fig. 482) is

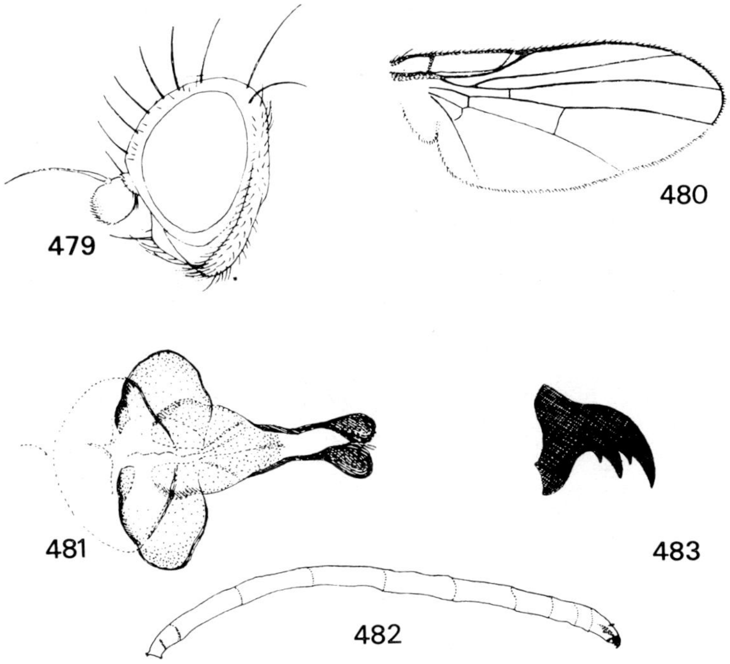

Figs. 479–483. *Phytobia cambii* (Hd.): 479, head; 480, wing; 481, aedeagus; 482, larva; 483, larval mouth-hooks (482 after BARNES, 483 after SCHIMITSCHEK).

unusually long and slender, up to 18 mm in length when boring in its mine channel but on removal contracts and becomes more normally cylindrical.

The mouth-hooks each have one conspicuously strong tooth and below two substantially shorter ones (Fig. 483). The anterior spiracles are on minute projections and normally have 10 pores, although this number can vary between 8 and 12. The posterior spiracles consistently have three well-defined pores. The puparium is dark reddish-brown, 4–5 mm long (from BARNES, 1933; SCHIMITSCHEK, 1935).
HOST-PLANTS. *Salix alba* L. (=*caerulea* L.) (Cricket Bat Willow), *S. purpurea* L., *S. triandra* L., *S. viminalis* L. (Basket Willow). Larval feeding has also been noted in *S. fragilis* L.

BIOLOGY. There is only a single generation and the adults emerge from the end of May until the end of June. BARNES, working with basket willows, observed that oviposition takes place in young shoots in the first year of growth about 30 cm above the ground. The egg stage lasts from 7–14 days. The larva initially feeds downwards, remaining throughout its life in the cambium tissue. Feeding continues throughout the summer and the larvae pupate between the end of July and

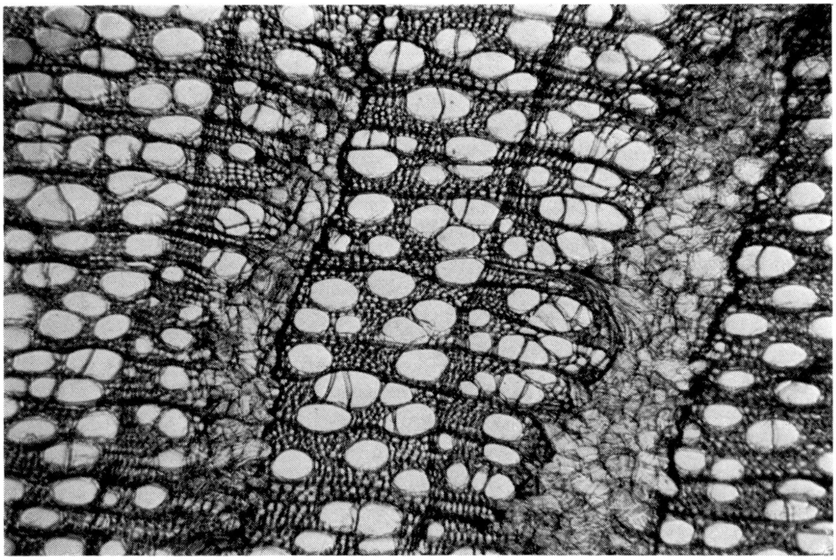

Plate 10: *Phytobia cambii:* micro-photograph of *Salix alba*, showing occluded mine channels (C. F. YEATES, Norwich).

mid-September. An exit slit is cut through the epidermis of the rod and the larva falls to the ground where actual pupation takes place.

A section of an infested bat willow from Norfolk which I have examined shows that larval feeding has taken place in all 18 annual growth rings but the infestation has diminished appreciably after ten years' growth, with the heaviest infestation occurring during the first seven years. A micro-photograph showing occluded mine channels in *Salix alba* is shown in Plate 10.

Salix alba, the bat willow, has a diameter of some 30 cm after 15 years' growth and it is not clear how the females lay their eggs in trees of this size. It is possible that eggs are laid directly in the soft, last year's growth through cracks in the bark or alternatively they may be laid in young, low, lateral branches, from which the larvae could without difficulty tunnel into the main trunk.

It has not been established whether the larva remains in the year's growth of wood into which the egg was laid or penetrates deeper into earlier years' growth, as has been reported for *Phytobia cerasiferae* (p. 311).

DAMAGE CAUSED. In basket willows the larvae tunnel up and down the basal portion of the rods. When the rods are grown for sets the most valuable portion is useless. For basket work the thick end of such rods has to be discarded. The larval tunnels become occluded with fresh tissue after a year or so but nevertheless such rods are weakened throughout and this is particularly important when the rods are used as stakes (BARNES, 1933).

In the older wood which is used for making cricket bats all except the latest channels have been filled with callus tissue but the channels remain readily visible on the surface of the bats. Apart from the obvious reduction in quality, there must be also some weakening of the wood.

DISTRIBUTION. England, widespread including also Scotland; Austria, Netherlands, Finland.

PARASITES. *Symphya hians* Nees, *S. ringens* Haliday (Braconidae). BARNES suggests that the rate of parasitism is high but inadequate material has been studied for any accurate assessment to be made.

Discussion

TAXONOMY. HENDEL described this species twice, both as *cambii*, 1931, and later as *barnesi* (in BARNES, 1933). He considered the specimens sent to him by BARNES from England to be distinct from *cambii* described from the Netherlands, in view of the lack of the mid-tibial bristle in the single specimen he saw from Holland. However, I have confirmed that this bristle is in fact present in other specimens of the type series, the male genitalia are identical in the two species and *barnesi* was therefore synonymized with *cambii* (SPENCER, 1971a).

ECONOMIC IMPORTANCE. BARNES (1933) indicated significant commercial damage to basket willows in England and SCHIMITSCHEK (1935) has discussed similar damage in Austria. In recent years the 'pith flecks' giving evidence of larval feeding have become increasingly numerous in the wood of the bat willow. Serious economic loss is now being experienced by growers in Norfolk, one of the main areas of willow cultivation in England. For example, a recent consignment of 2000 cricket bat blanks was rejected by the manufacturers owing to 'pith flecks' and could only be sold by the timber merchants at a loss.

Cricket bats are manufactured in Australia from blanks supplied from England and I have seen such blanks which have been exported to Australia and subsequently returned as rejects for the same reason.

The situation in the cricket bat industry is thus serious. However, British salesmanship is rising to the occasion and a virtue is being made of necessity by utilizing a white nylon covering for low quality bats, which would certainly be otherwise rejected. These are now being sold as 'Supertuf' bats, with all blemishes conveniently obscured, and the implication is made that they are superior to bats made from healthy wood.

PEACE (1952) discusses a species damaging poplar in England (which he also misidentified as *Agromyza carbonaria* Zett.). This has not yet been reared but it seems probable that the species represents *P. cambii* (Hendel). He writes: 'The

Poplar Cambium Borer causes damage of great economic importance. This insect is widely distributed, and numerous in many localities. The eggs are deposited singly in punctures made in the bark during late spring, and the larvae bore long irregular burrows in the cambium of the stem throughout the summer months.

'The excavated larval tunnels become infested with bacteria, and the frass and injured tissues become discoloured and turn brown. As the growth of the stem proceeds, the larval tunnels are covered with a layer of new wood and ultimately become imbedded in the timber, a new series of larval tunnels being formed each year while the infestation lasts.

'If the log is sawn or split lengthwise, the larval tunnels appear on the radial surface as narrow streaks, varying in length according to the angle of incidence. When logs are peeled as veneers, the larval tunnels appear on the surface as brown marks about one-twelfth of an inch in width, of irregular pattern and varying length. These frass-filled tunnels with their damaged cell tissues constitute lines of weakness in the wood structure and render the material unsuitable for the manufacture of matches or plywood.'

JOBLING and YOUNG (1964) discuss bacterial canker in *Populus* spp. in England associated with the tunnels of Agromyzid larvae (probably of *P. cambii*). They state that 'Agromyzid damage previously considered to be of relatively minor importance, could lead to major economic loss through stem canker'. Similar damage has been observed in France.

Phytobia carbonaria (Zetterstedt, 1848) – Europe

ADULT. Very large species, with wing length about 4 mm, last and penultimate sections of vein M_{3+4} equal, mesonotum black rather than grey.

HEAD. Frons broad, twice width of eye, conspicuously projecting above eye in profile, normally 2 largely reclinate ors and 4 more incurved ori, orbital setulae reclinate; orbits broad, lunule semicircular; jowls deepest at rear, $\frac{1}{4}$ height of eye, cheeks forming broad ring below eye; third antennal segment large, rounded, arista only finely pubescent; mouth-margin normal, without epistoma.

MESONOTUM. 3 + 1 strong dc, acrostichals irregularly in 8 rows.

WING. Length 4–4.2 mm, costa extending to vein M_{1+2}, last and penultimate sections of M_{3+4} approximately equal, first cross-vein slightly beyond midpoint of discal cell.

COLOUR. Frons mat black, normally slightly brownish in front, orbits more shining blackish; third antennal segment blackish-brown; mesonotum black, with slight sub-shine, abdomen more shining black; squamae pale grey, margin and fringe black; wing base reddish-brown.

MALE GENITALIA. Aedeagus distinctive, as in Fig. 484.

EARLY STAGES. Larva with posterior spiracles in form of two short, closely-adjoining projections, each with 3 minute pores and below, a single pointed protuberance above the anus (Fig. 485); mouth-parts with 1 strong and 1 smaller tooth, lower arm of cephalo-pharyngeal skeleton conspicuously short (Fig. 486); puparium reddish-brown, up to 5.5 mm long.

HOST-PLANTS. *Malus* species (apple); also *Crataegus* species (Hawthorn); possibly *Cydonia* (quince).
BIOLOGY. Oviposition normally takes place in young twigs, at the end of May or early in June, in the previous year's wood. The young larva initially feeds upwards into increasingly tender tissues and later turns to feed downwards, sometimes penetrating up to 20 cm below ground into the root. On completion of feeding in August or early September, the larva bores through the bark and falls to the ground, where pupation takes place just beneath the surface. There is only a single generation (based on discussion with Mr. NIELSEN, Denmark).
DAMAGE CAUSED. Larval feeding has some effect on the growth of young apple shoots and budding can be delayed by up to three weeks. There can also be some

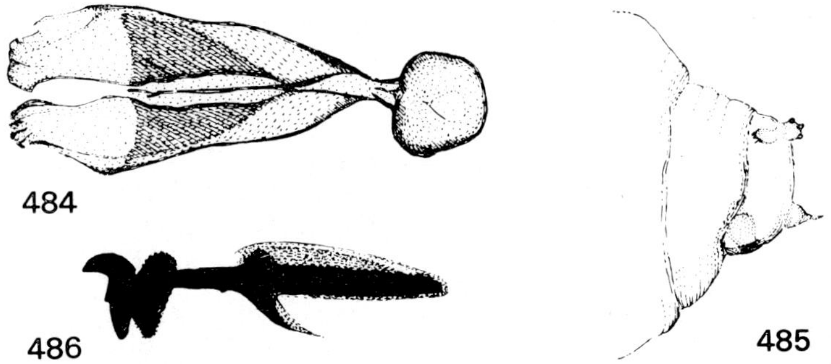

Figs. 484–486. *Phytobia carbonaria* (Zett.): 484, aedeagus; 485, posterior end of larva; 486, cephalo-pharyngeal skeleton.

wilting of affected shoots. The exit hole of the larva increases in size with the growth of the twigs and facilitates fungal attack.
DISTRIBUTION. Norway, Denmark; Britain, including Scotland; Austria. Probably widespread, at least in northern Europe.
PARASITES. None have yet been obtained.

Discussion

TAXONOMY. The name *carbonaria* has been widely used in the literature and has been applied rather indiscriminately to a number of species in this group. NIELSEN (1906) identified as *carbonaria* specimens reared from puparia obtained from the soil below *Alnus* trees. However, in view of the positive identification of *carbonaria* from *Malus*, it seems virtually certain that another species feeds in *Alnus*.
HOST-RANGE. Species in this genus are highly host-specific and it is noteworthy that in Europe even within the family Rosaceae there are two distinct species – *carbonaria* probably limited to the section Pomoideae and *cerasiferae* probably

limited to the Prunoideae. LEE (1953) recorded pith-flecks in apple, and also *Cydonia vulgaris* Pers. (quince) in S.E. England. *Cydonia* is of course closely related to *Malus* and it seems virtually certain that the species concerned was *carbonaria*.

ECONOMIC IMPORTANCE. The natural hosts of this species in Europe are probably *Crataegus* and *Malus* but cultivated apple has been readily colonized. An apple orchard of some 16 hectares in Denmark at Hornsharred on the west coast of the main island Zealand is now so heavily infested that almost every branch of every tree has been attacked. The Manager, Mr. NIELSEN, whom I visited on 27 September, 1971, is considerably worried by this infestation which has significantly increased in recent years. So far there has been no detectable reduction in the crop yield but there is a real danger of a serious secondary fungal infection affecting the tunnelled branches.

The situation at this particular orchard is exacerbated by the protective hedges surrounding the entire area consisting of *Crataegus* which is as heavily infested as the apple trees themselves. The increase in population of *P. carbonaria* which has been noted in recent years is probably linked with a reduction in the numbers of parasites resulting from frequent spraying with 'Midol', a preparation based on the persistent organo-chlorine compound, Lindane (BHC). Up to 14 applications per year are now normal, directed primarily against saw-flies and lepidoptera.

P. carbonaria is known to be present in other apple orchards in Denmark but apart from LEE's (1953) record of feeding tracks in apple in S.E. England there are no other reports of damage to this crop. The situation in Denmark, however, may indicate a general build-up of the population levels of this species and its significance as a pest clearly deserves further study.

Phytobia cerasiferae (Kangas, 1955) – England, Corsica

ADULT. Large black species, with last section of vein M_{3+4} $1\frac{1}{2}$ times length of penultimate.

HEAD. Frons broad, from 2–$2\frac{1}{2}$ times width of eye, distinctly projecting above eye in profile; 2 strong ors, 3 ori, the lowest entirely incurved; orbital setulae sparse, reclinate, jowls about $\frac{1}{5}$ height of eye; third antennal segment round, large, arista only finely pubescent.

MESONOTUM. 3 + 1 strong dc, acrostichals irregularly in 8 rows.

WING. Length from 2.9 mm in male to 3.5 mm in female; costa extending to vein M_{1+2}, last section of M_{3+4} just less than $1\frac{1}{2}$ times penultimate, first cross-vein at anterior third of discal cell (Fig. 487).

COLOUR. Frons mat black, orbits and ocellar triangle weakly shining; lunule silvery-brownish; all antennal segments black; mesonotum greyish-black, all pleura and legs black; squamae grey, fringe black.

MALE GENITALIA. Aedeagus as in Fig. 488.

EARLY STAGES. Eggs ranging in size from 0.66 × 0.20 mm to 0.83 × 0.29 mm; larva up to 18 mm long, mouth-hooks with one very strong tooth, arrangement on other side not fully understood; posterior spiracles each with 3 bulbs. The size of the puparium is 4 × 1.5 mm (KANGAS, 1955; PITCHER, 1956).

HOST-PLANTS. *Prunus cerasifera* Ehrh., *P. domestica* L. (plum), *P. insititia* L. (damson), *Prunus* sp. (cherry).

BIOLOGY. Oviposition takes place from the end of June into early July. 'Caged flies, offered cut shoots of one-year-old and current year's wood, laid all their eggs in the former. Eggs were inserted, usually through a lentical, into the periderm, and lay directly on the phellogen. Usually only one egg was laid through each puncture but groups of up to four did occur. . . . Larval feeding takes place from July to May in the cambial layer of the host. . . . Movement is in general downward, although recurrent upwards tracks are common. After hatching, the larva

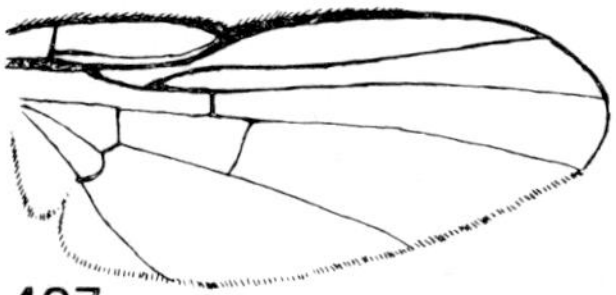

487

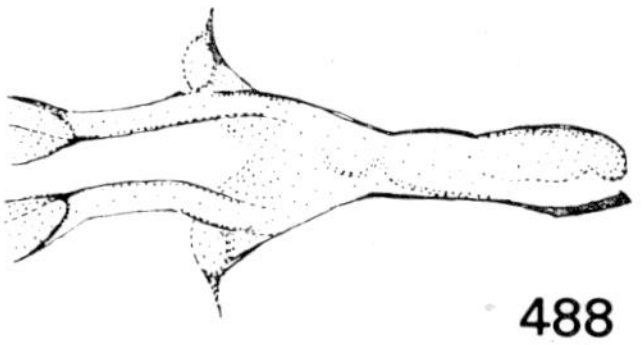

488

Figs. 487–488. *Phytobia cerasiferae* (Kangas): 487, wing; 488, aedeagus.

loses little time in leaving the one-year wood, and most of the feeding occurs in two- to four-year wood. Mining in five-year wood is infrequent.' The larvae feed slowly during the summer and autumn, apparently hibernate in the tree throughout the winter and the following spring feed actively, 'nearly trebling their length in six weeks. From mid-May onwards each larva cuts a narrow vertical slit in the bark through which it leaves the tree, usually at night, to pupate in the soil'. The pupal period lasts about 30 days (PITCHER, 1956).

DAMAGE CAUSED. The feeding tracks in the cambium, frequently referred to in the literature as 'pith flecks', apparently have little effect on tree growth. 'During the budding season, when cambial disturbance may affect the percentage take, the larvae are either not yet present or are so small that they are of little consequence. The proportion of fleck to normal wood is negligible in plum, and it is unlikely that the presence of flecks would lead to branch breakage' (LEE, 1953).

DISTRIBUTION. England: Kent; Corsica: Calvi (E. Lindner).

PARASITES. *Symphya mandibularis* Nees (Braconidae). PITCHER reports that up to 80% of puparia are parasitized and suggests that the thinly covered eggs or very young larvae are the stage attacked.

Discussion

TAXONOMY. This is one of the smaller species in this genus, which comprises the largest known species in the family. It is similar to *carbonaria* (Zett.) in having the mesonotum black rather than grey, but is immediately distinguishable by the longer last section of vein M_{3+4} and also by the smaller size.

DISTRIBUTION. The only record apart from the type series from southern England is of a single caught specimen from Corsica (SPENCER, 1972b).

ECONOMIC IMPORTANCE. This is apparently negligible at existing population levels but if populations increase, as in *Phytobia cambii* on *Salix* and *P. carbonaria* on *Malus*, significant damage might be expected. PITCHER (1956) refers to the possibility that larval feeding might influence the development of cankers caused by the bacterium *Pseudomonas mors-prunorum* Wormald.

Japanagromyza viridula (Coquillett, 1902) – N. America

ADULT. Medium-sized species, with black mesonotum and green abdomen.

HEAD. Frons not projecting above eye, the 2 ors equal, lower ori incurved; jowls

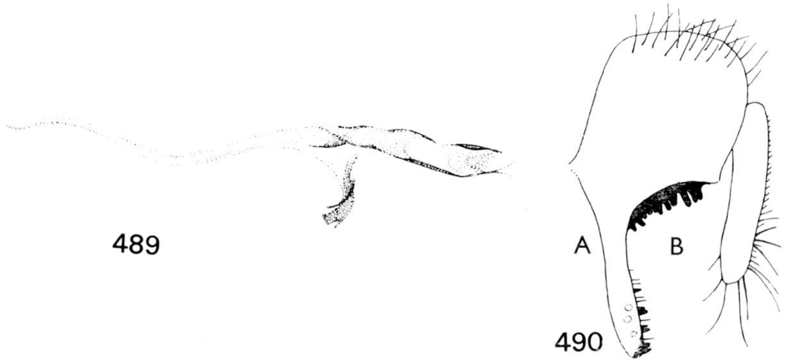

Figs. 489–490. *Japanagromyza viridula* (Coq.): 489, aedeagus; 490, epandrium with A, surstylus and B, cercus.

narrow, about $\frac{1}{12}$ height of eye; third antennal segment large, round, arista virtually bare.

MESONOTUM. 2 strong dc, pre-scutellars well developed.

LEGS. Fore-tibia with 1 strong lateral bristle, mid-tibia with 2.

WING. Length from 2.8–3 mm, last section of vein M_{3+4} short, little more than half length of penultimate.

COLOUR. Frons mat black, lunule conspicuously silvery; mesonotum moderately shining blackish, abdomen conspicuously green; squamae and fringe silvery, halteres white.

MALE GENITALIA. Aedeagus as in Fig. 489, with long, straight distiphallus and well-developed hypophallus; ninth sternite with elongated hypandrial apodeme; surstyli projecting downwards (Fig. 490A), long, straight, with about 10 strong bristles towards end and another group at upper corner; cerci (Fig. 490B) without bristles.

HOST-PLANTS. *Quercus alba* L., *Q. borealis* Michx. var. *maxima* Sarg., *Q. rubra* L. and probably other *Quercus* species.
BIOLOGY. The only detailed study known is that by LABONTE and LIPOVSKY (1967) who studied the species in Maine, U.S.A. By mid-June, 1965, feeding punctures made by the females on young leaves were numerous but the number of eggs laid was small. These punctures are made through the upper epidermis of the leaf and the female 'probes and macerates the inner tissues to one side of the puncture, forming a roughly oval to circular area . . . the flies attack only very tender juvenile leaves. This phenomenon occurs from the time the oak buds first open until the leaves are almost 2 inches long. Juvenile leaves are available through June and into July on oak sprouts'. In 1966 the first females were observed between 4–7 June and the first larval mines were seen during the second week of June. The larva forms a blotch-mine, pupating on the ground. There is only a single generation, with the puparia overwintering and producing adults the following spring.
DAMAGE CAUSED. The main damage to ornamental oak foliage is caused not by the larval leaf-mines but by so-called 'shot-hole' injury. In some seasons many leaves are found in southern Maine with 'neat, prominent holes'. Investigations by the two authors cited above traced this injury to the feeding punctures made in immature leaves by the females, which is a regular habit throughout the family. They describe the damage as follows: 'The injured leaf area soon becomes brownish and dries in the form of a small disc. As the young leaf continues to grow each necrotic disc gradually separates from the living tissues to leave a break completely through the leaf. This break, or hole, continues to expand in direct proportion to the enlargement of the leaf surface. The resulting hole in the mature leaf is about 6 times the size of the original injury made by the fly. Measurements of the holes in mature leaves averaged 4.4 mm in diameter but ranged from 2.3–10.1 mm. The final hole is characteristically neat appearing with definite margins bounded by a cicatrix. Occasionally juvenile leaves have so many punctures that holes coalesce to form irregular breaks and tattered edges.

The necrotic disc on the edge of the hole in the leaf is characteristic, but unfortunately it remains attached only for about 2 weeks. Thereafter the disc sloughs off leaving no further evidence to verify the cause of the injury. While the disc remains in place, shot-holes initiated by agromyzid flies can be identified by the typical ovipositor puncture, visible with a hand lens, within the disc. This typical round puncture, not visible to the naked eye, appears as a minute hole through the upper epidermis of the leaf. Thus, identification of this injury should be made by examining oak leaves when they are very young.'
DISTRIBUTION. Widespread in eastern U.S. from Gulf of Mexico northwards into Canada (Ontario, Nova Scotia).
PARASITES. None recorded.

Discussion

TAXONOMY. This species is larger than others in the genus known in the United States and the male genitalia are entirely distinctive. It is the only Agromyzid feeding on oaks in America.

ECONOMIC IMPORTANCE. The expanding hole phenomenon following initial damage by the female feeding punctures is quite unusual in the Agromyzidae. In some, although apparently not in all, seasons this effect leads to gross defects in the foliage of ornamental oaks. LABONTE and LIPOVSKY point out that the shot-hole condition is less serious to oaks than complete defoliation, browning of trees caused by skeletonizing or other leaf diseases. However, 'severe attacks, sufficient to cause tattering, browning of leaf areas and inability of leaves to develop to normal size, result in poor overall appearance of oaks and cause real consternation to owners of ornamentals'.

Phytomyza heringiana Hendel, 1922 – Europe

ADULT. Small black species, wing length up to 1.8 mm; second costal section short, less than $2\frac{1}{2}$ times length of fourth. Frons dark brownish-black, somewhat paler above. Mesonotum mat, greyish-black, acr in 4–5 rows. Legs black, only fore-knees distinctly yellowish.
MALE GENITALIA. Aedeagus as in Fig. 491.
EARLY STAGES. Larva with anterior spiracles distinctly divided (Fig. 492A), posterior spiracles elongated, each with some 17 pores (Fig. 492B); puparium pale brown.
HOST-PLANTS. *Malus sylvestris* L. and probably other *Malus* species.
BIOLOGY. The only detailed investigations have been made by VIGGIANI (1962) in Italy. Adults emerged at the end of March or early in April. Oviposition takes place on the lower leaf surface, where the young larva forms a short mine before moving to the upper surface where the main irregularly linear mine is formed (Fig. 493). The colour of the mine can vary from greenish to reddish. There are two complete and a third partial generation in a year.
DISTRIBUTION. Europe, known in England, Germany, Italy, apparently widespread.
PARASITES. In Italy *Achrysocharella formosa* Westw.; *Chryocharis pubens* Delucchi, *C. pubicornis* Zett.; *Closterocerus trifasciatus* Westw.; *Opius* sp. (VIGGIANI, 1962).

Discussion

TAXONOMY. The similarity in external characters and also in the male genitalia between this species and *P. persicae* Frick from North America feeding on *Prunus persica* clearly indicates relatively recent speciation.
ECONOMIC IMPORTANCE. The populations of *P. heringiana* appear to remain small in all localities and no damage has as yet been recorded on cultivated apple.

Phytomyza persicae Frick, 1954 – N. America

ADULT. Small, black species, wing length in male 1.9 mm; second costal section short, twice length of fourth. Frons black in front, paler brown above; 2 equal ors,

1 ori; third antennal segment small, round. Mesonotum mat, greyish-black, with acr in 4 rows. Legs black, fore-knees slightly yellowish, tibiae and tarsi paler, brownish-yellow; squamal fringe dark.

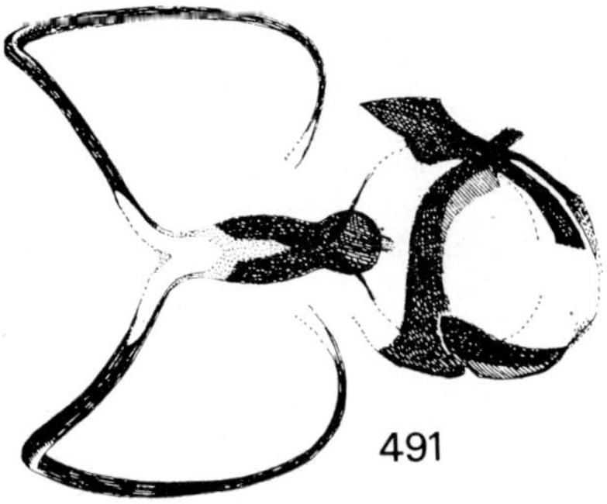
491

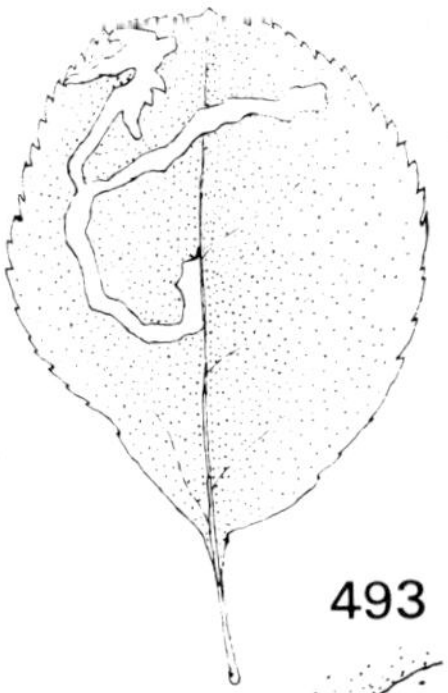
493

A 492 B

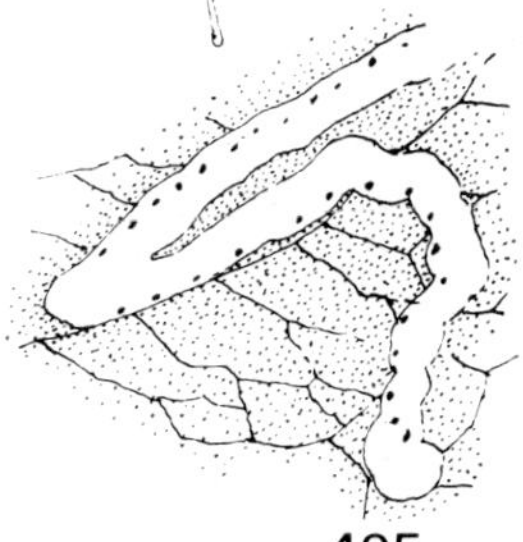
495

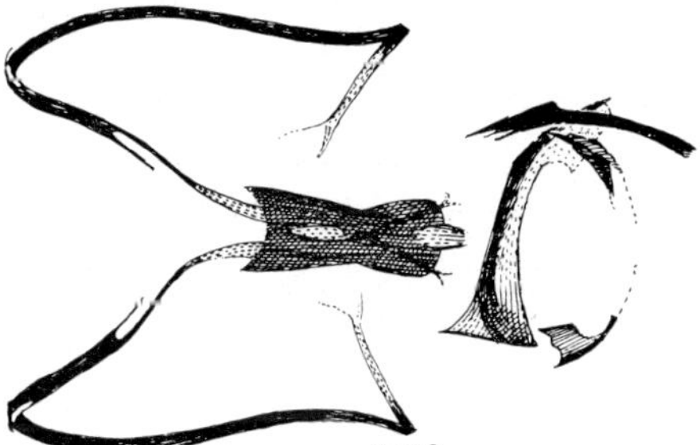
494

Figs. 491–493. *Phytomyza heringiana* Hd.: 491, aedeagus; 492A, anterior spiracles of larva; 492B, posterior spiracles; 493, leaf-mine on *Malus* (Figs. 492, 493 after VIGGIANI).

Figs. 494–495. *Phytomyza persicae* Frick: 494, aedeagus; 495, leaf-mine on *Prunus persica* (after FROST).

MALE GENITALIA. Aedeagus as in Fig. 494.
HOST-PLANT. *Prunus persica* L. (peach).
BIOLOGY. FROST (1924) discusses this species as *Phytomyza obscurella* var. *nigritella* Zett. The larva forms an upper surface linear mine, which is whitish or

greenish, with frass in distinct black grains (Fig. 495). The puparium remains in the leaf at the end of the mine. There are at least two generations.
DISTRIBUTION. U.S.A.: Connecticut, Ohio, Virginia (FRICK, 1959); Canada: Ontario (SPENCER, 1969a).
PARASITES. No species identified but FROST writes: 'Many parasites issue during the summer'.

Discussion

This species apparently only occurs in small numbers and is effectively controlled by its parasites.

XI. INTERNAL FEEDERS AND LEAF-DINERS ON ORNAMENTALS

This review of cultivated ornamentals attacked by Agromyzidae cannot deal with more than a selection of the most important hosts and species, in view of the almost infinite number of plants which can be found in gardens. Twenty plant families and 47 Agromyzid species are discussed, mainly from Europe. Information on infestation of ornamentals outside Europe is inevitably fragmentary but some references are given to species in North and South America, South Africa, India, Australia and New Zealand.

Polyphagous species such as *Tropicomyia polyphyta* Kleinschmidt in Australia, *Liriomyza bryoniae* (Kalt.) and *L. strigata* (Mg.) in Europe, *L. huidobrensis* (Blanchard) in North and South America, *Phytomyza horticola* Goureau in Europe, Africa and Asia, and *P. syngenesiae* (Hardy) in Europe, North America, Australia and New Zealand will certainly be found on many hosts additional to those mentioned. It is nevertheless hoped that identification will be possible from the information given on these species regarding their host range, normal distribution and the type of mine formed.

Botanical Gardens, with their concentration of a wide range of plants, frequently introduced from other Continents, always harbour a surprising number of Agromyzid species, and are a valuable source of information both on new hosts and conversely on the non-acceptability of apparently suitable hosts introduced from distant areas. Buhr (1932, 1941, 1954) has provided useful lists of hosts and their Agromyzid species from Mecklenburg, Germany, based to a large degree on a study of plants in the Botanical Gardens at Rostock; some records are also included from the Botanical Gardens at Bremen. Similar lists are given for Thuringia (1960) and Saxony (1964). Prof. Hering, who lived only a short distance from the Berlin Botanical Gardens at Dahlem, collected there regularly for many years and discovered many interesting species and new host records. A brief account of 30 species found in the Royal Botanic Gardens at Kew was given by Spencer (1961a). If similar intensive collecting in Botanical Gardens can be undertaken in other parts of the world, this will certainly greatly add to our knowledge of the host range of many species.

Ornamentals to suffer most from attack by Agromyzidae are without question chrysanthemums, both in Europe and in North America. Glasshouse cultivation provides an ideal environment, and the population build-up which so frequently occurs can prove highly destructive and lead to substantial economic loss. In Europe the leaf-mining species *Phytomyza syngenesiae* (Hardy) is a most serious and damaging pest, and in North America also proved a serious problem on cultivated chrysanthemums following its introduction to the New York area around 1880. However, at the present time the main pest of chrysanthemums in the United States is *Liriomyza trifolii* (Burgess).

Cinerarias have suffered heavy attacks by *P. syngenesiae* in England, and also in Australia and New Zealand. *Calendula* has been badly damaged by the stem-borer *Napomyza lateralis* (Fallén) in England and in Italy. Damage can occur both to the stems and also directly to the flower-heads. In Italy and Belgium *Dianthus caryophyllus* L. can be heavily attacked by the leaf-miner *Paraphytomyza dianthicola* (Venturi) and carnation growers regularly suffer appreciable commercial loss. Lilies, such as *Lilium martagon* L. and *L. candidum* L., are widely cultivated in central Europe and the blooms are periodically badly damaged by the seed-feeding species, *Liriomyza urophorina* Mik. Finally, the holly leaf-miners deserve mention, as the unsightly damage to the foliage is not only a source of concern in individual gardens but also leads to commercial loss in nurseries where the young plants are cultivated. In Europe, and also in British Columbia where it has been introduced, *Phytomyza ilicis* Curtis damages *Ilex aquifolium* L. In the United States, among several species now known in this complex, both *P. ilicicola* Loew feeding primarily on *I. opaca* Ait. and *P. vomitoriae* Kulp which is host-specific on *I. vomitoria* Ait. can be considered as serious pests.

The remaining species which are discussed below are less obviously injurious but will certainly come to the attention of the keen gardener, and on occasions may cause some damage to their host-plants. Of the 47 species referred to, eight are stem-feeders, one is a seed-feeder and the others are leaf-miners.

AQUIFOLIACEAE

Ilex species (Europe, North America)

Both the European holly, *Ilex aquifolium* L., and the American holly, *Ilex opaca* Ait. are invariably heavily attacked by leaf-miners which produce large, unsightly blotches. These can cover much of the foliage in bad infestations and cause commercial loss to growers.

Only a single holly miner is known in Europe, described by CURTIS in England in 1846 as *Phytomyza ilicis*. The larva of this species feeds in a characteristic way. Early in June the egg is laid near the base of the midrib on the underside of the leaf and the resulting oviposition puncture is readily visible. For two months or so the young larva feeds slowly towards the apex of the leaf inside the midrib and only at the end of the summer moves into the leaf-blade to start the formation of an irregular blotch-mine. The winter is passed in the larval stage within the mine and feeding can be resumed during spells of warm weather. Pupation takes place, also within the mine, by early April and the adult flies emerge at the end of May or early in June. The life history has been discussed by MIALL and TAYLOR (1907) and CAMERON (1939) in England and by DOWNES (1931) in British Columbia.

Ilex aquifolium has been introduced into British Columbia in western Canada and is also present in the N.W. American States of Washington and Oregon. *Phytomyza ilicis* is well established in these areas and was presumably introduced together with its host. It has proved very damaging to this holly grown in British

Columbia, particularly on Vancouver Island and the lower mainland, where it is unusual to find a holly bush which is not infested. Up to 80% of leaves can be affected (Downes, 1931).

A second species of miner, *Phytomyza ilicicola* Loew, 1863, is common on its main host, *Ilex opaca*, in the eastern United States. Somewhat similar blotches are formed to those of *P. ilicis* but the biology of this species differs in the egg being laid not in the midrib but directly in the leaf-blade, where the first instar larva forms a short linear mine before changing its feeding pattern to form an irregular blotch (Fig. 496). Additional hosts recorded in the Washington, DC area are *Ilex aquifolium*, *I. cumulicola* Small and the hybrid *I. opaca* X *I. myrtifolia* Walt. (Kulp, 1968).

A number of earlier workers in the United States had noticed that on *Ilex opaca* and other hollies entirely different linear mines were formed and it was suspected that additional species might be present. Kulp (1968) investigated this problem and was able to segregate and describe five further closely related species. Morphological and even genitalia differences between these species are in most cases slight but differences in the leaf-mines can be striking.

Phytomyza opacae Kulp forms a long, narrow linear mine (Fig. 499) on *Ilex opaca* and less frequently on *I. aquifolium* and *I. cumulicola*. *Phytomyza vomitoriae* Kulp forms blackish linear-blotch mines in the minute leaves of *Ilex vomitoria* Ait. In Florida and South Virginia infestations are frequently extremely heavy where this holly is cultivated in nurseries. Two further species recorded by Kulp are *P. ditmani* on *Ilex decidua* Walt. and *I. serrata* Thunb., and *P. glabricola* on *I. glabra* (L.) A. Gray. Kulp also records *P. verticillatae* as forming linear blotch-mines on the soft-leaved deciduous holly, *Ilex verticillata*; however, I have seen a long, entirely linear mine on this host from Florida which may represent yet another species in this complex.

The male genitalia immediately confirm the close relationship of the species in this complex. However, the small but consistent differences permit ready identification of the species concerned, as can be seen from illustrations of the aedeagus of *P. ilicicola* and *P. opacae* shown in Figs. 497, 498 and 500, 501.

Frost and Sasakawa (1954) described *Phytomyza jucunda*, a further species in this same complex from Japan, which forms irregular linear blotch-mines on *Ilex crenata* Thunb., *I. macropeda* Mq., *I. radicans* Ohwi and also on *Illicium religiosum* Sieb. & Zucc. The species is discussed further by Sasakawa (1961).

CAMPANULACEAE

Campanula species (Europe)

Cultivated *Campanula* spp. are favourite hosts for *Liriomyza strigata* and towards the end of the summer few garden-grown Campanulas in the London area are without the characteristic leaf-mines of this species (Fig. 502).

CARYOPHYLLACEAE

Dianthus caryophyllus L. (Europe, North America)

In Europe two species regularly attack Carnations. The commoner of these, *Amauromyza* (*Trilobomyza*) *flavifrons* (MEIGEN, 1830), forms conspicuous white

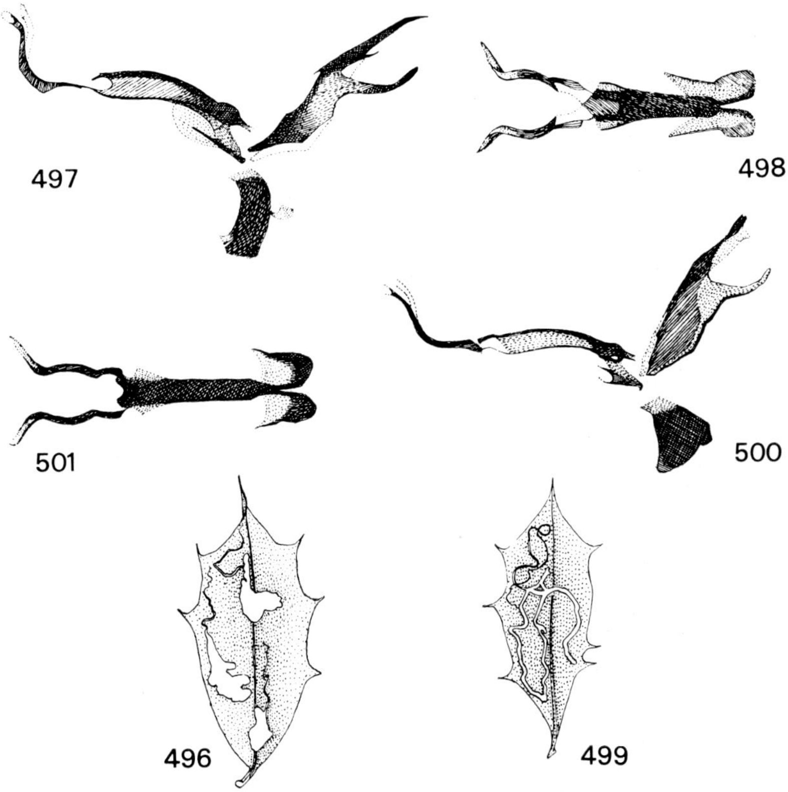

Figs. 496–498. *Phytomyza ilicicola* Loew: 496, leaf-mine on *Ilex opaca;* 497, aedeagus, side view; 498, same, ventral view.

Figs. 499–501. *Phytomyza opacae* Kulp: 499, leaf-mine on *Ilex opaca;* 500, aedeagus, side view; 501, same, ventral view.

blotch-mines in the leaves (Fig. 187), cf. Chapter 6, p. 132. The second, primarily Mediterranean species, *Paraphytomyza dianthicola* (VENTURI, 1949) forms narrow linear mines mainly on the underside of leaves (Fig. 503) and occasionally also in the stem. CIAMPOLINI (1949, 1952) has discussed in detail the biology of this species which causes considerable damage to cultivated carnations in Tuscany and Liguria in Italy. He observed no less than 7 generations between April and October and not infrequently 6–8 mines can occur in a single leaf. When infestations are heavy, the plants suffer considerable damage, resulting in a reduced number of flowers and these have less brilliant colours and last for a shorter time than normal. Plants with obvious attack on the leaves and stems depreciate substantially in commercial value (CIAMPOLINI, 1949). Severe damage has also been caused to glasshouse-grown carnations in Belgium (HEUNGENS, 1966). It is also known from Dalmatia and Hungary (SPENCER, 1966c). SÉGUY (1949), who was unaware of VENTURI's description of *dianthicola*, also described it as *Phytomyza jannonei*, after Prof. JANNONE who had discovered specimens at Genoa.

In the United States, leaf-mines on cultivated carnations first came to notice in San Mateo County, California, in 1956 (PRITCHARD, 1957). FRICK (1958) described the species as *Liriomyza dianthi* sp.n. The larvae form large white mines and several may be present in a single leaf. The species was considered at this time to be quite destructive.

After studying additional material from many other localities in California, FRICK (1964) synonymized *L. dianthi* with *L. langei* Frick. This species is itself now known to be synonymous with *L. huidobrensis* (Blanchard) described from Argentina, which is a highly polyphagous species causing damage to a wide range of hosts (see Chapter 7, p. 215).

COMPOSITAE

Aster (United States).

In the Los Angeles area the most important pest of asters is *Liriomyza huidobrensis* (Blanchard). Single mines run down the petiole but when many larvae feed together in a single leaf, a large blotch is formed. The life cycle can be completed in 3–4 weeks (JEFFERSON and EADS, 1952, as *Lir. langei* Frick).

In Europe *Phytomyza syngenesiae* (Hardy) is not infrequent on some cultivated asters but it rarely occurs on this host in sufficient numbers to cause appreciable damage.

Calendula officinalis L., also *Centaurea*, *Dimorphotheca* (Europe)

Cultivated marigolds have suffered serious damage in England and in Italy from *Napomyza lateralis* (FALLÉN, 1823b), the larvae of which bore in the stem or in the flower-head. *N. lateralis* has also been reported in England on cultivated *Centaurea* and *Dimorphotheca*, and feeds commonly on other wild Composites.

In Italy damage to marigold flowers was first reported by DELLA BEFFA (1939), who identified the species as *Phytomyza atricornis* Mg. It is clear from his paper that he was dealing with two species – the common leaf-miner in the Mediterranean area, *Phytomyza horticola* Goureau, and also the stem-borer, *Napomyza lateralis*, of which good photographs are given. Four illustrations are also included showing damage to the flower-heads of Calendula.

In Holland, while working on the pest of chicory, *Napomyza cichorii* (see Chapter 6, p. 141), VAN'T SANT (1961) also bred *N. lateralis* from one of its wild hosts, *Matricaria maritima*, ssp. *inodora* L.

The most detailed account of the life history of this species and of the damage caused to marigolds is that given by MORGAN (1954) in England. The damage first came to notice in the winter of 1949–1950 when cut flowers, grown in the mild S.W. counties of Devon and Cornwall, were found to be wilting on the way to market. It was then discovered that both the stem and receptacle had been eaten out by the larvae of *N. lateralis*.

Eggs are laid in the upper leaf surface and the young larva forms a short mine until it reaches a vein, which is then followed to the midrib and ultimately to the stem. Some larvae feed exclusively in the stem but others move up into the capitulum. The complete life cycle can take as little as 7 weeks in the summer or in winter 10–12 weeks. In S.W. England there is apparently no extended diapause and flies may emerge in January or February from eggs which were laid in November or December.

'Three types of damage occur which may be of economic importance. Attack in the seedling stage may cause death of the primary growing point, giving rise to a "many-necked" type of plant. This in itself may not be a disadvantage unless the secondary crowns are themselves attacked and killed out, so that the entire plant is crippled. Attack at a later stage of growth may weaken the stems, particularly at the nodes, which leads to breakdown of the plants under the stress of wind, mechanical cultivation or picking. Still later attack will cause damage to the flowers, either killing the bud before it opens or causing it to wilt after picking.' (MORGAN, 1954.)

Five parasites have been recorded from *N. lateralis*, namely *Rhizarcha pubescens* (Curtis) and *Dacnusa leptogaster* Hal. (Braconidae); *Sphegigaster aculeatus* (Walker) and *Stenomalus muscarum* L. (Pteromalidae); and *Halticoptera fusicornis* Walker (Eulophidae). These probably in many years exercise effective control but, if their populations are reduced by some natural causes or from indiscriminate use of insecticides, *N. lateralis* will certainly periodically reach pest proportions. Reservoirs of population exist on favourite wild hosts such as *Matricaria maritima* and *Senecio vulgaris*.

Chrysanthemum (Europe, N. America, New Zealand)

Chrysanthemums were first introduced into England in 1823 from China. In Japan these plants were being cultivated by the mid-nineteenth century and J. G. VEITCH who was there on a collecting expedition in 1860 wrote that he saw chrysanthemums 'which would not disgrace even a London exhibition'. (COATS, 1969).

Chrysanthemums imported into Western Europe were no doubt quickly colonized by *Phytomyza syngenesiae* (discussed earlier on p. 233) from local wild hosts such as *Senecio* and *Sonchus*. However, no serious damage was apparently reported in chrysanthemum nurseries prior to the early 1920s. From that time on serious infestations have been reported in most years throughout England and

Wales. For example, in 1923 there was a severe outbreak at Aberystwyth, Wales, with 90% of plants attacked; in 1930 young rooted cuttings in Leicestershire and Derbyshire were practically destroyed; in 1952 at Penrith, Cumberland, attacks were on a serious scale and caused considerable financial loss; in a nursery in Buckinghamshire in 1968 all leaves of every plant were reported to be attacked (M.A.F.F., unpublished reports).

At the present time, with *P. syngenesiae* present at least in small numbers throughout the country on a wide range of Composite hosts, cultivated chrysanthemums are quickly colonized and without control, the damage caused, particularly to young plants, frequently assumes serious proportions. The damage varies significantly from season to season but heavily mined leaves shrivel and die, the plant can be severely weakened and at best only inferior blooms will be produced. Breeding is continuous under glass and infestations may occur at any time of the year. Some varieties are less susceptible to attack than others, and detailed studies have indicated decreasing susceptibility in the following 14 varieties which have been tested: Galaxy, Vibrant, Tuneful, Long Island Beauty, Ice cap, Salmonchip, Silverstone, Criterion, Favourite, Corsair, Yellow Iceberg, Yellow Shasta, Dark Delmar, Cerise Shoesmith. (M.A.F.F., 1969.)

A photograph from the Glasshouse Crops Research Institute at Rustington, Sussex, shows the typical leaf-mining damage of this species (Plate 11). HUSSEY *et al.* (1969) report that as many as 20–30 larvae may be found in a single leaf.

A number of insecticides have been tested and are recommended for use against this species on chrysanthemums. Further testing is currently being carried out at the G.C.R.I. This is briefly discussed later in Chapter 12.

Biological control has also been under investigation for some years at the Glasshouse Crops Research Institute and is producing encouraging results (see Chapter 12, p. 352).

In the United States *P. syngenesiae* first came to notice in a greenhouse on Long Island, N.Y., in October, 1886. 'The leaves of some daisies (marguerites) were seen to show some wart-like specks, and irregular, whitish, linear markings, and soon afterward to shrivel up and die. Examination for the cause disclosed very small "worms" working within channels in the interior of the leaves' (LINTNER, 1888).

In 1887 infested marguerite leaves were recorded from Germantown, Pa. At the same time it had 'become troublesome' as a pest of chrysanthemums, and also *Cineraria* and *Eupatorium*, in glasshouses in the Boston area.

SMULYAN (1914) reports that the fly was originally considered to be a European species but it was nevertheless described by the European specialist, KOWARZ from Bohemia, as *Phytomyza chrysanthemi* sp.n. (in LINTNER, 1891). GRIFFITHS (1967c) formally synonymized *chrysanthemi* with *syngenesiae*, after examining a male from the original type series preserved in the Deutsches Entomologisches Institut, Eberswalde, G.D.R. It seems virtually certain that the species was introduced into N. America in the 1880s.

Two favourite foodplants at the time of these early infestations were apparently *Chrysanthemum frutescens* L. (marguerite) and *C. parthenium* (L.) Pers. (double white feverfew). The common greenhouse chrysanthemums, *C. indicum* L., *C. morifolium* Ramat. (= *C. sinense*) and also *C. leucanthemum* L. and *C.* (*Tanacetum*)

Plate 11: *Phytomyza syngenesiae:* leaf-mines on Chrysanthemum, England (Dr. N. W. Hussey, Glasshouse Crops Research Institute, Worthing).

vulgare were also regularly attacked. Other Composites known as hosts in the United States are referred to below under *Cineraria*.

Serious damage was caused to cultivated Composites, one informant reporting to SMULYAN that 'the infestation reduces the number of flowers and weakens the plants to a great extent'. There is also a tendency towards a reduction in the size of the flowers. Buds in many instances formed, but they dried up after reaching a certain size.

As early as 1890 one grower in Connecticut reported that 'he would be compelled to abandon the growing of such plants as marguerites and feverfews, owing to the attack of this pest'. Following an outbreak in Wisconsin, the growers faced 'an entire loss of their flowering plants caused by a complete infestation of the leaves' (SANDERS, 1912).

It was noted that this species appeared to be essentially a pest of greenhouses and over many years there were no reports of it as an outdoor pest (SMULYAN, 1914).

More recently the major pest of Chrysanthemums in America has been not *P. syngenesiae* but *Liriomyza trifolii* (Burgess) (cf. p. 226). This causes serious damage in nurseries in Florida and is considered a major pest in greenhouses in Delaware, Maryland, New Jersey, Ohio and Pennsylvania, and also around Washington, DC. The species was originally considered to be new (SMITH, BOSWELL and WAVE, 1962) and was later identified as *Liriomyza munda* Frick (WEBB and SMITH, 1969). I am now satisfied, however, that the species represents *Lir. trifolii* (Burgess), which is a polyphagous species, occurring on many genera of the Compositae.

SMITH *et al.* (1962) report that both the adults and larvae injure chrysanthemums but the larval injury is the more serious. 'When the mined leaves occur on the upper parts of the stem, they seriously reduce the salability of the crop.' As many as four generations have been recorded in a commercial greenhouse between the end of January and early May. Although all varieties can be attacked, 'the most severely infested in one greenhouse containing 13 varieties were Iceberg, Yellow Iceberg, Golden Boy, Indianapolis and Luyona'.

BRITTON (1935) reported that in Connecticut chrysanthemums in glasshouses suffered considerable damage from the larvae of *Amauromyza maculosa* (discussed earlier in Chapter 6, p. 151). This species has also damaged chrysanthemums in Florida. The larvae form a large greenish-black blotch (Fig. 504), with several feeding together, and this leaf-mining activity can be particularly destructive to young plants.

In Argentina *Agromyza guaranitica* Brèthes, 1920, was described attacking chrysanthemums in Rio Grande do Sul. This species was synonymized with *A. maculosa* by SPENCER, 1963a: 336.

Cineraria and other composites (Europe, North and South America, New Zealand)

Phytomyza syngenesiae attacks *Cineraria* almost as readily as chrysanthemums and it can be almost equally destructive on this host. It was first reported damaging

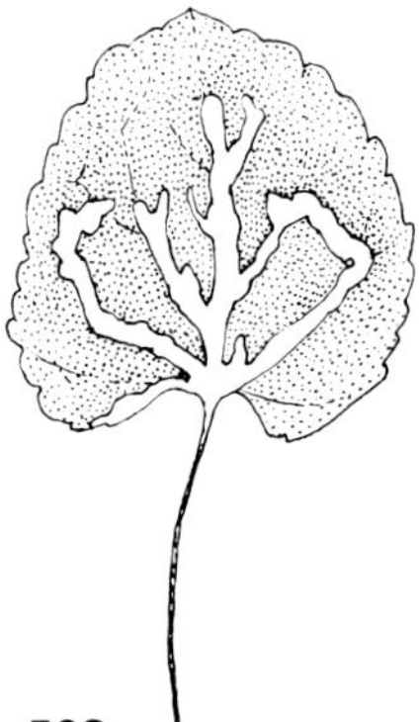

502

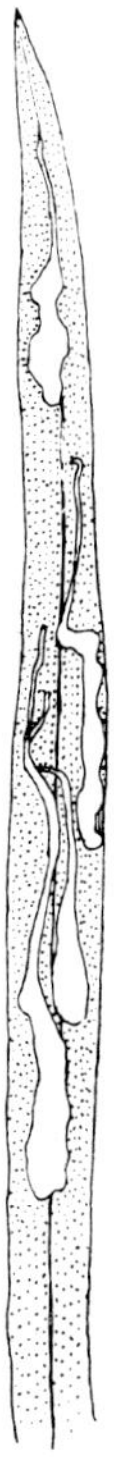

503

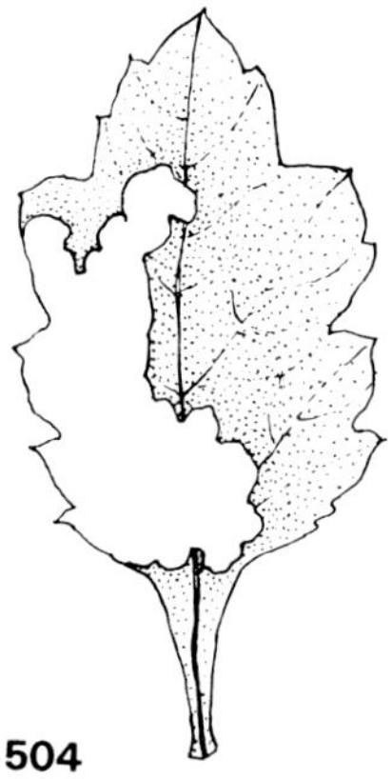

504

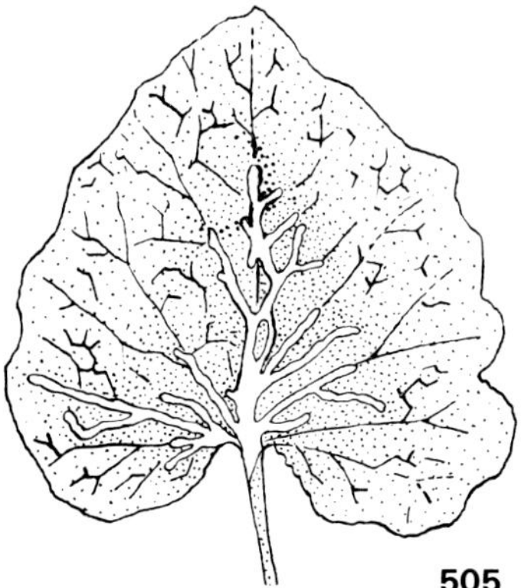

505

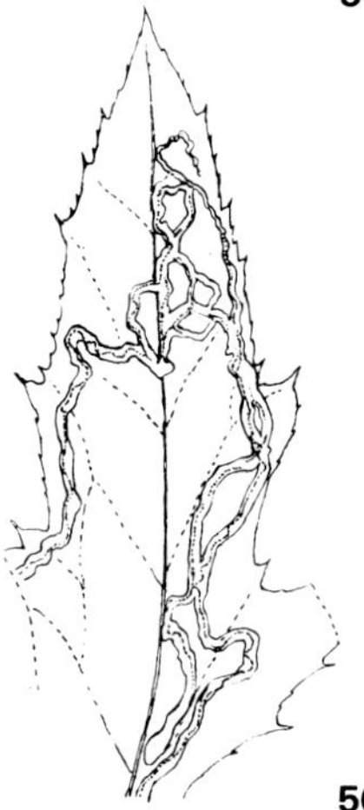

506

Fig. 502. *Liriomyza strigata* (Mg.): leaf-mine on *Campanula trachelium* (Dresden, Germany).

Fig. 503. *Paraphytomyza dianthicola* (Venturi): leaf-mines on *Dianthus caryophyllus* (Italy), after Ciampolini.

Fig. 504. *Amauromyza maculosa* (Mall.): leaf-mine on *Chrysanthemum*.

Fig. 505. *Liriomyza huidobrensis* (Bl.): leaf-mine on *Cineraria* (Argentina), after Blanchard.

Fig. 506. *Phytomyza bipunctata* Loew: leaf-mine on *Echinops*.

Cinerarias in England as early as 1844 by Curtis (as *Phytomyza nigricornis* Macq.). Curtis mentions that the larvae, in view of their feeding habit, have been denominated 'miners' and writes: 'Although they may not occasion much mischief, as regards the health of plants, yet their labours are by no means acceptable to the gardener in the greenhouse or conservatory, for the blighted and sickly aspect they cast over the foliage is certainly very unsightly. I have frequently seen the leaves of Cinerarias thus affected to a great extent . . ..'

In England in 1928 there were severe and persistent attacks at Quadring, Lincs., and in the same year a nursery at Usk in S. Wales suffered an unusually severe attack. The following year considerable damage was caused at Salford, Lancs.; other composites which have suffered damage in England are *Gazania, Pyrethrum* and *Zinnia* (M.A.F.F., unpublished reports).

In Tasmania and New Zealand *P. syngenesiae* is commonly known as the Cineraria Fly. LEA (1906) reports that in 1897 it was widely distributed in Western Australia and 'in Tasmania there is probably not a single garden in which it is not present'. Other composite hosts recorded from Tasmania and Australia are 'marguerite, sunflower, dahlia, chrysanthemum, marigold, lettuce, capeweed, dandelion and milkthistle'. WISE (1953) reports this species as the commonest leaf-miner in New Zealand. Marigold and dahlia are mentioned as additional hosts.

Liriomyza huidobrensis (BLANCHARD, 1926) was described in Argentina attacking *Cineraria* (Fig. 505). This has since been found to be a highly polyphagous species of wide distribution which has been subsequently described under a number of names (see Chapter 8).

SMULYAN (1914) records a number of Composites which are readily attacked in the United States, in addition to Chrysanthemums, including *Ambrosia artemisiifolia* L., *Antennaria plantaginifolia* (L.), *Bidens frondosa* L., *Helianthus annuus* L., *Solidago nemoralis* Ait. and *Taraxacum officinale* L.

In Florida the species primarily attacking cultivated Composites is *Liriomyza trifolii* Burgess (see above under *Chrysanthemum*). STEGMAIER (1966a) lists the following ornamentals as additional hosts: *Aster, Callistephus chinensis* (L.) Nees, *Dahlia, Guillardia aristata* Pursh., *Helianthus annuus* L. and *Zinnia* sp.

Amauromyza maculosa (Malloch) is a highly oligophagous species on Composites which occurs commonly on a number of ornamentals, in addition to Chrysanthemums (see above), and also on a wide range of wild hosts. STEGMAIER (1967b) reviews the known hosts of this species and among ornamentals records *Aster, Dahlia, Helianthus annuus* L., *Gaillardia aristata* Pursh. and *Tagetes*. At Tucuman in Argentina *Zinnia elegans* Jacq. is frequently attacked by *A. maculosa* and a wild species, *Z. peruviana* L. is an additional host (P. WYGODZINSKY, personal communication).

In Europe *Liriomyza strigata* (Meigen), a highly polyphagous species (see Chapter 8), occurs on many ornamental Composites and has been recorded on *Aster, Bellis, Calendula, Callistephus, Centaurea, Chrysanthemum, Coreopsis, Dahlia, Gaillardia, Helianthus, Tagetes* and *Zinnia*. The characteristic leaf-mine runs primarily along the midrib, with lateral offshoots into the leaf-blade (cf. Figs. 502, 514).

Echinops (Europe)

A number of *Echinops* species, such as *E. banaticus* Rochel, *E. commutatus* Juratzka and *E. sphaerocephalus* L. are known as hosts of *Phytomyza bipunctata*

Loew, 1858. It is remarkable how quickly isolated plants in gardens are colonized by this host-specific species. Plants in Botanical Gardens such as Kew (SPENCER, 1961a) are invariably heavily infested. The leaf-mine is long and narrow, with unusually long strips of black frass at alternate sides of the channel (Fig. 506).

Gerbera (S. Africa, India, Europe)

In South Africa two stem-borers are known on *Gerbera*, *Melanagromyza gerberae* Spencer, 1960a, and a related species, *M. gerberivora* Spencer, 1960b, which also occurs in India.

Stem-borers are not known on cultivated *Gerbera* in Europe but in England a severe leaf-mining attack occurred in Warwickshire in 1958 by *Phytomyza syngenesiae* (M.A.F.F., unpublished report).

Solidago canadensis L. (N. America, Europe)

A common leaf-miner on this plant, which is native to North America and is commonly cultivated in Europe, is *Nemorimyza posticata* (MEIGEN, 1830). The larva forms a large blotch with a characteristic pattern of feeding lines (Fig. 507). This is an isolated species of holarctic distribution. Considerable damage can be caused to the foliage of cultivated golden-rod in England.

In the United States this species occurs as far south as the Miami area in Florida. It is considered that this represents a relict Pleistocene population which moved south during the last glaciation and has now adapted to the sub-tropical climate of southern Florida (SPENCER and STEGMAIER, 1973).

CONVOLVULACEAE

Calonyction aculeatum (L.) House (Florida)

The characteristic circular blotch-mines (Fig. 508) with frass deposited in an almost regular spiral found by C. STEGMAIER at Miami immediately suggested that these must be distinct from those caused by the species forming irregular linear blotch-mines on *Ipomoea* and less commonly also on this host. Examination of the male genitalia confirmed this and the species which, as far as is known, is host-specific on Moon-Flower was recently described as *Calycomyza stegmaieri* Spencer, 1966d.

Ipomoea species (Florida, Central and South America).

Calycomyza ipomaeae (FROST, 1931) occurs commonly on *Ipomoea batatas* (L.) Lam. and *I. tiliacea* (Willd.) Choisy but appears consistently to avoid some other *Ipomoea* species. Wherever 'Morning Glory' is grown as an ornamental within its normal host range in the Neotropical Region the characteristically irregular linear blotch-mines (Fig. 509) of *C. ipomaeae* will probably be in evidence.

STEGMAIER (1967a) includes a number of photographs of *Cal. ipomaeae* on *I. tiliacea* in Florida.

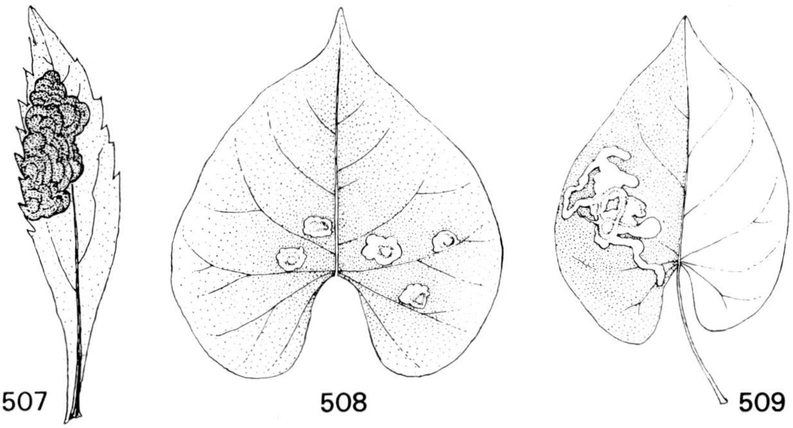
507 508 509

Fig. 507. *Nemorimyza posticata* (Mg.): leaf-mine on *Solidago canadense*.

Fig. 508. *Calycomyza stegmaieri* Sp.: leaf-mine on *Calonyctium aculeatum*.

Fig. 509. *Calycomyza ipomaeae* (Frost): leaf-mine on *Ipomoea* sp.

IRIDACEAE

Iris species (Europe)

Iris ochroleuca L. is commonly cultivated in England and the foliage of plants in a nursery at Enfield, Middlesex, was considerably damaged in 1952–1953 by *Cerodontha* (*Dizygomyza*) *iridis* (HENDEL, 1927), a species which had hitherto only been known to feed on the wild host *Iris foetidissima* L. As many as 12 larvae can feed together, forming a large greenish-mottled mine. Pupation is characteristic in this species, with the puparia lying side by side horizontally across the leaf.

Iris foetidissima is occasionally grown in gardens in Italy and its biology has been discussed in some detail by VENTURI (1946b).

Iris pseudacorus L. (Yellow Flag) is frequently cultivated around pools and has its specific leaf-miner. The species, *Cerodontha* (*Diz.*) *ireos* (GOUREAU, 1851), forms a conspicuous white mine (Fig. 510), with the larvae feeding singly and the puparium lying lengthwise in the leaf (contrast *Diz. iridis* above).

Iris versicolor L. (North America)

This and other *Iris* species are attacked in North America by *Cerodontha* (*Dizygomyza*) *thompsoni* (FRICK, 1952a). As a result of larval feeding the leaves become seriously deformed, developing gall-like swellings. The biology of the species has been discussed in detail by CLAASEN (1918).

Iris sp. (Florida)

A new species from Florida, *Cer.* (*Diz.*) *iridophora*, from an unidentified *Iris* sp. was described by SPENCER (in SPENCER and STEGMAIER, 1973). Little information is available on the distribution of this species on cultivated Irises.

LABIATAE

Salvia splendens Sellow (Argentina)

This plant, which is a native of Brazil, is a popular ornamental in Buenos Aires, and is apparently heavily attacked by a leaf-miner described in 1923 by BRÈTHES as *Phytomyza platensis*. The exact status of this species is not known and it would be of interest if fresh material could be obtained for detailed examination.

LEGUMINOSAE

Clianthus, *Sophora* (New Zealand)

WATT (1923) described as *Agromyza clianthi* a leaf-mining species on *Clianthus puniceus* Banks & Solander in New Zealand. HARRISON (1959: 322) transferred the species to *Liriomyza*.

WISE (1953), discussing leaf-mining pests in New Zealand, writes: 'A common and well-known example of leaf-mining damage in gardens is the mine of the native clianthus fly. Red Kowhai is attacked so severely by this species that it is most difficult to grow in some areas.' Red Kowhai, *Sophora tetraptera* F. Mill, is in a related genus to *Clianthus* but it would be desirable in due course to check that the fly on *Sophora* is conspecific with that on *Clianthus*. The leaf-mine of *Liriomyza clianthi* on *Clianthus* is shown in Fig. 511.

Laburnum (Europe)

Two species of Agromyzidae occur commonly in western Europe as leaf-miners of *Laburnum anagyroides* Medic., which is now widely used as a garden ornamental, particularly in Britain, Germany, the Netherlands and Sweden.

Agromyza demeijerei Hendel, 1920, forms an upper surface mine which is typical of a number of related species, in which the larva first feeds along the margin of the leaf towards the apex and this narrow channel finally widens into a broad blotch (Fig. 512A). The main generation of mines starts in mid-June and continues well into July. At least in southern England there is a partial second generation, when mines can be seen in large numbers in September or October. G. C. D. GRIFFITHS found a heavy infestation of this second generation in Yorkshire at the

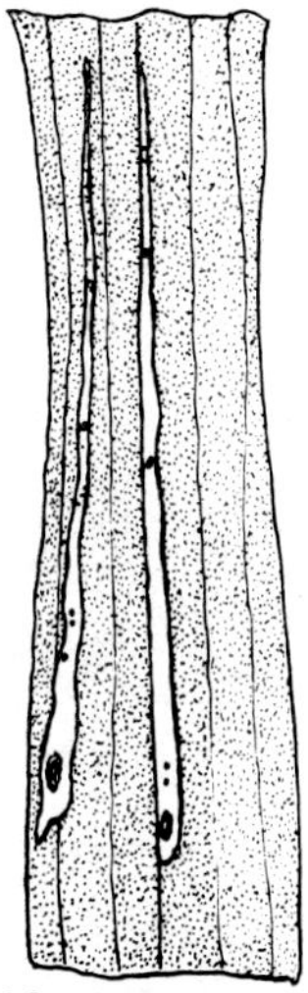

510

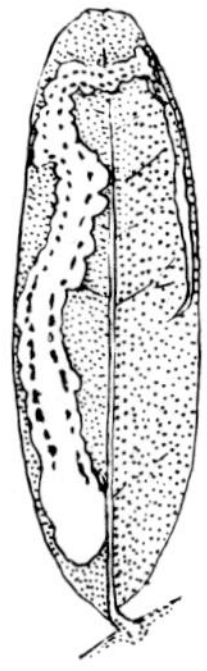

511

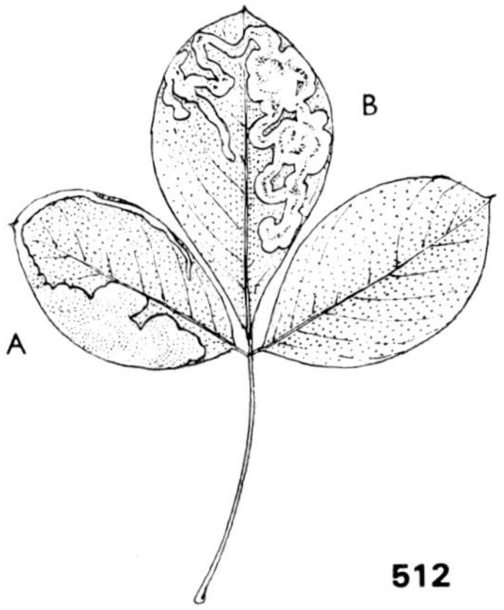

512

Fig. 510. *Cerodontha (Diz.) ireos* (Gour.): leaf-mines on *Iris pseudacorus* (Hering).

Fig. 511. *Liriomyza clianthi* (Watt): leaf-mine on *Clianthus puniceus*.

Fig. 512A. *Agromyza demeijerei* Hd.: leaf-mine on *Laburnum*.

Fig. 512B. *Phytomyza cystisi* Bri.: leaf-mine on *Laburnum*.

end of September, 1964. ASKEW (1968b) records that by July trees in the Manchester area can have a scorched appearance, due to the many mines, which by this time have turned brown with age.

Phytomyza cytisi Brischke, 1881, forms an entirely different type of mine, which is irregularly linear (Fig. 512B) and is frequently, at least partially, on the lower leaf surface.

Laburnum was introduced into Britain late in the 16th century from southern Europe. There are no records of either *A. demeijerei* or *P. cytisi* from areas where *Laburnum* is native but as they are both strictly monophagous on this host, they must occur, perhaps in small numbers, in southern Europe, and have presumably built up larger populations in areas further north where *Laburnum* is now widely cultivated.

Lathyrus latifolius L. (Europe)

This 'Sweet Pea' which is frequently grown in gardens in England is commonly attacked by *Agromyza varicornis* Strobl, 1900 (=*A. watersi* Spencer, 1957a). The larva initially forms a whitish linear mine in the leaves but the main feeding continues in the winged stem. In the London area there is apparently a large population of this species, based on individual plants scattered in hundreds of gardens. It is rare to find this host not attacked by this miner and any freshly grown plant invariably becomes infested within one or two seasons.

Lathyrus odoratus L. (Europe, S. America)

In Europe the white linear mines of *Phytomyza horticola* are not uncommon on Sweet Pea. At Tucuman in Argentina both *Liriomyza huidobrensis* (Blanchard) and *L. sativae* Blanchard have been found on this host. In the former the mine is generally associated with the midrib and lateral veins (cf. Fig. 505), while in the latter the mines are more irregularly linear, sometimes producing a secondary blotch.

Sarothamnus scoparius L. (Europe)

A common species in much of western Europe attacking cultivated Broom is *Agromyza johannae* de Meijere, 1924. The leaf-mine starts as a narrow channel running along the leaf-margin towards the apex, where it turns to develop into a substantial blotch in the area of the midrib (cf. *Agromyza demeijerei* Hd. on *Laburnum*, Fig. 512A). The mines are inconspicuous when fresh and can occur in the smallest leaves. Mined leaves are invariably destroyed by the larval mining and in due course dry up, and are then more readily visible.

A. johannae is common on wild broom and is regularly found in gardens in England on a wide range of cultivated varieties.

Wistaria chinensis DC. (Japan, North America).

In 1912 galls were found on twigs, generally near buds, on a shipment of Wistaria imported from Japan to Los Angeles. The flies which emerged were identified as *Agromyza schineri* (a gall-causer on *Populus*). Buds in the immediate vicinity of the galls invariably die (Amundsen, 1912).

Malloch (1913a) correctly appreciated that the Wistaria fly was distinct from that on poplar and described it as *Agromyza websteri*. It has also been discussed by Sasakawa (1961), as *Melanagromyza websteri*. Spencer (1966b) placed the gall-causing species in this group in the available genus, *Hexomyza* Enderlein, to which *websteri* is now transferred.

LILACEAE

Lilium martagon L. and other lilies (Europe)

MIK (1894) described *Liriomyza urophorina* as a new species feeding in the buds of *L. martagon* and *L. candidum* L. in Lower Austria. The species has since been reported as a major pest of cultivated lilies in the German Democratic Republic (RODE and VORSATZ, 1966).

Yellow and black *Liriomyza* species, as at present understood, had been included in the genus *Agromyza*, until MIK described *urophorina* in the new genus *Liriomyza*. He justified the erection of a new genus on two unusual characters – the ability of the larvae to jump, and the unusually long ovipositor in the female (Fig. 513). Although these characters are not necessarily of generic significance, subsequent genitalia studies fully confirm the validity of this genus.

Larval feeding by *L. urophorina* causes the flower-buds to rot and subsequently dry up and then fall. Slightly infested buds may produce flowers but these will be deformed, show holes or have rotting areas. The severity of the infestation is clearly shown by the following table relating to buds of *Lilium candidum* in gardens around Jena in 1964–1965:

Table IX.

Year	*Plants*	*Total buds*	*Buds damaged* *Number*	*Per cent*	*Buds totally damaged* *Number*	*Percent of total buds*	*Percent of damaged buds*
1964	969	5913	3373	57%	1821	30.8	54
1965	1185	7171	4217	58.8%	2021	28.2	47.9

Many species of lily may serve as hosts and the following have been confirmed, in addition to the favoured hosts, *L. candidum* and *L. martagon*; *L. bulbiferum* L., *L. croceum* Chaix, *L. davidii* var. *willmottiae* W. & S., *L. regale* W. & S., *L. thunbergianum* Schult and also several hybrids, such as *Lilium* × *aurelianense* W. & S., *L. testaceum* Lindl. and *L. umbellatum* Hort.

L. urophorina has been found on wild *L. martagon* in a number of areas in the south of the German Democratic Republic, including Jena, Erfurt, Waltershausen, Meiningen and Suhl (RODE and VORSATZ, 1966) and also reaches Berlin (THOMAS, 1893). It is not known in western Europe.

MALVACEAE

Althaea rosea Cav. (Europe, North America)

The characteristic mines of *Liriomyza strigata* are not infrequently found on hollyhocks in Europe (Fig. 514).

Hollyhocks were introduced into Europe from China in the mid-sixteenth century and later from Europe into North America. Distinctive long, greenish-yellow mines were first discovered by the author on hollyhocks in front gardens in Toronto in early October, 1965 (Fig. 515). These represented the second generation and earlier mines were found two years later in July. I have also noticed that similar mines were not uncommon in gardens at Amityville, Long Island, New

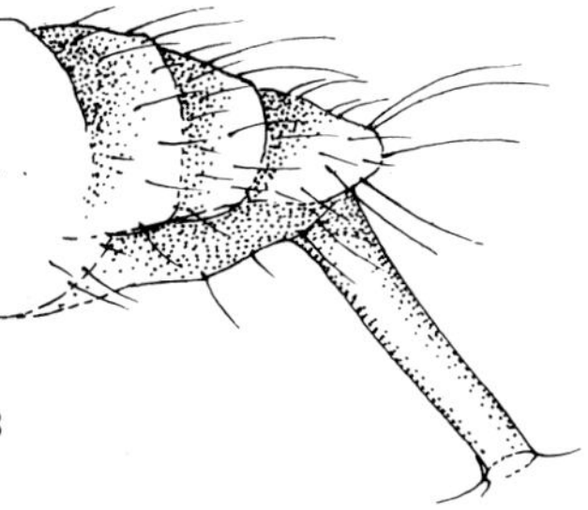

513

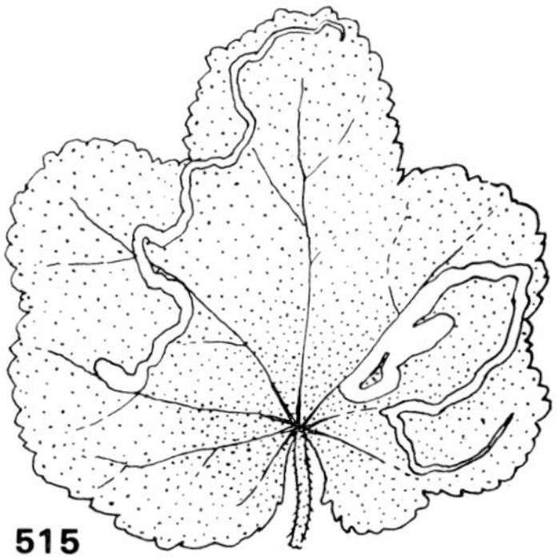

515

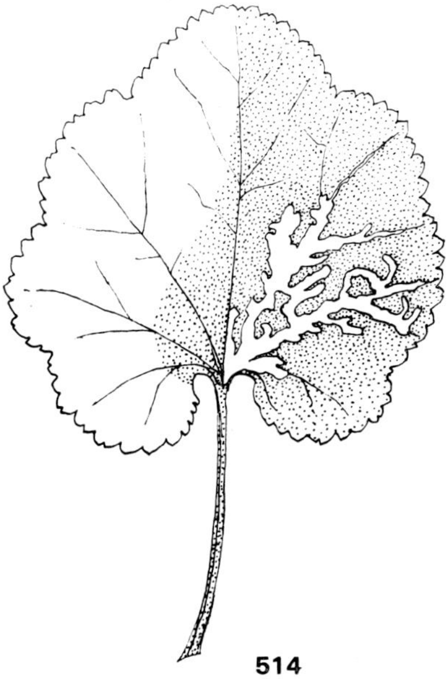

514

Fig. 513. *Liriomya urophorina* Mik: ovipositor of female.

Fig. 514. *Liriomyza strigata:* leaf-mine on *Althaea rosea* (Europe).

Fig. 515. *Calycomyza malvae* (Burg.): leaf-mines on *Althaea rosea* (Canada).

York, at the end of October, 1965. The species was at first thought to be undescribed but was later identified as *Calycomyza malvae* (Burgess), 1880 (SPENCER and STEGMAIER, 1973).

Leaf-stalks and stems of hollyhocks in southern Ontario can be heavily attacked by larvae of a species referred to as *Melanagromyza* sp. (Steyskal) by SPENCER (1969a). This remains an undescribed species which Mr. G. STEYSKAL was proposing to describe in 1967.

NYCTAGINACEAE

Bougainvillea spectabilis Willd. (Africa, S. America, Florida)

Two entirely unrelated species produce conspicuous, silvery epidermal mines on young leaves of this widely grown ornamental.

In Argentina and probably in many other areas of South America, in Central America and in Florida cultivated *Bougainvillea* is frequently attacked by *Liriomyza schmidti* (ALDRICH, 1929). This ornamental is widely grown around Tucuman in Argentina and older, empty mines cause some leaf deformation (P. WYGODZINSKY, personal communication). A wild *Bougainvillea* sp. is also attacked at Tucuman.

In East Africa *Bougainvillea spectabilis* is a favoured host of *Tropicomyia flacourtiae* (see Chapter 7, p. 185). Similar mines are not uncommon on *Bougainvillea* in India but it has not been established which *Tropicomyia* sp. is responsible.

PAPAVERACEAE

Papaver (Europe)

Cultivated poppies are rarely heavily infested by leaf-mines but are regularly used as hosts by *Phytomyza horticola* Goureau. BUHR (1964) records mines on *Papaver rhoeas* L. and *P. somniferum* L. in Saxony, East Germany.

POLEMONIACEAE

Phlox (Europe, South America)

In Europe *Phlox Drummondii* Hook. is occasionally attacked by *Liriomyza strigata* and damage noted in England is referred to below under Violaceae. In Argentina the related species, *Liriomyza huidobrensis*, has also been recorded attacking cultivated *Phlox*.

BUHR (1964) also records *Phytomyza horticola* (as *atricornis*) on this host in Saxony, East Germany.

PRIMULACEAE

Primula species (Europe)

The conspicuous white mines of *Phytomyza primulae* Hendel (Fig. 516) occur commonly on cultivated Primulas and *Polyanthus*. These mines were first discussed by HARDY (1849) on *Primula vulgaris* Huds. in England but the species was misidentified as *Phytomyza nigra* Meigen. HARDY gave a vivid, if somewhat unscientific, description of these mines: 'On picking up one of the leaves, sometimes the middle part of the upper surface will be found of nearly a pure white, which, where it is limited by the original green, presents a wavy and exceedingly fantastic outline; at other times small undulating bands issue from the colourless central area, like streams

"Devolving from their parent lakes:"

at times we have before us the representation of a serpent untwisting its many coils, and at others a congeries of minute worms, inextricably intertwined, of which we can trace a general source, but whose terminations are quite a maze.'

RANUNCULACEAE

Aconitum, Delphinium (Europe, North America)

Phytomyza aconiti Hendel, 1920, forms large blackish blotch-mines on these two hosts, with several larvae feeding communally, and infested leaves can be virtually destroyed (Fig. 517). This species is common in gardens in the south of England. Mines have also been found in Quebec on *Aconitum* (SPENCER, 1969a) and it is believed that this represents an introduction from Europe.

FROST (1928) described *Phytomyza delphiniae* from New York and GRISWOLD (1928) discusses the life history based on original specimens sent to FROST. He points out that the leaves are badly damaged not only by larval feeding but also by the female feeding punctures.

FRICK (1959) records also having seen specimens from Ohio and Pennsylvania reared from *Delphinium cultorum* Voss. This species was synonymized with *P. aconiti* by SPENCER (1969a: 223).

At Edmonton, Alberta, cultivated Delphiniums can be heavily infested with short white linear mines (Fig. 518) produced by *Phytomyza delphinivora* Spencer (1969a). This is so far the only known locality of this species.

Aquilegia, Thalictrum (Europe, North America)

These genera are favourite hosts of Agromyzidae and they are attacked by stem-borers and seed-feeders, in addition to leaf-miners.

In Europe *Phytomyza krygeri* Hering, 1949b, is known from Denmark and England, the larvae feeding and pupating in the seed-heads of *Aquilegia vulgaris* L. A parallel species occurs in seed-heads of *Thalictrum aquilegiaefolium* L., *Phytomyza thalictri* Escher-Kundig (in DE ROUGEMONT, 1912), which is only known from Switzerland. (No males of this species are known and the possibility that it is identical with *krygeri* cannot be excluded.) *Phytomyza aquilegiae* Hardy, 1849, is a common species in western Europe on cultivated *Aquilegia* and the large

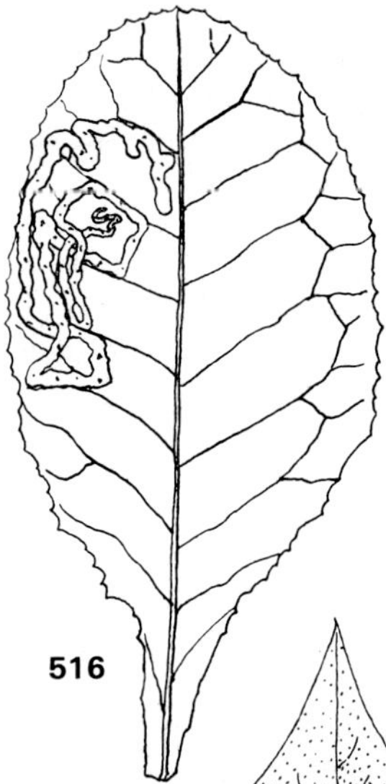

516

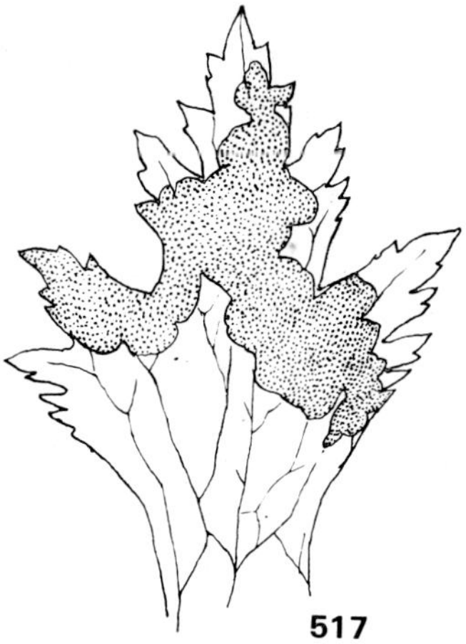

517

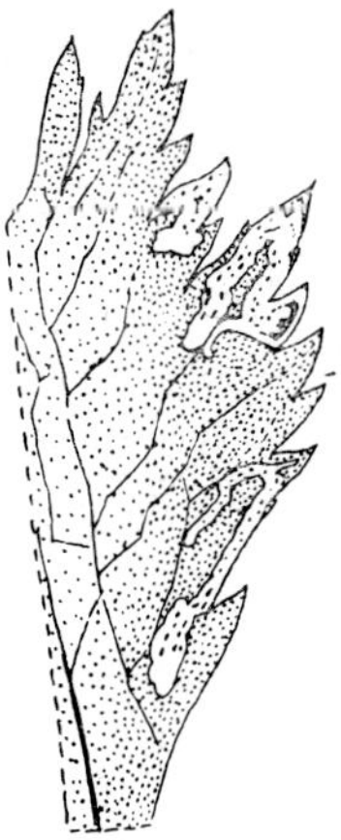

518

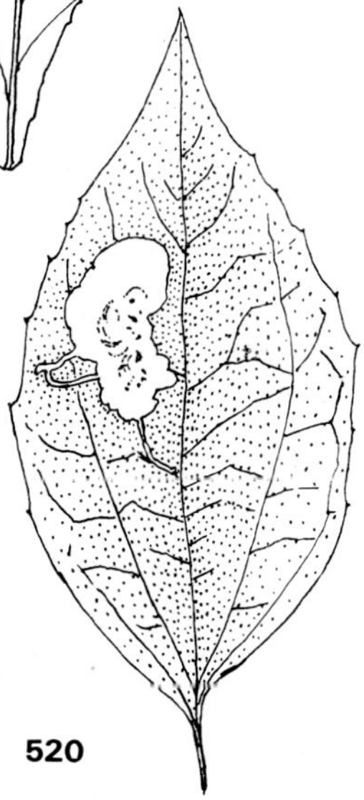

520

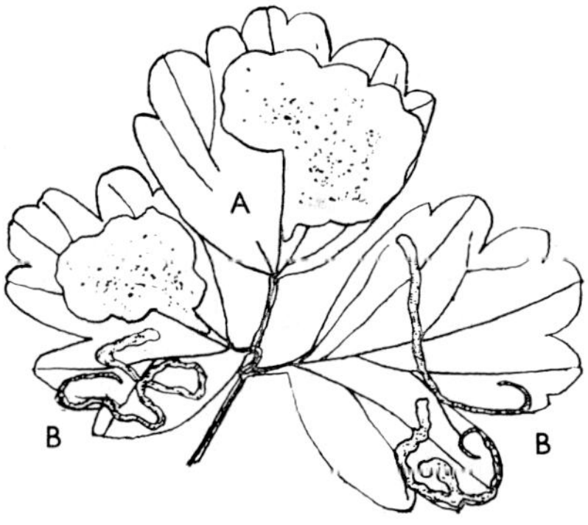

519

Fig. 516. *Phytomyza primulae* Hd.: leaf-mines on *Primula* (Hering).

Fig. 517. *Phytomyza aconiti* Hd.: leaf-mine on *Aconitum* sp. (England).

Fig. 518. *Phytomyza delphinivora* Sp.: leaf-mines on *Delphinium* sp. (Canada).

Fig. 519A. *Phytomyza aquilegiae* Hardy: leaf-mines on *Aquilegia*.

Fig. 519B. *Phytomyza minuscula* Gour.: leaf-mines on *Aquilegia* (both Hering).

Fig. 520. *Liriomyza philadelphivora* Sp.: leaf-mines on *Philadelphus coronarius* (Canada).

blotch mines (Fig. 519A) formed by the larvae can considerably weaken the plants. A smaller species, *Phytomyza minuscula* Goureau, forms characteristically irregular linear mines (Fig. 519B) on both *Aquilegia* and *Thalictrum* and the ovoid, shining brown puparia can frequently be found adhering to the leaves near the end of the mines.

In North America four species are known to feed on cultivated *Aquilegia*. Of these, two form blotch-mines, one linear mines and the other feeds in the stem and root. The blotch-miners are *Phytomyza aquilegiana* Frost, 1930, known from the eastern United States and Manitoba and Ontario in Canada; and *P. columbinae* Sehgal, 1971, from Edmonton, Alberta. *P. aquilegiophaga* Spencer, 1969a, is known from Ottawa, Ont. and Edmonton, Alta. and feeds in stems and bores into the root. *P. aquilegivora* Spencer, 1969a, forms short linear mines and has been recorded only in Ontario. The male genitalia show that it is closely related to the European *P. minuscula*.

Both in Europe and North America further species are known on wild *Aquilegia* or *Thalictrum* species, and of the species recorded on ornamentals in these genera, all certainly feed on related wild hosts.

In North America there has in the past been understandable confusion in the identification of species on these hosts, and European names have also been incorrectly applied to Nearctic species. In particular, the papers by Cory (1916), Frost (1924 and 1930) will be found misleading in their use of names.

ROSACEAE

Agrimonia, Geum, Potentilla, Rubus odoratus L. and other genera (Europe)

A common miner on many genera of Rosaceae which is encountered frequently on ornamentals is *Agromyza spiraeae* Kaltenbach, 1867. This species occasionally occurs as a pest on strawberries and has been discussed earlier in Chapter 6, p. 173. *A. spiraeae* forms a characteristic mine which is at first linear but after the first instar develops into an elongate blotch (Fig. 251).

SAXIFRAGACEAE

Hydrangea macrophylla Ser. (Australia)

This cultivated *Hydrangea* has been found with mines of *Tropicomyia polyphyta* (Kleinschmidt, 1960) at Toowong, Queensland (Kleinschmidt, 1970), and in New South Wales at Lisarow and at the Botanical Gardens, Sydney (Spencer, 1963b). A photograph of the silvery, epidermal mines is shown in Plate 12.

The only other record of Agromyzidae on *Hydrangea* is of *Phytomyza hydrangeae* Sasakawa, 1956, found on the wild host, *Hydrangea paniculata* Sieb. in Japan at Mt. Hira, Shiga Pref.

Plate 12: *Tropicomyia polyphyta:* mines and puparia in *Hydrangea macrophylla*, Toowong, Queensland (Dept. of Primary Industries, Brisbane, Queensland).

Philadelphus coronarius L. (Canada)

Mock Orange is widely cultivated and in Ontario it is heavily mined by a recently described species, *Liriomyza philadelphivora* Spencer, 1969a. The larvae form conspicuous white blotch mines (Fig. 520).

SCROPHULARIACEAE

Antirrhinum, *Digitalis* (Europe)

Napomyza scrophulariae Spencer, 1966h, has been bred both from stems and flower-heads of ornamental *Digitalis purpurea* L. in England, and puparia of what is probably the same species have been found in stems of *Antirrhinum*.

In the Angers district of western France where *Digitalis* is cultivated for medical purposes, the main damage was found to occur at the base of the stem, caused by secondary fungal attack.

N. scrophulariae is discussed in greater detail earlier in Chapter 5, p. 116.

SOLANACEAE

Nicotiana (Europe)

BUHR records mines of *Liriomyza bryoniae* (Kalt.) on a number of *Nicotiana* species in Mecklenburg (1941) and also on *N. alata* Link and Otto var. *grandiflora* Comes in Saxony. Cultivated *N. tabacum* L. was apparently not attacked (1964).

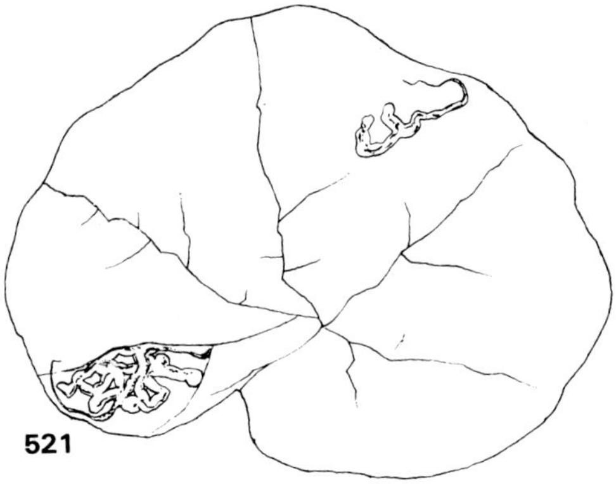

521

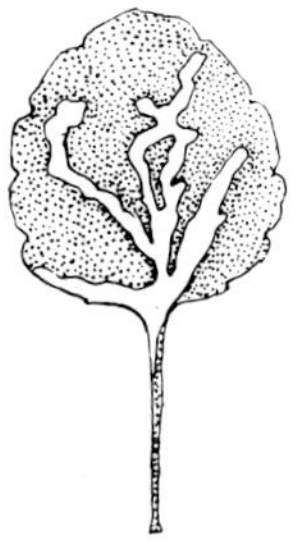

522

Fig. 521. *Liriomyza brassicae* (Ril.): leaf-mines on *Tropaeolum* (Australia).

Fig. 522. *Liriomyza huidobrensis* (Bl.): leaf-mine on *Viola* sp. (Argentina).

In England mines of *Phytomyza horticola* occur not infrequently on *N. tabacum* grown as an ornamental.

Petunia sp. (Europe, S. America)

In W. Europe cultivated *Petunia* is occasionally attacked by *Phytomyza horticola*.

In Argentina *Petunia* is a common ornamental in gardens around Tucuman. A heavy attack by *Liriomyza huidobrensis* was noted by Dr. P. WYGODZINSKY (personal communication), with about 50% of all leaves mined and with six or more mines per leaf.

TROPAEOLACEAE

Tropaeolum (Europe, S. America, Australia)

In Europe *Liriomyza strigata* occurs not uncommonly on this host.

In Argentina Nasturtium is a frequent ornamental in Tucuman gardens and is attacked by two leaf-miners – *Liriomyza huidobrensis* (posterior spiracles of larva each with 6 pores) and *L. sativae* (posterior spiracles each with 3 pores).

In Australia Nasturtium serves as a common host for *Liriomyza brassicae* (RILEY, 1884). The irregular white linear mines, with conspicuous black frass (Fig. 521) were particularly common in the Sydney area in January, 1958 (K.A.S.).

VIOLACEAE

Viola species (Europe, South America)

Cultivated violets or pansies are not normally attacked by Agromyzidae in Europe but one notable case occurred in England in the Manchester area in 1844 when serious damage was caused to the blossoms of pansies by the females making numerous feeding punctures in the petals, a procedure which usually only takes place in leaves.

The species was described as *Agromyza violae* Curtis, 1844, but this has since been synonymized with *Liriomyza strigata* by HERING (1957c).

In view of its interest and inaccessibility, the following extract from the original description of *Agromyza violae* (the Pansy-fly) in the *Gardners' Chronicle* (1844) may be quoted: 'During the last year a small fly made its appearance in the gardens about Manchester, which completely spoiled the more delicate flowers, by discharging the colours, and spotting the blossoms. The flowers I noticed which suffered the most, were those of the Pansy and *Phlox Drummondii*; whole beds of the former were so extensively attacked that I could not gather a perfect bloom after an hour of sunshine. The appearance thus caused is not dissimilar to that of the Thrips, which was also plentiful that year. For a long time I was unable to find out how the mischief was effected, so insignificant in itself is the culprit. I first noticed these flies on a flower of the Carlo Dolce Pansy, which has the upper petals of a dark blue, and the lower ones white; it had just passed from the blue to the white ground, when it secreted a blue liquid, upon seeing which I immediately suspected that the mischief was owing to the operation of the flies, and subsequent examination proved I was correct. The manner in which the fly effects its purpose is by puncturing the flower with a small sting situated at the tail, and, afterwards drawing its head to the hole it has made, it sucks up the juice, and by this means completely extracts the colouring matter'.

In Argentina cultivated violets are attacked by *Liriomyza huidobrensis* (P. WYGODZINSKI, personal communication) and a typical leaf-mine is shown in Fig. 522.

XII. CONTROL MEASURES*

The control of Agromyzidae is complicated by their internal feeding habits and by differences in the life histories of individual species. Insecticides which may be highly effective against adults resting on foliage or against free-living larvae may be quite ineffective against eggs or larvae protected within the tissue of the leaf or stem. Even systemics which may be effective with small herbaceous plants could not be used in a similar way against larvae boring internally in trees. Foliar sprays may be successful against species having only a single generation and a limited flight period, such as *Phytomyza ilicis*, but will be economically impractical against a species with six or more overlapping generations when adults can be found from spring until late autumn, such as *Paraphytomyza dianthicola*. In a number of species the puparium remains within the leaf or stem, in others it overwinters in the soil, and obviously a different control approach is required in each case.

A detailed knowledge of the life history of all Agromyzid pests, and ideally also of their parasites, is thus a pre-requisite for obtaining the best results from any programme of chemical control. It follows that accurate timing of applications, which will of course vary from year to year and is vital for the success of such programmes, can only be properly worked out with the necessary knowledge of the life cycle of the species. The spectacularly successful initial results obtained with DDT and some other insecticides in the immediate post-war years led many growers to ignore the effect of these powerful toxicants on the overall ecology of the pest species, with the now familiar result of the resurgence of the pest owing to the greater toxicity of DDT to its hymenopterous parasites. Examples of this effect have been mentioned in Chapter 1. However, it appears that the necessary lesson has still not been fully learnt and insecticides are still on occasions being used indiscriminately and therefore ineffectually in overall terms. The desirability of not attempting total kill of the pest but rather of reinforcing the natural control of parasites in an integrated programme is now generally accepted, but this approach is not always easy to work out in practice and has found little application with Agromyzid pests.

Before discussing modern chemical control methods in greater detail, it may not be without interest briefly to consider some of the first attempts at control of Agromyzidae. Many of the earliest control recommendations were limited to purely mechanical methods. CURTIS (1844), discussing damage to pansies and phlox in England, in the Manchester area, by *Liriomyza strigata* (as *Agromyza violae*, cf. p. 341), admits that no effective control can be applied until it is known where the eggs are deposited. In the meantime it is suggested that 'prize pansies might perhaps be preserved from their attacks by gauze-frames to cover over the pots, similar to those used for protecting meat against flesh-flies'. Discussing the

* See also Appendix A.

holly fly, *Phytomyza ilicis*, Curtis (1846) makes two suggestions: 'I expect no remedy could be found for preventing the mischief occasioned by the Holly-leaf fly, unless any trap could be invented to decoy the winged insects. If it were very important to protect the plants from their attacks, dusting the Hollies with soot and lime, when they are damp with dew might deter the flies from depositing their eggs in the leaves'. Lintner (1891) describing the considerable damage to chrysanthemums in the New York area caused by *Phytomyza chrysanthemi* (=*P. syngenesiae*, cf. p. 323), recommends 'plucking and destroying every infested leaf'.

Tryon (1895), in one of the first discussions of damage caused by *Ophiomyia phaseoli* in Australia, writes: 'it is recommended that whenever the presence (of the pest) is detected the entire growing crop should be destroyed, and the ground occupied by it planted for a time with other vegetables'. Jack (1913) first described damage by this species in Africa and his main suggestion is to sow a trap crop with the first rains which could be ploughed under after 3 weeks' growing and this would serve to reduce the initial infestation. He noted the differing resistance of some strains to this pest and emphasized the botanical side to the problem, urging investigations to discover the most resistant varieties. Moutia (1945) tried several novel repellents against *O. phaseoli* in Mauritius, such as diesel oil clay emulsion, tobacco decoction and even cow dung water. However, all these proved ineffective.

One of the first accounts of damage by *Liriomyza* species in America is given by Webster and Parks (1913); referring to damage to alfalfa they make these recommendations: 'doubtless cutting the crop for hay at once as soon as the depredations were observed would prevent a recurrence. Deep fall ploughing would bury the pupae so deep in the ground as to put them beyond the possibility of emerging as adults. Keeping down volunteer growth along ditch banks and in waste lands would greatly diminish the number of pupae which yearly enter hibernation.' Picard (1914b) also records damage to alfalfa in France by '*Agromyza nigripes*' (=*Agromyza* ?*nana*) and he suggests that the crop should be cut 'as soon as the white blotches appear'.

Nicotine in various forms has been widely used and has proved largely satisfactory as a control agent. In the United States Sanders (1912) reported that 'Black Leaf 40', a nicotine solution used as a spray with or without whale oil soap, 'proved a complete and satisfactory control' of '*Napomyza chrysanthemi*' (=*Phytomyza syngenesiae* Hardy). Sanders found this effective not only against adults but also against larvae – 'it is evident that the nicotine affects the larvae through the leaf epidermis by osmosis'. Smulyan (1914) also recommended nicotine, either as 'Black Leaf 40', 'Nico-Fume' or 'Nicoticide'. He added that picking the leaves 'is in most cases neither adequate nor satisfactory'. In British Columbia nicotine dust and lime proved highly successful against the holly fly, *Phytomyza ilicis*. As many as 3000 dead flies could be found beneath trees 10 ft high. However, emulsions had little effect against larvae owing to the tough consistency of the leaves. Mesnil and Marcel (1935) in France found that immersing chicory plants for 10 minutes in a 1% nicotine solution proved completely effective. In England Miles and Cohen (1936) also found that routine spraying of chrysanthemums with a nicotine solution at intervals of not more than 10 days gave effective control of *Phytomyza atricornis* (=*syngenesiae* Hardy). Nicotine dust and sulphur dust were also recommended where spraying might not be convenient.

In Australia MORGAN (1938) confirmed that nicotine sulphate plus white oil emulsion provided an effective means of controlling *Ophiomyia phaseoli* by spraying the leaves. MORGAN emphasized the critical importance of accurate timing. He had correctly noticed that the eggs of this stem-feeder are laid in the leaves and that normally three days are required before the young larvae reach the stem, where they would be relatively immune to nicotine spray. It was therefore recommended that treatment should begin on the third day after the plants commence to show above ground. If spraying was delayed until the fourth day after eggs were laid 'a sufficient number of larvae sometimes escaped treatment to cause serious damage to the plants'. It was found that few eggs hatched which had been laid within 48 hours of spraying. A four day interval between spray applications was therefore recommended. This method of control became standard practice in Australia until the introduction of DDT in 1947.

In Japan KUWAYAMA (1950) after field trials recommended derris emulsion against *Agromyza oryzae*, although in laboratory experiments pyrethroid insecticides had proved most effective.

Various cultural practices have been recommended and these maintain their validity, in combination with chemical control. In the Netherlands it was found that damage to carrots could be greatly reduced by earthing up the rows of young plants immediately the first foliage broke through the ground, thus preventing oviposition at this early stage (VAN'T SANT *et al.*, 1961). VAN DEN BRUEL (1938) in Belgium found that larvae of the two chicory flies *Ophiomyia pinguis* and *Napomyza cichorii* could be effectively killed by soaking the plants prior to transplantation to the forcing beds in hot water at 40°C for $1\frac{1}{2}$ hours. A temperature of 39°C was markedly less effective. Despite the simplicity of this procedure it failed to find favour with growers and has never been widely adopted.

The cleaning up and burning of infested plant remains after harvesting can substantially reduce populations in following generations. This is particularly recommended against *Ophiomyia pinguis* attacking chicory in Italy (SÜSS, 1970).

The new chlorinated hydrocarbon and organophosphorus compounds which became available at the end of World War II were used initially as dusts or sprays. Later, soil drenches, seed dressings and systemic granules were widely used in appropriate circumstances and have many advantages, although conventional DDT as a spray or dust is still significantly cheaper.

In Australia HELY (1947) tested DDT against *Ophiomyia phaseoli* and found that 0.05% sprays gave effective control. Dusts were not satisfactory. HELY considered that DDT was superior to the standard nicotine treatment owing to 'the lower cost of the diluted spray, smaller quantity required per acre, ease of mixing, control of associated pests and apparent safety of application. However, it was noted that DDT treatments resulted in an increase of red spider infestation.

SPEYER and PARR (1948) tested DDT and BHC in England against *Liriomyza bryoniae* (as *L. solani* Hering). BHC was found to be the more effective but the really significant finding was the greater toxicity of DDT to hymenopterous parasites than to the adult Agromyzids. 'A recently emerged male which had moved about on the surface of a dusted leaflet for about two minutes was removed to a fresh leaflet: the insect gradually lost control of its limbs, became completely paralysed in four hours, and subsequently died. Adult Braconid parasites (*Opius* and *Pachysema*) died much more rapidly. The movements of a female *Opius* which had walked for a few seconds over a lightly dusted surface, became uncoordinated in 13 minutes, after one hour the insect could make use neither of its legs nor wings, and in three hours had become practically motionless'. This work has been largely overlooked but was confirmed later in America by HILLS and TAYLOR (1951) and WENE (1953, 1955), who explained increases in leaf-miner infestations following the use of DDT by the greater sensitivity of the parasites.

WOLFENBARGER (1958), in a useful review of a decade of control measures against *Liriomyza* species in Florida, showed that DDT was only partially effective against leaf-miners and definitely less effective than aldrin, dieldrin and endrin, and also a number of organophosphorus compounds. KELSHEIMER (1947) reported that in Florida DDT gave apparent control of *Liriomyza sativae* (as *pusilla* Mg.) but added: 'Populations build up at an alarming rate upon cessation of DDT treatments'. HILLS and TAYLOR (1951), working in Arizona on the *Liriomyza* sp. attacking cantaloups, found that 'repeated applications of DDT were ineffective against the leaf-miners but reduced the parasite population, resulting in increased leaf-miner populations'. WILCOX and HOWLAND (1952) confirmed that DDT was relatively ineffective against *Liriomyza* species in California. On the other hand various combinations of DDT both with other organochlorine and with organophosphorus compounds were significantly more effective. WILCOX and HOWLAND (1955) again confirmed that DDT gave only fair to poor results in controlling *Liriomyza huidobrensis* (as *langei* Frick) on peas in California. WENE (1953 and 1955), investigating control of *Liriomyza sativae* (as *subpusilla* Frost) attacking peppers in Texas, found that DDT increased the number of mined leaves by reducing the parasites. This effect of accentuating the leaf-miner problem was also found with aldrin, TDE and methoxychlor. MAYEUX and WENE (1950) report that good control of *Liriomyza sativae* (as *pusilla* Mg.) was obtained by a 5% dust mixture of chlordane-DDT. Control is effected by killing adults before oviposition, not by killing larvae in the mines. They warn that 'miners are likely to become very heavy when dusting is discontinued or when irregular applications are made'.

In India VENUGOPAL and VENKATARAMANI (1954) obtained good results against *Melanagromyza hibisci* (as *obtusa* Mall.) by using either BHC or DDT (0.05 or 0.01%) as prophylactic sprays on apparently healthy plants of *Hibiscus esculentus* at fortnightly intervals. In Egypt MAHER ALI (1957) found that DDT was inferior both to chlordane and diazinon in controlling *Ophiomyia phaseoli*, and ALY (1971), also in Egypt, found that DDT gave only moderate protection against *Pseudonapomyza spicata* attacking maize.

The limited effectiveness of DDT alone against Agromyzidae appears to be established but other organochlorine compounds are recorded as having greater success. WOLFENBARGER (1947) found that chlordane gave very satisfactory control over *Liriomyza sativae* (as *pusilla*) on potatoes. MICHELBACHER *et al.* (1951) tested aldrin and dieldrin in California against *Liriomyza sativae* (as *subpusilla* Frost) on melons. Both sprays and dusts were used by ground and air. 'Dieldrin proved to be more effective than aldrin irrespective of formulation, and sprays were better than dusts'. KERR (1951) found that aldrin and dieldrin were significantly more effective than Lindane and chlordane as a larvicide used against *Phytomyza opacae* Kulp (as *ilicis* Curt.) on Rhode Island, U.S.A. MICHELBACHER, BACON, MIDDLEKAUF and ERWIN (1953) found dieldrin to be somewhat more effective than aldrin and heptachlor in controlling *Liriomyza sativae* (as *subpusilla* Frost) on tomatoes in California.

WILCOX and HOWLAND (1952) reported that camphechlor (Toxaphene), BHC, Lindane and chlordane were more effective than DDT in controlling *Liriomyza* species on tomatoes in California. It was later found that dieldrin and EDN gave even better control.

THEVASAGAYAM and CANAGASINGHAM (1960) carried out detailed investigations into the effectiveness of insecticides against *Melanagromyza obtusa* in Ceylon. Seven insecticides were tested, namely dieldrin, Dipterex, endrin, parathion (Folidol), Gusathion, malathion and Phosdrin. Best results were obtained with dieldrin. These authors conclude that pests of dhal could be controlled economically by insecticides and write: 'The total yield of 2132 lb per acre from the entire trial area shows that treatment with insecticide significantly increases yields. This compares favourably with yields at Maha Illuppallama which have been in the neighbourhood of 300–500 lb per acre in the past few years. The seed weight in all the treatments has been about 65–75% of the bulk pod weight and in one case it was as high as 87% of bulk weight. Normally the seed weight is about 50% or less of the bulk weight. This is due to heavy damage to seed by pests.'

Of the various organophosphorus insecticides used against Agromyzidae, parathion and diazinon have had considerable success and have found general application in some areas, but the most consistently satisfactory results appear to have been achieved with dimethoate. Several workers have emphasized the desirability of using these compounds rather than chlorinated hydrocarbons owing to their lesser persistence.

In Florida as early as 1948 parathion proved to be 'outstanding' in preventing leaf-mines (WOLFENBARGER, 1948). JEFFERSON and EADS (1953) also found that parathion was most effective in controlling *Liriomyza langei* in California on asters, and it was better than chlordane, Toxaphene, BHC (benzene hexachloride) and EPN. In the Netherlands DE BROUWER and VAN OFFEREN (1966) found parathion both as dusts and sprays 100% successful against larvae of *Liriomyza bryoniae* (=*solani* Hering) attacking tomatoes and it was almost equally effective in killing eggs. Dichlorvos (DDVP) and mevinfos (Phosdrin) were also 100% effective against larvae but significantly less so against eggs. WOLFENBARGER (1958), however, reported that there is some evidence in Florida that resistance had developed to parathion and that it was 'not so effective after 10 years as it was initially'. Nevertheless GETZIN (1960) reported that in Texas parathion was one of the most effective chemicals and in the Rio Grande Valley there was no evidence of parathion-tolerant leaf-miners.

Several workers have confirmed that the best control could be obtained with diazinon. It was formally recommended for use against leaf-miners in Florida, following its approval by Federal regulatory agencies (WOLFENBARGER, 1958).

However, only a few years later WOLFENBARGER (1961) admitted that this, like parathion, was proving no longer as effective as earlier. Combinations of parathion and diazinon could often be used to provide satisfactory control. In Egypt MAHER ALI (1957) found that diazinon was superior to parathion in controlling the bean fly, *Ophiomyia phaseoli*. SMITH, BOSWELL and WAVE (1962) recommended diazinon, and also Dibrom, against the chrysanthemum leaf-miner, *Liriomyza trifolii* (as *Liriomyza* sp.n.) in Maryland, U.S.A. SHELL (1970) and SALINAS (1966) also recommend diazinon for use against the tomato leaf-miner, *Liriomyza sativae*, in Venezuela. In Egypt ALY (1971) found diazinon gave satisfactory control of *Pseudonapomyza spicata* on maize. AUDENARD and D'AGUILAR (1969) tested several organophosphorous compounds against *Liriomyza bryoniae* in France on lettuce, tomato and melon. Diazinon, and also dimethoate, ethion (=diethion) and malathion all gave 100% control.

Among this group of insecticides dimethoate has been shown to provide the most satisfactory control by many workers. GETZIN (1960) tested a number of compounds against *Liriomyza sativae* (as *munda* Frick) on peppers and cantaloups in Texas. Most highly recommended were dimethoate, ethion and Delnav, with residual toxicity lasting for 9 days. Parathion and diazinon were satisfactory but lacked this residual toxicity and had the further disadvantage, together with ethion, of being highly toxic to one of the main parasites, *Derostenus variipes*. WOLFENBARGER (1961) in Florida found dimethoate the most effective foliar spray against larvae. WOLFENBARGER and GETZIN (1963) found that weekly applications of dimethoate gave complete control of *Liriomyza sativae* (as *munda*) on peppers in Texas and such treatment resulted in a quantifiable increase in yields as follows:

	Weekly applications	*One application*	*No. applications*
Yield (ton/acre)	6.29	5.80	4.75

In Belgium LOUNSKY (1961, 1963a, 1966b) confirmed that dimethoate gave consistently good results against the chicory flies, *Ophiomyia pinguis* and *Napomyza cichorii*. Soil drenches applied after the roots have been moved to the forcing beds have now become standard practice in Belgium and give complete control. Earlier work on control measures in Belgium has also been discussed by VAN DEN BRUEL (1938, 1941) and VAN DEN BRUEL, LOUNSKY and BERNARD (1958). In the Netherlands VAN'T SANT (1961) initially obtained good results from parathion and diazinon but finally recommended dimethoate (Rogor) as giving virtually 100% control for use on chicory in the field but owing to relatively high residues, parathion and diazinon were still recommended for use in the forcing beds. Current practice is to spray plants in the field with dimethoate in the latter half of August and also twice in September. After removal to the forcing beds dimethoate is again sprayed on top of the roots (VAN'T SANT, personal communication, 1972).

SÜSS (1970) also recommends dimethoate against *Ophiomyia pinguis* in northern Italy. Both SALINAS (1966) and SHELL (1970) suggest that best control of the tomato leaf-miner in Venezuela can be obtained with dimethoate.

Experiments recently carried out with various insecticides in Cuba against *Liriomyza sativae* attacking tomatoes and potatoes showed that by far the best control was obtained with omethoate (Folimat) but dimethoate also gave acceptable results. Twelve further preparations, including diazinon and phosphamidon were significantly less effective, clearly indicating the difficulties in controlling this leaf-miner. A detailed paper on this work is in course of preparation (H. HOFFMANN, personal communication).

HUSSEY *et al.* (1969) recommended both diazinon and gamma-BHC for control of the chrysanthemum leaf-miner, *Phytomyza syngenesiae*, in England but HUSSEY (1969) confirmed that strains resistant to BHC and to diazinon had developed, though considerable variation in this respect could be detected in different populations.

WOLFENBARGER and GETZIN (1963), in a series of interesting experiments in Texas, investigated a number of phosphates and other new chemicals, and also various insecticide-surfactant combinations, for control of *Liriomyza sativae* (as *munda* Frick). Among organophosphorous compounds Bayer 29493, dimethoate and ethion were the most effective materials. Dibrom, Tedion, Kepone and Ronnel were ineffective after 3 days. All insecticides added to the surfactants B-1956 and X-100 gave significant initial control. Diazinon and parathion in combination with B-1956 gave increased control as the surfactant concentration was increased. WOLFENBARGER (1964) tested further insecticides and found that the phosphates Bidrin, Bayer 25141 and Monsanto 40294 gave good control at all dosages but Toxaphene and endosulfan in various oil and surfactant combinations were not effective toxicants.

In India SOHI and VERMA (1967) tested several organophosphorous compounds against the pea leaf-miner, *Phytomyza horticola* (as *atricornis* Mg.), working with *Tropaeolum*, another of its favourite hosts. Using foliar sprays dimethoate, phosphamidon and methyl demeton gave protection for a period of two weeks. Substantially better results were obtained using seed dressings and with this technique methyl demeton and phosphamidon gave protection for six weeks, but dimethoate was slightly less effective and infestations started after only four weeks.

Carbamates have not been widely used against leaf-miners. WOLFENBARGER and GETZIN (1963) tested four compounds in Texas – Sevin, Zineb, Bayer 39007 and Zectron but considered that these were not effective. WOLFENBARGER (1964), however, found that carbaryl in various oil and surfactant combinations was superior to Toxaphene and endosulfan. In India SOHI and VERMA (1967) considered that 0.2% carbaryl, as a foliar spray, provided as good control as phosphamidon, methyl demeton and dimethoate. GURNEY and SCOPES (1970) report that in England conflicting results have been obtained with carbaryl used against *Phytomyza syngenesiae*. Carefully timed experiments showed that while initial control is good, the larvae become immune to this chemical as they mature.

Seed dressings have proved conspicuously successful in controlling *Ophiomyia phaseoli* in Tanganyika and Rhodesia. TAYLOR (1958) suggested that best results are obtained with dieldrin, using a wet, emulsifiable concentrate dressing. Protection was obtained for 30 days after germination. Of dry dressings aldrin appeared to give the best results. In further trials in Rhodesia TAYLOR (1959) reported that wet seed dressings of dieldrin and endrin gave 100% protection, while aldrin was little less effective, with only a 0.5% infestation. Chlorthion gave some protection but gamma-BHC, malathion and dimethoate (Rogor) had little effect. The rate of infestation of the untreated control was 80%. TAYLOR indicates that it is not known exactly how these treatments control *O. phaseoli* but suggests that they have a systemic effect 'by killing the larvae at some stage before they descended to soil level to pupate'. Similar trials were carried out in Tanganyika by WALKER (1960) with both dry and wet dressings. Wet dressings generally proved superior to dry dressings. The best results were obtained with endrin and it was again confirmed that with this type of treatment the chlorinated hydrocarbons gave better results than organophosphorous insecticides. SWAINE (1969) also confirmed these findings after yet further trials in Tanganyika.

Similar results had been obtained in Ceylon by WICKRAMASINGHE and FERNANDO (1962) who carried out detailed comparative tests with seed dressings, soil treatments and foliar sprays for the control of *Ophiomyia phaseoli*. Adequate control was obtained with sprays of DDT, endrin and parathion. Soil drenches, prior to sowing, with the chlorinated hydrocarbons Shell WL 1650, aldrin and gamma-BHC, were all highly effective in that order. Used as seed dressings the organochlorine compounds were definitely superior to the organophosphates. The best results of all were obtained with endrin, with an infestation rate of only 0.7% at the end of six weeks, compared with 63% or more attack in the untreated controls. It is concluded that the chlorinated hydrocarbons used as seed dressings in these trials may have been acting systemically and it was confirmed that endrin has a larvicidal effect when used as a pre-sowing treatment.

In Thailand SEPSAWADI and MEKSONGSEE (1970) carried out trials against *O. phaseoli* using seed dressings of Furadan and also granules of carbofuran (Furadan), aldicarp (Temik) and phorate (Thimet) and obtained good control.

HARDING and WOLFENBARGER (1963) investigated the effectiveness of a number of granular systemic insecticides against *Liriomyza sativae* (as *munda* Frick) attacking southernpeas and cucumbers in Texas. They emphasize the increasing interest in this form of control for three good reasons: '(1) many of these materials have demonstrated insect control 30–50 days after application while foliar applications generally provide only 5–7 days control; (2) these materials can usually be applied as easily as a spray application; and (3) the predator-parasite host relationship may not be unbalanced.' On southern peas, phorate and Di-Syston were more effective than dimethoate and methyl demeton; on cucumbers, phorate gave markedly the best results.

A considerable variation in the degree of control obtained with a number of insecticides used by different workers against different pests in different parts of the world has been noted. The following examples may be cited relating to Dipterex, Phosdrin and malathion. In Ceylon Dipterex was the least effective of seven insecticides tested against *Melanagromyza obtusa*; both malathion and Phosdrin gave medium to poor control (THEVASAGAYAM and CANAGASINGHAM, 1960). Phosdrin gave only poor control of *Napomyza cichorii* (as *lateralis* Fall.) in the Netherlands, while Dipterex gave better control than diazinon but was markedly inferior to Rogor (VAN'T SANT, 1961). In Ceylon Dipterex as a soil soak against *Ophiomyia phaseoli* gave negligible control but as a foliar spray was reasonably effective at weekly intervals but ineffective at fortnightly intervals (WICKRAMASINGHE and FERNANDO, 1962). In Texas Phosdrin used as a spray against *Liriomyza sativae* (as *munda* Frick) gave only moderate control (WOLFENBARGER and GETZIN, 1963). In Venezuela SALINAS (1966) found that Dipterex gave poor control of the tomato leaf-miner but SHELL (1970) recommended this as an effective treatment. In the Netherlands DE BROUWER found Phosdrin 100% effective as a larvicide used against *Liriomyza bryoniae* (as *solani* Hering) on tomatoes. HUSSEY (1969) reports that in England in 1958 malathion was 100% effective against *Phytomyza syngenesiae* but in 1966 gave very little control.

These varying results may only in part be explainable by differing environmental conditions or methods of application, or by differences in the life histories of the pests concerned. Constant re-assessment under varying conditions even of established insecticides thus seems desirable. Entirely predictable results with specific compounds can still not be guaranteed and the complex of factors affecting the desired level of control are far from being properly understood. In England further investigations on a number of insecticides are currently being carried out at the Glasshouse Crops Research Institute. These include pirionicarb (Pirimor), carbophenothion (Trithion), trichlorphon (Dipterex), dimethoate (Rogor), dioxothion (Delnav) and bromophos as sprays, and malathion as a dust. Very effective results are being obtained with all the sprays but malathion is proving relatively ineffective (GURNEY, personal communication).

Current recommendations of leading manufacturers of insecticides for leaf-miner control cover mainly organophosphorous compounds such as parathion, diazinon, phosphamidon, ethion, Bidrin and Birlane; but BHC, gamma-BHC and the combination of dimethoate and DDT are also included. Preparations recommended against leaf-miners by one leading manufacturer include Disyston, Folimat, Folithion, Lebaycid, Tamaron and Dipterex. The latter is specifically recommended against *Liriomyza cicerina* (p. 90) and *L. flaveola* (p. 259). Several still recommend nicotine preparations, at least for horticultural use. One supplier suggests that Methomyl, a carbamate with systemic properties, will prove effective against insects showing resistance to organophosphates. Phosdrin is particularly recommended, in view of its rapid breakdown and thus limited toxicity against parasites, which will survive applications while in the pupal stage; by the time they emerge this insecticide will have broken down to harmless components. It is appreciated that very accurate timing will be necessary to obtain such optimum results.

Application of biological control programmes against Agromyzid pests has been limited but nevertheless this approach has achieved some success in isolated cases. Several programmes have recently been initiated but their effectiveness cannot yet be assessed.

The European holly, *Ilex aquifolium*, was introduced to the coastal areas of southern British Columbia and with it its host-specific Agromyzid, *Phytomyza ilicis*, which by 1931 was damaging 80% of the leaves in commercial holly plantations (TURNBULL and CHANT, 1961). Nicotine sprays had considerable success in controlling this leaf-miner but it was decided in 1935 to supplement this by introducing European parasites. Between 1936 and 1939 numbers of five species were sent to Canada (McLeod, 1954), comprising two Euloplids, two Pteromalids and one *Opius* sp., as follows:

	Vancouver Island	*Mainland*
Chrysocharis gemma (Walk.)	34,564	11,393
Chrysocharis pubicornis (Walk.)	1,978	179
Cyrtogaster vulgaris Walk.	2,227	471
Sphegigaster flavicornis Walk.	6,359	852
Opius ilicis Nixon	33	10

The first specimens to be received in B.C. consisted of 1941 adults of *Chrysocharis gemma* Walk. (Downes and Andison, 1940). The species became well established on Vancouver Island and there accounted for 90% of the parasitism of *P. ilicis*. Despite the far smaller numbers of *Opius ilicis* released, this species has become fully established in Vancouver and is responsible for 80–90% of the parasitism there (McLeod, 1954). The two Pteromalids, *Sphegigaster flavicornis* (Walk.) and *Cyrtogaster vulgaris* Walk. successfully established themselves but recoveries have been small and these two species appear to be of minor importance. The second Eulophid, *Chrysocharis pubicornis* (Walk.) (=*syma* Walk.), has apparently failed to become established (McLeod, 1962).

Turnbull and Chant (1961) consider that this project has had only limited success. In commercial plantations, where holly is grown for the Christmas trade the unsightly mines still occur and spraying is still necessary. In this area therefore the project has failed. Holly is also widely used as an ornamental in gardens, parks and boulevards. These trees are rarely sprayed and parasitism usually keeps the population of *Phytomyza ilicis* at a level where the trees are not seriously disfigured. To this extent therefore biological control is proving successful.

In work which appears to have been largely overlooked, Kelsey (1937) in New Zealand clearly demonstrated the potential of a parasite to control a leaf-mining pest in the restricted environment of a glasshouse. He was working with *P. syngenesiae* (as *atricornis* Mg.) attacking ragwort and with a Braconid parasite, *Dacnusa areolaris* Nees. The results achieved are closely parallel to those obtained at the G.C.R.I. in England with chrysanthemums (see below). Kelsey summarizes his work as follows: 'Several generations of the fly and parasite were kept under close watch during their natural development in the insectary containing ragwort, and it was found that the parasite was gradually subduing the fly, but not rapidly enough for our purposes. In order to hasten this control, leaves containing the puparia of the fly were collected in the insectary and placed in an emergence box, and when emergences had ceased the flies, which amounted to 56% of the total emergences, were destroyed, and the parasites returned to the hothouse. At the completion of the third generation of the fly and parasite close on 100% of the fly puparia contained parasites, and the leaf-miner infestation was completely checked. This method of assisting parasites could be easily adopted by glasshouse-owners who are faced with similar problems, providing that a parasite of the pest is present.'

SCOPES and GURNEY (1969) are currently working on the biological control of *Phytomyza syngenesiae* attacking chrysanthemums in England. The original work used the endoparasitic Eulophid, *Chrysocharis melaensis* (Walker). Currently, however, even better results are being achieved with the ectoparasite, *Diglyphus isaea* (Walker). It has been found that females of this species 'sting' and kill many larvae, in addition to those on which they oviposit. The rapidity and effectiveness of control with this species have been spectacular. In one case an estimated population of half a million *P. syngenesiae* was virtually destroyed within a month by 500 *D. isaea*. Effective, though somewhat unscientific control, can be achieved by introducing into an infested greenhouse a few heavily mined leaves of some wild host, such as *Sonchus oleraceus* L., on the reasonable assumption that a significant proportion of the puparia will contain parasites of some kind (SCOPES and GURNEY, 1969).

Preventive biological control can be practised successfully in glasshouses where chrysanthemums have no infestation, by first introducing a small number of *P. syngenesiae* and then, with this host base for the parasite and before it has built up a damaging population, introducing a known effective control agent, such as the Eulophid, *Diglyphus isaea* Walk. In this way small, stable populations of both host and parasite can be established, and any risk of the leaf-miner becoming injurious can be forestalled (HUSSEY and SCOPES, personal communication).

Further parasites which might be used successfully for biological control of *Phytomyza syngenesiae* are referred to earlier in Chapter 8, p. 234.

The possibilities of biological control of *Ophiomyia phaseoli* have recently been investigated by the East Africa Station of the Commonwealth Institute of Biological Control. However, GREATHEAD (1969) concludes that in spite of the existing heavy rate of parasitism, sufficient flies survive to cause substantial damage and the parasites cannot prevent, and can at most reduce, the level of infestation. It is nevertheless suggested that certain parasites might with benefit be introduced to areas where they are known not to be present.

An attempt is currently being made to control *O. phaseoli* in Hawaii with the parasite *Opius melanagromyzae* Fischer, imported from East Africa. Four shipments of 100–200 puparia were sent to Hawaii in April–May, 1969. Releases were made in June and a recovery was made one month later. No further information has been obtainable on this project.

The possibilities of sterilizing males of *Phytomyza syngenesiae* are also being investigated at the G.C.R.I. in England. However, difficulties are being encountered in inducing the males to imbibe the chemosterilants. Work is continuing but at the moment no way has been found of overcoming this problem (B. GURNEY, personal communication).

One of the main difficulties in the management of control measures lies in the impossibility of predicting in advance whether or not a serious infestation will develop. As with most other groups of insects, the population dynamics of the Agromyzidae are imperfectly understood. Information on the most important factor of all in determining population levels, hymenopterous parasites, is very limited and further study of this group is an urgent requirement. Of the 156 species of Agromyzidae discussed here, no parasites at all are known in 50 species, and information on most of the other species is certainly incomplete.

As early as 1913 WEBSTER and PARKS appreciated the significance of parasites in effecting natural control of one of the most serious Agromyzid pests in America, a *Liriomyza* sp. attacking alfalfa in California, when they wrote: 'excessive parasitism has prevented it from becoming destructively abundant.' MILES and COHEN (1936) recorded 98% of puparia of *Phytomyza syngenesiae* attacking chrysanthemums in England as being parasitized. In Kenya in October, 1971, 100% parasitism was found by the author in 40 puparia of *Tropicomyia flacourtiae* collected on coffee at Nairobi and Ruiru.

Few detailed studies of the life histories of parasites of Agromyzidae have been carried out but three valuable contributions may be mentioned. CAMERON (1939) discussed the life history of *Opius ilicis* parasitizing the holly leaf-miner, *Phytomyza ilicis*, and included a fascinating account of the competition between *Opius ilicis* and a Eulophid, *Chrysocharis gemma*. ASKEW (1968b) made a comprehensive study of the parasite complex of the two leaf-miners on *Laburnum* in England, *Agromyza demeijerei* and *Phytomyza cytisi*. GREATHEAD (1969) in East Africa described the life histories of the parasites of the bean flies. *Ophiomyia centrosematis*, *O. phaseoli* and *O. spencerella*.

Despite the detailed studies of a number of authors in England and America showing beyond doubt the destruction caused to the parasites of Agromyzidae by the use of insecticides and the correlation between this and the increase of leaf-miner infestations (SPEYER and PARR, 1948; WOLFENBARGER, 1958; HILLS and TAYLOR, 1953; WENE, 1955; GETZIN, 1960), many modern insecticides have continued to be used with a total lack of discrimination and with no positive effort to safeguard parasite populations. At the present time, however, this aspect of control is appreciated by most manufacturers and probably also by the more enlightened growers. Useful guidelines for sensible use of insecticides taking into account the inherent disadvantages in the use of chemicals are given by SHENTON (1971): 1. Choose materials that are particularly toxic to the pests; 2. Choose pesticides with residues which are broken down quickly; 3. Use materials at dosages that give less than 100% kill of the pests. 4. Place pesticides so that, as far as possible, they reach the target pest with the least disturbance to natural predators and parasites.

In the absence of any possibility of predicting serious infestation levels in advance, some recommendations have at least been made regarding quantitative assessment of incipient damage as an indication of the need for chemical control measures. MICHELBACHER *et al.* (1951) suggested that 'the efficiency of a melon leaf is considerably reduced if it contains 20 or more miners and a leaf containing over 10 miners is approaching the point where some control methods should be employed'. WOLFENBARGER (1961) made suggestions based on the number of flies seen on the plants. He wrote: 'The number of flies which it is suggested would determine the application or non-application is hereby suggested as two to three per plant. It is suggested that two flies per plant may be an insufficient number for which to apply an insecticide. If the number is three flies or more per plant it is suggested that sufficient toxicant be applied to kill larvae in the leaves. Plant size, season and number of mines in the leaves are factors which may justifiably alter the decision to apply a toxicant. Two flies per seedling plant with mines in evidence is obviously justification for applying a killing treatment. Three flies per plant on a large tomato plant with but few mines in the leaves are regarded as insufficient to apply treatments.' WOLFENBARGER and WOLFENBARGER (1966) proposed a more scientific method of assessing whether or not control treatments are necessary. They suggested a sequential sampling plan, from which a graph can be drawn with two decision lines, indicating whether or not control measures are required.

This brief review of the various methods which have been used for controlling Agromyzidae – mechanical, cultural, chemical and biological – indicates that these differing approaches have been applied in a largely arbitrary and unscientific manner, frequently with a total disregard of the most important single control factor, the pest's own parasite complex. VARLEY (1971) has emphasized the generally unscientific approach which has been adopted in the use of insecticides, with rarely any attempt being made to establish and achieve the optimum degree of kill of the pest, or to decide whether persistent or non-persistent compounds should be employed. VARLEY and GRADWELL (1968) and HASSEL and VARLEY (1969), working with the winter moth, *Operophtera brumata* (L.) in England, proposed the construction of mathematical models for calculating and predicting population levels of a parasite and its host. It would clearly be of the greatest value if comparable models for the integration of biological and insecticidal control of the main Agromyzid pests could be worked out.

SMITH and HAGEN (1959) make a particularly lucid plea for integrated control programmes, combining natural or manipulated biological with chemical control. Such programmes are an obvious requirement for effective control of Agromyzid pests. Although information both on pest species of Agromyzidae and their parasites is far from complete, sufficient knowledge is now available for a more considered effort to be made in applying rational, integrated control programmes. As early as 1957 LANGE *et al.* showed the way when they wrote: 'The best solution to the leaf-miner control programme seems to lie in the judicious use of insecticides in such a way as to more fully utilize the natural enemies.' However, as WOLFENBARGER (1961) correctly remarked: 'There is no quick, simple nor easy solution to the leaf-miner problem.'

XIII. BENEFICIAL SPECIES

Very few Agromyzidae have been used or even actively considered for possible control of weeds. One species, *Phytomyza orobanchia* Kaltenbach, has been conspicuously successful in controlling *Orobanche* attacking a variety of crops in the U.S.S.R.; another, *Ophiomyia lantanae* (Froggatt), has been introduced to many parts of the world where Lantana has become a problem but its effectiveness in controlling Lantana has been very limited. Three further species have been or are being investigated for possible use against *Convolvulus arvensis* L., *Cuscuta* and *Striga*; these are *Melanagromyza convolvuli* Spencer, *M. cuscutae* Hering and *Ophiomyia strigalis* Spencer. Furthermore, it has been discovered that another species feeding in stems of *Convolvulus* in Egypt and Israel represents *Mel. albocilia* Hendel, described in 1931 from Budapest and Vienna. KELSEY (1937) suggested that *Phytomyza syngenesiae* (Hardy) might have some value in controlling *Senecio jacobaea* L. in New Zealand. These seven species are discussed in greater detail below.

Many serious weeds are attacked by either stem-boring, seed-eating or leaf-mining Agromyzids and the possibilities of using these for biological control appear to deserve further study. In a number of cases they are host-specific, and introduction into a continent outside the range of their normal distribution where they might escape natural control by their normal parasites could lead to a significant population build-up which might prove destructive to their particular host. Alternatively, if the parasite complexes of these species are better known, critical application of selective insecticides might substantially reduce the parasite populations and achieve a useful increase in the Agromyzid populations. There are many common species with which it would be possible to conduct experiments on these lines. Some weeds and their Agromyzids for which these possible approaches might be considered are also discussed below.

A further interesting possibility would be to attempt to colonize weeds which are not known to have any specific Agromyzids of their own with species feeding on closely related genera. For example, *Striga* has no Agromyzid feeding in the seed-heads but related parasitic genera, such as *Euphrasia*, *Melampyrum*, *Odontites*, *Pedicularis* and *Rhinanthus*, each have seed-feeding *Phytomyza* species which conceivably might be persuaded to transfer to *Striga* in the absence of their normal host. If any of the five species concerned – *Phytomyza affinis* Fallén (*Euphrasia*), *P. rostrata* Hering (*Melampyrum*), *P. isais* Hering (*Odontites*), *P. tenella* Meigen (*Pedicularis*) and *P. varipes* Macquart (*Rhinanthus*) could be established experimentally on *Striga*, a few generations might suffice for complete adaptation to the artificial but closely-related host. There are certainly other comparable cases which might repay detailed investigation.

1. **Melanagromyza convolvuli** Spencer, 1971c – West Pakistan

This species was confirmed as new during the course of this study and was described in a separate, earlier paper in order to provide it with a name, in view of its interest to the Commonwealth Institute of Biological Control.

M. convolvuli is one of the large complex of greenish species which occur throughout the Oriental and Ethiopian Regions. Superficially, it closely resembles the common and widespread species, *M. metallica* (Thomson), and a positive identification of caught specimens will be possible only from the male genitalia. Both the head (Fig. 523) and the wing (Fig. 524) are typical of many species in the genus. The wing length varies from 1.9–2.2 mm. The aedeagus (Figs. 525, 526) is entirely distinctive. The puparial spiracles (Fig. 527), which can be of great assistance in identifying species in this group, are separated by twice their own diameter, each with an irregular ellipse of about 9 pores around a central chitinized area.

Oviposition takes place near ground level in the soft stems of the only known host, *Convolvulus arvensis* L. (bindweed). The larvae bore downwards mostly along the peripheral parts of the stem; pupation takes place in the lower stem. Larval feeding in the stems or lateral branches considerably weakens the plant and in non-irrigated areas of limited rainfall, the combined effect of the drought and larval attack causes the infested branches to dry up. There is no apparent effect on the growth of plants in irrigated fields (GHANI, personal communication).

Convolvulus arvensis is a cosmopolitan weed which has spread widely in the United States and is now causing concern to agriculturists in many areas. Possibilities of biological control are being actively studied and *Melanagromyza convolvuli*, which is certainly host-specific on *Convolvulus*, could be a useful factor in such a programme.

M. convolvuli is known only from West Pakistan, where it occurs both in the hills and on the plains. At the time of its description it was the only Agromyzid known on *Convolvulus*. Recently, however, specimens bred from *Convolvulus arvensis* have also been seen from Egypt and Israel. Externally the flies agree virtually completely with those from Pakistan but the male genitalia (Figs. 528, 529) clearly indicate that this is a distinct species, which I have been able to identify as *M. albocilia* Hendel.

2. **Melanagromyza albocilia** Hendel, 1931 – Austria, Hungary, Israel, Egypt

This species was described from three caught specimens from the vicinity of Vienna and of Budapest. No further specimens have since been found.

Recently I have seen four specimens from the Giza region of Egypt reared from stems of *Convolvulus arvensis* and also six from the same host from four different localities in Israel – Chof Ashkelon, 3 km East of Ashdod, Kffar Sirkin and 5 km S.W. of Raffid. The species is not distinguishable on external characters from *M. convolvuli* but differences in the aedeagus clearly indicated that the species on *Convolvulus* in Pakistan is distinct from that in the eastern Mediterranean area.

I have now found that this species beyond doubt represents *M. albocilia* Hendel, the host of which has not hitherto been known. The aedeagus of the lectotype was illustrated in a revision of European *Melanagromyza* species (SPENCER, 1966b:

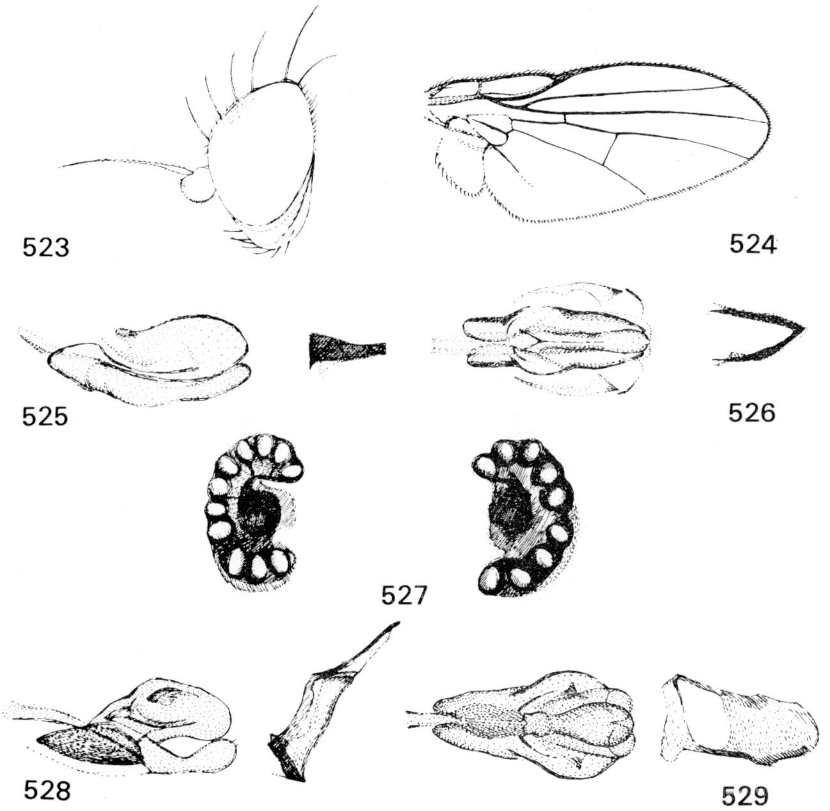

Figs. 523–527. *Melanagromyza convolvuli* Sp.: 523, head; 524, wing; 525, aedeagus, side view; 526, same, ventral view; 527, posterior spiracles of puparium.

Figs. 528–529. *Melanagromyza albocilia* Hd.: 528, aedeagus, side view; 529, same, ventral view.

Figs. 5, 6). Two further illustrations of a male from Egypt are shown in Figs. 528, 529.

It is not known to what extent larval feeding damages *Convolvulus* in Egypt, and eastern Europe. However, the possible value of *M. convolvuli* for use in biological control must now be re-assessed in the light of this second *Convolvulus* feeder. Little is so far known about its habits and a detailed study of its life history, either in Israel or in Egypt, is now required.

3. **Melanagromyza cuscutae** Hering, 1958 – Europe, Kazakhstan, Pakistan, India, Burma

M. cuscutae is a typical member of this genus, with over 80 species in the area of its distribution. Identification is difficult and is frequently only possible from the male genitalia. Larval (also puparial) characters are distinctive and, in view of the high degree of host specificity encountered in the genus, these together with a knowledge of the host-plant permit ready identification of bred material.

This species was described from seed-heads of *Cuscuta europea* L. in Germany. The adult is entirely black, without any metallic coloration of the mesonotum or abdomen, and the squamal fringe is also dark. In European specimens the wing length is up to 2.4 mm, in those from Pakistan up to 2.8. Despite this greater size and other slight differences, I am not inclined to treat these as of specific significance. The aedeagus of a specimen from Burma is shown in Fig. 530.

In Europe *M. cuscutae* is only known to feed in seed-heads but in West Pakistan at Rawalpindi the larvae were found feeding exclusively in young stems of *C. reflexa* Roxb. Elsewhere in the plains of West Pakistan other species of *Cuscuta* have been recorded as hosts, including *C. approximata* Bab., *C. hyalina* Roth. and *C. planiflora* Tenore (Baloch *et al.*, 1967). Sehgal (1965) records *M. cuscutae* at Namkum, India on *Cuscuta reflexa*, feeding both in stems and flower-heads. This habit of feeding in these different parts of the plant is unusual in the Agromyzidae but also occurs with *Phytomyza orobanchia* discussed below (p. 364).

Baloch *et al.* (1967) report that a larva can feed along a 9–11 cm length of stem and 'the attacked section gradually dries up. If the vine is not attached to its host plant beyond the attacked part, the entire section is killed'. These authors carried out host specificity trials with plants from 36 different families. Single eggs were laid only on *Tradescantia* sp. (Commelinaceae) and *Ipomoea batatas* (Convolvulaceae). The eggs hatched but the larvae failed to develop.

In West Pakistan one important factor in the failure of *M. cuscutae* effectively to control *Cuscuta* spp. is the high degree of parasitism to which it is subject. Baloch *et al.* (1967) record six species of parasites, namely *Bruchophagus* sp. (Eurytomidae), *Opius* sp., *Dacnusa* sp. and *Bracon* sp. (Braconidae), *Eupelmus* sp. (Eupelmidae) and *Eucoilidea* sp. (Cynipidae). They found that parasitism varied at different periods but could be as high as 60–70%. They nevertheless feel that *M. cuscutae*, which is confirmed as host-specific on *Cuscuta* and is 'an effective destroyer of the weed', could prove useful in the biological control of *Cuscuta*, if introduced to areas without its parasites. In Kazakhstan in the U.S.S.R. *M. cuscutae* is already being used on a small scale against several species of *Cuscuta*. Larval-infested seed-heads are collected and stored during the winter and the adults are released the following spring (Ivannikov, 1969).

4. **Ophiomyia lantanae** (Froggatt, 1919)

In Hawaii *Lantana camara* had spread to very large areas by 1902 and it was at this time that the possibility of biological control by the introduction of insect enemies from Mexico, one of its natural habitats, was actively investigated.

Koebele, an entomologist from Hawaii, spent most of that year in Mexico studying and collecting insects attacking Lantana and no less than 23 different species were identified and their introduction to Hawaii was attempted. The Agromyzid was considered as among the potentially most useful of these. This had been identified as belonging to the Agromyzidae but was not named until it was later discussed as *Agromyza lantanae* by Froggatt in 1919 from material which had in the meantime been introduced into Queensland, Australia from Hawaii

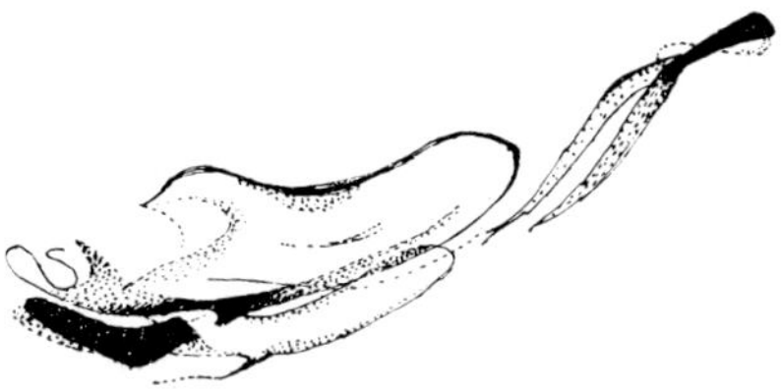

530

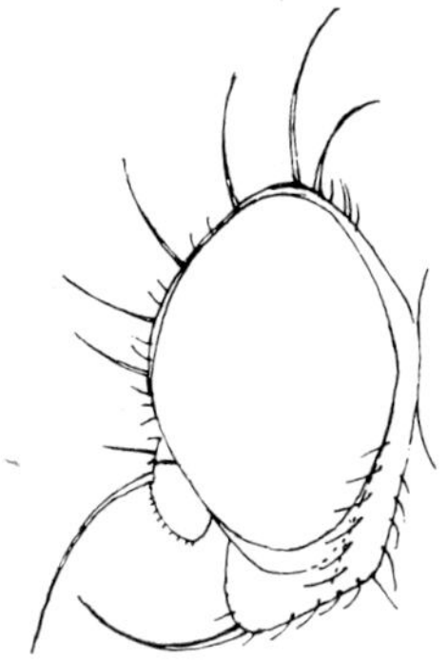

531

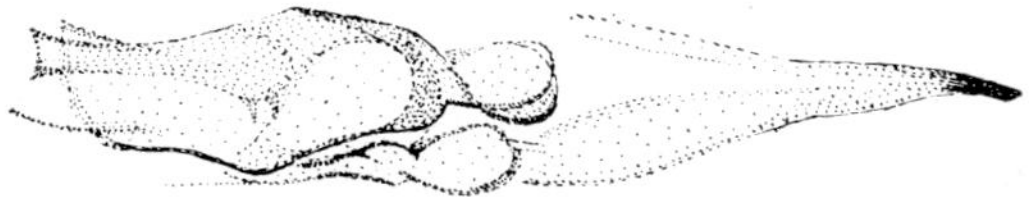

532

Fig. 530. *Melanagromyza cuscutae* Hg.: aedeagus (Burma).

Figs. 531–532. *Ophiomyia lantanae* (Frogg.): 531, head; 532, aedeagus.

in 1914. Froggatt did not give a detailed description, mistakenly believing that it had been described earlier, but he included in his paper a number of excellent illustrations and the name *lantanae* is now correctly ascribed to him. Aldrich (1923) redescribed the species from specimens bred from Lantana in Hawaii.

Ophiomyia lantanae belongs to a complex of small black species, in which a positive identification is frequently only possible from examination of the male genitalia. Females of many of these species can still not be satisfactorily identified. Important characters of the males are the form of the distinctive vibrissal fasciculus, the shape of the jowls and of the raised keel dividing the base of the antennae. *O. lantanae* has a wing length from 1.9–2.2 mm. It is entirely black, with the mesonotum somewhat mat. The jowls are approximately one-fifth the vertical eye height, not greatly projecting forwards and forming an angle in front of 80°; the vibrissal fasciculus of the male is long, narrow and regularly curving (Fig. 531). The keel dividing the antennae is pronounced, broad and raised but somewhat variable. The highly asymmetrical aedeagus of a specimen from Kenya is shown in Fig. 532.

O. lantanae occurs naturally, apart from Mexico, in Central America where it is known from Panama, Honduras and Costa Rica; throughout the Caribbean, with records from Cuba, Puerto Rico and Trinidad; probably throughout much of tropical South America, with a confirmed record from Porto Alegre, Brazil; and in southern areas of the United States, where it has been identified in southern California, Texas and Florida (FRICK, 1952a; SPENCER and STEGMAIER, 1973).

Deliberate introductions have taken place in Hawaii, New Caledonia, Fiji, Australia, India, Hong Kong, Kenya and South Africa. The species is also present in Burma, Ceylon, Malaya and Singapore. It thus seems certain that larvae or puparia must have been introduced inadvertently together with seeds for growing the plant, not only in areas where no deliberate introductions have been made for control purposes but probably in all or most areas where the plant is now established. This view was first expressed by SUBRAMIAM (1934) who records that the initial release of flies at Bangalore, India in 1921 was very limited and during adverse weather conditions. No trace of larval feeding could be found during the two seasons following the introduction and it was concluded that the fly had not been able to establish itself. Some ten years later *O. lantanae* was confirmed at Bangalore and was also found to be present at a number of other localities in the state of Mysore. SUBRAMIAM suspected that the fly might have been present in India before the deliberate introduction in 1921 and seeds obtained from localities ranging from Cochin in the south to Bombay and Saharanpur in the north, and also from Burma and Ceylon, all contained larvae and many adults were reared. GREATHEAD (1968) confirms that the fly was present in East Africa before its deliberate introduction in 1958. In South Africa also OOSTHUIZEN (1963) came to the conclusion that *O. lantanae* was present both in Natal and the Transvaal prior to its introduction to Natal in 1961. On the other hand, during a visit to the Philippines in January and February, 1960, the writer could find no trace of larval feeding on Lantana at several localities on Luzon.

The larvae of *O. lantanae* feed within the developing seed-head. The exact area of feeding will depend to some extent on the timing and site of oviposition. Eggs can be laid either directly in the soft tissues of the endocarp surrounding the young seeds or lower in the peduncle. KUNHI KANNAN (1924), referring to the larval feeding, writes: 'the damage it did was more to the pulp of the fruit in which the egg was laid and the larvae developed for pupation. The larva usually sought the pithy hollow to be found between the germ cells in every seed. It is obvious that all that is vital to the full grown seed is left untouched.' A few infested flower-heads examined by the author at Brisbane, Australia in February, 1961, showed scarcely any damage to the actual ovule and puparia were imbedded in the receptacle below. SWEZEY (1921) tested the germination of infested Lantana seeds in Hawaii. From 100 infested seeds retained until after emergence of flies, 82 seedling plants were obtained. From the 100 non-infested control seeds 95 seedlings were obtained.

On the other hand, SWEZEY (in PERKINS and SWEZEY, 1924) later found that a significant proportion of seeds may be damaged. 'In one lot of 200 berries, 70%

of which were infested, the infested berries were examined and 51 % of them had the embryo entirely destroyed, while 48 % had one remaining sound embryo each. In other lots of infested berries those with embryos entirely destroyed ranged from 65–79 %, and those containing at least one good embryo were from 21–32 %. These observations would show that the seed-maggot is not thoroughly effective in destroying lantana seed.' Infested seeds dry up prematurely and thus become unattractive to birds. He suggests that 'the prevention of lantana spread by birds has been in a great measure accomplished, even though the maggots do not always destroy the seeds'.

Between 1902 and 1961 *O. lantanae* was deliberately introduced to eight different areas and the following brief review of these introductions indicates the limited success achieved in controlling Lantana.

KOEBELE and PERKINS were enthusiastic about the possibilities of introducing *O. lantanae* to Hawaii. In a letter from Mexico to R. C. L. PERKINS on 23 November, 1902, KOEBELE wrote: 'The Agromyzid fly is becoming still more numerous. I find that the berries before becoming ripe turn a pale colour. This is where the Agromyzid larvae have been at work eating the pericarp, and pupating in the seeds. Let us establish the insects destroying flowers and seeds and lantana will not be any longer a pest.' RAMACHANDRA RAO (1920: 300) quotes PERKINS as stating the success of the Agromyza to be 'instant and phenomenal'. In a report on his work in Mexico addressed by KOEBELE early in 1903 'To the Honourable Board of Agriculture and Forestry, Honolulu', he wrote: 'It goes without saying that a large percentage of the seeds will be destroyed by this little fly alone, especially if established without its parasites, which in Mexico destroy the largest part of the maggots.' (PERKINS and SWEZEY, 1924.)

KUNHI KANNAN (1924), reporting on his visit to Honolulu in 1913, found that the areas where Lantana had to some extent been checked had a low rainfall and the effect achieved was due to the joint efforts of physical eradication by man, together with varying degrees of natural control by the various insects which had been imported from Mexico. His final conclusion was that 'the further set-back it (Lantana) has received from insect pests should be looked upon as no more than supplemental'.

One introduction of *O. lantanae* to New Caledonia was made from Hawaii in 1908–1909. TRYON (1912) reports that by 1912 the fly was well established, following its introduction to 'La Belle Isle' in 1911.

The Government Entomologist of Fiji was impressed by the apparent effectiveness of *O. lantanae* in Hawaii and following a visit there in 1910, the fly was introduced to Fiji in the following year. The release of relatively few flies led to successful colonization and JEPSON (1916) was also enthusiastic about the results achieved. 'By the agency of these small flies the berries are prevented from giving rise to new lantana plants which was the case before these flies were introduced. In this way the spread of the weed is to a large extent prevented.' The degree of infestation was unusually high on Fiji, JEPSON recording that '95 % of the berries are attacked by the larvae of this useful fly'.

O. lantanae was introduced to Queensland, Australia in 1917 and VEITCH (1935) reported that 'it is considered highly improbable that there is any lantana-infested area in which it does not now occur'. VEITCH, however, was cautious about the results achieved by the seed-fly, suggesting that its overall effect was merely to slow down the spread of lantana. He concluded that 'the establishment of the lantana seed fly has by no means solved the problem of the biological control of that serious weed pest.'

By 1919 Lantana had become a serious problem in some areas of New South Wales. W. W. FROGGATT, Government Entomologist, refused to recommend the introduction of *O. lantanae*, in view of the potential dangers to native plants from the introduction of foreign insects. No deliberate introductions were therefore ever made in New South Wales but nevertheless the fly was discovered in 1919 to be well established in many areas. At least one observer considered that it was 'greatly checking' the spread of Lantana (FROGGATT, 1919).

Introduction of *O. lantanae* to India was originally delayed through fears that the larvae might find an alternative host in the seeds of teak, *Tectona grandis* L., also belonging to the Verbenaceae. After it was established that the hard seed, with little fleshy covering suitable for feeding by dipterous larvae, would almost certainly be immune from attack by *O. lantanae*, the fly was introduced to Bangalore in 1921 (KUNHI KANNAN, 1924; SUBRAMIAM, 1934). The fly is now present throughout India but there is little evidence to suggest it has been in any way effective in limiting the spread of Lantana.

GREEN (1934) reports the introduction of *O. lantanae* to Hong Kong in 1933. No information is available on the degree of success of this importation.

Introductions were made to East Africa, where flies were liberated in the Mombasa and Nairobi areas in 1958. SIMMONDS (1960) reported the fly to be present at Dar-es-Salaam (Zanzibar) and Serere (Uganda), far from the original liberation sites, by early 1960. There is no evidence from this area that the deliberate introduction achieved any positive results in checking the spread of Lantana.

OOSTHUIZEN (1963) gives a detailed review of the biological control of Lantana in South Africa. Lantana had spread over large areas of Natal and was proclaimed a weed in 1946 (NAUDE, 1956). After some years of hesitation approval was given in 1960 for a three year assessment of the possibilities and limitations of biological control and it was decided to introduce six insects from Hawaii. *O. lantanae* was introduced in 1961 but OOSTHUIZEN writes: 'It is considered that this species is of very little importance as a controlling factor.' Infestation rates in South Africa appear to be lower than noted elsewhere, ranging from 4% at Durban, to a maximum of 31% at Shakaskraal (OOSTHUIZEN, 1963).

It seems clear from the foregoing that *Ophiomyia lantanae* has some but only limited value in controlling Lantana, and the early enthusiasm shown over the first introductions to Hawaii can be seen to have been without real foundation. Furthermore, there is strong evidence to suggest that in many, if not in all, areas where the fly was deliberately introduced, it had already become established earlier together with importations of Lantana seeds for horticultural purposes.

In Indonesia it is not without interest that importation of *O. lantanae* was rejected in 1953, in view of the value of *Lantana camara* for erosion control on mountain slopes. Furthermore, in Ceylon Lantana has been regarded as helpful in regenerating old coffee plantations (RAMACHANDRA RAO, 1920) and in

Queensland it has proved useful in regenerating old banana plantations (OOSTHUIZEN, 1963).

5. **Ophiomyia strigalis** Spencer, 1963c – East Africa

This species belongs to the small group in *Ophiomyia* in which the fused vibrissal fasciculus of the male is lacking (Fig. 533) and the squamae and fringe are silvery-white. The colour is otherwise entirely black, without any metallic reflections on the mesonotum or abdomen. It is a small species with wing length from 1.9–2.3 mm. The aedeagus is shown in Fig. 534.

The parasitic genus *Striga* (Scrophulariaceae) causes damage mainly to cereal crops in many parts of the world, including America, Africa, the Middle East,

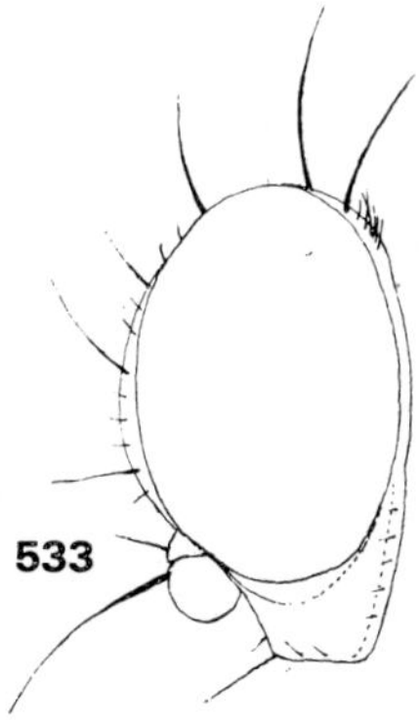
533

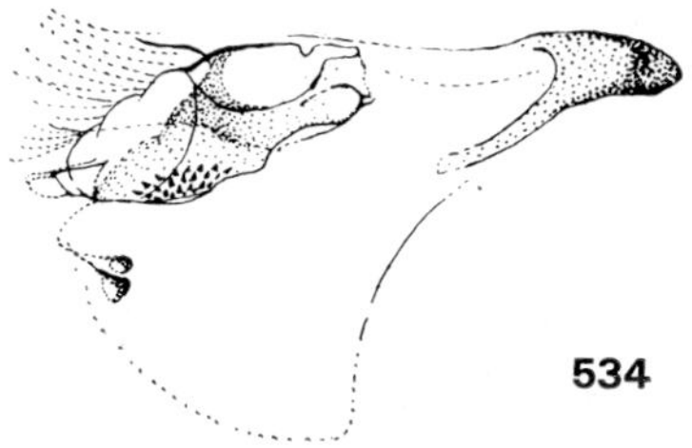
534

Figs. 533–534. *Ophiomyia strigalis* Sp.: 533, head; 534, aedeagus.

Asia and Australia. Four species attack cultivated crops in East Africa, *Striga hermontheca* (Del.) Benth. (*S. senegalensis* Benth), *S. asiatica* (L.) O. Kuntze (*S. lutea* Lour), *S. forbesii* Benth. and *S. gesnerioides* (Willd.) Vatke (*S. orobanchioides* Benth.). The commonest host plant of *S. hermontheca* and *S. asiatica* is sorghum but maize, sugar-cane and finger millet are also attacked. *S. forbesii* does not normally parasitize cultivated plants but it has been found on maize both in Tanzania and Uganda. *S. gesnerioides* has been recorded from several genera of wild Papilionaceae in East Africa and may also be found on cultivated legumes in addition to graminaceous hosts. Plots of sorghum at many localities in northern and eastern Uganda and in the Lake Region of Tanzania have been almost totally ruined by *Striga* infestations. In some parts of the Sudan sorghum cultivation has been abandoned owing to persistent and heavy infestation by *Striga* (GREATHEAD and MILNER, 1971).

DAVIDSON (1963) published a brief survey of insects attacking *Striga hermontheca* in the Lake Victoria area of western Kenya in 1960 and 1962. The commonest species proved to be the Agromyzid, described from material obtained by DAVIDSON as *Ophiomyia strigalis*. He found up to ten puparia per plant but the mean number found in August, 1962, at Kibos was only two per plant. Attack commences when *Striga* begins to flower in late June and by the beginning of August more than 90% of plants have been attacked.

GREATHEAD and MILNER (1971), whose survey between 1963 and 1967 covered wide areas of Tanzania, Uganda and Kenya, report rearing up to 28 individuals from a single plant of *S. hermontheca* but one to three was more normal. The highest infestations observed were 49% (*S. asiatica*) and 35% (*S. hermontheca*) in western Tanzania, while in northern and eastern Uganda (*S. hermontheca*) over 25% of plants were infested.

O. strigalis lays its eggs in the axils of leaves at or just above ground level. The young larva forms a dark purplish-black mine beneath the epidermis of the stem running downwards to the root to a depth of 8–10 cm. Prior to pupation the larva returns to ground level and the puparium is visible beneath the epidermis of the stem in a distinct 'pupal blister' (DAVIDSON, 1963).

The main larval feeding by *O. strigalis* takes place when the plants are robust and mature. This attack 'causes the root epidermis to rot and peel off and reduces the number of lateral roots, but no debilitation of the aerial parts of the plant or reduction of numbers of flowers or fruits has been observed'. (DAVIDSON, 1963.) GREATHEAD and MILNER (1971) confirm DAVIDSON's observations, despite secondary infestation by other insects, mites and fungi of the damaged root epidermis. 'The damage caused by it is very slight and the infestation only builds up by the end of the season, by which time mature seeds have already been produced.'

In East Africa several hymenopterous parasites have been recorded from *O. strigalis*. In the Serere area of northern Uganda parasitism was as high as 31.5% by a *Eucoilidea* sp. (Cynipidae). The rate of parasitism by an *Opius* sp. was 13% in eastern and northern Uganda (GREATHEAD and MILNER, 1971). Introduction of *O. strigalis* to other areas where *Striga* is a problem might possibly lead to increased populations, in the absence of its normal parasites. However, in view of the life history of this species, with only mature plants being attacked and the damage caused not being sufficient to reduce production of seeds, its potential for use in biological control does not seem promising.

6. **Phytomyza orobanchia** Kaltenbach, 1864 – Europe, W. Asia, Ethiopia

Biological control of *Orobanche* with *Phytomyza orobanchia* was first considered early in this century in Italy where beans were heavily parasitized by *Orobanche speciosa* (MARTELLI, 1933). A similar approach is being considered in Egypt (HAMMAD *et al.*, 1967) and the possible introduction of *P. orobanchia* into southern India is currently being studied. However, really effective results have now been achieved both in Yugoslavia and in the U.S.S.R. where this Agromyzid is proving conspicuously successful as a control agent of *Orobanche*.

KALTENBACH described this species from flies bred from larvae on *Orobanche* sp. in Germany. He stated that the larvae feed in the seed-heads and then proceed down to the stem to pupate. This is not entirely accurate, as this is one of the few species in which the larvae can feed indiscriminately either in the seed-heads or in the stem. On occasions they also move from one part of the plant to the other. In my experience the main feeding takes place in the seed-head, where substantial quantities of seeds are destroyed. Damage to the plant from stem-feeding can produce premature desiccation of the seed-heads and also lead to destruction of the plant from secondary fungal rot.

The adult fly is distinctive. The third antennal segment (Fig. 535) is brownish and conspicuously elongate, the frons is yellowish-brown and mesonotum and pleura are mat grey; the legs are black, with yellow knees. The size is somewhat variable, with females having a wing length up to 2.5 mm.

The male genitalia (Fig. 536) clearly indicate a very close relationship between *P. orobanchia* and *P. affinis* Fallén (cf. SPENCER, 1969a: Fig. 394) which feeds in the

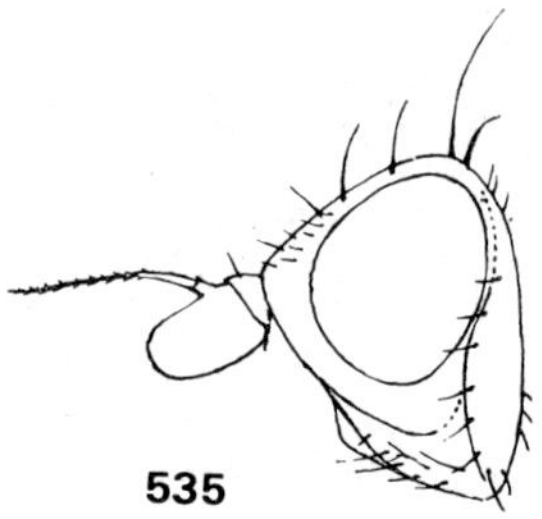

535

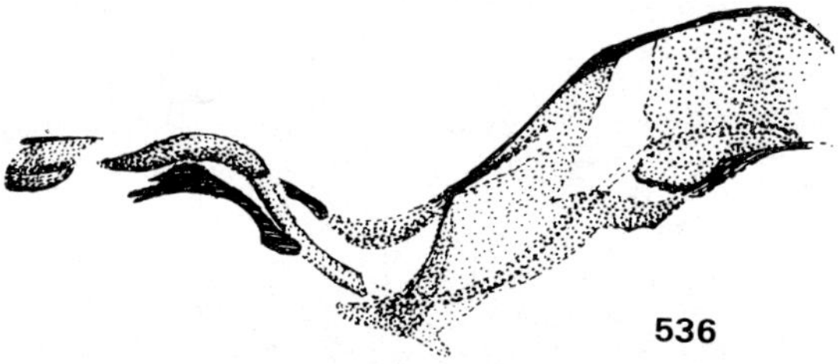

536

Figs. 535–536. *Phytomyza orobanchia* Kalt.: 535, head; 536, aedeagus.

seed-heads of *Pedicularis*, a parasitic genus in the Scrophulariaceae. Botanists are agreed that the Orobanchaceae are directly derived from the Scrophulariaceae and this suggests the probable evolutionary development of *P. orobanchia* either directly from *P. affinis* or one of its close relatives among feeders on Scophulariaceae.

P. orobanchia is particularly common throughout the Mediterranean area and is known from Spain, Italy, Malta, Egypt and Israel. It has also been recorded in Ethiopia, near Addis Ababa (SPENCER, 1961c: 429). It is present in N.W. Europe, for instance in Germany and England, but in smaller numbers and is not known in Scandinavia. Its eastwards distribution extends through the Balkans (Yugoslavia, Bulgaria) to the Crimea, the Ukraine and as far as Samarkand in the Uzbek Republic of the U.S.S.R. (Weed Research Organization, 1970). It is almost certainly present in Afghanistan but is not known in India.

KALTENBACH (1864: 265) records obtaining a Braconid parasite from *P. orobanchia* in Germany and MARTELLI (1933) reports a Braconid and also two species of Chalcids in Italy. One Pteromalid, *Sphegigaster orobanchiae* Kurdjumov, has been described from Russia and has also been reported in Iraq (AL-AZAWI, 1967).

The genus *Orobanche* has some 90 species mainly in the northern temperate areas of Europe and Asia, but with three also present in western Canada. There are

40 species in the U.S.S.R. It is present in smaller numbers in India and in the New World, for example in California and Cuba, and as an introduced weed from Europe in Florida and Australia. Some species are host specific but many parasitize a wide range of hosts, including many crops. The main damage is caused by withdrawal of water from the host and a serious reduction in sugar content can also result.

Crops particularly affected are tobacco, tomatoes and to a lesser extent eggplant and potatoes (Solanaceae), melon, water-melon, cucumber and pumpkin (Cucurbitaceae), sunflowers and safflower (Compositae), horse-beans, broad beans and

Plate 13: *Orobanche minor:* parasitizing tomatoes (L. KASASIAN, Weed Research Organization, Yarnton, England).

clover (Leguminosae), cabbage (Cruciferae) and carrots (Umbelliferae). A photograph of *Orobanche minor* attacking tomatoes is shown in Plate 13.

In Egypt severe reduction in the yield of beans, carrots and tomatoes has been reported by HAMMAD *et al.* (1967). In Yugoslavia, particularly in Vojvodina, tobacco, tomato, sunflower and hemp all have their yields reduced by *O. cumana* Walter and *O. ramosa* L. (LEKIĆ, 1970). In Bulgaria the seed yields of sunflower have suffered drastically with up to 85% of plants being infested. In the U.S.S.R. damage to many crops is serious and widespread in the Central Asian Republics of Uzbekistan, Kazakhstan and Turkmenistan. Cucumber yields have been reduced by half, up to 50% of the sugar content of musk-melons has been lost. A survey of 100,000 hectares of tomatoes, cabbage and sunflower in the Ukraine in 1969 showed that 8000 hectares were infested with *Orobanche* at an average density of 10–22 shoots per square metre. In California *Orobanche ramosa* is a serious parasite of tomato, lettuce, safflower and other crops. Yield losses of tomatoes increase with infestation to 100% at 50 parasites per tomato plant (W.R.O., 1970). Recently *Orobanche* has become a serious problem on the tobacco crop in the Guntur area of southern India.

Phytomyza orobanchia has had conspicuous success in controlling *Orobanche* both in Yugoslavia and the U.S.S.R. Lekić (1970) reports that in the Vojvodina the primary role in reducing the damage caused by *Orobanche* to sunflower, tomato and tobacco crops has been played by the larvae of this Agromyzid. 'On test plots and in nature they destroyed 96.03% of the seed capsules. They often damage young plant tops and in this way render further growth impossible.'

In one area of the Ukraine 96% of the *Orobanche* shoots were attacked and 94% of the peduncles perished; damaged seeds lost viability. On another farm 1000 puparia per hectare were applied to 80 hectares infested with 46 *Orobanche* shoots per square metre; 74% of peduncles died compared with 18% in the untreated control. Over a 3- to 4-year period use of *P. orobanchia* reduced parasitization of melon crops. Also in the Ukraine cabbage yields were increased from 7–10 tons to 27–29 tons per hectare and *Orobanche* was practically eliminated from tomato crops. In the Crimea in 1968 *P. orobanchia* was used successfully for the control of *Orobanche* spp. on 1500 hectares of tobacco and tomatoes. The numbers of broomrape shoots were greatly reduced and most of the remaining shoots did not set seed (W.R.O., 1970).

Introduction of the fly to new areas is relatively simple, as emergence of the adults normally coincides with fresh growth of *Orobanche* plants, onto which the emerging flies can be released. Lekić (1970) reports that overwintering puparia can suffer severe mortality from excessive cold and also from deep ploughing, as the adults are unable to penetrate through more than a shallow layer of earth. The rate of parasitism can also be high. However, with proper management *Phytomyza orobanchia* can be employed as a highly efficient control agent against *Orobanche* and there are no reasons inhibiting its introduction to any areas where *Orobanche* is proving a serious problem.

7. **Phytomyza syngenesiae** (Hardy) – New Zealand

Ragwort (*Senecio jacobaea* L.) is an important weed in many temperate areas, including New Zealand and Australia. Kelsey (1937), during investigations into a number of insects as potential agents for biological control of ragwort in New Zealand, found that in a number of seasons damage to ragwort by the leaf-miner *P. syngenesiae* (as *atricornis* Mg.) was so severe that the experiments being conducted with these specially imported insects were severely hampered. He therefore carried out subsidiary experiments on the degree of control of ragwort exercised by *P. syngenesiae*, and also on the control of *P. syngenesiae* by its Braconid parasite, *Dacnusa areolaris* Nees.

Firstly, a series of four experiments were carried out under laboratory conditions in the absence of the parasite. Details of these experiments and their discussion are of sufficient interest to merit reproduction verbatim.

'EXPERIMENT 1. – For the first experiment of this series ten flies, of which five were females, were placed in a cage containing a large healthy ragwort plant. The leaves were extensively mined by the larvae, but were not killed. When all the larvae had pupated the material was collected and the flies emerging from this were returned to the same cage. This second generation infestation took place before the plant had had time to recover from the previous attack, and it was killed before many of the larvae had reached their full development.

EXPERIMENT 2. – For the second experiment thirty-six female flies with males were caged in a section of an insectary with seven strong plants. In this instance none of the plants was killed by the first generation larvae, and although those of the second generation killed all the leaves and three of the plants, the remaining four sent up new growths, these, with the stocks, being destroyed as a result of the mining of the third generation larvae before they had recovered their original vigorous growth, and, indeed, before the majority of the larvae had completed their development.

EXPERIMENT 3. – Ten female flies with males were caged with a single plant, which was apparently killed by the first generation of fly larvae: later, however, the stock produced a sickly growth of new leaves which did not attain any great size, and which were eventually killed by the second generation fly larvae.

EXPERIMENT 4. – Two plants and twenty-four female flies with males were used in experiment 4: the leaves were all killed by the first generation larvae, the stocks later producing fresh leaves, which, as in the last case, were destroyed by the second fly generation. Both stocks remained green for some considerable time, but no new leaves were produced.

From the above experiments it will be seen that the larvae produced by five female flies are not able to kill a healthy ragwort plant, but the second generation of larvae has no trouble in destroying the same plant under caged conditions. In the field, of course, the second generation flies would not concentrate on the same plant or plants, since it has been observed frequently that the miners choose for oviposition healthy plants in preference to those in a weakened condition; thus, under such circumstances, it is very probable that plants attacked by the first flies to emerge in any year would recover sufficiently during the second and succeeding generations of the flies to produce flowers and seeds. In a case of this kind the only benefits derived from the miner infestation would lie in the stunting of plant-growth, with consequent reduction in the quantity of the leaves, and in the numbers of seed-heads produced. But where there are small isolated patches of ragwort, such as those occurring on farms on steep sidings where they may be overlooked during the general treatment of pastures, the fly would be, perhaps, of more value in controlling the weed. In these small patches of ragwort the later broods of flies, unless they migrated, would be compelled to lay their eggs in the weakened growths produced by plants attacked by the earlier emerging miners, and as shown in the foregoing experiments, such plants are not able to survive heavy successive infestations following closely on each other, and, taking into consideration the numbers of generations of the leaf-miner, it is possible that the flowering of these small areas, of ragwort, even though many plants remained alive, would be retarded.'

It must be emphasized that KELSEY's results were obtained in the absence of any parasites. In a further series of experiments, it was found that introduction of parasites in the restricted conditions of a glasshouse quickly destroys the leaf-miner population. This has recently been confirmed in work on *P. syngenesiae* as a pest of chrysanthemums in England (see Chapter 12, p. 352). In the field, where in New Zealand a rate of parasitism of between 40 and 65 % is normal, a considerable degree of natural control will be obtained, but with the periodical fluctuations in populations resulting from imbalance in the parasite-host relationship which is a familiar phenomenon. KELSEY's final conclusion is that 'a weed-infesting insect, if freed of its parasites, may become a powerful beneficial agent'.

Some further serious weeds attacked by Agromyzidae

Any possible use of Agromyzids for biological control of weeds must obviously satisfy the essential criteria of such programmes concerning host specificity. This would be particularly difficult in the case of the many serious annual and perennial grasses, as the host specificity within the Gramineae is less marked than in most families of dicots, and furthermore for most species is imperfectly understood. Information on grass-mining species, both in the Old and New World tropics, is still fragmentary but it can at least be recorded that there are many Agromyzidae, mostly still unidentified, feeding on close relatives of serious weeds, which might offer some possibilities of biological control, particularly if introduced to areas outside their normal range.

AMARANTHACEAE

Alternanthera philoxeroides (Mart.) Standley occurs widely in South America and the southern States of the U.S.A. It is particularly troublesome in southern Louisiana. *Melanagromyza marellii* (Brèthes) forms conspicuous stem-galls on this species (Fig. 537). It was originally known from Argentina and Brazil and has recently been recorded from the Bahamas and Florida (SPENCER and STEGMAIER, 1973).

COMMELINACEAE

Commelina benghalensis L. and other *Commelina* spp. are widespread and can be a problem in Africa and Asia, especially among low-growing crops such as ground-nuts (GREATHEAD, 1967). No Agromyzid is known to feed on this genus in Africa or Asia but *Liriomyza commelinae* (FROST, 1931) is a well-known leaf-mining species in the Neotropical Region and S. Florida. The only host known apart from *Commelina* is the related genus *Tradescantia* and the species is certainly host-specific on the Commelinaceae.

COMPOSITAE

Bidens pilosa L. is now found as a troublesome weed throughout the tropics. *Melanagromyza floris* Spencer, 1963a, appears to feed primarily on this host, in the seeds, but has also been reared from *Calendula* sp.; it is a neotropical species and has been found particularly commonly in S. Florida (SPENCER and STEGMAIER, 1973).

Mikania cordata (Burm. f.) B. L. Robins and *M. micrantha* H.B.K. are serious weeds in Ceylon, India and Malaya, particularly among plantations of tea, rubber and oilpalm. Legume cover crops can be completely smothered (KASASIAN, 1971); *Calycomyza mikaniae* Spencer, 1973, was recently described from Florida from *Mikania batatifolia* DC. and is almost certainly host-specific on *Mikania*. The larvae form a large white blotch mine (Fig. 538). Irregular, linear-blotch mines of an undescribed species (Fig. 539) were also found by the author on *Mikania* sp. at Singapore.

Wedelia trilobata (L.) Hitch., a native of the West Indies, occurs as a serious weed throughout the American tropics and subtropics and has also been introduced into West Africa. A neotropical species, *Melanagromyza minima* (Malloch), feeds in the seeds of this species and the female ovipositor is specially elongated for oviposition deep within the immature seed-head (Fig. 540).

M. minima has been reared from *Senecio confusus* Britton in Florida. I have also found females apparently ovipositing in the seed-heads of *Eupatorium odoratum* L. in Jamaica, which has colonized large areas in Ceylon and India, and is considered as a serious weed in these areas.

A recently described species, *Melanagromyza wedeliae* Spencer, 1973, was obtained from flower-heads of *Wedelia paludosa* L. at Miami, Florida. It seems probable that this species would also feed on *Wedelia trilobata*.

CYPERACEAE

Cyperus rotundus L. is considered as perhaps the most serious of tropical weeds (HOLM and HERBERGER, 1970) and occurs widely throughout the world. In temperate areas of Europe and N. America many Agromyzidae in the genus *Cerodontha*, sub-genus *Dizygomyza*, are known as host-specific to the family Cyperaceae. Very little is known about these species in the tropics. An unidentified species was found near Caracas, Venezuela (K.A.S.) in December, 1960, on *Cyperus tenuis* Sw., with many larvae feeding together in a leaf, which was largely eaten out and destroyed. The pale brown puparium of this species bears a distinctive pair of long black thoracic spiracles (Fig. 541). Mines were also found by the author on *Cyperus ajax* C.B. Cl. in the Aberdare Mountains, Kenya in October, 1971. No adults were reared but it is clear from the empty puparia that this is one of the largest species in the genus and a single larva forms a substantial mine (Fig. 542).

GRAMINEAE

It will have been seen earlier in Chapter 9 that many pest species attack with equal frequency the main cereals *Avena*, *Hordeum*, *Secale* and *Triticum*. *Agromyza* species which attack weeds such as *Agropyron* and *Alopecurus*, even if populations could be built up to effective levels, could be expected to colonize any cultivated

537

538

539

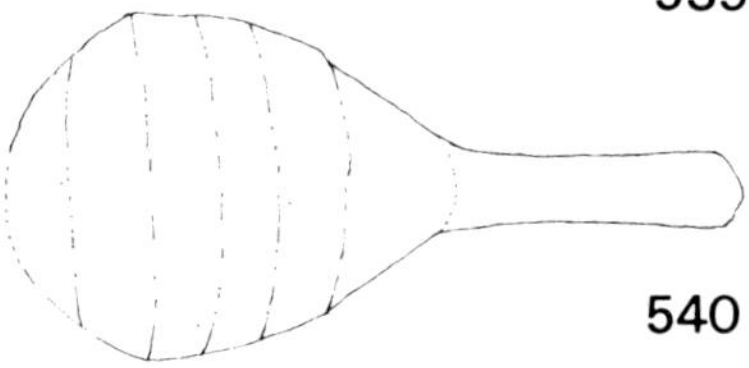

540

541

542

543

Fig. 537. *Melanagromyza marellii* (Brèthes): galls on *Alternanthera philoxeroides* (Argentina).

Fig. 538. *Calycomyza mikaniae* Sp.: leaf-mine on *Mikania batatifolia* (Florida).

Fig. 539. *Calycomyza* sp.: leaf-mines on *Mikania* sp. (Singapore).

Fig. 540. *Melanagromyza minima:* elongated ovipositor sheath of female.

Fig. 541. *Cerodontha* (*Diz.*) sp.: puparium ex *Cyperus tenuis* (Venezuela).

Fig. 542. *Cerodontha* (*Diz.*) sp.: leaf-mine on *Cyperus ajax* (Kenya).

Fig. 543. *Pseudonapomyza philippinensis:* leaf-mine on *Rottboelia exaltata* (Philippines).

cereals which might be growing in the vicinity. However, if such grasses were a problem among non-cereal crops in areas where cereals are not habitually cultivated, such reservations would not apply.

Cynodon dactylon (L.) Pers. is readily attacked by *Pseudonapomyza asiatica* Spencer (see Chapter 9, p. 264) and this was the host from which the type series was obtained at Singapore. Tips of young leaves are entirely eaten out by larvae feeding singly (Fig. 396). An important annual, *Rottboellia exaltata* L., can be heavily mined by another species in the same genus, *Ps. philippinensis* Spencer, 1961b. In this case a number of larvae feed together to form a substantial blotch mine (Fig. 543). In the southern United States *Paspalum* is a favourite host genus of *Liriomyza marginalis* (Malloch). Although *Paspalum conjugatum* Berg can be a serious weed in some tropical areas, it and other species in the genus have importance as fodder crops and *L. marginalis* is discussed earlier in this context in Chapter 9, p. 260.

Unidentified leaf-mines have been found by the author on a number of genera of Gramineae in Africa and Asia, as follows: *Digitaria*, Bombay; *Echinochloa*, Pretoria; *Eragrostis*, Bombay; *Ischaemum*, Ceylon; *Oplismenus*, Singapore; *Panicum*, Victoria Falls; *Setaria*, Bulawayo and Pietermaritzburg; *Sporobolus*, Nairobi; *Urochloa*, Pretoria. A number of these genera include important weeds, such as *Digitaria scalarum* (Schweinf.) Chiov. in East Africa and *D. sanguinalis* Royal in Brazil where coffee yields can be seriously affected; *Echinochloa crus-galli* (L.) Dulac and *E. colonum* (L.) Link, which are a problem wherever rice is cultivated; *Ischaemum rugosum* Salisb., another rice weed; *Panicum maximum* Jacq., and finally, *Setaria* spp., which can be particularly troublesome with soy bean. (HOLM, 1969; KASASIAN, 1971).

APPENDIX A

Visit to Venezuela from 11 April to 12 May 1972

I. Introduction

For some years Agromyzid leaf-miners on a number of crops have been causing increasing concern in Venezuela. The Servicio Shell para el Agricultor at Cagua, Aragua investigated this problem and since 1965 have been recommending chemical control of leaf-miners on various crops, particularly tomatoes, beans and melons. Specimens sent to me for identification represented mainly the two polyphagous species, *Liriomyza huidobrensis* and *L. sativae*. However, it was not clear whether other species were present on some of the crops and the ecology of these pest species was improperly understood. My visit to Venezuela followed an invitation from Servicio Shell jointly with the Faculty of Agriculture of Caracas University at Maracay and additional support was provided in England by the Overseas Development Administration of the Foreign and Commonwealth office. I would like to take this opportunity of thanking everyone who made this visit possible.

During my stay in Venezuela I was able to examine crops in six states: Aragua, Miranda, Dto. Federal, Lara, Trujillo and Merida. My stay coincided with the end of the dry season and the beginning of the rains and this provided a useful opportunity for observing changing stages of crop growth and associated changes in insect populations.

Despite the limited duration of my visit and the relatively small area of the country I was able to see, I found 8 of the 9 Agromyzid species of economic importance known in Venezuela and obtained much additional information concerning their ecology and biology.

II. Agromyzid species of actual or potential economic importance

1 *Melanagromyza sp.n.* (Tomato) (cf. Chapter 5, p. 102)

Examination of three specimens from the University Collection at Maracay bred from stems of tomato at Sanaré, La., 14.ix.53 (R. Prieto) has confirmed that they represent an undescribed species, distinct from the two further species on tomato known from Colombia and Ecuador. The formal description of this new species has now been prepared and will be published in the Revista de la Facultad de Agronomia, Maracay, together with the descriptions of some 30 further new species collected during my visit.

It is possible that this species may also feed on wild *Solanum* spp. from which it has colonized the extensive areas of cultivated tomato.

2. *Japanagromyza sp.n.*

The undescribed species discussed in Chapter 4, p. 83 was found on *Phaseolus vulgaris* at three localities: La Villa, Ar., Rancho Grande, Ar. and Timotes, Me.

In addition, specimens were seen in the University collection from El Valle, Dto. F., 24.ii.50 (R. Hernandez) and from Caripe, M., 15.iii.72 (R. Cásares M.). This species was not previously known in Venezuela.

A few isolated plants found by Prof. FERNÁNDEZ YÉPEZ at the roadside near Rancho Grande on 18.vi.1972 (obviously resulting from beans discarded from a picnic some weeks before) were heavily attacked, owing to the non-availability of additional host-plants in the vicinity on which oviposition could take place normally. Each leaf was mined by several larvae and was in the process of being destroyed; it looked doubtful whether the plants could survive.

3. *Amauromyza maculosa* (Malloch) (cf. Chapter 6, p. 151)

This widespread neotropical feeder on Compositae was found at several localities in Aragua and Merida. Its normal hosts are among wild Compositae and it has only once been found damaging lettuce in Venezuela.

4. *Liriomyza braziliensis* (Frost) (cf. Chapter 5, p. 118)

This high altitude species is restricted in Venezuela to the Andes and is known from Tachira and Merida. It has been recorded both from *Solanum tuberosum* and *S. andigenum*. It may well also occur on *S. columbianum* which grows locally in the Paramos but there is no evidence to suggest that it feeds on other groups of *Solanum*.

Many specimens were seen on *Solanum tuberosum* above La Venta, Me. (*c*. 3000 m.), 6.v.72, and a smaller population was present on the same host in El Valle Grande, Me. (*c*. 2400 m.), 11.v.72. It has previously been bred from potato tubers at La Erita, Paramo El Rosal, Ta., 2600 m. on 5.xii.64 (N. Angeles and R. Vegas).

The life history of this species is still inadequately known and further investigations seem called for to assess its economic significance in Venezuela.

5. *Liriomyza huidobrensis* (Blanchard) (cf. Chapter 8, p. 215)

Very large populations of this species were seen at Las Minas, Mi., 27.iv.72, on a farm beside the road above Burbusay, Tr. and at Timotes, Me., 5.v.72, and specimens were also present above La Venta, Me., 6.v.72.

It became apparent from my collecting that this is a more temperate species than *Liriomyza sativae*. It appears to be normally limited to altitudes above the rain forest and was clearly at its upper limit at La Venta (3000 m.). It was absent from the Cagua area and was not present at Rancho Grande, Ar.

The population at Las Minas was heavily infesting *Beta vulgaris* and was also present on the same farm on *Beta vulgaris*, var. *cicla*, *Cichorium endivia*, *Lactuca vulgaris*, *Cucurbita pepo* and *Phaseolus vulgaris*. The area of beet in particular was badly overgrown with the two weeds *Amaranthus viridis* and *Galinsoga caracasana*, both of which were heavily mined. These common weeds obviously provide a base from which cultivated crops can be readily colonized.

Above Burbusay the main infestation was on *Galinsoga caracasana* which was growing in profusion in a field of *Allium schoenoprasum*. However, only few leaf-mines were found on the garlic. On potatoes above La Venta, Me. (*Solanum tuberosum* (no *S. andigenum* was seen) only a few flies were present; at this altitude *L. huidobrensis* is replaced on Solanaceae by *L. quadrata* (see below). Isolated specimens were caught in El Valle Grande, above Merida, *c.* 2000 m but no infestation of crops was noted.

The leaf-mines of *L. huidobrensis* are frequently associated with the midrib or stronger lateral veins. This feeding pattern can be obscured when several larvae are feeding in the same leaf and the mines then appear irregularly linear. It has not previously been recorded that larval feeding occurs not only on the upper surface of the leaf but with almost equal frequency on the lower surface. This was observed on *Amaranthus* and *Beta* at Las Minas, on *Galinsoga* at Burbusay and on *Chrysanthemum* and *Dahlia* at Timotes.

6. *Liriomyza marginalis* Malloch (cf. Chapter 9, p. 260)

This species has not previously been recorded from Venezuela but was present in the University collection at Maracay. It had been reared from *Zea mays* at Pozo del Diablo, Maracay, 26.vi.62 and a large population had been found at Caicara de Maturin, Mo., 14.vi.68, attacking *Digitaria decumbens* (pesto pengola), an important fodder crop. I obtained isolated specimens in Aragua at Cagua and La Villa, bred ex *Panicum*, and also at Merida.

7. *Liriomyza sativae* Blanchard (cf. Chapter 8, p. 219)

L. sativae is probably the most serious Agromyzid pest in Venezuela. It is highly polyphagous and populations are frequently very large. I noted heavy infestations on beans, tomatoes and cucurbits in the experimental plots at Fundación Shell, Cagua. A small plot of *Phaseolus vulgaris* at La Villa, Ar. had many mines per plant on 13.iv.72. At El Rodeo, near Quibor, La. both adults and mines were present in moderate numbers on beans, tomatoes and cucurbits on 3.v.72. On the large farm visited on 4.v.72 at La Goajira, near El Tocuyo, potatoes were heavily infested and both larvae and adults were present; smaller numbers were also present on alfalfa.

Leaf-mines of this species were found on a number of wild hosts, frequently in the immediate proximity of cultivated crops. The most important of these are *Solanum nigrum*, *Ricinus communis*, *Amaranthus viridis* and *Cleome spinosa*. *L. sativae* occurs only rarely on Composites.

Specimens were bred from *Ricinus communis* at Cagua with a pupal period of only 9 days; at the higher altitude at El Tocuyo (*c.* 1500 m.) the pupal period of specimens bred from potato lasted 14 days. This would indicate a complete life cycle of 18–24 days. It is not know how many annual generations there are under natural conditions nor in the artificial environment of almost continuous fresh vegetation found with irrigated crops. It seems probable that breeding may be almost continuous under favourable conditions. Parasites could almost certainly not reproduce as rapidly and match the numerous generations of *L. sativae* induced by the artificial conditions of intensive agriculture. This may well be a significant factor in the population increases regularly occurring in this species and resulting in substantial economic damage.

8. *Liriomyza quadrata* Malloch (cf. Chapter 6, p. 176)

This is a high altitude species restricted in its host-plants to Solanaceae and in Venezuela it has only been recorded from the Andes, where it appears to be not uncommon between 2000–3000 m. At Timotes it was present on practically all plants of *Solanum nigrum* examined. From this wild host it is everywhere available at high altitudes to transfer to *Solanum tuberosum* and almost certainly also *S. andigenum*. These may well represent its original hosts. It is not known whether it also occurs on *S. columbianum*, which is found at even higher altitudes in the Paramos.

Above La Venta (*c.* 3000 m.) a field of potatoes was found to be very heavily infested. Virtually all the lower leaves of most plants had five or more mines and these leaves were effectively destroyed. However, with the abundant rainfall, growth of the plants was rapid and adequate fresh foliage was no doubt produced after the initial infestation on the young leaves to prevent any significant weakening of the plants.

L. quadrata is a large species, with wing length up to 3.6 mm. The larva first forms a linear mine, later producing a large white blotch mine, feeding both on the upper and lower surface of the leaves. The early linear mine is distinguishable from that of *L. huidobrensis* by the frass being deposited in distinct, roundish black pellets (in *huidobrensis* narrow strips).

III. Control measures

In all the main areas where crops were examined (Las Minas, Mi.; Quibor and El Tocuyo, La.; La Venta, Me.), it was found that current practice involves regular, weekly spraying with insecticides. On smaller farms this is done manually but I was informed that larger areas are sprayed by aircraft. The three preparations most frequently used appear to be Dipterex, malathion and dimethoate.

The use of foliar sprays is accepted as an essential part of crop management. Applications are carried out in a routine way, with little or no appreciation of the significance of critical timing in relation to parasites or distinction as to whether the aim is to kill larvae or adults. Several farmers I met not only did not realise that the larvae found in the leaf-mines were in any way associated with the flies which could be seen on the foliage but positively denied that there could be any such association. In view of such ignorance of elementary insect metamorphosis, the concept of integrated control or any sophisticated approach to the timing of insecticide applications would clearly be beyond the comprehension of many farmers.

Insecticide salesmen no doubt do what they can to educate their customers but their primary objective is to sell their products and in this they have been conspicuously successful. When I asked one farmer whether he used insecticides, the immediate reply was: 'Of course'. In many, and probably most, cases the use of insecticides is no doubt beneficial and essential, particularly against serious pests such as *Phtorimaea operculella* (which was seriously damaging the potato crop together with *Liriomyza sativae* on the farm visited near El Tocuyo, La.). However, it has been demonstrated by American workers that infestation by *Liriomyza sativae* can significantly increase with the use of insecticides, if through inaccurate timing of the applications the parasites suffer more than the leaf-miner. It seems certain that this effect has also on occasions, if not regularly, occurred in Venezuela, although it has not been proved experimentally. The overall correlation between the increase of leaf-mining damage and the widespread use of chlorinated hydrocarbons and organophosphorus insecticides has nevertheless been noted.

APPENDIX B

New taxa, new synonyms and new combinations

1. New genus and new species

Name	Host	Area/country
Tropicomyia gen. n.	Many families	Old World Tropics, N. Australia
Melanagromyza azawii	*Vicia faba*, *Sesamum*	Iraq
M. fabae	*Vicia faba*	England
M. phaseolivora	*Phaseolus*	Ecuador
Ophiomyia vignivora	*Vigna*	Pakistan
Liriomyza nietzkei	*Allium*	Germany
Haplomyza capsici	*Capsicum*	Argentina
Pseudonapomyza balkanensis	unknown	Yugoslavia
Ps. europea	unknown	Central Europe
Ps. hispanica	*Sorghum halepense*	Spain, Israel
Ps. hungarica	unknown	Hungary
Ps. malayensis	*Cynodon dactylon*	Malaya
Ps. mongoliensis	unknown	Mongolia
Ps. spinosa	*Hordeum*, *Triticum*. *Eleusine indica*	Africa, India, Australia, Pacific
Ps. strobliana	unknown	Yugoslavia
Ps. vota	unknown	Spain
Ps. zeae	*Zea mays*	Ghana
Cerodontha milleri	unknown	Australia, New Zealand

2. New synonyms and new combinations

Original status	Present status	Area/country
Melanagromyza alustrata	*Tropicomyia vigneae*	Cape Verde Is.
M. atomella	*T. atomella*	Formosa, India, Ceylon
M. capeneri	*T. capeneri*	Africa
M. cassinis	*T. cassinis*	South Africa
M. ceratiosicyi	*T. ceratiosicyi*	South Africa
M. citri	*T. flacourtiae*	Africa
M. clutiae	*T. clutiae*	South Africa
M. coffeae	*T. coffeae*	Java
M. crotonella	*T. crotonella*	Ethiopia
M. dianellae	*T. dianellae*	Australia
M. dicksoni	*T. dicksoni*	South Africa
M. dioscoreae	*T. flacourtiae*	Madagascar
M. dolichi	*T. vigneae*	India
M. flacourtiae	*T. flacourtiae*	Madagascar, Africa
M. gossypii	*T. flacourtiae*	Nigeria
M. gymnosporiae	*T. gymnosporiae*	South Africa
M. haemanthi	*T. haemanthi*	South Africa
M. indigoferae	*T. indigoferae*	Australia
M. jonesi	*T. vigneae*	Ethiopia
M. kikuyana	*Melanagromyza chalcosoma*	East Africa
M. laburnifoliae	*T. laburnifoliae*	Ethiopia

Original status	Present status	Area/country
M. lappivora	*Ophiomyia lappivora*	Japan
M. nigriclava	*T. nigriclava*	Rodrigues
M. passiflorae	*T. flacourtiae*	Africa
M. phaseoli	*Ophiomyia phaseoli*	Australia, Pacific, Asia, Africa
M. philocroton	*T. philocroton*	South Africa
M. pisi	*T. pisi*	Australia
M. polyphaga	*T. polyphaga*	Ceylon, India
M. polyphyta	*T. polyphyta*	Australia
M. shibatsuji	*Ophiomyia shibatsuji*	Japan
M. spencerella	*Ophiomyia spencerella*	East Africa
M. styricicola	*T. styricicola*	Japan
M. theae	*T. theae*	Ceylon
M. thunbergivora	*T. thunbergivora*	South Africa
M. vigneae	*T. vigneae*	Cape Verde Is. to Indonesia
M. wikstroemiae	*T. wikstroemiae*	Australia
Agromyza graminivora	*Agromyza penniseti*	West, East and South Africa
Cephalomyza crucifericola	*Liriomyza crucifericola*	Germany
Cephalomyza chinensis	*Liriomyza chinensis*	China, Japan, Malaya
Liriomyza alliovora	*Liriomyza trifolii*	U.S.A., Venezuela
Liriomyza canomarginis	*Liriomyza sativae*	Hawaii
Liriomyza cucumifoliae	*Liriomyza huidobrensis*	Argentina
Liriomyza dianthi	*Liriomyza huidobrensis*	California
Liriomyza guytona	*Liriomyza sativae*	Alabama
Liriomyza langei	*Liriomyza huidobrensis*	California
Liriomyza minutiseta	*Liriomyza sativae*	Hawaii
Liriomyza munda	*Liriomyza sativae*	California
Liriomyza pullata	*Liriomyza sativae*	Hawaii
Phytobia (Poemyza) orbitona	*Cerodontha (Poemyza) orbitona*	South and West Africa

REFERENCES

Abul-Nasr, S. and M. A. H. Assem, 1961. A leaf-miner, *Liriomyza bryoniae* (Kalt), attacking cucurbitaceous plants in Egypt. *Bull. Soc. Ent. Egypte* 45: 401–403.

Ahmad, T., 1938. The Tur-pod fly, *Agromyza obtusa* Mall., a pest of *Cajanus cajan. Indian J. Agric. Sci.* 8: 63–76.

Ahmad, T. and R. L. Gupta, 1941. The pea leaf-miner, *Phytomyza atricornis* (Meigen), in India. *Indian J. Ent.* 3 (1): 37–49.

Al-Azawi, A. F., 1967. Agromyzid leafminers and their parasites in Iraq. *Bull. Ent. Res.* 57: 285–287.

Al-Azawi, A. F., 1971. Parasites of Agromyzid leafminers in Iraq. *Bull. Iraq. nat. Hist. Mus.* V (1): 35–37.

Aldrich, J. M., 1918. Seasonal and climatic variation in *Cerodontha. Ann. Ent. Soc. Am.* 11: 63–66.

Aldrich, J. M., 1923. Descriptions of Lantana Gall-Fly and Lantana Seed-Fly (Diptera). *Proc. Hawaii Ent. Soc.* 5 (2): 261–263.

Aldrich, J. M., 1929. Three new acalyptrate Diptera. *Proc. Ent. Soc. Wash.* 31 (5): 89–91.

Ali, A. M., 1957. On the bionomics and control of the bean-fly, *Agromyza phaseoli* Coq. *Bull. Soc. Ent. Egypte* 41: 551–554.

Aly, A. S., 1971. Studies on the biology, ecology and control of the maize leaf-miner *Pseudonapomyza spicata* (Mall.) (Diptera, Agromyzidae) in U.A.R. Unpublished *M.Sc. Thesis*, Faculty of Agriculture, Cairo University.

Amundsen, E. O., 1912. Wistaria gall fly, *Agromyza schineri* (Giraud). *Calif. Month. Bull.* 1 (10): 730–733.

Anon., a,1898. A serious bean pest at Gosford. *Agric. Gaz. N.S.W.* 9: 560–561.

Anon., b,1937. Diseases, Insects and other Pests injurious to plants. Bienn. Rep. Kansas agric. exp. Stn 8 (1934–36): 90–106.

Anon., c,1947. A Silver Beet Stalk Miner (*Haplomyza imitans*). *Agric. Gaz. N.S.W.* 58: 158.

Arru, G. M., 1967. I più importanti Insetti minatori delle foglie di Pioppo nell'Italia settentrionale. *Boll. Zool. Agric. Bachic.* (Ser. 2) 8: 41–74.

Askew, R. R., 1968a. Hymenoptera, Chalcidoidea. R. Ent. Soc. Lond. *Handbk Ident. Br. Insects* **8**, Part 2(b): 1–39.

Askew, R. R., 1968b. A survey of leaf-miners and their parasites on laburnum. *Trans. R. ent. Soc. Lond.* 120 (1): 1–37.

Assem, M. A. H., 1966. Studies of vegetable Leaf Miners. Unpublished *Ph.D. Thesis*, Faculty of Agriculture, Cairo.

Audenard, H. et J. d'Aguilar, 1969. Un diptère ravageur des cultures légumières sous serre: *Liriomyza bryoniae* Kalt. (Dipt. Agromyzidae). Acad. Agr. France 2 July 1969: 896–901.

Bailey, W. K. and H. K. Plank, 1940. An Agromyzid Fly Infesting sweet-potato seed in Puerto Rico. *J. Econ. Ent.* 33(4): 704–705.

Baloch, G. M., A. I. Mohyuddin and M. A. Ghani, 1967. Biological control of Cuscuta sp. II. Biology and host-plant range of *Melanagromyza cuscutae* Hering (Dipt. Agromyzidae). *Entomophaga* 12 (5): 489.

Barnes, H. F., 1933. A cambium miner of basket willows (Agromyzidae) and its inquiline gall midge (Cecidomyidae). *Ann. App. Biol.* 20 (3): 498–519.

Barnes, H. F., 1937. The asparagus miner (*Melanagromyza simplex* H. Loew) (Agromyzidae; Diptera). *Ann. App. Ent.* 24: 574–588.

Becker, T., 1903. Ägyptische Dipteren. *Mitt. zool. Mus. Berlin* 2 (2): 1–195.

Becker, T., 1907. Die Ergebnisse meiner dipterologischen Frühjahrsreise nach Algier und Tunis, 1906. *Z. Hym. U. Dipt.* 7: 406.

Becker, T., 1908. Dipteren der Kanarischen Inseln. *Mitt. zool. Mus. Berlin.* 4: 166–171.

Beckmann, J. C., 1680. Bericht von denen auff den Blättern der Bäume in diesem 1680tem Jahre häufig gefundenen Schlangen-Gestalten. Frankfurt/O.

Beffa, G. della, 1939. La *Phytomyza atricornis* Meig. in Piemonte e danni arrecati alla *Calendula arvensis. Boll. Lab. sper. Oss. Fitopatologia* 16: 1–18.

Berg, W., 1956. Die Blumenkohl-Minierfliege, ein wenig beachteter Schädling. Rhein. Monatschr. Gemüse-, Obst und Gartenbau 5.

BERI, S. K., 1971a. Immature stages of Agromyzidae (Diptera) from India IV. Taxonomy and Biology of three species of *Japanagromyza* Sasakawa. *Oriental Insects Suppl.* 1: 19–28.

BERI, S. K., 1971b. Immature stages of Agromyzidae (Diptera) from India VI. Taxonomy and Biology of thirteen species of Genus *Agromyza* Fallén. *ibid.* 1: 53–84.

BERNARD, J., 1959. Essais préliminaires de la lutte estivale contre les mouches de la chicorée. *Meded. Landbouwhogesch. & Opzoek Stn Gent* 24 (314): 994–1004.

BEZZI, M. and C. G. LAMB, 1926. Diptera (excluding Nematocera) from the Island of Rodrigues. *Trans. Ent. Soc. Lond.* 1925 (314): 537–573.

BEZZI, M., 1928. Diptera Brachycera and Athericera of the Fiji Islands based on material in the British Museum. British Museum (Nat. Hist.) pp. 163–167.

BLANCHARD, E., 1926. A dipterous leaf-miner on *Cineraria*, new to Science. *Revta Soc. Ent. Argent.* 1: 10–11.

BLANCHARD, E., 1938. Descripciones y anotaciones de dipteros argentinos. Agromyzidae. *An. Soc. Cient. Argent.* 126: 352–359.

BLANCHARD, E., 1954. Sinopsis de los Agromyzidos argentinos (Diptera, Agromyzidae). Min. Agr. Ganaderia (A) 56: 1–50.

BOLLOW, H. VON, 1953. Ein Massenauftreten der Minierfliege *Agromyza frontanella* Rond. an Luzerne. *Pflschutz* 7: 141–143.

BÖRNER, C., 1907. Zwei neue Möhrenschädlinge aus den Gattungen *Ceutorrhynchus* und *Phytomyza*. *Arb. Bundesanst. f. Land- u. Forstw.* 5: 289–292.

BOUČEK, A. and R. R. ASKEW, 1968. Index of palearctic Eulophidae. Index of entomophagous Insects. Le François, Paris.

BRÈTHES, J., 1920. Insectos útiles y dañinos de Rio Grande do Sul y de la Plata. *An. Soc. Rur. Argent.* 54: 281–290.

BRÈTHES, J., 1923. Sur un diptère mineur des feuilles de *Salvia splendens* et deux Hymenoptères, ses parasites. *Rev. Zool. Agric. Bordeaux* 22 (6): 154.

BRISCHKE, C. G. A., 1881. Die Blattminierer in Danzig's Umgebung. *Schr. naturf. Ges. Danzig* 5: 233–290.

BRITTON, W. E., 1935. Connecticut State Entomologist, Thirty-fifth Report 1935. *Bull. Conn. Agric. Exp. Stn* 383: 245–366.

BROUWER, WA. M. TH. J. DE, and OFFEREN, A. L. VAN, 1966. De mineervlieg (*Liriomyza solani* Her.) bij tomaat. *Jversl. Proefstn Groenten en Fruitteelt Glas Naaldwijk.*

BROWN, L. R. and C. O. EADS, 1969. Unnamed and Little-known Insects attacking Cottonwood in Southern California. *J. Econ. Ent.* 62 (3): 667–677.

BRUEL, W. E. VAN DEN, 1938. A propos de la Lutte contre les Mouches de la Chicorée de Bruxelles. 7th Congr. Ent., Berlin, pp. 2512–2521.

BRUEL, W. E. VAN DEN, 1939. Note sur le mode d'hibernation des Mouches de la Chicorée de Bruxelles. *Bull. Inst. Agron. et Stn Rech. Gx.* 8: 223–230.

BRUEL, W. E. VAN DEN, 1941. A propos de la Lutte contre les Mouches de la Chicorée de Bruxelles. *ibid.* 10: 26–52.

BRUEL, W. E. VAN DEN, 1961. Quel est l'état des recherches effectuées en vue de prévenir les dégâts des Mouches de la Chicorée de Bruxelles: '*Napomyza lateralis* Fall.' et '*Ophiomyia pinguis* Fall.' (Diptères-Agromyzidae). *Rev. Agr.* 14: 507–515.

BRUEL, W. E. VAN DEN, J. LOUNSKY et J. BERNARD, 1958. Essais de destruction des larves de *Napomyza lateralis* Fall. dans les racines non forcées de la chicorée de Bruxelles. *Meded. Landbouwhogesch. en Opzoekstn Gent* 23: 738–744.

BUHR, H., 1932. Mecklenburgische Minen. *Stettin Ent. Z.* 93: 57–115.

BUHR, H., 1941. Mecklenburgische Minen IV. Nachtrag zu den Dipteren-Minen mit Einschluss der in den Rostocker Botanischen Gärten festgestellten. *Archiv Ver. Freunde Naturg. Mecklenburg N.F.* 15: 21–101.

BUHR, H., 1953. Die Frassbilder und weiteren Nahrungspflanzen der an Vicia faba L. lebenden Minierinsekten. *Beitr. Ent.* 3 (3): 258–278.

BUHR, H., 1954. Mecklenburgische Minen. V. Ueber neue und wenig bekannte Dipteren. Minen. *Archiv. Ver. Freunde Naturg. Mecklenburg N.F.* 29: 238–288.

BUHR, H., 1955. Minierfliegen als Gallenerzeuger. *Nachrichtenblatt bay. Ent.* 4 (9): 81–87.

BUHR, H., 1960. Bemerkenwerte oder neue Gallen und Minen aus Thüringen. *Mitt. Thüring. Bot. Ges.* 2 (1): 56–150.

BUHR, H., 1964. Sächsische Minen. *Abh. u. Ber. Naturkunde Mus. Görlitz* 39 (3): 1–72.

BURGESS, E., 1880. The clover Oscinis (*Oscinis trifolii* Burgess n. sp.). *Rep. U.S. Dept. Agr.* 1879: 200–201.

CAMERON, A. E., 1945. Insect Pests of 1944. *Trans. Highland Agric. Soc.* 57: 54–72.

CAMERON, E., 1939. The holly leaf-miner, *Phytomyza ilicis* Curt. and its parasites. *Bull. Ent. Res.* 30: 173–208.

Cañizo, J. del, 1934. Dos Agromícides perjudiciales al garbanzo. *Bol. Pat. Veg. Ent. Agric.* 7: 91–103.
Cañizo, J. del, 1945. Notas sobre la mosca del garbanzo. *ibid.* 13: 15–26 (1944).
Carden, P. W., 1962. The cabbage leaf miner attacking calabrese. *Plant Path.* 11 (1): 36–37.
Chittenden, F. H., 1907. The Asparagus miner. *U.S. Dep. Agr. Bull.* 66: 1–5.
Ciampolini, M., 1949. Osservazioni sull'etologia della *Pseudonapomyza dianthicola* Venturi (Diptera Agromyzidae). *Redia* 34: 289–301.
Ciampolini, M., 1952. La *Pseudonapomyza dianthicola* Venturi. *Redia* 37: 69–120.
Claasen, P. W., 1918. Observations on the life history and biology of *Agromyza laterella* Zetterstedt (Diptera). *Ann. Ent. Soc. Am.* 11: 9–18.
Coats, A. M., 1969. The Quest for Plants. London, pp. 1–400.
Cohen, M., 1936. The biology of the Chrysanthemum leaf-miner, *Phytomyza atricornis* Mg. (Diptera: Agromyzidae). *Ann. Appl. Biol.* 23: 612–632.
Collin, J. E., 1911. A new fly of the genus *Agromyza* from Egypt. *Bull. Ent. Res.* II: 79–80.
Commonwealth Institute of Entomology, 1961. Distribution Maps of Pests: *Melanagromyza phaseoli* (Tryon). London.
Coquillett, D. W., 1899. Description of *Agromyza phaseoli*, a new species of leaf-mining fly. *Proc. Linn. Soc. N.S.W.* 24: 128.
Coquillett, D. W., 1902. New acalyptrate Diptera from North America. *J. N.Y. Ent. Soc.* 10: 177–191.
Cory, E. N., 1916. The Columbine Leaf-miner. *J. Econ. Ent.* 9: 419–424.
Cotes, E. C., 1896. An account of the insects and mites which attack the tea plant in India. *Indian Mus. Notes* 3 (4): 1–74.
Couden, F. D., 1908. A gall-maker of the family Agromyzidae. *Proc. Ent. Soc. Wash.* 9: 34–36.
Cronquist, A., 1968. The Evolution and Classification of Flowering Plants. T. Nelson, London.
Crowe, T. J. and D. J. Greathead, 1970. Parasites of *Leucoptera* spp. (Lepidoptera, Lyonetiidae) on coffee in East Africa. *E. Afr. Agric. For. J.* 35 (4): 364–371.
Cumber, R. A. and R. A. Harrison, 1959. The insect complex of sown pastures in the North Island. III. The Diptera as revealed by summer sweep-sampling. *N.Z. J. Agric. Res.* 2 (4): 741–762.
Curran, C. H., 1931. Four new Diptera in the Canadian National Collection. *Can. Ent.* 63: 93–98.
Curtis, J., 1844. Agromyza violae (*the pansy fly*). *Gardeners' Chron.* April 20: 244–245.
Curtis, J., 1845. Phytomyza nigricornis (*the black-horned leaf-miner*). *ibid.* Feb. 22, 1845: 117.
Curtis, J., 1846. The holly-leaf-fly (*Phytomyza Ilicis*). *ibid.* July 4: 444.
Davidson, A., 1963. Insects attacking *Striga* in Kenya. *Nature* 197, No. 4870: 923–924.
Deshusses, A. and J. Deshusses, 1929. Parasites nouveaux. Les mouches de l'endive (*Ophiomyia pinguis* Fallén et *Phytomyza continua* Hendel). C.R. Acad. Agr. France: 533–537.
Deshusses, A., 1933. Insectes Nuisibles aux Cultures. 3. Les Mouches de l'Endive, *Ophiomyia pinguis* Fall. et *Phytomyza continua* Hend. *Mitt. Schweiz. Ent. Ges.* 15: 474–483.
Dovnar-Zapolski, D. P., 1969. Mining flies on plants of Kirghizia and neighbouring territories. Akad. Nauk. Kirghiz S.S.R., Frunze (in Russian).
Downes, W., 1931. Holly Insects. *Proc. Ent. Soc. Br. Columb.* 28: 25–28.
Downes, W. and H. Andison, 1940. The establishment in British Columbia of Parasites of the Holly Leaf Miner, *Phytomyza ilicis* Curtis. *J. Econ. Ent.* 33 (6): 948–949.
Duthoit, C. M. G., 1968. Cereal Leaf Miner in the south-east Region. *Pl. Path.* 17: 61–63.
Eichler, W., 1950. Zwiebelminierschädlinge in Mitteldeutschland (1949). *NachBl. Dtsch. Pflsch. Dienst. Berlin, N.F.*, 4: 71–73.
Eichmann, R. D., 1943. Asparagus miner really not a pest. *J. Econ. Ent.* 36: 849–852.
Elberg, K., 1964. Über einige Funde der Agromyziden (Brachycera, Acalyptrata) aus der estnischen SSR. *Eesti NSV Taed. Avad. Toim.* 13: 319–321.
Elmore, J. C. and C. A. Ranney, 1954. Injury to Pepper Plants by the Pea Leaf Miner. *J. Econ. Ent.* 47 (2): 357–358.
Enderlein, G., 1936. 22. Ordnung. Zweiflügler, Diptera. *In* Brohmer *et al.* Die Tierwelt Mitteleuropas. Insekten. Teil 3. Diptera 6.
Esterberg, L. K., 1931. On the mass increase of *Dizygomyza lateralis* Mg. in the year 1929. *Plant Protection* 8 (2): 187–189 (in Russian).
Fallén, C. F., 1820. Diptera Sueciae. Vol. 2. Oscinides sueciae, p. 1–10, Lund.
Fallén, C. F., 1823a. Agromyzides. *Diptera suecica* 2 (37): 1–10.
Fallén, C. F., 1823b. Phytomyzides. *Diptera suecica* 2 (41): 1–10.
Fink, D. E., 1913. The Asparagus miner. *Bull. Cornell Agric. Exp. Stn* 331: 411–421.

Fischer, M., 1962. Beitrag zur Kenntnis der Wirte von *Opius*-Arten (Hymenoptera, Braconidae). *Entomophaga* 7 (2): 79–90.

Fischer, M., 1971. Index of world Opiinae. Index of entomophagous Insects. Le François, Paris.

Freeman, C. C., 1958. *Liriomyza guytona:* a new species of Agromyzid leaf-miner. *Ann. Ent. Soc. Am.* 51: 344–345.

Frey, W., 1951. Über das Auftreten der Minierfliege *Phytomyza rufipes* Mg. on Raps. *Z. Pflkrankh.* 58: 10–20.

Frick, K. E., 1951. *Liriomyza langei*, a new species of leaf-miner of economic importance in California. *Pan-Pacif. Ent.* 21: 81.

Frick, K. E., 1952a. A generic revision of the family Agromyzidae (Diptera) with a catalogue of new world species. *Univ. Calif. Publ. Ent.* 8: 339–452.

Frick, K. E., 1952b. Four new Hawaiian *Liriomyza* species and notes on other Hawaiian Agromyzidae (Diptera). *Proc. Hawaii Ent. Soc.* 14 (3): 509–518.

Frick, K. E., 1953. Further Studies on Hawaiian Agromyzidae (Diptera) with Descriptions of Four New Species. *Bull. Wash. St. Agric. Exp. Stn* 1182: 207.

Frick, K. E., 1954. Three North American Phytomya species closely related to *P. nigritella* Zetterstedt (Agromyzidae: Diptera). *Ann. Ent. Soc. Am.* 47: 367–374.

Frick, K. E., 1955. Nearctic species in the *Liriomyza pusilla* complex, No. 3. *L. alliovora*, new name for the Iowa onion miner. *J. Kansas Ent. Soc.* 28 (3): 88–92.

Frick, K. E., 1957a. Nearctic species in the *Liriomyza pusilla* complex, No. 2. *L. munda* and two other species attacking crops in California. *Pan-Pacif. Ent.* 33 (2): 59–70.

Frick, K. E., 1957b. Nomenclatural changes and type designations of some New World Agromyzidae (Diptera). *Ann. Ent. Soc. Am.* 50: 198–205.

Frick, K. E., 1958. *Liriomyza dianthi* n. sp., a new pest of carnations in California. *Proc. Ent. Soc. Wash.* 60 (1): 1–5.

Frick, K. E., 1959. Synopsis of the species of Agromyzid leaf miners described from North America (Diptera). *Proc. U.S. Natn. Mus.* 108: 347–465.

Frick, K. E., 1964. *Liriomyza dianthi*, a New Synonym of *L. langei* Frick (Diptera: Agromyzidae). *Pan-Pacif. Ent.* 40 (1): 11–12.

Froggatt, W. W., 1919. The Lantana fly (*Agromyza lantanae*). *Agric. Gaz. N.S.W.* 30: 665–668.

Frost, S. W., 1924. A study of the leaf-mining Diptera of North America. *Mem. Cornell Univ. Agric. Exp. Stn* 78: 1–228.

Frost, S. W., 1928. Notes on *Phytomyza* with a description of a new species. (Diptera). *Can. Ent.* 60 (4): 77–78.

Frost, S. W., 1930. The leaf-miners of *Aquilegia*, with a description of a new species. *Ann. Ent. Soc. Am.* 23: 457–460.

Frost, S. W., 1931. New Species of West Indian Agromyzidae (Diptera). *Ent. News* 42: 72–75.

Frost, S. W., 1936. New Central American Agromyzidae. *Ann. Ent. Soc. Am.* 29: 298–318.

Frost, S. W., 1939. Two New Species of *Agromyza* from South America (Dipt.: Agromyzidae). *Ent. News* 50: 97–100.

Frost, S. W., 1943. Three new species of Diptera related to *Agromyza pusilla* Meig. *J. New York Ent. Soc.* 51: 253–260.

Frost, S. W. and M. Sasakawa, 1954. New Agromyzidae from Japan. II (Diptera). *Mushi* 27 (7): 49–52.

Gadd, C. H., 1945. Report of the Entomologist. *Ann. Rep. 1945 Tea Res. Inst. Ceylon, Bull.* 27: 30–31.

Gangrade, G. A., 1962. The biology and morphology of immature stages of *Euderus agromyzae* Gangrade (Eulophidae: Hymenoptera). *Indian J. Ent.* 24 (4): 265–273.

Gangrade, G. A., 1963. Assessment of damage to Tur (*Cajanus cajan*) in Madhya Pradesh by the Tur-pod fly, *Agromyza obtusa* Malloch. *Indian J. Agric. Sci.* 8: 63–76.

Gangrade, G. A., 1964. Losses to tur (*Cajanus cajan*) by *Melanagromyza obtusa* Mall. *Indian J. Ent.* 26 (3): 364–365.

Geigenmüller, M., 1966. Beobachtungen bei einem Massenauftreten der Minierfliege *Agromyza megalopsis* Hering an Gerste. *Anz. Schädlingsk.* 39 (4): 57–60.

Geoffrion, R. et A. Hérisset, 1967. Une 'maladie' nouvelle des digitales cultivées. *Plantes médicinales et Phytothérapie* 1: 99–107.

Gersdorf, E. and A. von Kraft, 1965. Die 'falsche Möhrenfliege' auch in Deutschland. *Gesunde Pflanzen* 17 (3): 49–52.

Getzin, L. W., 1960. Selective insecticides for vegetable leaf miner control and parasite survival. *J. Econ. Ent.* 53 (5): 872–875.

Giraud, J., 1861. Fragments entomologiques. *Verh. Zool. Bot. Ges. Wien* 11: 447–494.

GOIDANICH, A., 1928. Contributi alla conoscenza dell'Entomofauna della Canapa. *Boll. Lab. Ent. R. Ist. Sup. Agr. Bologna* 1: 37–64.
GOODLIFFE, F. D., 1941. Studies on Insects bred from Barley, Wheat, Maize and Oats. *Bull. Ent. Res.* 32 (4): 309–325.
GOOT, P. VAN DER, 1930. De *Agromyza* vliegjes der inlandsche Katjanggewassen op Java. *Meded. Inst. Plziekt., Buitenz.* 78: 1–97.
GOUREAU, C. C., 1851. Mémoire pour servir à l'histoire des Diptères dont les larves minent les feuilles des plantes. *Annls. Soc. Ent. Fr.* 9: 131–176.
GRAHAM, M. W. R. DE V., 1969. The Pteromalidae of North-Western Europe (Hymenoptera: Chalcidoidea). *Bull. Brit. Mus. (Nat. Hist.) Ent. Suppl.* 16: 1–908.
GREATHEAD, D. J., 1967. A list of the more important weeds of East Africa. C.I.B.C. unpublished report.
GREATHEAD, D. J., 1968. Biological control of Lantana – a review and discussion of recent developments in East Africa. *PANS* (c) 14 (2): 167–175.
GREATHEAD, D. J., 1969. A study in East Africa of the bean flies (Dipt., Agromyzidae) affecting *Phaseolus vulgaris* and of their natural enemies, with the description of a new species of *Melanagromyza* Hend. *Bull. Ent. Res.* 59 (3): 541–561 (1968).
GREATHEAD, D. J., 1971. A new species of *Melanagromyza* Hendel (Dipt. Agromyzidae) from pods of the bonavist bean (*Lablab niger*) in Uganda. *Bull. Ent. Res.* 60: 463–465.
GREATHEAD, D. J. and J. E. D. MILNER, 1971. A Survey of *Striga* spp. (Scrophulariaceae) and their insect Natural Enemies in East Africa with a discussion on the possibilities of Biological Control. *Trop. Agric.* (Trinidad) 48 (2): 111–124.
GREEN, H., 1934. Report of the Botanical and Forestry Department (Hong Kong) for the Year 1933.
GREENE, C. G., 1914. The cambium miner in river birch. *J. Agric. Res.* 1 (6): 471–474.
GREENE, C. G., 1917. Two new cambium miners (Diptera). *ibid.* 10 (6): 313–317.
GRIFFITHS, G. C. D., 1963. A revision of the palaearctic species of the *nigripes* group of the genus *Agromyza* Fallén (Diptera, Agromyzidae). *Tijdschr. Ent.* 106 (2): 113–168.
GRIFFITHS, G. C. D., 1964a. The Alysiinae (Hym. Braconidae) parasites of the Agromyzidae (Diptera). – III. *Beitr. Ent.* 16: 775–951.
GRIFFITHS, G. C. D., 1964b. The Agromyzid fauna of Iceland and the Faroes, with appendices on the *Phytomyza milii* and *robustella* groups. *Ent. Meddr.* 32: 393–450.
GRIFFITHS, G. C. D., 1967a. A stem-mining fly (*Ophiomyia*) (Dipt., Agromyzidae) attacking peas (*Pisum sativum* L.). *Entomologist's Mon. Mag.* 102: 135–137 (1966).
GRIFFITHS, G. C. D., 1967b. The Alysiinae (Hym. Braconidae) parasites of the Agromyzidae (Diptera) – IV. *Beitr. Ent.* 17: 653–696.
GRIFFITHS, G. C. D., 1967c. Revision of *Phytomyza syngenesiae* group (Diptera: Agromyzidae), including species hitherto known as '*Phytomyza atricornis* Meigen'. *Stuttg. Beitr. Naturk.* 177: 1–28.
GRIFFITHS, G. C. D., 1968. The Alysiinae (Hym. Braconidae) parasites of the Agromyzidae (Diptera) – VI. Parasites of *Cerodontha* Rondani s.l. *Beitr. Ent.* 18: 63–152.
GRIFFITHS, G. C. D., 1972a. Comment on STEYSKAL's note on *Phytomyza horticola* Goureau and *P. syngenesiae* (Hardy) (Diptera, Agromyzidae). *Ent. News* 83: 39–40.
GRIFFITHS, G. C. D., 1972b. Studies on the phylogenetic classification of Diptera Cyclorrhapha, with special reference to the structure of the male postabdomen. Dr. W. Junk, The Hague.
GRISWOLD, G. H., 1928. A new leaf miner injurious to larkspur (*Phytomyza delphiniae* Frost). *J. Econ. Ent.* 21: 855–857.
GROSSENBACHER, J. G., 1915. Medullary spots and their cause. *Bull. Torrey bot. Club* 42: 227–239.
GÜNTHART, E., 1949. Beiträge zur Lebensweise und Bekämpfung von *Ceuthorrhynchus quadridens* Panz. und *Ceuthorrhynchus napi* Gyll. mit Beobachtungen an weiteren Kohl- und Rapsschädlingen. *Mitt. Schweiz. Ent. Ges.* 22: 441–591.
GURNEY, B., 1960. Agromyzid Leaf-miners. *In* Rep. Glasshouse Crops Res. Inst., p. 65.
GURNEY, B. and N. W. HUSSEY, 1967. Chrysanthemum leaf-miner. *ibid.* 1967, p. 82.
GURNEY, G. and N. E. A. SCOPES, 1970. Chrysanthemum pests. *ibid.* 1970, pp. 124–125.
HALIDAY, A. H., 1833. Catalogue of Diptera occurring about Holywood in Downshire. *Ent. Mag.* 1: 147–180.
HAMILTON, A., 1947. Ryegrass attacked by wheat sheath miner. *N.Z. J. Agric.* 75: 447.
HAMMAD, S. M., 1955. On some dipterous leaf-miners from Egypt. *Bull. Soc. Ent. Egypte* 39: 391–393.
HAMMAD, S. M., 1956. Further additions to the knowledge of some leaf-miners from Egypt. *Bull. Soc. Ent. Egypte* 11: 21–27.

HAMMAD, S., H. EL-AROSI and O. AL-MENOUFI, 1967. *Phytomyza orobanchia* Kalt. feeding on *Orobanche crenata* Forsk., in Egypt. *Bull. Soc. Ent. Egypte* 51: 141–144.
HARDING, J. A., 1965. Parasitism of the leaf miner *Liriomyza munda* in the Winter Garden Area of Texas. *J. Econ. Ent.* 58 (3): 442–443.
HARDING, J. A. and D. A. WOLFENBARGER, 1963. Granulated systemic insecticides for vegetable Insect control in South Texas. *J. Econ. Ent.* 56 (5): 687–689.
HARDY, J., 1849. On the primrose-leaf miner; with notice of a proposed new Genus, and characters of three Species of Diptera. *Ann. Mag. Nat. Hist.* (2) 4 (24): 385–392.
HARRIS, H. M. and H. D. TATE, 1933. A Leaf Miner attacking the cultivated onion. *J. Econ. Ent.* 26: 515–516.
HARRISON, R. A., 1959. Acalyptrate Diptera of New Zealand. *N.Z. Dep. Sci. Industr. Res. Bull.* 128. 382 pp.
HASSAN, A. J., 1947. The Beanfly *Agromyza phaseoli* Coq. in Egypt. *Bull. Soc. Fouad I Ent.* 31: 217–224.
HASSELL, M. P. and G. C. VARLEY, 1969. New inductive population model for insect parasites and its bearing on biological control. *Nature* 223: 1133–1137.
HELY, P. C., 1947. Control of French bean fly. Experiments with D.D.T. *Agric. Gaz. N.S.W.* 58: 85–89.
HENDEL, F., 1920. Die paläarktischen Agromyziden (Prodromus einer Monographie): *Arch. Naturgesch.* A 84 (7): 109–174.
HENDEL, F., 1922. Blattminierende Fliegen. *Wien. Ent. Ztg* 39: 65–72.
HENDEL, F., 1923. Blattminierende Fliegen. *Dt. Ent. Z.* 1923: 394.
HENDEL, F., 1927. Beiträge zur Systematik der Agromyziden. *Zool. Anz.* 69: 248–271.
HENDEL, F., 1931–1936. Agromyzidae in Lindner. *Flieg. Pal. Reg.* 59: 1–570.
HENNIG, W., 1941. Verzeichnis der Dipteren von Formosa. *Ent. Beihefte* 8: 1–239.
HENNIG, W., 1965. Die Acalyptratae des Baltischen Bernsteins. *Beitr. Ent.* 8: 505–688.
HERING, M., 1924. Zur Kenntnis der Blattminenfauna des Banats. *Z. Wiss. InsektBiol.* 19: 31–41.
HERING, M., 1927a. Agromyzidae in F. DAHL, Die Tierwelt Deutschlands 6: 1–172.
HERING, M., 1927b. Eine neue gallenerzeugende Agromyzide (Dipt.) *Z. Wiss. InsektBiol.* 22: 319–326.
HERING, M., 1927c. Die Minenfauna der Canarischen Inseln. *Zool. Jb.* B., 53: 405–486.
HERING, M., 1930. Eine Minierfliege als Schädling an Erbsenpflanzen. *Anz. Schädlingsk.* 6 (6): 61–64.
HERING, M., 1932. Die Minenfauna von Tighina (Bender) in Bessarabien. *Bull. Sci. Acad. Roum.* 15 (1/2): 1–22.
HERING, M., 1933. Neue Agromyziden (Dipt.). *Konowia* 12: 33–40.
HERING, M., 1935. Minen von Shanghai, gesammelt von H. HOEHNE. I. *Mitt. Dt. Ent. Ges.* 5 (3/4): 23–28.
HERING, E. M., 1940. *Melanagromyza coffeae*, sp.n., a Leaf-Miner of the Coffee-Plant. *Ann. Mag. Nat. Hist.* 1940: 471–472.
HERING, E. M., 1942. Ein neuer Spargelfeind. *Z. PflKrankh.* 52 (12): 529–533.
HERING, E. M., 1949a. Die Nelken-Minierfliege *Dizygomyza flavifrons* Mg. (Dipt.). *Ent. Z., Frankf. a. M.* 59 (1/2): 2–5.
HERING, E. M., 1949b. Neue palaearktische Agromyziden. *Notul. Ent.* 29: 18–32.
HERING, E. M., 1951a. Ein neuer Sellerie-Feind, *Melanagromyza apii*, sp.n. (Dipt. Agromy.). *Ann. Mag. Nat. Hist.* 12 (IV): 736–745.
HERING, E. M., 1951b. Ein neuer Getreideschädling. *Z. Angew. Ent.* 32 (4): 604–608.
HERING, E. M., 1951c. Biology of the Leaf Miners. Dr. W. Junk, The Hague, 420 pages.
HERING, E. M., 1951d. Neue paläarktische und nearktische Agromyziden (Dipt.). *Notul. Ent.* 31: 31–45.
HERING, E. M., 1953. *Agromyza nigrociliata* Hendel als Getreideschädling (Dipt.). *Tijdschr. Plziekt.* 59: 188–191.
HERING, E. M., 1956. Schnelle Unterartbildung bei *Cephalomyza cepae* (Hg.) (Dipt. Agromyz.). *Dt. Ent. Z.* (*N.F.*) 3: 258–261.
HERING, E. M., 1957a. Zur Blattminenkunde von Südafrika. *Ann. Trans. Mus.* 23: 59–80.
HERING, E. M., 1957b. Die Larven der Agromyziden (Diptera) III. *Tijdschr. Ent.* 100 (1): 73–94.
HERING, E. M., 1957c. Bestimmungstabellen der Blattminen von Europa. Vols. 1–2: 1–1185, Vol. 3: 1–221. Dr. W. Junk, The Hague.
HERING, E. M., 1958. *Melanagromyza cuscutae* sp.n., eine neue fruchtfressende Agromyzide der Seide. *Dt. Ent. Z. N.F.* 5: 217–220.
HERING, E. M., 1960. Neue Blattminen-Studien. *Dt. ent. Z. N.F.* 7: 119–145.

Hering, E. M., 1962. Neue Blattminen-Studien II. *Dt. ent. Z. N.F.* 9: 30–65.
Hering, E. M., 1963. Neue Blattminen-Studien III. *Dt. ent. Z. N.F.* 10: 221–250.
Heungens, A., 1966. Voorkomen van de mineervlieg *Phytagromyza dianthicola* Venturi in de Anjerteelt. *Meded. Rijks Faculteit Landbw Gent.* 31: 107–116.
Hills, O. A. and E. A. Taylor, 1951. Parasitization of dipterous leaf-miners in cantaloups and lettuce in the Salt River Valley, Arizona. *J. Econ. Ent.* 44 (5): 759–762.
Hochapfel, H., 1937. Stärkeres Auftreten von *Phytomyza rufipes* in Schlesien. *Anz. Schädlingsk.* 13: 114–115.
Holm, L., 1969. Weed problems in developing countries. *Weed Sci.* 17 (1): 113–118.
Holm, L. and J. Herberger, 1970. Weeds of tropical crops. Proc. 10th Br. Weed Control Conf. 1132–7.
Hussey, N. W., 1969. Differences in susceptibility of different strains of Chrysanthemum leaf miner (*Phytomyza syngenesiae*) to BHC and Diazinon. Proc. 5th Br. Insecticide and Fungicide Conf., pp. 93–97.
Hussey, N. W. and B. Gurney, 1962. Host selection by the polyphagous species *Phytomyza atricornis* Meigen (Dipt., Agromyzidae). *Entomologist's Mon. Mag.* 98: 42–47.
Hussey, N. W., W. H. Read and J. J. Hesling, 1969. The pests of protected cultivation. London.
International Trust for Zoological Nomenclature, 1961. International Code of Zoological Nomenclature. London.
Ipe, I. M., 1966. A detailed morphological study of external and internal genital organs of female *Melanagromyza obtusa* (Malloch), a serious pest of *Cajanus indicus* L. (Agromyzidae: Diptera). *Indian J. Ent.* 28 (3): 287–298.
Ipe, I. M., 1967. A detailed morphological study of the external and internal genital organs of male *Melanagromyza obtusa* (Malloch), a pest of *Cajanus indicus* L. *ibid.* 29 (1): 1–10.
Isitani, H., 1938. The Leaf-miners of Barley and Wheat found near the Environs of Tokyo (in Japanese). *Oho Kontyû* 1 (3): 101–109.
Ivannikov, A. I., 1969. [The fly *Melanagromyza cuscutae* Hg. – an enemy of dodder in Kazakhstan] Vest. selkhoz. *Nauki, Alma-Ata* 12 (1): 85–88 (in Russian).
Jack, R. W., 1913. The bean stem maggot. *Rhodesia Agric. J.* 10: 545–553.
Jefferson, R. N. and C. O. Eads, 1952. Control of Leaf Miners and other Insect Pests of Aster. *J. Econ. Ent.* 45 (3): 476–481.
Jensen, G. L. and C. S. Koehler, 1970. Seasonal and Distributional Abundance and Parasites of Leaf Miners of Alfalfa in California. *J. Econ. Ent.* 63 (5): 1623–1628.
Jepson, F. P., 1916. The Lantana Seed Fly (Agromyzidae). Dep. Agric., Fiji. Pamphlet 21, 1 page.
Jobling, J. and C. W. T. Young, 1964. Apparent variations in the resistance of poplar clones to bacterial canker. Report on Forest Research, H.M.S.O., London.
Kaltenbach, J. H., 1858. Die deutschen Phytophagen aus Klasse der Insekten (B). *Verh. naturh. Ver. preuss. Rheinl.* 15: 77–161.
Kaltenbach, J. H., 1864. Die deutschen Phytophagen aus der Klasse der Insekten (M-P). *Verh. Naturh. Ver. Preuss. Rheinl.* 21: 228–404.
Kaltenbach, J. H., 1867. Die deutschen Phytophagen aus der Klasse der Insekten (Q-R). *Verh. Naturh. Ver. Preuss. Rheinl.* 24: 21–117.
Kangas, E., 1935. Die Braunfleckigkeit des Birkenholzes und ihr Urheber *Dendromyza* (*Dizygomyza*) *betulae* n.sp. *Comm. Inst. Forest. Fenn.* 22: 1–31.
Kangas, E., 1937. Ueber die Braunfleckigkeit der Laubhölzer, *Ann. Ent. Fenn.* 3: 33–39.
Kangas, E., 1949. Die Dendromyza-Arten von Eberesche, Espe und Salweide und ihre Biologie. *Ann. Ent. Fennici* 14: 106–116.
Kangas, E., 1955. *Dendromyza cerasiferae* n. sp. (Dipt., Agromyzidae) from England. *Ann. Ent. Fenn.* 21 (4): 165–170.
Kasasian, L., 1971. Weed Control in the Tropics, London. pp. 1–307.
Kato, S., 1949. A preliminary report on a survey of agricultural insect pests in Chahar, Suiyuan and northern Shansi. *Peking nat. Hist. Bull.* 18 (1): 11–36.
Kato, S., 1956. A taxonomic study of the Rice Leaf-miner, *Agromyza oryzae* (Munakata) (Dipt., Agromyzidae). *Bull. Natn. Inst. Agric. Sci.* (Japan). Series C, 6: 25–34.
Kato, S., 1961. Taxonomic studies on Soy Bean Leaf and Stem Mining Flies (Diptera, Agromyzidae) of Economic Importance in Japan, with Descriptions of Three New Species. *Bull. Natn. Inst. Agric. Sci.* (C) 13: 171–206.
Kelsey, J. M., 1937. The Ragwort leaf-miner (*Phytomyza atricornis* Mg.) and its parasite (*Dacnusa areolaris* Nees). *N.Z. J. Sci. Tech.* (A) 18 (10): 762–767.
Kelsheimer, E. G., 1948. Insecticidal value of DDT and related synthetic compounds on vegetable crop insects in Florida. Fla agric. exp. Stn a. Rep. 1947: 46–47.

Kerr Jr., K. W., 1951. The chemotherapeutic value of several insecticides for larvae of certain leaf mining insects. *J. Econ. Ent.* 44: 493–498.

Kerrich, G. J., 1970. On the taxonomy of some African Eulophidae (Hym., Chalcidoidea) associated with oil palm, coffee and mango. *ibid.* 60 (2): 327–331.

Kleinschmidt, R. P., 1960. New species of Agromyzidae from Queensland. *Qd. J. agric. Sci.* 17: 321–337.

Kleinschmidt, R. P., 1970. Studies of some Agromyzidae in Queensland. *Qd. J. agric. Sci.* 27: 341–384.

Koehler, C. S. and P. N. Metha, 1971. Effects of attack by *Melanagromyza chalcosoma* Spencer on germination and development of cowpea. *E. Afr. agric. For. J.* 36 (1): 83–87.

Koizumi, K., 1953. *Melanagromyza lappivora*, a new species of root miner of Burdock in Japan (Diptera, Agromyzidae). Okayama Univ., *Sci. Rep. Fac. Agr.* 3: 46.

Koningsberger, J. C., 1897. De dierlijke vijanden van het Koffiekultuur op Java. Meded. Lands Plantentuin, Batavia, pp. 25–26.

Kulp, L. A., 1968. The taxonomic status of the dipterous Holly Leaf Miners (Diptera: Agromyzidae). Univ. Maryland *Agric. Exp. Stn Bull.* A-155: 1–42.

Kunhi Kannan, K., 1924. Lantana flies (*Agromyza lantanae* Frogg.) in Hawaii. *Agric. J. India* 19: 504–508.

Kuroda, M., 1958. Studies on the spiracles and cephalopharyngeal sclerites of the larvae of the Agromyzid flies (Report II). *Kontyû* 26: 148–152.

Kuroda, M., 1960. Studies on the spiracles and cephalopharyngeal sclerites of the larvae of the Agromyzid flies (Report IV). *Kontyû* 28: 172–176.

Kuwayama, S., 1950. The Rice leaf-miner: some observations on its ecology and experiments along the control measures. Proc. 8th int. Congr. Ent. (Stockholm, 1950), pp. 662–671.

LaBonte, G. A. and L. J. Lipovsky, 1967. Oak-Leaf Shot-Hole caused by *Japanagromyza viridula*. *J. Econ. Ent.* 60 (5): 1266–1270.

Lamb, C. G., 1912. The Percy Sladen Trust Expedition to the Indian Ocean in 1905. XIX. Diptera. *Trans. Linn. Soc. Lond.* 15: 344–347.

Lange, W. H. Jr., 1949. Notes on the occurrence of Agromyzid flies during 1948, and a record of two unreported species in Calif. *Pan-Pacif. Ent.* 25: 91–92.

Lange, W. H., A. A. Grigarick and E. C. Carlson, 1957. Serpentine Leaf Miner Damage. *Calif. Agr.* 11: 3–5.

Lea, A. M., 1906. Notes on some summer beetles and two leaf-attacking flies. *Agr. Stock Dep. Tasm. Bull.* 9: 1–6.

Lee, N. R., 1953. Note on a plum cambium miner (Agromyzidae). A. Rep. East Malling Res. Stn 1952: 78–79.

Lehmann, H., 1934. Luzerneschädlinge. 2. Diptera. *Z. Pflanzenkrankh.* 44: 331–348.

Lekić, M., 1971. The Role of the Dipteron *Phytomyza orobanchia* Kalt. (Agromyzidae) in reducing parasitic phanerogam populations of the *Orobanche* Genus in Vojvodina. *Savremena Poljoprivreda* 18 (7–8): 627–637.

Le Pelley, R. H., 1968. Pests of coffee. Longmans, London, pp. 1–590.

Lever, R. J. A. W., 1942. Pests of the Vegetable Garden and their Control. *Agric. J. Fiji* 13 (4): 109–225.

Lever, R. J. A. W., 1944. Division of Entomology *Ann. Rep.* 1943. *Agric. J. Fiji* 15 (3): 73–75.

Lindemann, K., 1886. Über *Agromyza lateralis* Macq. und ihre Verwandlungen. *Bull. Soc. Imp. Natur. Moscou*, 62 (3): 9–14.

Lintner, J. A., 1888. *Phytomyza lateralis* (The Marguerite Fly). Fourth a. Rep. St. Entomologist N.Y., pp. 73–79.

Lintner, J. A., 1891. Seventh Report on the injurious and other insects of New York State. *Rep. N.Y. Mus.* 44: 199–246.

Lodos, N., 1961. Egede nohutlara zarar veren iki sinek turu: *Liriomyza cicerina* Rond. ve *Phytomyza atricornis* Meigen. *Bitki Koruna Büll.* 2 (10): 44–48.

Loew, H., 1858. Zwanzig neue Diptern. *Wien. Ent. Monatschr.* 2 (3): 65–79.

Loew, H., 1863. Diptera Americae septentrionalis indigena. *Berl. Ent. Z.* 7: 1–55.

Loew, H., 1869. Diptera Americae septentrionalis indigena. *ibid.* 13: 1–52.

Lotto, G. De, 1948. Gli insetti dannosi alle piante coltivate e spontanae dell'Eritrea – Elenco delle specie riscontrate nel 1947. *Boll. Soc. ital. Med. e Ig. trop.* (Sez. Eritrea) 8 (1/2): 84–90.

Lounsky, J., 1961. Traitement en couche des racines de la chicorée witloof dans la lutte contre les mineuses. *Napomyza lateralis* Fall. et *Ophiomyia pinguis* Fall. *Meded. Landbouwhogesch. e. Opzoek Stn Gent* 26 (3): 1451–1470.

Lounsky, J., 1963a. Nouvelles observations sur le traitement estival en tant que moyen de lutte contre les dégâts des mineuses de la chicorée witloof. *ibid.* 28 (3): 691–702.
Lounsky, J., 1963b. Les mineuses de la chicorée Witloof. *Parasitica* 19 (2): 81–85.
Lounsky, J., 1964. Les mineuses de la chicorée de Bruxelles. Nouvelles observations sur l'importance du décolletage. *Meded. Landbouwhogesch. e Opzoek Stn Gent* 29 (3): 1056–1062.
Lounsky, J., 1965a. Les mineuses de la chicorée de Bruxelles. De l'intervention respective de différentes espèces. *Parasitica* 21 (4): 113–123.
Lounsky, J., 1965b. Les mineuses de la chicorée witloof. De l'évaluation des dégâts en couche de forçage. *Meded. Landbouwhogesch. & Opzoek Stn Gent* 30 (3): 1488–1495.
Lounsky, J., 1966a. Les mineuses de la chicorée de Bruxelles. De l'intervention respective de différentes espèces. II. *Parasitica* 22 (1): 12–16.
Lounsky, J., 1966b. Observations pratiques diverses dans la lutte contre les mineuses de la chicorée de Bruxelles. *Meded. Rijksfac. Landbouw-Wetenschappen Gent* 31 (3): 810–814.
Luginbill, P. and T. D. Urbahns, 1916. The spike-horned leaf-miner, an enemy of grains and grasses. *Bull. U.S. Dep. Agr.* 432: 1–18.
McGregor, E. A., 1914. The serpentine leaf-miner on cotton. *J. Econ. Ent.* 7: 447–454.
McLeod, J. H., 1954. Statuses of some introduced parasites and their hosts in British Columbia. *Proc. ent. Soc. Br. Columb.* 50: 19–27.
McLeod, J. H., 1962. A review of the biological control attempts against insects and weeds in Canada. Part 1. C.A.B., England.
Macquart, J., 1835. Histoire Naturelle des Insectes. *Diptères.* 2: 605–620. Paris.
Malloch, J. R., 1913a. A revision of the species in *Agromyza* Fallén and *Cerodontha* Rondani. *Ann. Ent. Soc. Am.* 6: 269–336.
Malloch, J. R., 1913b. A synopsis of the genera of Agromyzidae, with descriptions of new genera and species. *Proc. U.S. Natl. Mus.* 46: 127–154.
Malloch, J. R., 1914a. Formosan Agromyzidae. *Annales Hist.-Nat. Mus. Hung.* 12: 306–336.
Malloch, J. R., 1914b. Description of a new species of *Agromyza* from Porto Rico. *Proc. ent. Soc. Wash.* 16: 89–90.
Malloch, J. R., 1916. A new species of Agromyzid destructive to beans in the Philippines. *Proc. Ent. Soc. Wash.* 18: 93.
Malloch, J. R., 1925. Notes on Australian Diptera VI. *Proc. Linn. Soc. N.S.W.* 50: 88–91.
Malloch, J. R., 1934a. Agromyzidae. In: Diptera of Patagonia and South Chile, pp. 465–486. London, British Museum (Nat. Hist.).
Malloch, J. R., 1934b. Notes on Australian Diptera 34. *Proc. Linn. Soc. N.S.W.* 59: 1–8.
Marchal, P., 1911. Sur quelques Insectes récemment observés comme nuisibles aux cultures. *Bull. Soc. Ent. France* 87: 261–262.
Martelli, G. M., 1933. Nota preliminare sui parassiti animali dell'*Orobanche* della fava '*Orobanche speciosa*' D.C. *Riv. Pat. veg.* 23 (5–6): 1–8 (Pavia).
Matsumura, S., 1915. '*Dainippon Gaichu Zensho* [Manual of the injurious insects in Japan] 2: 53 (in Japanese).
Matsumura, S., 1916. *Konchû sekai* [Insect World] 20: 443–444 (in Japanese).
Mayeux, H. and G. P. Wene, 1950. Control of Serpentine Leaf Miner on Pepper. *J. Econ. Ent.* 43 (5): 732–733.
Meigen, J. W., 1830. Systematische Beschreibung der bekannten europäischen zweiflügeligen Insekten 6: 166–196. Hamm.
Meigen, J. W., 1838. Systematische Beschreibung der bekannten europäischen zweiflügeligen Insekten 7: 396–407. Hamm.
Meijere, J. C. H. de, 1922. Zur Kenntnis javanischer Agromyzinen. *Bijd. Dierk.* 22: 17–24.
Meijere, J. C. H. de, 1924. Verzeichnis der holländischen Agromyzinen. *Tijdschr. Ent.* 67: 119–155.
Meijere, J. C. H. de, 1925. Die Larven der Agromyzinen. *ibid.* 68: 195–293.
Meijere, J. C. H. de, 1926. Die Larven der Agromyzinen (Fortsetzung und Schluss). *ibid.* 69: 227–317.
Meijere, J. C. H. de, 1940a. Über *Melanagromyza centrosematis* n. sp. aus Java nebst Bemerkungen über andere tropische Melanagromyzen (Dipter. Agrom.). *ibid.* 83: 128–131.
Meijere, J. C. H. de, 1940b. Die Larven der Agromyzinen. *ibid.* 83: 160–188.
Meijere, J. C. H. de, 1943. Die Larven der Agromyzinen. Siebenter Nachtrag. *ibid.* 86: 61–76.
Melis, A., 1935. Contributo alla conoscenza morfologica e biologica della *Phytomyza atricornis* Meig. *Redia* 21: 205–262.
Mendes, L. O. T., 1940. O minador da batatinha. *J. Agron. Piracicaba* 3 (3): 207–220.
Mesnil, L., 1934. *Ophiomyia pinguis* Fall. (Dipt. Agromyzidae) nuisible aux Endives. *Bull. Soc. Ent. France* 39: 131–136.

MESNIL, L. and C. MARCEL, 1935. Une méthode de destruction de la mouche des endives. *C.R. Acad. Agr. France*: 75–80.
MIALL, L. C. and T. H. TAYLOR, 1907. The structure and life history of the holly fly. *Trans. Ent. Soc. Lond.* 1907: 259–283.
MICHELBACHER, A. E., O. G. BACON, W. W. MIDDLEKAUFF and W. ERWIN, 1953. Tomato Insect Investigations in Northern California in 1951. *J. Econ. Ent.* 46 (1): 73–76.
MICHELBACHER, A. E., O. G. BACON and J. UNDERHILL, 1953. Leaf-miner on tomato. *Calif. Agr.* 7 (7): 15.
MICHELBACHER, A. E., W. W. MIDDLEKAUF and L. C. GLOVER, 1951. Studies with Aldrin and Dieldrin against Melon Insects. *J. Econ. Ent.* 44 (3): 390–393.
MIK, J., 1894. Ueber eine neue Agromyza, deren Larven in den Blüthenknospen von *Lilium Martagon* leben. *Wien. Ent. Ztg* 13: 284–290.
MILES, H. W. and M. COHEN, 1936. The chrysanthemum leaf miner and its control. *J. Min. Agr.* 43: 256–261.
MILLER, D. E. and G. L. JENSEN, 1970. Agromyzid Alfalfa Leaf Miners and their parasites in Massachusetts. *J. Econ. Ent.* 63 (4): 1337–1338.
MINISTRY OF AGRICULTURE, FISHERIES AND FOOD, Fisheries and Food, 1969. Chrysanthemum Leaf Miner. Advisory Leaflet 550, 4 pages, London.
MINISTRY OF AGRICULTURE, FISHERIES AND FOOD, 1923–1970. Unpublished Reports.
MORGAN, H. G., 1954. *Napomyza lateralis* (Fall.): a new pest of Calendulas. *Pl. Path.* 3: 85–86.
MORGAN, W. L., 1938. Control of bean fly. *Agric. Gaz. N.S.W.* 49: 501–503.
MORRISON, L., 1938. Surveys of the insect pests of wheat crops in Canterbury and North Otago during the summers of 1936–1937 and 1937–1938. *N.Z. J. Sci. Tech.* (A) 20 (3): 142–155.
MOUTIA, A., 1932. Entomological Division. A. Rep. Dept. Agric. Mauritius 1931.
MOUTIA, A., 1945. Entomological Division 5. Miscellaneous Pests. A. Rep. Dep. Agric. Mauritius, 1944: 1–30.
MUESEBECK, C. F. W., 1963. A new Hawaiian *Opius* from a leaf-mining pest of beans (Hymenoptera: Braconidae). *Proc. Hawaii. Ent. Soc.* 1962, 18 (2): 289–290.
MÜHLE, E., 1971. Die Krankheiten und Schädlinge der zur Samengewinnung angebauten Futtergräser. S. Hirzel Verlag, Leipzig.
MUNAKATA, E., 1910. '*Konchu Sekai*' [Insect World] 14 (9): 7–116 (in Japanese).
NAUDE, C. P., 1956. Eradication of Lantana and Thorn trees. *Farming in South Africa*, March, 1956.
NECZYPOR, W., 1962. *Phytomyza* (*Napomyza*) *lateralis* Fallén, ein neuer Schädling an *Digitalis lanata* Ehrh. und *Digitalis purpurea* L. *Pharmazie* 17: 780–782.
NIELSEN, J. C., 1905. Über die Entwicklung von *Agromyza carbonaria* Zett., der Urheber der 'Markflecken'. *Zool. Anz.* 29: 221–222.
NIELSEN, J. C., 1906. Zoologische Studien über die Markflecke. *Zool. Jb.* 23 (Abt. Syst.): 725–738.
NIETZKE, G., 1943. Rassendifferenzierung bei *Dizygomyza cepae* Her. *Arch. Hydrobiol.*, Aug. – Thienemann Festband, pp. 48–56.
NIETZKE, G., 1953. Beiträge zur Biologie und Epidemiologie der Zwiebelminierfliege *Dizygomyza cepae* Her. *Z. Angew. Ent.* 35: 249–270.
NISHIJIMA, Y., K. HONMA and S. OKYUAMA, 1963. Mortality and its factors in the immature stages of the black wheat leaf miner, *Agromyza albipennis* Meigen [in Japanese, with English summary]. *Res. Bull. Obihiro Zootech. Univ.* (1) 3: 487–504.
NOLTE, H.-W., 1960. Schäden an Porree im Herbst 1959. *NachrBl. Dtsch. PflSchDienst, Berlin*, N.F., 14: 190–193.
NOLTE, H.-W., 1964. Zur Epidemiologie der Zwiebelminierfliege (*Phytobia cepae* Hering). *Beitr. Ent.* 14 (1/2): 119–124.
NOWAKOWSKI, J. T., 1962. Introduction to a systematic revision of the family Agromyzidae (Diptera) with some remarks on host plant selection by these flies. *Ann. Zool., Warsz.* 20 (8): 67–183.
NOWAKOWSKI, J. T., 1967. Vorläufige Mitteilung zu einer Monographie der europäischen Arten der Gattung *Cerodontha* Rond. (Diptera, Agromyzidae). *Polskie Pismo Ent.* 37: 633–661.
NOWAKOWSKI, J. T. In press. Monographie der europäischen Arten der Gattung *Cerodontha* Rond. (Diptera, Agromyzidae). Warsaw.
OATMAN, E. R., 1959a. Natural control studies of the Melon Leaf Miner, *Liriomyza pictella* (Thomson). *J. Econ. Ent.* 52 (5): 895–898.
OATMAN, E. R., 1959b. Host range studies of the Melon Leaf Miner, *Liriomyza pictella* (Thomson) (Diptera: Agromyzidae). *Ann. Ent. Soc. Am.* 52 (6): 739–741.

Oatman, E. R., 1960. Parasitism of the overwintering pupae of the Melon Leaf Miner, *Liriomyza pictella*. *J. Econ. Ent.* 53 (4): 682.
Oatman, E. R., 1961. Crossbreeding studies with two closely related species of *Liriomyza*. *Pan-Pacif. Ent.* 37 (1): 53–57.
Oatman, E. R. and A. E. Michelbacher, 1958. The Melon Leaf Miner, *Liriomyza pictella* (Thomson) (Diptera: Agromyzidae). *Ann. Ent. Soc. Am.* 51 (6): 557–566.
Oatman, E. R. and A. E. Michelbacher, 1959. The Melon Leaf Miner *Liriomyza pictella* (Thomson) (Diptera: Agromyzidae) II. Ecological Studies. *Ann. Ent. Soc. Am.* 52 (1): 83–89.
Oatman, E. R. and G. R. Platner, 1969. An Ecological Study of Insect Populations on Cabbage in Southern California. *Hilgardia* 40 (1): 1–40.
Oosthuizen, M. J., 1963. The biological control of *Lantana camara* L. in Natal. *J. Ent. Soc. Sth Afr.* 27: 3–16.
Otanes y Quesales, F., 1918. The Bean Fly. *Philipp. Agr.* 7: 2–31.
Pandey, N. D. and N. S. Agrawal, 1962. Morphology and anatomy of *Agromyza obtusa* Mall. (Diptera: Agromyzidae). *Indian J. Ent.* 24 (3): 149–166.
Panzer, G. W. F., 1806. Faunae Insectorum Germanicae Initia, No. 104, table 22.
Peace, T. R., 1952. Poplars. *Forestry Commission Bull.* 19: 1–50, H.M.S.O., London.
Perkins, R. C. L. and O. H. Swezey, 1924. The Introduction into Hawaii of Insects that attack Lantana. *Bull. Exp. Stn Hawaii.* Sugar Planters' Ass. 16: 1–83.
Phillips, W. J., 1914. Corn-leaf blotch miner. *J. Agric. Res.* 2: 15–31.
Picard, F., 1914a. A propos de l'action du froid sur les insectes. *Progrès agricole et viticole, Montpellier* 31: 332–333.
Picard, F., 1914b. Les insectes de la luzerne. *ibid.* 31: 555–561.
Pitcher, R. S., 1956. A Further Note on a Cambium Miner of *Prunus* spp. (*Dendromyza cerasiferae* Kangas, Agromyzidae, Dipt.). *A. Rep. East Malling Res. Stn* 1955: 139–140.
Porter, C. E., 1936. Notas breves de Entomologia agricola. *Revta Chil. Hist. Nat.* 40: 427–428.
Porter, C. E., 1938. Notas breves de Entomologia agricola. *ibid.* 42: 172.
Porter, C. E., 1939. Notas breves de Entomologia agricola. *ibid.* 43: 139–140.
Pritchard, A. E., 1957. New Carnation Pests. *California Agr.* 11 (3): 5.
Ramachandra Rao, R. S. Y., 1920. Lantana insects in India. *India Dep. Agr. Mem., Ent. Series*, 5: 239–314.
Ratzeburg, J. T. C., 1868. Die Waldverderbnis, Berlin.
Réaumur, M. de, 1738. Mémoires pour servir à d'Histoire des Insectes Vol. 3, Amsterdam.
Ricchello, A., 1928. Contributo alla conoscenza della Mosca del carciofo (*Agromyza andalusiaca* Strobl) e dei suoi parassiti. *Boll. Zool. Gen. agric.* 22: 81–147.
Riley, C. V., 1884. The Cabbage Oscinis. *Ann. Rep. U.S. Dep. Agric.* 1884, p. 322.
Rode, H. and E. Vorsatz, 1966. Zur Biologie und Bekämpfung der Lilienfliege (*Liriomyza urophorina* Mik). *NachrBl. Dt. PflSchDienst, Berlin* 20 (1): 107–112.
Roesler, R., 1937. *Phytomyza rufipes* Mg. (Dipt.) als Blumenkohl-Schädling. *Anz. Schädlingsk.* 13: 87–88.
Rohdendorf, B. B., 1950. A new pest of water-melon – the mining fly *Liriomyza citrulli* Rohdendorf, sp.n. (Diptera, Agromyzidae). *Rev. Ent. URSS* 31: 82–84 (in Russian).
Rohdendorf-Holmanová, E. B., 1958. Die Minierfliegen (Diptera, Agromyzidae) des Moskauer Gebiets. 1. Agromyzinae. *Rev. Ent. USSR* 37: 380–391.
Rondani, C., 1875. Dipterologiae Italicae Prodromus. *Bull. Soc. Ent. Italia* 7: 174.
Rougemont, F. de, 1912. Détails biologiques sur la *Phytomyza* de *Thalictrum*. *Mitt. Schweiz. Ent. Ges.* 12: 82–87.
Ruhmann, M. H., 1935. Report of Provincial Entomologist. *Rep. Dep. Agr. Brit. Columb.* 1934 29: 37–39. Victoria, B.C.
Sajó, K., 1896. Ein bisher unbekannter Feinde des Spargels. *Ill. Wochenschrift Ent.* 1: 597–598.
Sanders, J. G., 1912. A Remedy for Chrysanthemum Leaf Miner. *J. Econ. Ent.* 5 (6): 472.
Sant, L. E. van't, H. E. Vijzelman and J. G. C. Bethe, 1961. Enkele gegevens over de witlofmineervlieg (*Napomyza lateralis* Fall.) en haar bestrijdingsmogelijkheden. Inst. voor Plantenziektenkg. Onderzoek, Wageningen, No. 268: 1–36.
Sasakawa, M., 1953. The ecological studies on the corn-linear leaf-miner, *Phytomyza nigra* Meigen. (1). *Sci. Rep. Saikyo Univ., Agr.* 5: 106–116.
Sasakawa, M., 1954a. Ecological Studies on the corn-linear Leaf-miner, *Phytomyza nigra* Meig. II. Differences among the varieties of barley and wheat to the damage of the corn-linear leaf-miner. *ibid.* 6: 131–138.
Sasakawa, M., 1954b. Neue Agromyzidae aus Japan. *Trans. Shikoku Ent. Soc.* 4 (3): 35–49.
Sasakawa, M., 1956. A new species related to *Phytomyza saxifragae* Hering (Diptera, Agromyzidae). *Insecta Matsumarana* 19: 105–108.

Sasakawa, M., 1958. The female terminalia of the Agromyzidae, with description of a new genus (1). *Sci. Rep. Saikyo Univ., Agr.* 10: 138–139.
Sasakawa, M., 1961. A study of the Japanese Agromyzidae (Diptera), 2. *Pacif. Insects* 3: 307–472.
Sasakawa, M., 1963a. Oriental Agromyzidae (Diptera) in Bishop Museum, Part 1. *ibid.* 5 (1): 23–50.
Sasakawa, M., 1963b. A Revision of Polynesian Agromyzidae (Diptera). *ibid.* 5 (3): 489–506.
Sasakawa, M., 1963c. Papuan Agromyzidae (Diptera). *ibid.* 5 (4): 797–835.
Sasakawa, M., 1964. A Revision of the Hawaiian *Liriomyza. Proc. Hawaii. Ent. Soc.* 18 (3): 429–433 (1963).
Sasakawa, M., 1966. Host preference of the pea leaf miner, *Phytomyza atricornis* Mg. in relation to the leguminous plants. *Sci. Rep. Kyoto Pref. Univ., Agr.* 18: 57–62.
Sasakawa, M. and N. Fukuhara, 1965a. A new Mulberry Leaf-miner in Japan (Diptera: Agromyzidae). Japan. *J. Appl. Ent. Zool.* 9 (2): 121–124.
Sasakawa, M. and N. Fukuhara, 1965b. Emendation for the Original Spelling of the specific name of the Mulberry Leaf-miner (Diptera: Agromyzidae). Japan. *J. Appl. Ent. Zool.* 9 (3): 190.
Sasakawa, M., A. Yoshida and A. Kanayama, 1970. Observations on oviposition and adult survival of the pea leaf-miner, *Phytomyza horticola* Goureau (Diptera, Agromyzidae). *Bull. Exp. Farm, Fac. Agr. Kyoto Pref. Univ.* 3: 7–14.
Scherney, F., 1966. Zur Biologie der an Gerste minierenden Agromyzidenart *Agromyza nigrella* Rondani (Dipt.), *Z. PflanzKr. u. Pflschutz* 73: 141–146.
Scherney, F. and G. Weigand, 1967. Biologie und Bekämpfung der an Gerste auftretenden Minierfliege *Agromyza nigrella* Rondani. Bayr. Landesanstalt f. Bodenkultur, Pflanzenbau und Pflanzenschutz – *Pflanzenschutzinformationen* 14: 1–4.
Schimitschek, E., 1935. *Dendromyza*-Larven als Korbweidenschädlinge. *Dendromyza cambii* Hendel (Agromyzidae); Die Weiden-Kambiumminierfliege *Anz. Schädlingsk.* 11 (11): 121–126.
Scopes, N. E. A. and B. Gurney, 1969. Biological control of chrysanthemum leaf-miner. *In* A. Rep. Glasshouse Crops Res. Inst. 1969, pp. 107–108.
Séguy, E., 1949. Un nouveau *Phytomyza* parasite des oeillets cultivés (Dipt. Agromyzidae). *Ann. Mus. Civ. Storia Nat. Genova* 64: 56–61.
Séguy, E., 1951. Diptères mineurs de Madagascar. *Mém. Inst. Sci. Madagascar* (A) 5 (2): 309–321.
Sehgal, V. K., 1965. Studies on Indian Agromyzidae (Diptera) – 2. *Beitr. Ent.* 15: 3–10.
Sehgal, V. K., 1968. Descriptions of new species of flies of the family Agromyzidae from Alberta, Canada (Diptera). *Quaest. Ent.* 4: 57–88.
Sehgal, V. K., 1971. Biology and host-plant relationships of an oligophagous leaf miner *Phytomyza matricariae* Hendel (Diptera: Agromyzidae). *Quaest. Ent.* 7: 255–280.
Sepsawadi, P. and B. Meksongsee, 1970. A testing of Insecticides to control the Bean Fly (*Agromyza phaseoli* Coq.). Yearly Rep., Dep. Agr., Bangkok (in Thai).
Shanab, L. M. and S. Bognar, 1970. Some biological and ecological investigations on Onion leaf-miner *Dizygomyza cepae* Her., *Act. Agron. Acad. Sci.* Hung. 19: 37–45.
Shenton, T., 1971. Phosdrin: new perspectives. *Span* 14 (2): 1–3.
Shevtchenko, M., 1937. *Liriomyza cicerina* Rond. and its importance to cultivation of *Cicer arietinum. Plant Protection* (*Leningrad*) 10: 159–170 (in Russian, with English summary).
Shibatsuji, T., 1949. Ecological investigations on the soy-bean root-miner, *Ophiomyia* sp.l. The life history, 2. On the diurnal activity. *Oya-Kontyû*: 5 (3): 113 (in Japanese).
Simmonds, F. J., 1960. Report on a tour of Commonwealth countries, April–November, 1959. Comm. agric. Bur. pp. 1–108.
Singh, S. and S. K. Beri, 1971. Studies on the immature stages of Agromyzidae (Diptera) from India. Part I. Notes on the biology and descriptions of immature stages of four species of *Melanagromyza* Hendel. *J. Nat. Hist.* 5: 241–250.
Singh, S. and G. S. Sohi, 1957. A list of insect pests of economic importance in the Punjab. *Govt. Agric. Coll. Mag.*, Ludhiana, 5: 19–35.
Smith, F. F., A. L. Boswell and H. E. Wave, 1962. New Chrysanthemum Leaf Miner Species. *Florists' Review*, June, 1966, 2 pp.
Smith, K. M., 1927. A study of *Hylemyia* (*Chortophila*) *brassicae* Bouché, the cabbage root fly and its parasites. With notes on some other dipterous pests of cruciferous plants. *Ann. Appl. Biol.* 14: 312–330.
Smith, R. F. and K. S. Hagen, 1959. Integrated Control Programs in the Future of Biological Control. *J. Econ. Ent.* 52 (6): 1106–1108.

SMULYAN, M. T., 1914. The Marguerite fly or chrysanthemum leaf-miner. *Bull. Mass. Agric. Exp. Stn* 157: 21–52.
SOHI, G. S. and G. C. VERMA, 1967. Studies on the chemical control of pea leaf-miner *Phytomyza atricornis* (Meigen) (Diptera: Agromyzidae). *Indian J. Ent.* 29 (3): 281–284.
SPENCER, K. A., 1957a. Two new species of Agromyzidae (Dipt.) of the genus *Agromyza* Fallén. *Comm. Biol. Helsinki* 16 (3): 1–3.
SPENCER, K. A., 1957b. Notes on the British species of *Melanagromyza* Hendel (Diptera: Agromyzidae), with the description of four new species, and also of three new species from Germany. *Proc. R. Ent. Soc. Lond.* (B) 26: 179–188.
SPENCER, K. A., 1957c. Records of Agromyzidae in Spain including 3 species new to science. *Eos, Madrid* 33: 113–122.
SPENCER, K. A., 1959. A synopsis of the Ethiopian Agromyzidae (Diptera). *Trans. R. Ent. Soc. Lond.* 111: 237–329.
SPENCER, K. A., 1960a. Records of further Ethiopian Agromyzidae (Diptera), mainly from South Africa, including eighteen species new to science. *ibid.* 112: 15–361.
SPENCER, K. A., 1960b. A new species of Agromyzidae (Diptera) from South Africa. *J. Ent. Soc. Sth Afr.* 23 (2): 314–316.
SPENCER, K. A., 1961a. Some leaf-mining Agromyzidae (Diptera) noted at Kew. *In* Additions to the Wild Fauna and Flora of the Royal Botanic Gardens, Kew. 24. *Kew Bull.* 15 (2): 185–188.
SPENCER, K. A., 1961b. A synopsis of the Oriental Agromyzidae (Diptera). *Trans. R. Ent. Soc. Lond.* 113 (4): 55–100.
SPENCER, K. A., 1961c. Notes on the African Agromyzidae – 2. *Dt. Ent. Z.* (N.F.) 8 (5): 415–430.
SPENCER, K. A., 1961d. Notes on the African Agromyzidae – 3 (Diptera). *J. Ent. Soc. Sth Afr.* 24 (2): 322–344.
SPENCER, K. A., 1963a. A synopsis of the Neotropical Agromyzidae. *Trans. R. Ent. Soc. Lond.* 115 (12): 291–389.
SPENCER, K. A., 1963b. The Australian Agromyzidae (Diptera, Insecta). *Rec. Aust. Mus.* 25 (15): 305–354.
SPENCER, K. A., 1963c. Notes on the African Agromyzidae (Diptera) – 4. *J. Ent. Soc. Sth Afr.* 26 (1): 94–124.
SPENCER, K. A., 1963d. Diptera: Agromyzidae. *Insects Micronesia* 14 (5): 135–162.
SPENCER, K. A., 1964a. A revision of the palaearctic species of the genus *Ophiomyia* Brashnikov (Diptera: Agromyzidae). *Beitr. Ent.* 14 (7/8): 773–822.
SPENCER, K. A., 1964b. Notes on the African Agromyzidae (Diptera) – 5. *Dt. Ent. Z. N.F.* 11 (1/2): 15–41.
SPENCER, K. A., 1964c. Some Agromyzidae (Diptera) from the Juan Fernandez Islands. *Pacif. Insects* 6 (2): 251–255.
SPENCER, K. A., 1965a. A clarification of the status of *Liriomyza trifolii* (Burgess) and some related species. *Proc. Ent. Soc. Wash.* 67 (1): 32–40.
SPENCER, K. A., 1965b. Notes on the African Agromyzidae (Diptera) – 6. *J. Ent. Soc. Sth Afr.* 28: 233–276.
SPENCER, K. A., 1965c. Agromyzidae *in* Diptera from Nepal. *Bull. Brit. Mus. (Nat. Hist.), Ent.* 16 (1): 25–31.
SPENCER, K. A., 1965d. A clarification of Fallén's types of Agromyzidae in Stockholm and Lund. *Ent. Tidskr.* 86: 249–259.
SPENCER, K. A., 1966a. Notes on the Oriental Agromyzidae – 4. *Stuttg. Beitr. Naturk.* 147: 1–15.
SPENCER, K. A., 1966b. A revision of European species of the genera *Melanagromyza* Hendel and *Hexomyza* Enderlein, with a supplement on the genus *Ophiomyia* Braschnikov. *Beitr. Ent.* 16: 3–60.
SPENCER, K. A., 1966c. Notes on European Agromyzidae – 1. *Beitr. Ent.* 16: 285–309.
SPENCER, K. A., 1966d. New and interesting Agromyzidae from Florida. *Stuttg. Beitr. Naturk.* 158: 1–20.
SPENCER, K. A., 1966e. Notes on the Neotropical Agromyzidae – 1. *Papéis Dep. Zool. S. Paulo* 19: 141–150.
SPENCER, K. A., 1966f. Agromyzidae (Diptera) from the Bismarck Archipelago. *Ent. Meddr* 34: 489–520.
SPENCER, K. A., 1966g. Some Agromyzidae from Sicily. *Entomologist's Mon. Mag.* 101: 172–177.
SPENCER, K. A., 1966h. A clarification of the genus *Napomyza* Westwood (Diptera, Agromyzidae). *Proc. R. Ent. Soc. Lond.* (B) 35: 29–40.

SPENCER, K. A., 1967. Some Agromyzidae (Diptera) from Morocco. *Entomologist's Mon. Mag.* 103: 126–130.

SPENCER, K. A., 1968. Ed. ERICH M. HERING: Briefe über Blattminierer. Letters on Leaf Miners pp. 1–450, Dr. W. Junk, The Hague.

SPENCER, K. A., 1969a. The Agromyzidae of Canada and Alaska (Diptera). *Mem. Ent. Soc. Canada* 64: 1–311.

SPENCER, K. A., 1969b. Notes on European Agromyzidae (Diptera) – 2. *Beitr. Ent.* 19: 5–26.

SPENCER, K. A., 1971a. Notes on a revision of the British Agromyzidae (Diptera), including the description of 14 new species. *Entomologist's Gaz.* 22: 141–195.

SPENCER, K. A., 1971b. Notes on European Agromyzidae (Dipt.) – 3. *Beitr. Ent.* 21: 249–265.

SPENCER, K. A., 1971c. A new Agromyzid (Diptera) from Pakistan feeding as a stem-borer in *Convolvulus arvensis*. Bull. Ent. Res. 61: 369–371.

SPENCER, K. A., 1972a. Diptera from Southern Spain. Agromyzidae. *Steenstrupia* 2: 91–104.

SPENCER, K. A., 1972b. Diptera, Agromyzidae. R. ent. Soc. Lond. Handbk Ident. Br. Insects 10, Part 5(g): 1–136.

SPENCER, K. A., 1973. Some Agromyzidae (Diptera) from the Maltese Islands. *Entomologist's Mon. Mag.* 108 (1972): 190–2.

SPENCER, K. A. and C. E. STEGMAIER, 1973. The Agromyzidae of Florida with a Supplement on species from the Caribbean. Arthropods of Florida and neighboring Land Areas, Vol. 7: 1–204.

SPEYER, E. R., 1936. A Leaf-miner of the Tomato Plant. *Rep. Exp. Res. Stn*, Cheshunt (1936) 28: 70.

SPEYER, E. R., 1938. Leaf-miners of Glasshouse Plants. *ibid.* (1938) 24: 67–73.

SPEYER, E. R. and W. J. PARR, 1944. Leaf-miners of tomato (*Liriomyza* spp.). *ibid.* (1944) 30: 41–52.

SPEYER, E. R. and W. J. PARR, 1946. Tomato Leaf-miner (*Liriomyza strigata* Macq.). *ibid.* (1946) 32: 52–58.

SPEYER, E. R. and W. J. PARR, 1948. Tomato Leaf-miner (*Liriomyza solani* Hering). *ibid.* (1948) 34: 43–51.

SPEYER, E. R. and W. J. PARR, 1949. Tomato Leaf-miner (*Liriomyza solani*, Hering). *ibid.* (1949) 35: 48–56.

SPEYER, E. R. and W. J. PARR, 1951. Tomato Leaf-miner (*Liriomyza solani* Her.). *ibid.* (1951) 37: 51–54.

STEGMAIER, C. E. JR., 1966a. Host plants and parasites of *Liriomyza trifolii* in Florida (Diptera: Agromyzidae). *Fla Ent.* 49 (2): 75–80.

STEGMAIER, C. E. JR., 1966b. Host plants and parasites of *Liriomyza munda* in Florida (Diptera: Agromyzidae). *ibid.* 49 (2): 81–86.

STEGMAIER, C. E. JR., 1966c. Host plants and parasites of *Liriomyza schmidti* in Florida (Diptera: Agromyzidae). *ibid.* 49 (2): 119–122.

STEGMAIER, C. E. JR., 1967a. Notes on the biology and distribution of Florida leaf-mining flies of the genus *Phytobia* Lioy, sub-genus *Calycomyza* Hendel (Diptera: Agromyzidae). *ibid.* 50 (1): 13–26.

STEGMAIER, C. E. JR., 1967b. Some new host plant records and parasites of *Phytobia* (*Amauromyza*) *maculosa* in Florida (Diptera: Agromyzidae). *ibid.* 50 (2): 99–101.

STEGMAIER, C. E. JR., 1967c. Notes on new host plant records and parasites of *Liriomyza sorosis* in Florida (Diptera: Agromyzidae). *ibid.* 50 (2): 133–136.

STEGMAIER, C. E. JR., 1967d. Host plants of *Liriomyza brassicae*, with records of their parasites from south Florida (Diptera: Agromyzidae). *ibid.* 50 (4): 257–261.

STEYSKAL, G. C., 1964. Descriptive and Synonymical Notes on *Liriomyza munda* (Diptera: Agromyzidae). *Ann. Ent. Soc. Am.* 57 (3): 388–389.

STEYSKAL, G. C., 1969. Distinction between *Phytomyza horticola* Goureau and *P. syngenesiae* (Hardy) (Diptera, Agromyzidae). *Ent. News* 80 (11): 301–303.

STROBL, P. G., 1900a. Spanische Dipteren. IX. Theil. *Wien Ent. Ztg* 19: 61–70.

STROBL, P. G., 1900b. Dipterenfauna von Bosnien, Hercegovina und Dalmatien. *Wiss. Mitt. Bosnien u. Hercegovina* 7: 552–670.

STROBL, P. G., 1902. Neue Beiträge zur Dipterenfauna der Balkanhalbinsel. Glasnik. Muz. Bosn. Herz. 14: 504.

STROBL, P. G., 1906. Spanische Dipteren. II. BEITRAG. *Mem. R. Soc. Esp. Hist. Nat.* 3 (1905): 380.

STROBL, P. G., 1909. Neue österreichische Muscidae Acalypterae II. *Wien. Ent. Ztg* 28: 283–301.

SUBRAMIAM, T. V., 1934. The Lantana Seedfly in India, *Agromyza* (*Ophiomyia*) *lantanae* Froggatt. *Indian J. agric. Sci.* 4: 468–470.

Süss, L., 1970. *Ophiomyia pinguis* Fall. (Diptera Agromyzidae) in Lombardia. Osservazion biologische e morfologiche. *Boll. Zool. Agric. e Bach.* II: 43–84.
Swaine, G., 1969. Studies on the biology and control of pests of seed beans (*Phaseolus vulgaris*) in northern Tanzania. *Bull. Ent. Res.* 59 (1968): 323–338.
Swezey, O. H., 1921. Notes and Exhibitions. *Agromyza lantanae. Proc. Hawaii Ent. Soc.* 5: 187.
Swezey, O. H., 1941. Observations on Insect Pests in Samoa which are not yet known to occur in Hawaii. *Hawaii. Planters' Rec.* 45: 25–38.
Swezey, O. H., 1946. Some miscellaneous Diptera of Guam. *In* The Insects of Guam 2. *Bull. Bishop Mus.* 189: 195–200.
Swezey, O. H., 1948. Insect Invaders in Hawaii During and Since World War II. *J. Econ. Ent.* 41 (5): 669–672.
Takhtajan, A., 1969. Flowering plants – Origin and Dispersal. Oliver Boyd, Edinburgh. 310 pages.
Taylor, C. E., 1958. The bean stem maggot. *Rhodesia Agric. J.* 55: 634–636.
Taylor, C. E., 1959. Control of the bean stem maggot by insecticidal dressings. *Rhodesia Agric. J.* 56: 195–196.
Thevasagayam, E. S. and L. S. C. Canagasingham, 1960. Some observations on the insect pests of dhal (*Cajanus cajan*) and their control. *Tropical Agriculturist* 116: 287–298.
Thomas, F., 1893. Cecidiologische Notizen. 10. Schädigung der Lilium-Blüthen durch eine Muscidenlarve. *Ent. Nachrichten* 19: 300–302.
Thomson, C. G., 1869. Eugenies Resa. Diptera: p. 609. Stockholm.
Tryon, H., 1895. The Bean maggot. *Trans. Nat. Hist. Soc. Qd.* 1: 4–7.
Tryon, H., 1912. Insects for checking growth on dissemination of Lantana. *Qd. Agric. J.* 29: 493–497.
Tschirnhaus, M. von, 1969. Zur Kenntnis der Variabilität, Eidonomie und Verwandtschaft bemerkenswerter Agromyzidae (Diptera). *Senckenbergiana biol.* 50 (3/4): 143–157.
Tubeuf, C. F. von, 1897. Die Zellgänge der Birke und anderer Laubhölzer. *In:* Forst- und naturwiss. Z. Vol. 6.
Turnbull, A. L. and D. A. Chant, 1961. The Practice and Theory of biological control of insects in Canada. *Can. J. Zool.* 39: 697–753.
Uphoff, J. C. Th., 1968. Dictionary of Economic Plants. *J. Cramer*, Lehre, Germany.
Utech, L., 1962. Blattminen und Pflanzengallen aus Albanien und dem Kaukasus. *Dt. Ent. Z. (N.F.)* 9: 229–235.
Vappula, N. A., 1965. Pests of cultivated plants in Finland. *Acta Ent. Fenn.* 19: 1–239.
Varley, G. C., 1971. The ecology of insect parasites and predators. *In* Presidential Addresses of Swansea Meeting of British Association for the Advancement of Science, Section D, pp. 57–64.
Varley, G. C. and G. R. Gradwell, 1968. Population models for the winter moth. *In* Southwood, T. R. E., (Ed.) Insect Abundance. Symposia of the Royal Entomological Society of London, No. 4: 132–142.
Vecchio, C. del, 1917. *Phytomyza flavicornis*, a Dipteron injurious to the Milan cabbage in Lombardy. Int. Rev. Sci. Pract. Mon. Bull. Agric. Intell. et Pl. Dis. 8: 1069. Roma 7 (7).
Veitch, R., 1935. Insect Enemies of Lantana. *Qd. agric. J.* 44: 142–144.
Venturi, F., 1935. Contributo alla conoscenza dell'Entomofauna del Frumento. II. *Dizygomyza lateralis* Macquart. *Boll. Lab. Ent. Bologna* 8: 1–25.
Venturi, F., 1936. Contributo alla conoscenza dell'Entomofauna delle Graminacee coltivate a spontanee III. *Boll. Ist. Bologna* 8: 141–149.
Venturi, F., 1939. Contributi alla conoscenza dell'Entomofauna delle Graminacee VI. *Redia* 26: 27–70.
Venturi, F., 1946a. Studio biologico del genere *Cerodonta* Rond. (Diptera Agromyzidae). *Redia* 31: 191–226.
Venturi, F., 1946b. Note biologiche sulla '*Dizygomyza iridis*' Hendel (Diptera, Agromyzidae). *Boll. Ist. Ent. Univ. Bologna* 15: 203–216.
Venturi, F., 1949. *Pseudonapomyza dianthicola* n.sp. (Dipt. Agromyzidae) minatrice delle foglie di garofano. *Redia* 34: 161–164.
Venugopal, S. and K. S. Venkataramani, 1954. An Agromyzid Insect pest of 'Bhendi'. *J. Madras Univ.*, B, 24 (3): 335–340.
Viggiani, G., 1962. Contributo alla conoscenza degli insetti minatori e loro simbionti. I. La *Phytomyza heringiana* Hendel (Dipt. Agromyzidae) nuovo minatore del Melo per l'Italia. *Boll. Lab. Ent. Portici* 20: 31–69.
Walker, F., 1853. Insecta Britannica. Diptera, Vol. 2, pp. 240–247. London.
Walker, P. T., 1960. Insecticide studies on East African agricultural pests III – Seed dressings

for the control of the bean fly, *Melanagromyza phaseoli* (Coq.), in Tanganyika. *Bull. Ent. Res.* 50: 781–793.
WALLACE, G. B., 1939. French bean diseases and bean fly in East Africa. *East Afr. Agric. J.* 5: 170–175.
WATT, M., 1923. The Leaf-mining Insects of New Zealand: Part III – Species belonging to the Genera *Agromyza* (Fallén) and *Phytomyza* (Fallén) (Diptera). *Trans. Proc. N Z. Inst.* 54: 465–489.
WATT, M., 1924. The Leaf-mining Insects of New Zealand: Part 5 – The Genus *Nepticula* (Lepidoptera), and the Agromyzidae (Diptera) continued, and *Gracilaria selonitis* Meyr. (Lepidoptera). *ibid.* 55: 674–687.
WEBB, R. E. and F. F. SMITH, 1969. Effect of Temperature on Resistance in Lima Bean, Tomato, and Chrysanthemum to *Liriomyza munda*. *J. Econ. Ent.* 62 (2): 458–462.
WEBB, R. E. and F. F. SMITH, 1970. Survival of Eggs of *Liriomyza munda* in Chrysanthemum during cold storage. *J. Econ. Ent.* 63 (4): 1359–1361.
WEBSTER, F. M. and T. H. PARKS, 1913. The Serpentine Leaf-miner. *J. Agric. Res.* 1 (1): 59–87.
WEED RESEARCH ORGANIZATION, 1970. Selected references on broomrape (*Orobanche* spp.), 1964–1970. *Annot. Bibl.* 23: 1–15.
WEIGAND, G., 1966. Zur Problematik der Bekämpfung der an Gerste vorkommenden *Agromyza nigrella* Rondani. *Gesunde Pflanzen* 18: 43–48.
WENE, G. P., 1953. Control of the serpentine leafminer on peppers. *J. Econ. Ent.* 46 (5): 789–793.
WENE, G. P., 1955. Effect of some organic insecticides on the population levels of the serpentine leafminer and its parasites. *J. Econ. Ent.* 48 (5): 596–597.
WICKRAMASINGE, N. and H. E. FERNANDO, 1962. Investigations on insecticidal seed dressings, soil treatments and foliar sprays for the control of *Melanagromyza phaseoli* (Tryon) in Ceylon. *Bull. Ent. Res.* 53 (2): 223–240.
WIESMAN, R., 1961. *Phytomyza lateralis* Fall., ein wenig beachteter Möhren- und Karottenschädling. *Mitt. Ent. Ges. Basel* 11: 39–62.
WILCOX, J. and A. F. HOWLAND, 1952. Control of a Dipterous Miner on Tomatoes in California. *J. Econ. Ent.* 45 (4): 634–639.
WILCOX, J. and A. F. HOWLAND, 1955. Control of the Pea Leaf Miner in Southern California. *J. Econ. Ent.* 48 (5): 579–581.
WISE, K. J., 1953. Leaf-mining Pests in New Zealand. *N.Z. J. Agr.* 87: 75–76.
WOLFENBARGER, D. A., 1964. The effect of Insecticides, Rates, Intervals between, and number of applications and Insecticide-Oil and Surfactant combinations for Insect Control on Southernpeas. *J. Econ. Ent.* 57 (6): 966–969.
WOLFENBARGER, D. A. and L. W. GETZIN, 1963. Selective toxicants and toxicant-surfactant combinations for leafminer, *Liriomyza munda* Frick, control and parasite survival. *Fla Entomologist* 46 (3): 251–265.
WOLFENBARGER, D. A. and D. O. WOLFENBARGER, 1966. Tomato yields and Leaf Miner Infestations and a sequential sampling Plan for Determining Need for Control Treatments, *J. Econ. Ent.* 59: 279–283.
WOLFENBARGER, D. O., 1947. The serpentine leaf miner and its control. *Fla Agr. Exp. Stn. Press Bull.* 639: 1–6.
WOLFENBARGER, D. O., 1948. Control studies on the serpentine leaf miner on potato and tomato. *Proc. Fla St. hort. Soc.* 61: 181–186.
WOLFENBARGER, D. O., 1949. Border effects of serpentine leaf miner abundance in potato fields. *Fla Entomologist* 31 (1): 15–20.
WOLFENBARGER, D. O., 1958. Serpentine Leaf Miner: Brief History and Summary of a decade of Control Measures in South Florida. *J. Econ. Ent.* 51 (3): 357–359.
WOLFENBARGER, D. O., 1961. Leaf mining Insects, especially the serpentine miners on vegetable crop plants and their control. *Proc. Fla St. Hort. Soc.* 74: 131–133.
YUASA, H. and T. KOYAMA, 1938. '*Shokubutsu oyobi Dobutsu*' [*Botany and Zoology*] 6 (4): 820.
ZEHNTNER, L., 1900. De kedeleeboorder. *Ind. Natuur* 11: 113–137.
ZETTERSTEDT, J. W., 1838. Insecta Lapponica. Diptera: 1–867. Leipzig.
ZETTERSTEDT, J. W., 1848. Diptera Scandinaviae 7: 2728–2844. Lund.

LIST OF PLATES

INDEX

Part 1

GENERAL

The main reference to pest species and genera is set in bold. Synonyms are inset.

B

F

M

N

O

P

T

W

X

Y

Z

INDEX

Part 2

PARASITES

Parasites are indexed alphabetically under families, sub-families and genera.